Meat Biotechnology

Fidel Toldrá
Editor

Meat Biotechnology

 Springer

Editor
Fidel Toldrá
Department of Food Science
Instituto de Agroquímica y Tecnología
  de Alimentos (CSIC)
46100 Burjassot (Valencia), Spain
ftoldra@iata.csic.es

ISBN: 978-0-387-79381-8   e-ISBN: 978-0-387-79382-5

Library of Congress Control Number: 2008935902

© 2008 Springer Science+Business Media, LLC
All rights reserved. This work may not be translated or copied in whole or in part without the written permission of the publisher (Springer Science+Business Media, LLC, 233 Spring Street, New York, NY 10013, USA), except for brief excerpts in connection with reviews or scholarly analysis. Use in connection with any form of information storage and retrieval, electronic adaptation, computer software, or by similar or dissimilar methodology now known or hereafter developed is forbidden.
The use in this publication of trade names, trademarks, service marks, and similar terms, even if they are not identified as such, is not to be taken as an expression of opinion as to whether or not they are subject to proprietary rights.

Printed on acid-free paper

springer.com

Preface

The main goal of this book is to provide the reader with the recent developments in biotechnology for its application in the meat processing chain. To achieve this goal, the book is divided into four sections. The first part deals with the production systems towards an improved meat quality through the use of modern biotechnology applied to farm animals. This section includes chapters dealing with transgenic farm animals, genetic control of quality traits and traceability based on DNA. The second part is focused on the recent biotechnological developments in starter cultures to improve meat fermentation. The chapters cover the molecular identification of microorganisms, its characterization and the genetics of lactic acid bacteria, yeasts and molds. The third part presents the current approaches employed to improve the quality and nutritional properties of meat. This section includes chapters on flavor generation, probiotics and bioactive compounds. The final part deals with latest advances for the protection against foodborne pathogens and other recent trends in the field. The 9 chapters of this section cover biotechnological-based methods for the control of spoilage and detection of pathogens, GMOs, veterinary drugs, as well as recent developments in bioprotective cultures, bacteriocins, smart packaging and safety and regulatory aspects.

This book, which is written by distinguished international contributors with solid experience and reputation, brings together all the advances in such varied and different biotechnological topics related with meat. I thank the production team at Springer and wish to express my gratitude to Susan Safren (Editor) and David Parsons (Editorial assistant) for their kind assistance in this book.

Fidel Toldrá, Ph.D.
Editor

Contents

Part I Animal Biotechnology for the Enhancement of Meat Quality

1. **Transgenic Farm Animals** 3
   Morse B. Solomon, Janet S. Eastridge, and Ernest W. Paroczay

2. **Genetic Control of Meat Quality Traits** 21
   John L. Williams

3. **DNA-Based Traceability of Meat** 61
   G.H. Shackell and K.G. Dodds

Part II Biotechnology of Starter Cultures for Meat Fermentation

4. **Molecular Methods for Identification of Microorganisms
   in Traditional Meat Products** 91
   Luca Cocolin, Paola Dolci, and Kalliopi Rantsiou

5. **Characteristics and Applications of Microbial Starters in Meat
   Fermentations** 129
   Pier Sandro Cocconcelli and Cecilia Fontana

6. **Genetics of Lactic Acid Bacteria** 149
   Monique Zagorec, Jamila Anba-Mondoloni, Anne-Marie Crutz-Le Coq,
   and Marie-Christine Champomier-Vergès

7. **Genetics of Yeasts** 167
   Amparo Querol, Mª Teresa Fernández-Espinar, and Carmela Belloch

8. **Characteristics and Applications of Molds** 181
   Elisabetta Spotti, Elettra Berni, and Cristina Cacchioli

Part III Biotechnology for Better Quality and Nutritional Properties of Meat Products

9 Biotechnology of Flavor Generation in Fermented Meats............ 199
   Fidel Toldrá

10 Latest Developments in Probiotics 217
    Frédéric Leroy, Gwen Falony, and Luc de Vuyst

11 Bioactive Compounds in Meat 231
    Keizo Arihara and Motoko Ohata

Part IV Biotechnology for Safer Meat and Meat Products

12 Biocontrol of Pathogens in the Meat Chain 253
    Catherine M. Burgess, Lucia Rivas, Mary J. McDonnell,
    and Geraldine Duffy

13 At-Line Methods for Controlling Microbial Growth and Spoilage
    in Meat Processing Abattoirs 289
    Daniel Y.C. Fung, Jessica R. Edwards, and Beth Ann Crozier-Dodson

14 The Detection of Genetically Modified Organisms: An Overview..... 319
    Jaroslava Ovesná, Kateřina Demnerová, and Vladimíra Pouchová

15 Biosensors for Detecting Pathogenic Bacteria in the Meat Industry ... 335
    Evangelyn C. Alocilja

16 Immunology-Based Techniques for the Detection of Veterinary
    Drug Residues in Foods 361
    Milagro Reig and Fidel Toldrá

17 Antimicrobial Activity of Bacteriocins and Their Applications 375
    Eleftherios H. Drosinos, Marios Mataragas, and Spiros Paramithiotis

18 Bioprotective Cultures 399
    Graciela Vignolo, Silvina Fadda, and Patricia Castellano

19 Smart Packaging Technologies and Their Application
    in Conventional Meat Packaging Systems 425
    Michael N. O'Grady and Joseph P. Kerry

20 Meat Safety and Regulatory Aspects
    in the European Union 453
    Ron H. Dwinger, Thomas E. Golden, Maija Hatakka, and Thierry Chalus

Contributors

Evangelyn C. Alocilja

Department of Biosystems and Agricultural Engineering, Michigan State University, 213 Farrall Hall, East Lansing, MI 48824-1323, USA, e-mail: alocilja@egr.msu.edu

Jamila Anba-Mondoloni

Unité Flore Lactique et Environment Carné, UR309, INRA, Domaine de Vilvert, 78 350 Jouy-en-Josas, France

Keizo Arihara

Department of Animal Science, Kitasato University, 35-1 Higashi-23-Bancho, Towada-shi, 034-8628, Japan, e-mail: arihara@vmas.kitasato-u.ac.jp

Carmela Belloch

Department of Biotechnology, Instituto de Agroquímica y Tecnología de Alimentos (CSIC), PO Box 73, 46100 Burjassot (Valencia), Spain, e-mail: belloch@iata.csic.es

Elettra Berni

Stazione Spermentale per lí Industria delle Conserv Alimentari, Viale F. Tanara 31/A, Parma 43100, Italy

Catherine M. Burgess

Ashtown Food Research Centre, Teagasc, Ashtown, Dublin 15, Ireland

Cristina Cacchioli

Stazione Spermentale per lí Industria delle Conserv Alimentari, Viale F. Tanara 31/A, Parma 43100, Italy

Patricia Castellano

Centro de Referencia para Lactobacillos, CERELA, CONICET, Chacabuco 145, San Miguel de Tucumán, T 4000 ILC Tucumán, Argentina

Thierry Chalus
Unit of Hygiene and control measures, Health and Consumer Protection DG, European Commission, Office: Room 4/10, Rue Belliard 232, B-1049 Brussels, Belgium, e-mail: thierry.chalus@ec.europa.eu

Marie Champomier-Vergès
Unité Flore Lactique et Environment Carné UR309, INRA, Domaine de Vilvert, 78350 Jouy-en-Josas, France, e-mail: Marie-Christine.Champomier-Verges@jouy.inra.fr

Luca Cocolin
Dipartimento di Valorizzazione e Protezione delle Risorse Agroforestali, University of Turin, Faculty of Agriculture, via Leonardo da Vinci 44, 10095 Grugliasco – Turin, Italy, e-mail: lucasimone.cocolin@unito.it

Pier Sandro Cocconcelli
Istituto di Microbiologia, Centro Ricerche Biotecnologiche, Universita Cattolica del Sacro Cuore, via Emilia Parmense 84, 29 100 Piacenza-Cremona, Italy, e-mail: pier.cocconcelli@unicatt.it

Beth Ann Crozier-Dodson
Department of Animal Sciences and Industry, Kansas State University, Manhattan, KS 66506, USA, e-mail: bethann@ksu.edu

Anne-Marie Crutz-Le Coq
Unité Flore Lactique et Environment Carné, UR309, INRA, Domaine de Vilvert, 78350 Jouy-en-Josas, France, e-mail: Anne-Marie.LeCoq@jouy.inra.fr

Katerina Demnerová
Institute of Chemical Technology, Technická, 16000 Prague 6, Czech Republic, e-mail: Katerina.Demnerova@vscht.cz

Luc de Vuyst
Research Group of Industrial Microbiology and Food Biotechnology, Department of Applied Biological Sciences and Engineering, Vrije Universiteit Brussel, Pleinlaan 2, B-1050 Brussels, Belgium, e-mail: ldvuyst@vub.ac.be

Ken G. Dodds
Invermay Agr Center, Private Bag 50034, AgResearch, Mosgiel 9053, New Zealand, e-mail: ken.dodds@agresearch.co.nz

Paola Dolci
Dipartimento di Valorizzazione e Protezione delle Risorse Agroforestali, University of Turin, Faculty of Agriculture, via Leonardo da Vinci 44, 10095 Grugliasco – Turin, Italy, e-mail: paola.dolci@unito.it

Eleftherios H. Drosinos
Laboratory of Food Quality Control and Hygiene, Department of Food Science and Technology, Agricultural University of Athens, Iera Odos 75, GR-11855 Athens, Greece, e-mail: ehd@aua.gr

Geraldine Duffy
Ashtown Food Research Centre, Teagasc, Ashtown, Dublin 15, Ireland, e-mail: Geraldine.Duffy@teagasc.ie

Ronald H. Dwinger
Food and Consumer Product Safety Authority (VWA), Room 9A12, Prinses Beatrixlaan 2, P. O. Box 19506, 2500 CM The Hague, The Netherlands, e-mail: Ron.Dwinger@vwa.nl

Janet S. Eastridge
Food Technology and Safety Lab, USDA-ARS, Bldg 201, BARC-East, 10300 Baltimore Ave., Beltsville, MD 20705, USA

Jessica R. Edwards
Department of Animal Sciences and Industry, Kansas State University, Manhattan, KS 66506, USA

Silvina Fadda
Centro de Referencia para Lactobacillos, CERELA, CONICET, Chacabuco 145, San Miguel de Tucumán, T 4000 ILC Tucumán, Argentina, e-mail: sfadda@cerela.org.ar

Gwen Falony
Research Group of Industrial Microbiology and Food Biotechnology, Department of Applied Biological Sciences and Engineering, Vrije Universiteit Brussel, Pleinlaan 2, B-1050 Brussels, Belgium, e-mail: Gwen.Falony@vub.ac.be

Mª Teresa Fernández-Espinar
Department of Biotechnology, Instituto de Agroquímica y Tecnología de Alimentos (CSIC), PO Box 73, 46100 Burjassot (Valencia), Spain, e-mail: tfer@iata.csic.es

Cecilia Fontana
Istituto di Microbiologia, Centro Ricerche Biotecnologiche, Universita Cattolica del Sacro Cuore, via Emilia Parmense 84, 29 100 Piacenza-Cremona, Italy, e-mail: cecilia.fontana@unicatt.it

Daniel Y.C. Fung
Department of Animal Sciences and Industry, Kansas State University, Manhattan, KS 66506, USA, e-mail: dfung@ksu.edu

Thomas E. Golden
Unit of Hygiene and control measures, Health and Consumer Protection DG, European Commission, Office: Room 4/10Rue Belliard 232, B-1049 Brussels, Belgium, e-mail: thomas.golden@ec.europa.eu

Maija Hatakka
Evira. Mustialankatu 3, 00790, Helsinki, Finland, e-mail: maija.hatakka@evira.fi

Joseph P. Kerry

Department of Food & Nutrition Science, University College Cork, National University of Ireland, Cork, Ireland, e-mail: Joe.Kerry@ucc.ie

Frédéric Leroy

Research Group of Industrial Microbiology and Food Biotechnology, Department of Applied Biological Sciences and Engineering, Vrije Universiteit Brussel Pleinlaan 2, B-1050 Brussels, Belgium, e-mail: fleroy@vub.ac.be

Marios Mataragas

Laboratory of Food Quality Control and Hygiene, Department of Food Science and Technology, Agricultural University of Athens, Iera Odos 75, GR-11855 Athens, Greece

Mary J. McDonnell

Ashtown Food Research Centre, Teagasc, Ashtown, Dublin 15, Ireland

Motoko Ohata

Department of Animal Science, Kitasato University, 35-1 Higashi-23-Bancho, Towada-shi, 034-8628, Japan, e-mail: ohata@vmas.kitasato-u.ac.jp

Michael N. O'Grady

Department of Food & Nutrition Science, University College Cork, National University of Ireland, Cork, Ireland, e-mail: Michael.ogrady@ucc.ie

Jaroslava Ovesná

Crop Research Institute, Drnovska 507, 16106 Prague 6-Ruzyne, Czech Republic, e-mail: ovesna@vurv.cz

Spiros Paramithiotis

Laboratory of Food Quality Control and Hygiene, Department of Food Science and Technology, Agricultural University of Athens, Iera Odos 75, GR-11855 Athens, Greece

Ernest W. Paroczay

Food Technology and Safety Lab, USDA-ARS, Bldg 201, BARC-East, 10300 Baltimore Ave., Beltsville, MD 20705, USA

Vladimíra Pouchová

Crop Research Institute, Drnovska 507, 16106 Prague 6-Ruzyne, Czech Republic, e-mail: pouchova@vurv.cz

Amparo Querol

Department of Biotechnology, Instituto de Agroquímica y Tecnología de Alimentos (CSIC), PO Box 73, 46100 Burjassot (Valencia), Spain, e-mail: aquerol@iata.csic.es

Contributors

Kalliopi Rantsiou
Dipartimento di Valorizzazione e Protezione delle Risorse Agroforestali, University of Turin, Faculty of Agriculture, via Leonardo da Vinci 44, 10095 Grugliasco – Turin, Italy, e-mail: kalliopi.rantsiou@unito.it

Milagro Reig
Institute of Engineering for Food Development, Polytechnical University of Valencia, Ciudad Politécnica de la Innovación, edif 8E, Camino de Vera s/n, 46022 Valencia, Spain, e-mail: mareirie@doctor.upv.es

Lucía Rivas
Ashtown Food Research Centre, Teagasc, Ashtown, Dublin 15, Ireland, e-mail: lucy.rivas@teagasc.ie

Grant H. Shackell
Invermay Agr Center, Private Bag 50034, AgResearch, Mosgiel 9053, New Zealand, e-mail: grant.shackell@agresearch.co.nz

Morse Solomon
Food Technology and Safety Lab, USDA-ARS, Bldg 201, BARC-East, 10300 Baltimore Ave., Beltsville, MD 20705, USA, e-mail: msolomon@anri.barc.usda.gov

Elisabetta Spotti
Stazione Spermentale per lí Industria delle Conserv Alimentari, Viale F. Tanara 31/A, Parma 43100, Italy, e-mail: elisabetta.spotti@ssica.it

Fidel Toldrá
Department of Food Science, Instituto de Agroquímica y Tecnología de Alimentos (CSIC), PO Box 73, 46100 Burjassot (Valencia), Spain, e-mail: ftoldra@iata.csic.es

Graciela Vignolo
Centro de Referencia para Lactobacillos, CERELA, CONICET, Chacabuco 145, San Miguel de Tucumán, T 4000 ILC Tucumán, Argentina, e-mail: vignolo@cerela.org.ar

John L. Williams
Parco Tecnologico Padano, Via Einstein, Polo Universitario Lodi 26900, Italy, e-mail: john.williams@tecnoparco.org

Monique Zagorec
Unité Flore Lactique et Environment Carné, UR309, INRA, Domaine de Vilvert, 78 350 Jouy-en-Josas, France

Part I
Animal Biotechnology for the Enhancement of Meat Quality

Chapter 1
Transgenic Farm Animals

Morse B. Solomon, Janet S. Eastridge, and Ernest W. Paroczay

Introduction

Conventional science to improve muscle and meat parameters has involved breeding strategies, such as selection of dominant traits or selection of preferred traits by cross breeding, and the use of endogenous and exogenous hormones. Improvements in the quality of food products that enter the market have largely been the result of postharvest intervention strategies. Biotechnology is a more extreme scientific method that offers the potential to improve the quality, yield, and safety of food products by direct genetic manipulation. In the December 13, 2007 issue of the Southeast Farm Press, an article by Roy Roberson pointed out that biotechnology is driving most segments of U.S. farm growth. He indicated that nationwide, the agriculture industry is booming and much of that growth is the result of biotechnology advancements. For example, the United States produces over half the worldwide acreage of bio-engineered crops (GMO), and this growth is expected to continue worldwide. With respect to livestock, biotechnology is a more novel approach to the original methods of genetic selection and crossbreeding, or administration and manipulation of various hormones (i.e., growth).

Biotechnology in animals is primarily achieved by cloning, transgenesis, or transgenesis followed by cloning. Animal cloning is a method used to produce genetically identical copies of a selected animal (i.e., one which possesses high breeding value), while transgenesis is the process of altering an animal's genome by introducing (via gene transfer) a new or foreign gene (i.e., DNA) not found in the recipient species, or deleting or modifying an endogenous gene with the ultimate goal of producing an animal expressing a beneficial function or a superior attribute (e.g., adding a gene that promotes increased muscle growth). The gene or genes that are transferred or modified is called the transgene (TG). A combination of the two methods, i.e., transgenic cloning, is the process of producing a clone whose

---

Mention of brand or firm names does not constitute an endorsement by the United States Department of Agriculture over others of a similar nature not mentioned.

M.B. Solomon
USDA, ARS, BARC-East, Bldg. 201, Beltsville, MD 20705, USA

donor cells contain heritable DNA inserted by a molecular biology technique, as used in a transgenic event. The first to report on creating cloned animals was Hans Dreisch in the late 1800s. Dreisch's intent, however, was not to create identical animals but rather to prove that genetic material is not lost during cell division. His research experiments involved sea urchins, which he intentionally chose, since sea urchins have large embryo cells and grow independently of their mothers. A pioneering report by Palmiter et al. (1982) on the accelerated growth of transgenic mice that developed from eggs microinjected with a growth hormone (GH) fusion gene started the revolution in biotechnology of animals. Based on this research, many novel uses for biotechnology in animals were envisioned, beginning with the enhancement of production-related traits (yield and composition) and expanding into disease-resistance strategies and production of biological products (i.e., pharmaceuticals). The primary goal of transgenesis is to establish a new genetic line of animals, in which the trait is stably transmitted to succeeding generations. The past several years involving transgenic research has primarily focused on altering carcass composition, increasing milk production, enhancing disease resistance, and reducing excretion of phosphate by pigs. A significant amount of progress has been achieved. However, the success of this research is dependent upon improving the efficiency of the nuclear transfer technology, which will in turn reduce the cost of producing transgenic animals.

Early methods of cloning involved a technology called embryo splitting, but the traits of the resulting clone were unpredictable. Today's method of cloning, i.e., somatic (adult) cell nuclear transfer, became established in 1996 with the production of the world's first cloned farm animal, "Dolly" the sheep (Wilmut, Schnieke, McWhir, Kind, & Campbell, 1997), at the Roslin Institute in Scotland, and has since been used for cattle, goats, mice, and pigs. Cloning could be a promising method of restoring endangered, or nearly extinct, species and populations. Production of transgenic animals is carried out by a technique called pronuclear microinjection, reported first in mice (Gordon, Scangos, Plotkin, Barbosa, & Ruddle, 1980), and later adapted to rabbits, sheep, and pigs (Hammer et al., 1985). An excellent review on genome modification techniques and applications was published by Wells 2000).

Before 1980, applications for patents on living organisms were denied by the U.S. Patent and Trademark Office (USPTO), because anything found in nature was considered non-patentable subject matter. However, the U.S. scientist Amanda Chakrabarty, who wanted to obtain a patent for a genetically engineered bacterium that consumes oil spills, challenged the USPTO in a case that landed in the U.S. Supreme Court, which in 1980 ruled that patents could be awarded on anything that was human-made. Since then, some 436 transgenic or bio-engineered animals have been patented, including 362 mice, 26 rats, 19 rabbits, 17 sheep, 24 pigs, 20 cows, 2 chickens, and 3 dogs (Kittredge, 2005). Due to the steps specific to transgenic procedures, for instance the DNA construct, its insertion site, and the subsequent expression of the gene construct, animals derived from transgenesis have more potential risks than cloned animals. Based on a National Academy of Sciences, National Research Council (NRC) 2002 report, "Animal Biotechnology:

Science-Based Concerns," the U.S. FDA in 2003 announced that meat or dairy products from cloned animals are likely to be safe to eat, but to date has not yet approved these products for human consumption. More recently (2007 and 2008), the U.S. FDA has reported that meat and meat from cloned animals is as safe as those from their counterparts bred the old-fashioned way. However, progress in this area is very slow and has a long way to go before having an impact at a commercial usage level. It still will be years before many foods from cloned or transgenic animals reach the shelves in stores, mainly for economic reasons. At an estimated cost of $10,000–$20,000 for each bio-engineered animal, these technologically engineered animals are a lot more expensive than their ordinary bred counterpart. Thus, producers will be more inclined to use the bio-engineered offspring for meat and not the cloned or transgenic animal itself. The U.S. Department of Agriculture (USDA), however, recommended that the U.S. farmers should keep their cloned animals out of the market place indefinitely, even as FDA officials claim that food from cloned livestock is safe to eat.

Bio-engineered foods are regulated by three agencies: USDA, Food and Drug Administration (FDA), and the Environmental Protection Agency (EPA). The USDA has an oversight for meat and poultry, whereas seafood regulation falls under the FDA. The FDA Center for Veterinary Medicine (CVM) also regulates transgenic animals because any drug or biological material created through transgenesis is considered a drug and will have to undergo the same scrutiny to demonstrate safety and effectiveness (Lewis, 2001). The EPA has a responsibility for pesticides that are genetically engineered into plants. In the mid-1980s, federal policy declared that biotechnologically derived products would be evaluated under the same laws and regulatory authorities used to review comparable products produced without biotechnology. As stated on the FDA website, the CVM has asked companies not to introduce animal clones, their progeny, or their food products into the human or animal food supply until there is sufficient scientific information available on the direct evaluation of safety.

Characterization of Candidate Genes/Genetic Markers for Carcass and Meat Quality Traits

Animals vary widely in their genetic merit and commercial value. Classical selection techniques have been utilized, over the years, with great success for improving animal production traits, but the underlying genetic changes were elusive to researchers in the past. Technological advances in molecular biology in the early 1990s opened up a whole new area of investigations into the DNA genome. Presently, there is a lot of attention being paid to the identification and sequencing of chromosomal regions representing quantitative trait loci (QTL) influencing carcass traits, growth, and meat quality factors. Research aimed at elucidating potential candidate genes and characterizing their role on these important traits is an essential preliminary step to incorporate genetic manipulation into future biotechnology projects.

There are two proposed models for the genetic control of complex traits: the infinitesimal model and the major gene model. The infinitesimal model assumes that complex traits are controlled by large numbers of unlinked genes, of which each has only an infinitesimal effect on the trait. In contrast, the major gene model assumes that a small number of major genes contribute a substantial proportion of the genetic variation in the expressed trait. The results from QTL mapping reports suggest that modest numbers of QTL can explain some, but not all of the genetic variation in the complex traits.

In August of 2007, A Johns Hopkins University scientist (Se-Jin Lee) illustrated that the absence of the protein myostatin (MSTN) leads to oversized muscles in mice and reported that a second protein, follistatin, when triggered to overproduce in mice lacking the protein MSTN in turn quadruples the muscle mass (Lee, 2007). Transgenic mice expressing the MSTN pro-domain (Yang et al., 2001; Mitchell and Wall, 2004) also showed significantly increased muscle mass resulting in 22–44% heavier carcasses compared to the controls. They concluded that the lower percentage of fat in those mice was due to a higher proportion of lean mass, because the epididymal fat pad weight was not reduced. The dramatic muscular phenotype, observed throughout the whole carcass, was attributed to muscle hypertrophy since no change in fiber numbers between controls and transgenic mice were detected. Fast-twitch fibers were larger in transgenic mice. Thus, overexpression of the MSTN pro-domain could also be an alternative to MSTN knockout as a means of increasing muscle mass. Researchers at Adelaide University in Australia have identified a gene that they claim explains a large increase in the retail beef yield of edible tissue. While the gene, called MSTN F94L, is not the only gene that influence retail yield, they indicate that it has a tremendous effect on the retail yield.

Bovine

Information in this area is very limited and highly desired by federal agencies that regulate food safety issues. There have been some studies evaluating the meat of animals cloned from embryonic cells (Gerken, Tatum, Morgan, & Smith, 1995; Diles et al., 1996; Harris et al., 1997). Those results, however, do not correspond with the products from animals cloned from adult somatic cells. This is because embryonic animal clones are produced from blastomeres of fertilized embryos at a very early stage of development, and thus embryonic clones may undergo little gene reprogramming during their development. Consequently, they would not serve well as scientific evidence for assessing the food safety risks of somatically cloned food animals. A few reports which provide data on the composition of meat and dairy products derived from adult somatic cell clones indicate that these products are equivalent to those of normal animals. The first report on the chemical composition of bovine meat arising from genetic engineering was in cloned cattle (Takahashi & Ito, 2004). In the meat samples derived from cloned and non-cloned Japanese Black cattle, at the age of 27–28 months, data were collected for proximate analysis (water, protein, lipids, and ash) as well as fatty acids, amino acids, and cholesterol. The results of this study showed that the nutritional properties of meat from cloned cattle are similar to those of non-cloned animals, and were within the recommended values of the Japanese Dietetic Information guidelines. Also, based on the marbling score, the meat quality score of the cloned cattle in this study graded high (Class 4) according to the Japanese Meat Grading Standard (Class 1, poor to Class 5, premium). No other carcass characteristics were discussed in this report.

A comprehensive study designed specifically to provide the scientific data desired by U.S. regulatory agencies on the safety issue of the composition of meat and milk from animal cloning was recently published (Tian et al., 2005). All animals were subjected to the same diet and management protocols. They analyzed over 100 parameters that compare the composition of meat and milk from beef and dairy cattle derived from cloning, to those of genetic- and breed-matched control animals from conventional reproduction. The beef cattle, in this study, were slaughtered at 26 months of age and also examined for meat quality and carcass composition. A cross section between the sixth and seventh rib of the left side dressed carcass was inspected according to the Japan Meat Grading Association guidelines. Additional parameters of the carcass analyzed were organ or body part weights and the total proportion of muscle and fat tissue to carcass weight. The histopathology of seven organs was examined for appearance of abnormalities. Six muscles (infraspinatus (IS), longissimus thoracis, latissimus dorsi, adductor, biceps femoris (BF), and semitendinosus) were removed from the carcass and measured for the percentages of moisture, crude protein, and crude fat. Samples from these muscles for muscle fiber type profiling, however, were not performed. The fatty acid profile of five major fat tissues (subcutaneous fat, intra- and inter-muscular fats, celom fat, and kidney leaf fat) and the amino acid composition of the longissimus thoracis muscle was also determined. Out of more than 100 parameters examined, a significant difference was observed in 12 parameters for the paired comparisons (clone vs genetic comparator and clone vs breed comparator). Among these 12 parameters, 8 were related to the amount of fat or fatty acids in the meat/fat. The other four parameters that were found different between clones and comparators include yield score, the proportion of longissimus thoracis muscle to body weight, the muscle moisture, and the amount of crude protein in the semitendinosus muscle, all fall within the normal range of industry standards. Therefore, none of these parameters would be a cause for concern to product safety.

The mechanisms of regulation of muscle development, differentiation, and growth are numerous and complex. Meeting the challenge of optimizing the efficiency of muscle growth and meat quality requires a thorough understanding of these processes in the different meat-producing species. Application of biotechnology for livestock and meat production potentially will improve the economics of production, reduce environmental impact of production, improve pathogen resistance, improve meat quality and nutritional content, and allow production of novel products for food, agricultural, and biomedical industries.

In a recent article by Wall et al. (2005), the authors reported the success of genetically enhanced cows with lysostaphin to resist intra-mammary *Staphylococcus aureus* (mastitis) infection. Mastitis is the most consequential disease in dairy cattle and costs the U.S. dairy industry billions of dollars annually. Their findings indicated that genetic engineering of animals can provide a viable tool for enhancing resistance to the disease and thus improving the well-being of the livestock.

Ovine

Although the first mammalian species to be cloned using a differentiated cell (Wilmut et al., 1997) was ovine, continued development of cloning technology in this species has been in support of conserving endangered species (Loi et al., 2001; Ryder, 2002). About 5–10% of cloned sheep embryos result in offspring, but not all are healthy. Several groups have attempted transgenic introduction of growth hormone (GH) genes in sheep, but none have resulted in commercially useful transgenic animals. Growth promoting TG in sheep was first accomplished by Hammer et al. (1985) followed by Rexroad et al. (1989, 1991) where gene constructs inserted into the sheep produced a 10–20 times elevation of plasma GH level. Growth rates were similar to the control sheep early in life, but after 15–17 weeks of life, the over expression of GH was cited by Ward et al. (1989) and Rexroad et al. (1989) to be responsible for reduced growth rate and shortened life span. Ward et al. (1990) summarized their studies with transgenic sheep, noting reduced carcass fat, elevated metabolic rate and heat production, skeletal abnormalities, and impaired survival due to the unregulated production of GH in the transgenic sheep unless an all ovine construct was used.

The pattern of expression of the various growth hormones and growth-hormone releasing factor (GRF) TG in sheep could not be predicted (Murray and Rexroad, 1991), since circulating levels of GH and insulin-like growth factor I (IGF-I) levels did not correlate to expression of the TG. Transgenic sheep that were non-expressing had transgenic progeny that also failed to express the TG (Murray and Rexroad, 1991). Transgenic lambs which expressed either GH or GRF had growth rates similar to non-transgenic controls, even though the transgenic lambs had elevated plasma levels of IGF-I and insulin. Early literature on transgenic sheep expressing GH indicated similar growth rates and feed efficiency (Rexroad et al., 1989) as non-transgenic controls; however, all transgenic sheep displayed pathologies and shortened life span. Further, transgenic sheep expressing GH, were noted to have significantly reduced amounts of body and perirenal fat (Ward et al., 1990; Nancarrow et al., 1991), and were also susceptible to developing chronically elevated glucose and insulin levels of diabetic conditions.

Progress in overcoming the health problems of GH transgenic sheep was made by switching to an ovine GH gene with an ovine metallothionein promoter (Ward and Brown, 1998). They encountered no health problems through, at least, the first four years of life; although Ward and Brown (1998) noted increased organ sizes and noticeably reduced carcass fat in the G1 generation. Twenty transgenic lambs of the G2 generation (Ward and Brown, 1998) grew significantly faster than the controls, with differences detected between rams and ewes. Growth rate of transgenic rams was greater than controls from birth onwards; whereas, increased growth rate in transgenic ewes were not noted until 4 months of age. No difference in feed conversion from 4–7 months of age was observed between control and transgenic lambs (Ward and Brown, 1998). In the G3 generation, Brown and Ward (2000) reported the average difference in body weight between transgenic and controls at 12 months of age was 8 and 19% heavier for rams and ewes, respectively. Their results were consistent with the increased circulating levels of GH in the transgenics compared to controls.

Piper, Bell, Ward, and Brown (2001) evaluated the effects of an ovine GH TG on lamb growth and the wool production performance using 62 transgenic Merino sheep. The G4 transgenic lambs were from a single transgenic founder ram and were compared to 46 sibling controls. Pre-weaning body weights were similar for transgenics and controls, but began to diverge and were significantly different from 7 months of age onward. Transgenic lambs were about 15% larger than the controls at 12 months of age and had a very low amount of subcutaneous fat. Major wool production traits, greasy fleece weight and mean fiber diameter, were not different from the controls.

Adams, Briegel, and Ward (2002) also examined the effects of a TG encoding ovine GH and an ovine metallothionein promoter, in the progeny of 69 Merino and 49 Poll Dorset lambs from ewes inseminated by G4 transgenic rams heterozygous for the gene construct. As seen in earlier research using mouse-derived GH transgenes, the effects of the ovine construct varied according to the active expression of the TG. The TG failed to be expressed in some progeny (Adams et al., 2002) despite a positive status for the TG. The ovine GH produced negligible health problems, similar to that reported by Ward and Brown (1998). Among the progeny with active TG expression, plasma GH levels were twice those of the controls. Those sheep also grew faster to heavier weights and were leaner, but had higher parasite fecal egg counts compared to the non-transgenic sheep. Females at 18 months of age had decreased longissimus muscle depth compared to males. Adams et al. (2006) concluded that phenotypic effects of genetic manipulation of sheep may depend on age, breed, and sex of the animal and that modification to the fusion genes is required to meet the species-specific requirements to enhance expression in the transgenic sheep while maintaining the long-term health status.

Callipyge sheep have muscle fiber hypertrophy determined by a paternally inherited polar overdominance allele (Cockett et al., 1994), which is a result of a single base change (Freking et al., 2002; Freking, Smith, & Leymaster, 2004). This naturally occurring mutation that alters the muscle phenotype in sheep was described by Jackson and Green (1993) and Cockett et al. (1994), and since has been subject of much research. The callipyge phenotype is a post-translational effect (Charlier et al., 2001), in which the dam's normal allele suppresses the synthesis of at least four proteins that form muscle tissue. The phenotype is characterized by hypertrophy in certain muscles, vis., longissimus thoracis et lumborum (LTL), gluteus medius, semimembranosus, semitendinosus, adductor, quadriceps femoris, BF, and triceps brachii, while other muscles, such as IS, and supraspinatus (SS), are unaffected. The hypertrophy is caused by increased size of the fast-twitch fibers rather than increased fiber numbers (Carpenter, Rice, Cockett, & Snowder, 1996). Lorenzen et al. (1997) measured the elevated protein/ DNA ratio in callipyge LTL and BF but not in IS and SS muscles. Fractional protein accretion rate did not differ among those muscles, and protein synthesis rate was decreased by 22% in callipyge LTL and by 16% in callipyge BF muscles. Since the protein degradation rate was also decreased by 35% in callipyge compared to the controls, Lorenzen et al. (1997) concluded that callipyge-induced muscle hypertrophy was due to decreased muscle protein degradation. Reduced tenderness in callipyge was also related to higher calpastatin (CAST) (Koohmaraie, Shackelford, Wheeler, Lonergan, & Doumit, 1995; Freking et al., 1999; Goodson, Miller, & Savell, 2001) and m-calpain activities (Koohmaraie et al., 1995) compared to the control sheep. Otani et al. (2004) presented an evidence in mice that overexpression of CAST contributes to muscle hypertrophy, although this has not been investigated in relation to the callipyge phenotype.

Busboom et al. (1994) indicated that callipyge lambs had less monounsaturated and more polyunsaturated fatty acids (PUFA) than the controls. Muscle hypertrophy in callipyge sheep was also at the expense of adipose tissue (Rule, Moss, Snowder, & Cockett, 2002), possibly from a decrease in differentiation of the adipocytes. Rule et al. (2002) measured lower lipogenic enzyme activities in adipose tissues of heterozygous callipyge lambs compared to the controls, but were unable to relate these differences to insulin or IGF-I levels. The callipyge locus has been mapped to a chromosome segment that carries four genes that are preferentially expressed in the skeletal muscle and are subject to parental imprinting, namely, Delta-like 1 (DLK1), gene-trap locus 2 (GTL2), paternal expressed gene 11 (PEG11), and maternal expressed gene 8 (MEG8). The same conserved order was found on human and mouse chromosomes. The causative mutation for callipyge is a single base transition from A to G in the inter-gene region between DLK1 and GLT2 (Bidwell et al., 2004). Charlier et al. (2001) demonstrated the unique and very abundant expression of DLK1 (involved in adipogenesis) and PEG11 (unknown function) in callipyge sheep; however, the authors were not able to explain how the over expression of these genes were related to muscle hypertrophy. They suggested that the callipyge mutation does not alter the imprinting of DLK1 or PEG11, but modifies the activity of a common regulatory element which could be an enhancer or silencer. Bidwell et al. (2004) similarly detected elevated DLK1 and PEG11 in the muscles of lambs with the callipyge allele and named them as candidate genes responsible for the skeletal muscle hypertrophy. PEG11 was 200 times higher in heterozygous and 13 times higher in homozygous callipyge sheep than in the controls. Freking et al. (2004) discussed expression profiles and imprint status of genes near the mutated region of the callipyge locus. Markers for polymorphic genes that control fatness and leanness, such as, thyroglobulin, or the callipyge gene, could be used for making genetic selection improvements in animals (Sillence, 2004).

The apparent advantages of higher carcass yield, increased lean and reduced fat content of callipyge sheep would benefit the meat industry except for the associated toughness in the hypertrophied muscles. In contrast to minimal tenderness

1 Transgenic Farm Animals

improvement using ante-mortem techniques to control growth rate, size, or fatness level (Duckett, Snowder, & Cockett, 2000) or treatment with dietary vitamin D_3 (Wiegand, Parrish, Morrical, & Huff-Lonergan, 2001), some success at improving the tenderness of meat from callipyge has been accomplished by various postmortem treatments. Tenderness was improved slightly by electrical stimulation (Kerth, Cain, Jackson, Ramsey, & Miller, 1999). Other post-mortem treatments effective for improving the tenderness in callipyge include prerigor freezing prior to aging (Duckett, Klein, Dodson, & Snowder, 1998), calcium chloride injection (Koohmaraie, Shackelford, & Wheeler, 1998), hydrodynamic pressure treatment (Solomon, 1999), and extended aging to 48 days (Kuber et al., 2003). The higher CAST level responsible for the hypertrophy of callipyge lambs (Koohmaraie et al., 1995; Freking et al., 1999; Goodson et al., 2001) is often cited as contributing to the lower tenderness of the meat because CAST interferes with the normal post-mortem proteolysis during aging, particularly the breakdown of troponin-T (Wiegand et al., 2001). The lack of tenderness associated with the callipyge gene must be addressed before the economic advantages can be realized.

Porcine

Among major livestock species, the pig was last to be cloned (Onishi et al., 2000; Polejaeva et al., 2000; Betthauser et al., 2000). There appears to be more interest in transgenesis and cloning of pigs as a model for studying human diseases, such as osteoporosis and diabetes, and for donor organs for xeno-transplantation rather than for improving meat production. Pigs, due to their vast numbers and similar organ size and function like that of humans, are desirable for xeno-transplantation. Hyperacute rejection of xeno-transplanted organs was a major concern until Prather, Hawley, Carter, Lai, and Greenstein (2003) accomplished genetic modification of the (1,3)-galactosyltransferase gene prior to nuclear transfer cloning. Nuclear transfer cloning efficiency rates for swine averages between 1 and 6% of embryos. This and other issues need to be solved with this technology. Cloned pigs appear to have inadequate immune systems (Carroll, Korte, Dowd, & Prather, 2004), display behavioral variations (Archer, Friend, Piedrahita, Nevill, & Walker, 2003), and could transmit viruses (van der Laan et al., 2000). In contrast, Carter et al. (2002) used green fluorescent protein TG and then cloned pigs to evaluate the phenotype and health status. They declared that cloned pigs can be normal and without impaired immune system.

Approximately 40% of the red meat consumed worldwide comes from pigs (FAO, 2004), and pork consumption has increased consistently with increasing world population. Continued improvements in pork production, therefore, are needed to meet future demands for red meat. Research in genomics is one avenue to increase production efficiency. Selection of pigs based on the ranodyne receptor (RyR) gene, muscle regulatory factor (MRF) gene family, hormones, or other potential candidate genes affecting growth and fattening traits are needed to increase production. QTL evaluation of factors associated with meat quality and growth are underway; however, in pigs, some quality traits are polygenic (Krzecio et al., 2004b) requiring evaluation of their interactions.

In pigs, halothane sensitivity is associated with malignant hyperthermia syndrome and reduced meat quality. Kortz et al. (2004) evaluated meat quality parameters like pH, water binding capacity, water-soluble protein content, and meat color, among other traits to determine the frequency of occurrence of normal vs PSE (pale, soft, exudative) meat quality. Pigs that were recessively homozygous (nn) for halothane sensitivity had higher amount of carcass lean and had higher frequencies of PSE than the dominant homozygous (NN) pigs. The heterozygous genotype (Nn) pigs had the leanest and a lower proportion of carcasses with partial or fully PSE meat. The NN genotype did not guarantee PSE free meat as PSE was also observed in NN carcasses. Milan et al. (2000) related the Rendement Napole (RN) allele, which originated in Hampshire breed of pigs, to 70% increased glycogen content in the muscle and poor water binding quality. Hedegaard et al. (2004) characterized proteome patterns related to the porcine RN− genotype and showed changes in the expression and activity of the key enzymes of glycolysis as well as down-regulation of an intracellular antioxidant enzyme. The RN− mutation likely leads to a loss of function resulting in the reduced degradation of glycogen, based on adenosine monophosphate-activated protein kinase (AMPK) activity which is approximately three times lower in RN− than in normal rn+ pigs (Hedegaard et al., 2004). The RN− allele is of interest to pig breeders because it is also associated with increased growth rate and lean content in the carcass. The negative outcome of this mutation, however, is lower 24 hours post-mortem muscle pH, reduced water binding capacity, and reduced cooked ham yields. The RN− was mapped to a mutation, coined PRKAG3, which is the third isoform identified of a mammalian AMPK. AMPK plays a central role in regulating energy metabolism through glucose transport into the cell and in fatty acid synthesis and oxidation. The muscle-specific expression of PRKAG3 is consistent with the fact that RN− pigs have high glycogen content in their muscles but not in the liver. The PRKAG3 mutation was identified by seven nucleotide differences between rn+/rn+ and RN−/RN− pigs. Analysis of the single nucleotide polymorphisms further identified the 200 codon region to be the causative polymorphism. This 200Q substitution was found in RN− pigs but not in any rn+ pigs. Functional characterization of the RN− mutation is complicated by its location in a regulatory subunit of AMPK and by the expression of several isoforms of AMPK in skeletal muscle. Completion of the porcine genome sequence will increase the identification of genes and interactions with other genes associated with controlling muscle and fat. Transgenesis to inhibit or increase the action of these genes may prove useful in increasing pork production.

QTL analysis of factors affecting tenderness and juiciness of the pork were mapped to chromosome 2, and based on that location the CAST gene was considered (Ciobanu et al., 2004) a likely candidate. Meat quality traits in pigs negative for the halothane sensitivity ryanodyne receptor (RyR1) and RN− alleles were evaluated for interactions with CAST (Krzecio, Kury, Kocwin-Podsiada, & Monin, 2004a). For stress-resistant RyR1 pigs, CAST polymorphisms using the Rsa1 restriction enzyme (CAST/Rsa1) were identified as AA, AB, and BB genotypes. These were found to affect water holding capacity (WHC), drip loss, and water and protein content of the muscle. CAST/Rsa1 AA genotype pigs had lower WHC, lower drip loss at 96 hours, less moisture and higher protein content in muscle compared to BB genotype. Stress resistant pigs (homozygous and heterozygous RyR1 resistant genotype) had highly significant lactate level, pH at 35 and 45 minutes post-mortem and on reflectance values. Homozygous stress resistant pigs produced the most desirable quality traits. The interaction of CAST/Rsa1 and RyR1 was significant for the longissimus lumborum muscle pH at 45 minutes post-mortem and drip loss at 48 h; however, no interactions were detected for carcass lean (Krzecio et al., 2004a, 2004b) or cooking yield. That CAST and RyR1 would interact is not surprising since CAST is an endogenous inhibitor of calcium-dependent cysteine proteases, the calpains, and a mutation in RyR1 is partly responsible for the disturbed regulation of intracellular Ca^{2+} in pig skeletal muscle (Kuryl, Krzecio, Kocwin-Podsiada, & Monin, 2004). These studies indicate that the quality of meat should be considered not only by each individual genotype, but also by the interactions with other genes.

Polymorphisms of the CAST gene and their association between genotypes at the porcine loci MSTN growth differentiation factor 8 were considered by Klosowska et al. (2005). Mutations in the MSTN gene are responsible for extreme muscle hypertrophy, or double muscling, in several breeds of cattle. MSTN is important for controlling the development of muscle fibers and is considered to be a negative regulator of muscle growth (McPherron, Lawler, & Lee, 1997). Since calpain activity is required for myoblast fusion, cell proliferation and growth, it may also affect the number of skeletal muscle fibers. The fusion of myoblasts to form fibers is accompanied by a dramatic change in the calpain/CAST ratio. Over expression of CAST, an endogenous calpain inhibitor in transgenic mice resulted in substantially increased muscle tissue (Otani et al., 2004). Klosowska et al. (2005) analyzed the interaction of MSTN and CAST in Piétrain × (Polish Large White × Polish Landrace) cross-bred pigs and the Stamboek line of Dutch Large White × Dutch Landrace pigs. The MSTN genotypes identified using the Taq1 restriction enzyme were CC or CT, and CAST/Rsa1 genotypes were identified as EE, EF, or FF. They reported that 79.5% of the Stamboek line was characterized as MSTN/Taq1 CC genotype. Interestingly, the FF genotype of CAST/Rsa1 was not detected in the Piétrain cross-bred pigs. Muscle fiber size and type distributions were not affected by the MSTN genotypes although there were breed differences. Piétrain crosses had larger mean fiber diameters in all the fiber types compared to Stamboek pigs. Proportion of fiber types in a bundle was higher for slow-twitch oxidative (SO) and lower for fast-twitch glycolytic (FG) fibers in Piétrain cross-bred pigs compared to Stamboek pigs. Of the multiple deletions or substitutions identified for MSTN, only one results in muscle hypertrophy seen in double muscle cattle and in mice. The C to T replacement in the MSTN gene does not result in an amino acid substitution (Stratil and Kopecny, 1999), thus it is probable that this genotype has no effect on the MSTN function in pigs. Muscle fiber diameters and the number of fibers per unit area were not different for CAST genotypes in Piétrain cross pigs, whereas, the CAST genotype had an effect in the Stamboek line. In all the fiber types, fiber diameters were larger in the CAST EE and EF genotypes and smallest in FF. Loin eye area of EE genotype also was significantly larger than for EF or FF genotypes. Because of the missing FF genotype in Piétrain cross pigs, the interaction of CAST and MSTN could not be assessed.

Transgenic pigs expressing a plant gene, spinach desaturase, for the synthesis of the essential PUFAs, linoleic and linolenic acids, have been produced (Saeki et al., 2004), marking the first time that a plant gene has been functionally expressed in mammalian tissue. This transgenesis could result in a significant improvement in pork quality beneficial to human health. They detected levels of linoleic acid in adipocytes that was about ten times higher in transgenic than in the control pigs. Niemann (2004) suggested that modifying the fatty acid composition of products from domestic animals may make this technology more appealing to the public. High levels of dietary PUFA were shown to improve processing and increased PUFA in pork muscle. Earlier work with transgenic pigs and with injected porcine somatotropin also led to reduced levels of saturated fatty acids in pork (Pursel and Solomon, 1993; Solomon, Pursel, & Mitchell, 2002).

Many reports have documented the effects on growth of pigs receiving additional GH by exogenous administration or endogenously through transgenesis (Vize et al., 1988; Wieghart et al.,1988; Pursel et al., 1988; Pursel and Rexroad, 1993; Pursel and Solomon, 1993; Pursel et al., 1997; Solomon, Pursel, Paroczay, & Bolt, 1994). Transgenic pigs expressing IGF-I, a regulator of GH, have been described in detail (Solomon et al., 2002; Mitchell and Pursel, 2003; Pursel et al., 2001a, 2001b, 2004). Pursel et al. (2004) summarized the advances made in pigs expressing a skeletal α-actinin-hIGF-I TG, namely, the expression of IGF-I in skeletal muscles gradually improved body composition in transgenic pigs without major effects on growth performance. Lean tissue accretion rates were significantly higher (30.3 and 31.6%), and fat accretion rates were 20.7 and 23.7% lower in transgenic gilts and boars, respectively, compared to controls. Body fat, bone, and lean tissue measurements by dual-energy X-ray absorptiometry confirmed that transgenic pigs had less fat and bone but higher lean tissue amount than the control pigs.

Dietary conjugated linolenic acid (CLA) and IGF-I TG had little or no effect on pork quality (Eastridge, Solomon,Pursel, Mitchell, & Arguello, 2001; Solomon et al., 2002). Carcass weight of IGF-I TG pigs was less than non-TG controls; however, TG pigs had a 16% larger loin eye area, 26–28% reduced back fat thickness, and 21% less carcass fat. Dietary CLA acted synergistically with the IGF-I TG in reducing back fat thickness. Muscle pH at 45 minutes (pH_{45}) was lower ($p < 0.01$) in TG than non-TG (6.0 vs 6.1), while dietary CLA resulted in significantly higher pH_{45} than for pigs fed with control diets (pH_{45} 6.1 vs 6.0). At 24 hours, muscle pH was not different, averaging pH 5.6, for all carcasses. Neither the gene status nor dietary CLA affected drip/purge loss during the 21 days refrigerated storage in a vacuum package, pork chop cooking yield, or thiobarbituric reactive substances measured in vacuum packaged loins stored for 5 and 21 days fresh and 6 months frozen. In pigs receiving the control diet, pork chop tenderness was improved significantly,i.e., lower shear force values, in IGF-I TG compared to non-TG (5.3 vs 7.0 kgf). Dietary CLA improved the tenderness in non-TG pigs equivalent to the tenderness of TG. Wiegand et al. (2001) detected no effects of CLA supplementation of swine diets on sensory attributes; although, it improved meat color, marbling, and firmness. Bee (2001) detected no effect of CLA on pig growth performance, carcass lean, or fat deposition, but there was a marked effect on fatty acid profiles. Saturated fatty acids, palmitic and stearic, were increased significantly while monounsaturated linoleic and polyunsaturated arachidonic acids were reduced. Activity of lipogenic enzymes in vitro was not altered by the dietary CLA suggesting that lipogenesis was not affected by CLA (Bee, 2001).

Directing IGF-I expression specifically to skeletal muscle appeared to overcome the problems encountered with GH transgenics or with daily injections of exogenous IGF-I (Pursel et al., 2004) and clearly had a major impact on carcass composition. Piétrain pigs have 5–10% more meat than comparable pigs of other breeds (Houba and te Pas, 2004), although the muscle hypertrophy phenotype in Piétrain pigs is not as strongly expressed as the double-muscle condition in cattle or callipyge in sheep. The mechanism of Piétrain pig hypertrophy is still unknown; however, it may be associated with changes to the CAST gene. Klosowska et al. (2005) did not detect a CAST polymorphism FF genotype in Piétrain cross-bred pigs. Pigs with the FF CAST genotype had smaller muscle fiber diameters compared to the EE and EF phenotypes. Linking the CAST genotype with phenotype to meat quality would benefit the meat industry, especially in pigs. The relationship between the genotype at the CAST and MSTN loci to phenotype remains to be elucidated.

Conclusions

The development of recombinant DNA technology has enabled scientists to isolate single genes, analyze and modify their nucleotide structure(s), make copies of these isolated genes, and insert copies of these genes into the genome of plants and animals. The transgenic technology of adding genes to livestock species has been widely adopted because it is technically straightforward, although it is not efficient. The primary goal of transgenesis is to establish a new genetic line of animals, in which the trait(s) of concern are stably transmitted to succeeding generations. Not all injected eggs will develop into transgenic animals and not all transgenic animals will express the TG in the desired manner. Eating quality and food safety must not be compromised as meat animals are designed and developed using these biotechnological approaches.

References

Adams, N. R., Briegel, J. R., Pethick, D. W., & Cake, M. A. (2006). Carcass and meat characteristics of sheep with an additional growth hormone gene. *Australian Journal of Agricultural Research, 57*, 1321–1325.

Adams, N. R., Briegel, J. R., & Ward, K. A. (2002). The impact of a transgene for ovine growth hormone on the performance of two breeds of sheep. *Journal of Animal Science, 80*, 2325–2333.

Archer, G. S., Friend, T. H., Piedrahita, J., Nevill, C. H., & Walker, S. (2003). Behavioral variation among cloned pigs. *Applied Animal Behaviour Science, 82*, 151–161.

Bee, G. (2001). Dietary conjugated linoleic acids affect tissue lipid composition but not de novo lipogenesis in finishing pigs. *Animal Research, 50*, 383–399.

Betthauser, J., Forsberg, E., Augenstein, M., Childs, L., Eilertsen, K., Enos, J., et al. (2000). Production of cloned pigs from in vitro systems. *Nature Biotechnology, 18*, 1055–1059.

Bidwell, C. A., Kramer, L. N., Perkins, A. C., Hadfield, T. S., Moody, D. E., & Cockett, N. E. (2004). Expression of PEG11 and PEG11AS transcripts in normal and callipyge sheep. *BMC Biology, 2*, 17–27.

Brown, B. W., & Ward, K. A. (2000). *14th International Congress on Animal Reproduction, 14*, 250 (Abstract No. 19:22).

Busboom, J. R., Hendrix, W. F., Gaskins, C. T., Cronrath, J. D., Jeremiah, L. E., & Gibson, L. L. (1994). Cutability, fatty acid profiles and palatability of callipyge and normal lambs. *Journal of Animal Science, 72* (Suppl. 1), 60.

Carpenter, C. E., Rice, O. D., Cockett, N. E., & Snowder, G. D. (1996). Histology and composition of muscles from normal and callipyge lambs. *Journal of Animal Science, 74*, 388–393.

Carroll, J. A., Carter, D. B., Korte, S., Dowd, S. E., & Prather, R. (2004). The acute-phase response of cloned pigs following an immune challenge. Retrieved from http://www.ars.usda.gov/research/publications/publications.htm?SEQ_NO_115=170690.

Carter, D. B., Lai, L., Park, K. W., Samuel, M., Lattimer, J. C., Jordan, K. R., et al. (2002). Phenotyping of transgenic cloned piglets. *Cloning and Stem Cells, 4*, 131–145.

Charlier, C., Segers, K., Karim, L., Shay, T., Gyapay, G., Cockett, N., et al. (2001). The callipyge mutation enhances the expression of coregulated imprinted genes in cis without affecting their imprinting status. *Nature Genetics, 27*, 367–369.

Ciobanu, D. C., Bastiaanseni, J. W. M., Lonergan, S. M., Thomsen, H., Dekkers, J. C. M., Plastow, G. S., et al. (2004). New alleles in calpastatin gene are associated with meat quality traits in pigs. *Journal of Animal Science, 82*, 2829–2839.

Cockett, N. E., Jackson, S. P., Shay, T. L., Nielsen, D. M., Moore, S. S., Steele, M. R., et al. (1994). Chromosomal localization of the callipyge gene in sheep (*Ovis aries*) using bovine DNA markers. *Proceedings of the National Academy of Sciences of the United States of America, 91*, 3019–3023.

Diles, J. J. B., Green, R. D., Shepard, H. H., Mathiews, G. L., Hughes, L. J., & Miller, M. F. (1996). Relationships between body measurements obtained on yearling brangus bulls and measures of carcass merit obtained from their steer clone-mates. *The Professional Animal Scientist, 12*, 244–249.

Duckett, S. K., Klein, T. A., Dodson, M. V., & Snowder, G. D. (1998). Tenderness of normal and callipyge lamb aged fresh or after freezing. *Meat Science, 49*, 19–26.

Duckett, S. K., Snowder, G. D., & Cockett, N. E. (2000). Effect of the callipyge gene on muscle growth, calpastatin activity, and tenderness of three muscles across the growth curve. *Journal of Animal Science, 78*, 2836–2841.

Eastridge, J. S., Solomon, M. B., Pursel, V. G., Mitchell, A. D., & Arguello, A. (2001). Dietary conjugated linoleic acid and IGF-I transgene effects on pork quality. *Journal of Animal Science, 79* (Suppl. 1). *Proceedings of the Reciprocal Meat Conference*, 54 (Vol. II), 20 (Abstract No. 85).

FAO (2004). *The state of agricultural commodity markets*. Food and Agriculture Organization of the United Nations. ISBN: 925105133X.

FDA (2003). *Executive summary of the assessment of safety of animal cloning. Food and Drug Administration*. Retrieved October 31, 2003, from http://www.fda.gov/bbs/topics/news/2003/new00968.html.

Freking, B. A., Keele, J. W., Shackelford, S. D., Wheeler, T. L., Koohmaraie, M., Nielsen, M. K., et al. (1999). Evaluation of the ovine callipyge locus: III. Genotypic effects on meat quality traits. *Journal of Animal Science, 77*, 2336–2344.

Freking, B. A., Murphy, S. K., Wylie, A. A., Rhodes, S. J., Keele, J. W., Leymaster, K. A., et al. (2002). Identification of the single base change causing the callipyge muscle hypertrophy phenotype, the only known example of polar overdominance in mammals. *Genome Research, 12,* 1496–1506.

Freking, B. A., Smith, T. P. L., & Leymaster, K. A. (2004). The callipyge mutation for sheep muscular hypertrophy – genetics, physiology and meat quality. In M. F. W. te Pas, M. E. Everts, & H. P. Haagsman (Eds.), *Muscle development of livestock animals: Physiology, genetics and meat quality* (pp. 317–342). Wallingford, UK: CABI Publishing.

Gerken, C. L., Tatum, J. D., Morgan, J. B., & Smith. G. C. (1995). Use of genetically identical (clone) steers to determine the effects of estrogenic and androgenic implants on beef quality and palatability characteristics. *Journal of Animal Science, 73,* 3317–3324.

Goodson, K. J., Miller, R. K., & Savell, J. W. (2001). Carcass traits, muscle characteristics, and palatability attributes of lambs expressing the callipyge phenotype. *Meat Science, 58,* 381–387.

Gordon, J. W., Scangos, G. A., Plotkin, D. J., Barbosa, J. A., & Ruddle, F. H. (1980). Genetic transformation of mouse embryos by microinjection of purified DNA. *Proceedings of the National Academy of Sciences of the United States of America, 77,* 7380–7384.

Hammer, R. E., Pursel, V. G., Rexroad, Jr., C. E., Wall, R. J., Bolt, D. J., Ebert, K. M., et al. (1985). Production of transgenic rabbits, sheep and pigs by microinjection. *Nature, 315,* 680–683.

Harris, J. J., Lunt, D. K., Smith, S. B., Mies, W. L., Hale, D. S., Koohmaraie, M., et al. (1997). Live animal performance, carcass traits, and meat palatability of calf- and yearling-fed cloned steers. *Journal of Animal Science, 75,* 986–992.

Hedegaard, J., Horn, P., Lametsch, R., Møller, H. S., Roepstorff, P., Bendixen, C., et al. (2004). UDP-glucose pyrophosphorylase is upregulated in carriers of the porcine RN-mutation in the AMP-activated protein kinase. *Proteomics, 4,* 2448–2454.

Houba, P. H. J., & te Pas, M. F. W. (2004). The muscle regulatory factors gene family in relation to meat production. In M. F. W. te Pas & E. Haagsman (Eds.), *Muscle development of livestock animals: Physiology, genetics and meat quality* (pp. 201–224). Wallingford, Oxfordshire, UK: Cambridge, MA.

Jackson, S. P., & Green, R. D. (1993). Muscle trait inheritance, growth performance and feed efficiency of sheep exhibiting a muscle hypertrophy phenotype. *Journal of Animal Science 71* (Suppl. 1), 14–18.

Kerth, C. R., Cain, T. L., Jackson, S. P., Ramsey, C. B., & Miller, M. F. (1999). Electrical stimulation effects on tenderness of five muscles from Hampshire × Rambouillet crossbred lambs with the callipyge phenotype. *Journal of Animal Science, 77,* 2951–2955.

Kittredge, C. (2005). A question of chimeras. *The Scientist, 19,* 54–55.

Klosowska, D., Kury, J., Elminowska-Wenda, G., Kapelanski, W., Walasik, K., Pierzchaa, M., et al. (2005). An association between genotypes at the porcine loci MSTN (GDF8) and CAST and microstructural characteristics of m. longissimus lumborum: A preliminary study. *Archiv für Tierzucht, 48,* 50–59.

Koohmaraie, M., Shackelford, S. D., & Wheeler, T. L. (1998). Effects of prerigor freezing and calcium chloride injection on the tenderness of callipyge longissimus. *Journal of Animal Science, 76,* 1427–1432.

Koohmaraie, M., Shackelford, S. D., Wheeler, T. L., Lonergan, S. M., & Doumit, M. E. (1995). A muscle hypertrophy condition in lamb (callipyge): Characterization of effects on muscle growth and meat quality traits. *Journal of Animal Science, 73,* 3596–3607.

Kortz, J., Rybarczyk, A., Pietruszka, A., Czarnecki, R., Jakubowska, M., & Karamucki, T. (2004). Effect of HAL genotype on normal and faulty meat frequency in hybrid fatteners. *Polish Journal of Food & Nutrition Science, 13,* 387–390.

Krzecio, E., Kocwin-Podsiada, M., Kury, J., Antosik, K., Zybert, A., Sieczkowska, H., et al. (2004b). An association between genotype at the CAST locus (calpastatin) and meat quality traits in porkers free of RYR1 SUP T allele. *Animal Science Papers Reports, 22,* 489–496.

Krzecio, E., Kury, J., Kocwin-Podsiada, M., & Monin, G. (2004a). The influence of CAST/Rsal and RYR1 genotypes and their interactions on selected meat quality parameters in three groups of four-breed fatteners with different meat content of carcass. *Animal Science Papers Report, 22,* 469–478.

Kuber, P. S., Duckett, S. K., Busboom, J. R., Snowder, G. D., Dodson, M. V., Vierck, J. L., et al. (2003). Measuring the effects of phenotype and mechanical restraint on proteolytic degradation and rigor shortening in callipyge lamb longissimus dorsi muscle during extended aging. *Meat Science, 63*, 325–331.

Kuryl, J., Krzecio, E., Kocwin-Podsiada, M., & Monin, G. (2004). The influence of CAST and RYR1 genes polymorphism and their interactions on selected quality parameters in four-breed fatteners. *Animal Science Papers Report, 22*, 479–488.

van der Laan, L. J. W., Lockey, C., Griffeth, B. C., Frasler, F.S., Wilson, C.A., Onlons, D. E., et al. (2000). Infection by porcine endogenous retrovirus after islet xenotransplantation in SCID mice. *Nature, 407*, 90–94.

Lee, S.J. (2007). Quadrupling muscle mass in mice by targeting TGF-β signaling pathways. *PLoSONE, 2*, 1–7.

Lewis, C. (2001). A new kind of fish story: The coming of biotechnology animals. *FDA Consumer January–February*. Retrieved from http://www.cfsan.fda.gov/~dms/fdbiofish.html.

Loi, P., Ptak, G., Barboni, B., Fulka, Jr., J., Cappai, P., & Clinton, M. (2001). Genetic rescue of an endangered mammal by cross-species nuclear transfer using postmortem somatic cells. *Nature Biotechnology, 19*, 962–964.

Lorenzen, C. L., Fiorotto, M. L., Jahoor, F., Freetly, H. C., Shackelford, S. D., Wheeler, T. L., et al. (1997). Determination of the relative roles of muscle protein synthesis and protein degradation in callipyge-induced muscle hypertrophy. *Proceedings 50th Reciprocal Meat Conference* (p. 175) June 29–July 2, Ames, IA. Savoy, IL: American Meat Science Association.

McPherron, A. C., Lawler, A. M., & Lee, S. J. (1997). Regulation of skeletal muscle mass in mice by a new TGF-β super family member. *Nature, 387*, 83–90.

Milan, D., Jeon, J.-T., Looft, C., Amarger, V., Robic, A., Thelander, M., et al. (2000). A mutation in PRKAG3 associated with excess glycogen content in pig skeletal muscle. *Science, 288*, 1248–1251.

Mitchell, A. D., & Pursel, V. G. (2003). Efficiency of energy deposition and body composition of control and IGF-I transgenic pigs. In W. B. Souffrant & C. C. Metges (Eds.), *Progress in research on energy and protein metabolism, EAAP scientific series* (pp. 61–64), 109.

Mitchell, A. D., & Wall, R. J. (2004). In vivo evaluation of changes in body composition of transgenic mice expressing the myostatin pro domain using dual energy x-ray absorptiometry *FASEB Journal 18*, A210.

Murray, J. D., & Rexroad Jr., C. E. (1991). The development of sheep expressing growth promoting transgenes. *NABC Report, 3*, 251–263.

Nancarrow, C. D., Marshall, J. T., Clarkson, J. L., Murray, J. D., Millard, R. M., Shanahan, C. M., et al. (1991). Expression and physiology of performance regulating genes in transgenic sheep. *Journal of Reproduction and Fertility, 43* (Suppl.), 277–291.

Niemann, H. (2004). Transgenic pigs expressing plant genes. *Proceedings of the National Academy of Sciences of the United States of America, 101*, 7211–7212.

NRC (2002). Animal biotechnology: Science-based concerns. *National research council committee on defining science-based concerns associated with products of animal biotechnology, committee on agricultural biotechnology, health, and the environment, board on agriculture and natural resources, board life sciences, division on earth and life studies* (p. 181). Washington, DC: National Academies Press.

Onishi, A., Iwamoto, M., Akita, T., Mikawa, S., Takeda, K., Awata, T., et al. (2000). Pig cloning by microinjection of fetal fibroblast nuclei. *Science, 289*, 1188–1190.

Otani, K., Han, D.-H., Ford, E. L., Garcia-Roves, P. M., Ye, H., Horikawa, Y., Bell, G. I., et al. (2004). Calpain system regulates muscle mass and glucose transporter GLUT4 turnover. *Journal of Biological Chemistry, 278*, 20915–20920.

Palmiter, R. D., Brinster, R. L., Hammer, R. E., Trumbauer, M. E., Rosenfeld, M. G., Brinbert, N. C., et al. (1982). Dramatic growth of mice that develop from eggs microinjected with metallothionein-growth hormone fusion genes. *Nature, 300*, 611–615.

Piper, L. R., Bell, A. M., Ward, K. A., & Brown, B. W. (2001). Effect of ovine growth hormone transgenesis on performance of merino sheep at pasture. 1. Growth and wool traits to 12 months

of age. *Proceedings of the Association for the Advancement of Animal Breeding and Genetics, 14,* 257–260.
Polejaeva, I. A., Chen, S.-H., Vaught, T. D., Page, R. L., Mullins, J., Ball, S., et al. (2000). Cloned pigs produced by nuclear transfer from adult somatic cells. *Nature, 407,* 86–90.
Prather, R. S., Hawley, R. J., Carter, D. B., Lai, L., & Greenstein, J. L. (2003). Transgenic swine for biomedicine and agriculture. *Theriogen, 59,* 115–123.
Pursel, V. G., Campbell, R. G., Miller, K. F., Behringer, R. R., Palmiter, R. D., & Brinster, R. L. (1988). Growth potential of transgenic pigs expressing a bovine growth hormone gene. *Journal of Animal Science, 66* (Suppl. 1), 267.
Pursel, V. G., Mitchell, A. D., Bee, G., Elsasser, T. H., McMurtry, J. P., Wall, R. J., et al. (2004). Growth and tissue accretion rates of swine expressing an insulin-like growth factor I transgene. *Animal Biotechnology, 15,* 33–45.
Pursel, V. G., Mitchell, A. D., Wall, R. J., Coleman, M. E., & Schwartz, R. J. (2001a) Effect of an IGF-I transgene on tissue accretion rates in pigs. *Journal of Animal Science, 79* (Suppl. 1), 29 (Abstract No. 121).
Pursel, V. G., Mitchell, A. D., Wall, R. J., Solomon, M. B., Coleman, M. E., & Schwartz, R. J. (2001b). Transgenic research to enhance growth and lean carcass composition in swine. In J. P. Toutant & E. Balazs (Eds.), *Molecular farming proceedings OECD conference on molecular farming* (pp. 77–86). Paris: INRA Editions.
Pursel, V. G., & Rexroad, Jr., C. E. (1993). Status of research with transgenic farm animals. *Journal of Animal Science, 71* (Suppl. 3), 10–19.
Pursel, V. G., & Solomon, M. B. (1993). Alteration of carcass composition in transgenic swine. *Food Reviews International, 9,* 423–439.
Pursel, V. G., Wall, R. J., Solomon, M. B., Bolt, D. J., Murray, J. D., & Ward, K. A. (1997). Transfer of an ovine metallothionein-ovine growth hormone fusion gene into swine. *Journal of Animal Science, 75,* 2208–2214.
Rexroad, Jr., C. E., Hammer, R. E., Boh, D. J., Mayo, K. E., Frohman, L. A., Palmiter, R. D., et al. (1989). Production of transgenic sheep with growth-regulating genes. *Molecular and Reproductive Development, 1,* 164–169.
Rexroad, Jr., C. E., Mayo, K., Bolt, D. J., Elsasser, T. H., Miller, K. F., Behringer, R. R., et al. (1991). Transferrin- and albumin-directed expression of growth-related peptides in transgenic sheep. *Journal of Animal Science, 69,* 2995–3004.
Rule, D. C., Moss, G. E., Snowder, G. D., & Cockett, N. E. (2002). Adipose tissue lipogenic enzyme activity, serum IGF-I, and IGF-binding proteins in the callipyge lamb. *Sheep Goat Research Journal, 17,* 39–46.
Ryder, O. A. (2002). Cloning advances and challenges for conservation. *Trends in Biotechnology, 20,* 231–232.
Saeki, K., Matsumoto, K., Kinoshita, M., Suzuki, I., Tasaka, Y., Kano, K., et al. (2004). Functional expression of a 12 fatty acid desaturase gene from spinach in transgenic pigs. *Proceedings of the National Academy of Sciences of the United States of America, 101,* 6361–6366.
Sillence, M. N. (2004). Technologies for the control of fat and lean deposition in livestock. *Veterinary Journal, 167,* 242–257.
Solomon, M. B. (1999). The callipyge phenomenon: Tenderness intervention methods. *Journal of Animal Science, 77* (Suppl. 2), 238–242; *Journal of Dairy Science, 82* (Suppl. 2), 238–242.
Solomon, M. B., Pursel, V. G., & Mitchell, A. D. (2002). Biotechnology for meat quality enhancement. In F. Toldra (Ed.), *Research advances in the quality of meat and meat products* (pp. 17–31). Kerala, India: Research Signpost.
Solomon, M. B., Pursel, V. G., Paroczay, E. W., & Bolt, D. J. (1994). Lipid composition of carcass tissue from transgenic pigs expressing a bovine growth hormone gene. *Journal of Animal Science, 72,* 1242–1246.
Stratil, A., & Kopecny, M. (1999). Genomic organization, sequence and polymorphism of the porcine myostatin (GDF8; MSTN) gene. *Animal Genetics, 30,* 468–469.
Takahashi, S., & Ito, Y. (2004). Evaluation of meat products from cloned cattle: Biological and biochemical properties. *Cloning and Stem Cells, 6,* 165–171.

Tian, X. C., Kubota, C., Sakashita, K., Izaike, Y., Okano, R., Tabara, N., et al. (2005). Meat and milk compositions of bovine clones. *Proceedings of the National Academy of Sciences of the United States of America, 102*, 6261–6266.

Vize, P. D., Michalska, A. E., Ashman, R., Lloyd, B., Stone, B. A., Quinn, P., et al. (1998). Introduction of a porcine growth hormone fusion gene into transgenic pigs promotes growth. *Journal of Cell Science, 90*, 295–300.

Wall, R. J., Powell, A. M., Paape, M. J., Kerr, D. E., Bannerman, D. D., Pursel, V. G., et al. (2005). Genetically enhanced cows resist intramammary *Staphylococcus aureus* infection. *Nature Biotechnology, 23*, 445–451.

Ward, K. A., & Brown, B. W. (1998). The production of transgenic domestic livestock: Successes, failures and the need for nuclear transfer. *Reproductive and Fertility Development, 10*, 659–665.

Ward, K. A., Nancarrow, C. D., Byrne, C. R., Shanahan, C. M., Murray, J. D., Leish, Z., et al. (1990). The potential of transgenic animals for improved agricultural productivity. *OIE Revue Scientifique et Technique, 9*, 847–864.

Ward, K. A., Nancarrow, C. D., Murray, J. D., Wynn, P. C., Speck, P., & Hales, J. R. S. (1989). The physiological consequences of growth hormone fusion gene expression in transgenic sheep. *Journal of Cellular Biochemistry*, Suppl. 13b, 164, Abstract F006.

Wells, K. D. (2000). Genome modification for meat science: Techniques and applications. *Proceedings of 53rd Annual Reciprocal Meat Conference* (pp. 87–93). Ohio State University.

Wieghart, M., Hoover, J., Choe, S. H., McGrane, M. M., Rottman, F. M., Hanson, R. W., et al. (1988). Genetic engineering of livestock – transgenic pigs containing a chimeric bovine growth hormone (PEPCK/bGH) gene. *Journal of Animal Science, 66* (Suppl. 1), 266.

Wiegand, B. R., Parrish Jr., F. C., Morrical, D. G., & Huff-Lonergan, E. (2001). Feeding high levels of vitamin D3 does not improve tenderness of callipyge lamb loin chops. *Journal of Animal Science, 79*, 2086–2091.

Wiegand, B. R., Parrish Jr., F. C., Swan, J. E., Larsen, S. T., & Baas, T. J. (2001). Conjugated linoleic acid improves feed efficiency, decreases subcutaneous fat, and improves certain aspects of meat quality in stress-genotype pigs. *Journal of Animal Science, 79*, 2187–2195.

Wilmut, I., Schnieke, A. E., McWhir, J., Kind, A. J., & Campbell, K. H. S. (1997). Viable offspring derived from fetal and adult mammalian cells. *Nature, 385*, 810–813.

Yang, J., Ratovitski, T., Brady, J. P., Solomon, M. B., Wells, K. D., & Wall, R. J. (2001). Expression of myostatin pro domain results in muscular transgenic mice. *Molecular and Reproductive Development, 60*, 351–361.

Chapter 2
Genetic Control of Meat Quality Traits

John L. Williams

Introduction

Meat was originally produced from non-specialized animals that were used for a variety of purposes, in addition to being a source of food. However, selective breeding has resulted in "improved" breeds of cattle that are now used to produce either milk or beef, and specialized chicken lines that produce eggs or meat. These improved breeds are very productive under appropriate management systems. The selection methods used to create these specialized breeds were based on easily measured phenotypic variations, such as growth rate or physical size. Improvement in the desired trait was achieved by breeding directly from animals displaying the desired phenotype. However, more recently sophisticated genetic models have been developed using statistical approaches that consider phenotypic information collected, not only from individual animals but also from their parents, sibs, and progeny. This combined information allows the genetic potential of individuals to be the better predicted. The predicted potential for several traits can then be combined into an index which provides a measure of the overall genetic merit of the individual. Using these statistical approaches animals are selected for breeding using the index of their estimated breeding value (EBV), rather than directly on their phenotype.

The results of these phenotype focused selection approaches have been highly successful, with dramatic improvements in the traits under selection. Modern broiler chickens used for meat production are eight times larger and grow much more rapidly than layer types that have been selected for egg production. Specialized beef cattle grow rapidly reaching a mature size in less than a year compared with the 24–36 months required to "finish" traditional breeds. Milk production from specialized dairy cows has also increased dramatically over the past 20 years under selection. However, these past selection choices have resulted in new problems,

---

J.L. Williams
Parco Tecnologico Padano, Via Einstein, Polo Universitario, Lodi 26900, Italy

such as a decrease in fertility, which in some breeds now threatens the viability of production. Fertility is currently a major cause for concern for dairy cattle and broiler poultry producers. Dairy farmers are also faced with increasing lameness in their herds while poultry breeders have to cope with birds that have brittle bones. These are major welfare problems as well as threatening productivity. In addition, the inadvertent selection for genetic defects linked to desirable production characteristics is a potential risk, especially when selection programs focus on a limited number of breeding individuals.

The traits that are routinely recorded are, through necessity, simple and focused on the commercially important traits. There has been little opportunity, and in some cases desire, to select other traits, many of which in the past were considered less relevant. However, for the livestock production to remain sustainable, it will be necessary to consider a wide range of traits in selection programs, particularly those that have an impact on the health and welfare of the animals. While there has been an increase in the quantity of production, arguably, until recently there has been little or no attention paid to the quality or composition of the products. With the growing awareness of the consumers with respect to choices, there is now an increasing demand for better quality, as well as lower cost products. Consumers are increasingly more conscious of their own health and also the welfare of animals in agricultural production systems. Over many years, there have been welfare concerns regarding, for example, battery farming for egg production. The frequent focus of the mass media on both human and animal health issues has drawn the public attention to health and welfare problems in, or arising from, agricultural production. This has resulted in increased pressures to change the associated production practices. Following on from the devastating BSE outbreak in the UK in the late 1980s, the public has seen *Salmonella* contamination in egg production, Escherichia coli contamination of beef products, and Foot and Mouth disease in cattle and sheep to name a few examples, all of which have had a major impact on the credibility and financial sustainability of the livestock industry. In addition to the public concerns over the risks from pathogens, there is now the desire to have food with healthy composition, in particular, meat products with lower fat content. In response to these demands the industry has turned its' attention to the quality as well as quantity of production. The major problem to date has been that improvement can only be made in traits where there are reliable measurements recoding variations in the phenotype. This information is required to develop selection strategies. For many health and quality related traits, this data is difficult and expensive to obtain. Additionally, the information can be difficult to apply in breeding programs: e.g., measurements of meat texture, composition or flavor are complicated to carry out, require specific samples, and can only be made after an animal is slaughtered. It is therefore difficult to collect these data in a routine way, and obviously post-slaughter there is no opportunity to breed from animals with superior characteristics. A further complication is that a large part of the variation, particularly in meat quality traits, results from differences in environmental conditions, in particular differences in feed and handling.

Approaches and Tools for Genetic Selection

Although the traits that are important for beef production are influenced by the environment, e.g., management conditions, nutritional status and handling pre- and post-slaughter, the genetically controlled variation (heritability) of the traits important for production, and product quality is relatively high, between 0.15 and 0.35 (e.g., Wheeler, Cundiff, Shackelford, & Koohmaraie, 2004). This suggests that an appreciable proportion of the variation is under genetic control, and hence could be improved by selection. Knowledge of the genes controlling the variation in a trait would open the opportunity to use genomic information in selection programs. By choosing to breed from the animals with the most favorable alleles at important genes, the rate of animal improvement could be significantly increased. Importantly, these gene-based methods have the potential to facilitate the improvement of traits that are difficult to select for by the traditional phenotype based methods. Information on polymorphisms in genes controlling particular traits, and understanding the biological effects of these polymorphisms, will allow genetic information to be used effectively in animal improvement programs. However, progress can be made even before these "trait genes" have been identified. Markers that are genetically linked to the traits genes can potentially be used in marker-assisted selection (MAS) programs (Dekkers, 2004).

As recombination events are relatively rare, large regions of chromosomes are passed intact from one generation to the next. Thus, polymorphisms in the DNA sequence that are close to the trait genes, if used with care, can be used to predict the alleles present at the trait loci. The MAS approach has received considerable attention as the results of numerous genome mapping studies have been published which identify markers linked to the genes controlling important production traits. The majority of traits that are important in livestock production are "complex", that is, they are under the control of several genes, each contributing to a part of the observed phenotypic variation. Therefore, these have been called quantitative traits and hence the genetic loci controlling them quantitative trait loci (QTL). By using a relatively small number or markers, it has been possible to crudely identify the chromosomal locations of QTL containing some of the major the genes controlling a number of important production traits in livestock species. Markers for these QTL could be used for MAS. However, the major drawback is that it is necessary to determine the alleles at the linked markers that are predictive of the favorable allele at the trait locus, which is known as "phase". Phase has to be determined in every population and family in which the MAS will be undertaken, as recombination at the population level means that the association between markers and the trait gene cannot be assumed. In addition, the phase may change from one generation to the next, because of recombination; thus the linked markers can only be used with confidence for a limited number of generations. The likelihood of recombination between the trait gene and the markers is dependent on the genetic distance separating them. Having markers either very close to the trait gene or ideally knowing the functional variation within the gene means that there is a very low probability of recombination between the markers and the gene. In such cases, the markers can be used directly without first having to determine the phase.

Over the last two decades there has been a considerable effort to develop several different types of maps that cover the whole genomes of many species, including all the major livestock species. Genetic maps (e.g., pig: Archibald et al., 1995; Rohrer et al., 1996; and cattle: Barendse et al., 1997; Kappes et al., 1997) have been produced using recombination to determine the relative position of polymorphic markers to each other based on the frequency of recombination between them. In addition, several types of physical maps have been produced, ranging from large fragment clone maps (Snelling, Chiu, Schein, & The International Bovine BAC Mapping Consortium, 2007) and maps produced by *in situ* hybridization (Hayes et al., 1995; Solinas-Toldo, Lengauer, & Fries, 1995; Chowdhary, Fronicke, Gustavsson, & Scherthan, 1996) to maps produced based on the probability of chromosomal breaks occurring between loci following irradiation (McCarthy, 1996). Together these genome wide maps have led to rapid advances in understanding the structure of genomes and have provided the markers required for genetic mapping studies to localize and identify the trait genes. Following the publication of the human genome sequence (Lander et al., 2001), several projects were initiated to sequence the genomes of a large number of other species. For livestock, the first sequence to be published was that of the chicken (International Chicken Genome Sequencing Consortium, 2004); a draft of the bovine sequence was also made available in 2004 and a more complete draft is soon to be published by the Bovine Genome Consortium (2008). Work is progressing rapidly on sequencing the pig genome, with a full draft sequence expected early in 2009. Work is also underway on the sequence of the sheep genome. Availability of the whole genome sequence provides information on the number and location of genes, and on gene regulation and genetic variations. Availability of the genome sequence also allows tools to be developed which can be used to identify the genes controlling target traits more rapidly.

This chapter will describe the approaches that have been followed to investigate the genetic control of meat production traits and provide examples of the identification of trait genes controlling variation in meat quality related traits.

Definition of Meat Quality

Meat quality can be defined in a number of ways, but the focus is on those factors that affect consumer appreciation of the product. The main sensory factors which influence purchase are color and visible fat, and the primary factors affecting the enjoyment of consuming meat are texture and flavor. However, consumers are increasingly concerned with food safety from the point of view of health implications, e.g., the composition of poly-unsaturated vs saturated fat, and microbiological contamination. The management of animals can influence these factors, for example, feed can affect fat composition and flavor, while the rate of growth and hence age at slaughter can affect the texture. Texture and color of meat are strongly influenced by the way an animal is handed prior to slaughter, and then the treatment and processing of the carcass post-slaughter. The genetics of the individual may have a smaller influence on many quality characteristics than these management aspects, but nevertheless, genetic variation can make a difference. Indeed, genetic variation may control the way different management practices affect meat quality, such as propensity for and animal convert food into fat vs muscle. Genetics also affects muscle composition and hence texture. In principle, molecular genetic approaches could be applied to defining the response of an individual to environmental factors, and help to establish the correct management conditions to optimize meat composition and physical characteristics. In addition to contributing to the control of microbiological infections, the genetic control of immune response and susceptibility to disease is important for improving animal health and reducing the use of antibiotics and other veterinary products that may contaminate food products. The reduction of pathogens through improved genetic resistance would also reduce the potential for bacterial contamination on carcasses and processed meat.

Traditional Genetic Selection

Genetic improvement of livestock has been achieved by selective breeding, which has been highly successful in improving some traits. However, to establish breeding programs it is necessary to have performance records for the traits that will be selected for. In a commercial context, only simple measurements and recording procedures are possible, without interfering with the management practices and hence adding significantly to the costs of production. Hence the traits routinely recorded are easily and quickly measured. At slaughter, basic information is routinely collected on carcass quality. Some more complex measurements are undertaken by some producers, such as the use of ultra-sound to measure back-fat or muscle depth. However, in general these measurements are not systematically recorded in a centralized way that would allow the data to be used effectively is selection programs. Measurement and routine recording of the more difficult-to-measure traits has not been attempted at a commercial level for obvious reasons: cost or because the measurements are imprecise. Many of the traits associated with, e.g., meat quality, or health, are subjective, dependent on the criteria set and the person carrying out the measurements. Without establishing standardized and detailed protocols for recording traits it is impossible to compare measurements taken in different places and at different times. To standardize trait recording it is often necessary to carry out complex measurements, which are difficult to apply in large populations, and certainly in a commercial setting. There is an increased interest by producers in developing selection criteria that are aimed at improving quality, efficiency, and health traits. In some countries, the recording of more complex traits has been centralized at a national level, e.g., centralized health recording was pioneered by the Scandinavian countries and more recently, Ireland has introduced a national animal health recording system. However, to be effective, the recording protocols should be simple and standardized, preferably at an international level. To achieve this, the international organization responsible for animal recording, ICAR, can play an important role.

For some species, especially cattle, artificial insemination (AI) has contributed significantly to breed improvement by allowing individual elite sires to produce large numbers of progeny. When improvement is required in a trait that is sex limited, such as milk-related traits, or that which can only be determined post-slaughter, such as meat quality or composition, testing of the progeny of a sire allows his genetic quality for target traits to be estimated. The genetic index of an animal calculated from the performance of daughters or sons can be used to select the highest merit sires for breeding. Sophisticated statistical methods have been developed to analyze this "progeny test" data and identify sires that are above average for the desired trait. Commercial progeny test schemes have maximized the genetic gain in traits such as milk yield and composition, which are easy to measure in a commercial setting compared with meat quality traits. Even with easily measured traits, the progeny test schemes are very expensive; especially considering that a large number of the sires tested will not be used for breeding. To recover the costs of testing, a large number of semen doses have to be sold for each elite bull. Unless carefully managed, this breeding strategy can risk high levels of inbreeding and the associated loss of vigor, and the concentration of deleterious recessive alleles. Progeny testing and artificial reproductive approaches have been used extensively in dairy cattle breeding, where AI is now used ubiquitously, and as a result milk yields of the Holstein breed have more than trebled over the past 25 years. However, this highly focused selection strategy has lead to an alarming reduction in the effective population size of the breed. The occurrence of bovine leukocyte adhesion deficiency (BLAD) was a dramatic example of the potential problems associated with using a limited pool of elite sires. BLAD is a result of a mutation in one gene (CD18) that originated in a single bull, probably about 60 years ago. The effects of the mutation are recessive; therefore while the mutated allele remained at low frequency the adverse affects were not observed. However, a carrier bull turned out to be highly productive and large numbers of his sons were used extensively for breeding as elite AI sires, and in their turn their sons were used to breed further elite sires. By 1990 the frequency of the mutated allele had reached 15% in some countries, and animals homozygous for the mutations started appearing in the Holstein population. The effects were observed as a disease of young Holstein calves characterized by pneumonia, delayed healing of wounds and death (Gilbert et al., 1993). The mutated allele has now been effectively removed from the Holstein population by genetic testing. This example illustrates why selection programs should be coupled with good breeding management to maintain the effective population size and hence the genetic diversity present in the population: to avoid inbreeding and the accumulation of recessive defects.

For progeny selection to be effective, a large number of sires have to be tested, which is very expensive. Therefore, approaches to identify potentially superior animals at an early age would help to reduce cost and would mean that a higher proportion of bulls selected for testing were of high quality. For slow-growing or late-maturing species, juvenile predictors of adult performance can be used to speed up selection and reduce costs (Meuwissen, 1998). Juvenile predictors would also allow animals with high potential to be selected before many of the rearing costs had been incurred. However, up to now few reliable juvenile predictors have been identified. DNA markers offer the potential to select breeding animals at a very early age, indeed as embryos, and to enhance the reliability in predicting the mature phenotype of the individual.

New Opportunities

There are undoubtedly genetic factors that affect meat quality, such as fatty acid composition, fat distribution, muscle fiber type, etc. Up to 35% of the variation seen in muscle composition is under genetic control (Wheeler et al., 2004). In addition, significant differences have been found in the organo-leptic properties and composition of meat produced from different breeds of cattle and pigs, which also suggests that genetics plays an important role in controlling variations in meat quality. Analysis of the genetic make-up of breeds has shown that, although there is genetic variation *within* breeds, this is small compared with the variation found between breeds (e.g., Blott, Williams, & Haley, 1999). Thus, genetic selection could be used to improve meat quality, and would be most effectively achieved using DNA markers. However, up to now few of the genes controlling variability in meat quality and composition have been identified, and specifically, few functional variations within the genes that control the phenotypic differences are known. With recent technological advances this situation may be about to change.

The application of simple phenotype-guided selection will inevitably be influenced by conflicting choices when considering the diverse range of traits that are important at different levels of the production chain. Some traits appear to be obligatorily in conflict: i.e., when alleles of a particular gene are beneficial for one trait but have negative effects on another. Molecular genetic approaches can be used to aid breeding decisions and may allow selection for a wide variety of traits. Another problem occurs when the genes controlling different traits are close together on a chromosome. In this case, it may appear that there is only one locus having an effect on both traits, as alleles at the closely linked genetic loci will generally be inherited together. However, even for very closely linked genes, by examining sufficient individuals (meiosis), some chromosomes will be identified where recombination has occurred, even between very closely linked loci. Knowledge of the alleles at particular genetic loci and their genetic effects will allow direct selection choices to identify individuals with the most beneficial combination of alleles. Therefore, in theory at least, a strategy to simultaneously select for improved performance in a number of traits could be devised using genetic markers, even when at the phenotypic level the traits may seem to be in conflict. If applied with care, the use of molecular information in selection programs has the potential to increase productivity, enhance environmental adaptation, and maintain genetic diversity.

Advances in Knowledge

Over the past 20 years there have been rapid advances in the development of molecular biological techniques, which have been applied to understanding the regulation of gene expression and function. The application of these techniques to the field of genetics has advanced the knowledge of the structure of the genome and the identification of sequence variations between individuals, some of which have known effects on gene function and phenotypic variation. The most significant advance in the past few years has been the completion of the human genome sequencing project (Lander et al., 2001). This project spurred the development of new technologies that are allowing biological problems to be addressed on a large scale. With these new technologies, it is now possible to analyze many thousands of DNA sequences in a single day. And instead of studying the expression of individual genes, it is now possible to examine the expression of all the genes in the genome simultaneously and to address the interactions between genes. The resources developed to sequence the human genome were subsequently used to sequence the genomes of many other species. The first draft of the bovine sequence was released in October 2004, and a more complete sequence including the annotation of the genes was made available in 2007 (http://www.hgsc.bcm.tmc.edu/projects/bovine/ and http://www.ensembl.org). A genome-sequencing project for pigs is currently underway, and the entire pig sequence is likely to be available in 2009. The sequences of these genomes, together with the information on genetic variations, gene structure, expression and regulation, together with the new technologies for rapidly sequencing the genomes of individuals, will facilitate the identification of the genes controlling variations in commercially relevant traits. Information on polymorphisms, within these genes, could then be used to enhance selection programs, or to develop improved management strategies. Information on large numbers of genetic polymorphisms together with the highthroughput methods to genotype them opens the possibility of genome-wide selection, rather than focusing on a limited number of loci.

For genetic studies, the most important developments arising from the genome sequencing projects has been the identification of large numbers of differences (polymorphisms) in the DNA sequence between individuals. Some of these polymorphisms may be functional, in so far as they alter levels of gene expression or the activity of the protein encoded by the sequence, e.g., changing the affinity of a receptor for its ligand, or the activity of an enzyme. Other variations may be neutral if they occur in inter-genic regions outwith regulatory regions, or within coding regions of genes but are conservative, i.e., do not change the amino acid in the protein. The functional polymorphisms may be involved in controlling variations in phenotypes including those relevant to meat quality, such as muscle composition or structure. The number of DNA polymorphisms known, and the way they are detected, is rapidly changing the way the identification of the genes controlling particular traits is carried out. Before discussing these advances, some examples of different genetic markers will be described.

Genetic Markers

The earliest form of DNA marker used to construct the first true genomic maps was the Restriction Fragment Length Polymorphism (RFLP). Bacterial "restriction" enzymes bind and cut DNA molecules at highly specific recognition sequences. Variation in the target restriction enzyme binding site results in differences in the size of the fragments generated, following digestion with a restriction enzyme. Initially, RFLPs at specific positions in the genome had to be identified individually. This was a laborious process that could only investigate one gene at a time. A special form of RFLP allowed variations in loci that are present in multiple copies throughout the genome to be investigated at the same time. These "Variable Number Tandem Repeat" VNTR markers were successfully used to identify familial relationships between individuals in wild populations by creating "genetic fingerprints" and were also used in genetic mapping studies (e.g., Jeffreys, Wilson, & Thein, 1985; Georges et al., 1990). A major breakthrough came with the identification of microsatellite sequences. These are loci in the genome that contain typically 5–20 copies of a short sequence motif 2 and 4 bp in length, repeated in tandem (e.g., CGCGCGCG). These sequences have a relatively high mutation rate, resulting from DNA replication errors, and so at a population level the number of repeat units at a locus can be highly variable, providing a large number of alleles that can be used as markers in genetic analyses. The number of alleles at these "microsatellite sequences" is approximately proportional to the number of repeat units present.

This high allele number and amenability of microsatellite loci to polymerase chain reaction (PCR) amplification make them excellent markers for use in genetic studies, and indeed most of the gene mapping studies carried out in the past 10 years have used this type of marker. Genotyping the microsatellite locus was initially achieved by PCR using primers that flanked the microsatellite repeat region. The PCR products were labeled by the incorporation of radioactive nucleotides in the reaction, and alleles identified by determining the sizes of PCR product by gel electrophoresis. More recently, the use of fluorescent dyes and automated DNA analyzers allowed the simultaneous analysis of, typically, 5–10 different microsatellite loci simultaneously (multiplexing). Nevertheless, the gel electrophoresis-based methods required to genotype this type of marker mean that it is difficult to automate the procedures and the cost of genotyping remains high.

Genetic variations fall into two classes: insertions or deletions of DNA sequence (indels), of which the microsatellite loci are a special type, or changes to the nucleotide sequence, often at individual bases. These single nucleotide polymorphisms (SNPs) are much more frequent than indels in the genome and occur in both coding and non-coding regions. Estimates from genome sequencing projects in different species suggest that SNPs occur at a frequency of one in every 200 bp, on average. Thus, there are potentially many millions of SNPs in the genome. SNPs within coding regions may have no effect on the protein coded by the gene (silent polymorphisms) or may result in a change in an amino acid. The latter are more likely to be the functional polymorphisms responsible for variations in traits, although in some cases the functional polymorphism may occur in intergenic regions (see, e.g., IGF2 below). There have been considerable technological advances to facilitate the high throughput genotyping of SNP variations, and current and future whole genome genetic studies are likely to use this type of marker.

Identifying the Genes Controlling Phenotypic Variations

There are several approaches that can be used to identify the genes controlling phenotypic variations. For simple monogenic traits, knowledge of the physiology of the trait may allow the biochemical pathways involved to be analyzed to identify the gene(s) likely to be responsible for the observed differences in the phenotype. If the pattern of expression of these genes among tissues is known, this information can be used to clone the gene(s). Once cloned, the genes from individuals displaying different phenotypes can be sequenced and compared to reveal if there are polymorphisms present that are responsible for the phenotypic variation. Some of the first genes controlling variations in monogenic traits were identified in this way. However, to be successful, a good *a priori* knowledge of the trait and the underlying physiology is clearly required. Another approach is to use comparative information between species; in some cases knowledge of the gene that controls a phenotype in one species may suggest a candidate gene that could be tested in another. Finding polymorphisms within the candidate genes that co-segregate with the phenotypic variation in the trait will indicate if the gene is responsible for the variation. This comparative approach requires a good knowledge of the phenotype, as even subtle differences may be the result of the action of different genes. Even with a good knowledge of the physiology of the trait, other genes, that are not obviously part of the biochemical pathways involved, may contribute to the variation and so would not be considered as candidates. Therefore for complex traits, it may be better to start with no prior assumptions regarding the physiology and to use a genetic, or linkage, mapping approach.

There are two requirements for identifying genes controlling particular traits by linkage mapping. The first are the genetic markers that are used to track inheritance of chromosomal regions in families segregating for the trait. The second are the families that are segregating for the target traits, for which both phenotype and pedigree information is available. Using statistical methods to correlate inheritance of the phenotype with inheritance of the genetic markers in the families, it is possible to localize the gene(s) controlling the trait to broad regions of chromosomes. The chromosomal location is then used as the starting point to identify the trait genes themselves. In practice, the identification of the "trait genes" starting from a chromosomal location is not easy, and the successful identification of trait genes has usually been achieved using a combination of genetic mapping, to localize the QTL region on a chromosome, followed by a candidate gene or positional cloning approaches to identify the trait gene within the QTL region.

Following an initial low resolution linkage mapping study, it is usual that several putative QTLs will be identified. The likely position of these QTL may span quite large chromosomal regions, typically a quarter or possibly half of a chromosome, which may contain between 200 and 400 genes. It is therefore necessary to refine the QTL localization by fine mapping before assessing putative candidate genes in the region. The precision of the localization of the QTL is dependent on the number of recombination events in the region, which will reduce the amount of flanking chromosomal region that is inherited with the QTL. To increase the possibility of recombination in the target region, additional individuals in the segregating families have to be studied, and additional markers will be required within the region to detect the recombination events. The fine mapping of a trait gene can be achieved most efficiently by examining the target chromosomal region inherited by different branches of a family arising from a common ancestor. The more generations that have separated the different branches of a family from the common ancestor, the greater the reduction in the ancestral genome inherited with the trait gene: i.e., the amount of the genome in linkage disequilibrium (LD). The genomic regions coming from the common ancestor are *identical by descent* (IBD) and can refine the location of a QTL to small chromosomal intervals (see Anderson and Georges, 2004). Once the QTL region has been fine mapped, the next step is to examine the region carefully for any genes, which from their known function, may be involved in controlling the trait (a candidate gene). Likely "candidate" genes will then be examined for variations that may alter their function and hence may be responsible for the phenotypic variation observed. In the absence of any good candidate genes it will be necessary to clone and sequence the refined QTL region from individuals carrying different QTL alleles in order to identify specific genetic polymorphisms that are associated with the variation in the trait.

Analysis of gene expression patterns may also indicate specific genes or biochemical pathways that are involved in the regulation of phenotypic variation: genes that have an increased or decreased level of expression associated with a specific phenotype may be directly responsible for the observed variation, or may be involved in the regulation of other genes that are. For many years, the assay for gene expression has been cumbersome and has focused on individual genes. The development of methods to create arrays of thousands of different DNA probes has facilitated the examination of changes in the expression of large numbers of genes simultaneously. The explosion in the amount of genomic information available and the development of these microarrays have resulted in large amounts of data on gene expression from different tissues and individuals with different genotypes to be gathered very rapidly. Analysis of such information may suggest those genes responsible for variations in phenotypes.

Application of the Techniques to Locating the Genes Controlling Meat Quality Traits

In this section, the different approaches to identify genes controlling meat quality traits are described and examples given for the successful application in each approach.

Candidate Genes

The most straightforward approach for identifying the genes controlling a particular trait is to use knowledge of the physiology of the trait to identify the biochemical pathways involved. This will then suggest genes that may be important for controlling key processing in the development of the phenotype. These genes are *"candidate"* genes that can be tested by identifying polymorphisms within the genes and observing whether the occurrence of the polymorphisms can account for some or all of the variation observed in the trait. In order to select the candidate genes, it is necessary to have a good a priori knowledge of the trait and the underlying physiology. However, even with a good knowledge of the trait, not all of the important genes will be identified and important candidates may be missed if they are not obviously involved in the known physiology. An additional drawback of testing candidate genes is that most traits that are important for livestock production, such as feed efficiency, disease resistance, growth rate, or muscle composition, are not under the control of a single gene, but are controlled by several genes that have an additive effect. For most of these "complex" traits, a very large number of genes could be considered as potential candidates. Testing all of these potential candidate genes is not practical; thus the candidate gene approach is better used in conjunction with other approaches that can refine the choice. A two-step approach, in which the chromosomal location of the gene is identified using a conventional linkage mapping approach, then this information used as the starting point to select *positional candidate genes* within the chromosomal location has proved successful. The application of these approaches is discussed below with reference to examples.

Examples of Genes and QTL Mapping in Livestock

Meat quality

In man, some individuals are know to have an adverse response to halothane anesthetic, which results in muscle spasm and a dangerous drop in muscle temperature. This condition is called malignant hypothermia and is a genetic disease controlled by a single major gene (HAL). A similar condition has been described in pigs. In addition to the adverse reaction to anesthetic, pigs carrying the halothane sensitivity gene (HAL) also respond badly to stressful situations, such as handling and transport. The physiological response can result in sudden death and is referred to as the porcine stress syndrome (PSS). Meat from pigs that carry the HAL gene is characterized by a pale color and a soft texture with a very high drip loss, which is referred to as pale, soft, exudative meat (PSE). The appearance of PSE meat is unacceptable to the consumer, and is associated with low yields of cooked and dry-cured ham. However, carcass traits of halothane sensitive pigs are heavier, shorter, and leaner than halothane negative pigs. HAL sensitive carcasses have better lean content, but are worse for pH, color, drip loss, intramuscular fat tenderness, and juiciness. Response to halothane anesthetic was used to identify pigs carrying the HAL mutation, and breeding programs have used this information to select boars and sows with different HAL carrier status. By crossing between the carriers and non-carriers, heterozygous production animals could be produced, thus taking advantage of the improved lean growth, without the problems of PSE associated with homozygosity. The gene responsible for the halothane response was initially identified in man as the muscle ryanodine receptor, which is a calcium release channel of the sarcoplasmic reticulum in the skeletal muscle (MacLennan et al., 1989). The HAL sensitivity locus was first mapped chromosome 6 in pigs (Fujii et al., 1991), and later the gene responsible was also shown to be porcine ryanodine receptor (RYR1), which in addition to the halothane sensitivity, causes PSS. After initial breeding programs to produce heterozygous animals, pig breeders decided that the detrimental effects of the HAL gene outweighed the positive effects and therefore the marker information has been used to eradicate the HAL-sensitive allele.

Fat, Obesity, and Feed Intake

The obese strain of mouse (*Ob*) has been used for many years as a model for human obesity. The obesity of the Ob strain was known to be the result if a single genetic mutation, and that the obesity of these mice was the result of a very high feed in take. The gene responsible was identified as leptin, using a positional cloning approach (Zhang et al., 1994). The LEP gene encodes a protein composed of 146 amino acids which is expressed in adipose tissue, but is released into the blood, and thus has a general endocrine effect. The protein acts as a hormone that induces satiety and hence regulates food intake and energy balance (Barb, Hausman, & Hoseknechtm, 2001). Studies in livestock have investigated leptin as a candidate gene that may affect carcass fat. Several polymorphisms have been reported in the bovine LEP gene, some of which have been associated to carcass fat levels and, feed intake in cattle (Buchanan et al., 2002; Lagonigro, Wiener, Pilla, Woolliams, & Williams, 2003; Barendse, Bunch, & Harrison, 2005; Liefers et al., 2005). The LEP gene has also been investigated in pigs and seven polymorphisms have been reported in the porcine leptin gene although none of these have been conclusively associated with fat levels in the carcass (Jiang & Gibson, 1999; Kennes, Murphy, Pothier, & Palin, 2001).

Some forms of genetically associated obesity in humans have been associated with non-functional (frame-shift) mutations in the melanocyte receptor hormone (MC4R). This gene is associated with pigmentation, and more than 30 amino acid variations have been found in different ethnic groups, some of which have been associated obesity and impaired melanocyte cell functions (Yeo et al., 2003; Kim, Reecy, Hsu, Anderson, & Rothschild, 2004). In pigs, a mutation in the MCR4 gene results in an amino acid substitution in the coded protein (Asp298Asn) that has been associated with growth, fatness, and feed intake traits (Kim, Larsen, Short, Plastow, & Rothschild, 2000; Huston, Cameron, & Rance, 2004). Different expression levels of MC4R and LEP genes have been reported in two pig breeds with significantly different levels of intra muscular fat (D'Andrea, Fidotti & Pilla, 2005). In cattle, several mutations have been reported in the MC4R gene and also in the pro-opiomelanocortin (POMC) gene that is the precursor of alpha-melanocyte stimulating hormone (αMSH), which is an agonist of MC4R (Buchanan, Thue, Yu, & Winkelman-Sim, 2005). But up to now these have not been associated with variation in fat traits. Thus, candidate genes selected because they show a major effect in one species may have less obvious effects in another species.

Meat Tenderization

The overriding factor that affects the liking of meat when consumed is the texture, or specifically tenderness. The structure of muscle has a major effect on meat tenderness, but post-slaughter aging and maturation of meat plays a major part in the quality achieved in the final meat product. This aging process includes the proteolytic degradation of proteins in the meat by calcium dependent proteases. Hence, the tenderization process is affected by the activity of the proteases and the availability of calcium. A family of photolytic enzymes, called calpains, is involved in the proteolysis of muscular proteins during meat aging; the Calpain 1 or μ-Calpain enzyme is encoded by the *CAPN1* gene (Smith, Casas, Rexroad III, Kappes, & Keele, 2000). As its name implies, this protease in activated in the presence of micromolar amounts of calcium and has been associated with degradation of myofibrillar protein in living muscle. A large number of polymorphisms have been described in the bovine *CAPN1* gene (e.g., Page et al., 2002), and two polymorphisms in particular, within exons 9 and 14 of this gene, result in amino-acid substitutions (A316G and I530V), and have been associated with variations in tenderness determined by the Warner Brazler shear force test.

The level of calpain proteolytic activity is regulated by the availability of calcium, and calpains are inhibited by calpastatin (CAST). The *CAST* gene codes for calpastatin, and several polymorphisms, have been found in the bovine *CAST* gene, some of which have been associated beef tenderness measured by the shear force test (Barendse, 2002a; Koohmaraie et al., 1995). There is a strong correlation between calpastatin activity and tenderness measured by Warner-Bratzler shear force (Shackelford et al., 1994). Two SNPs, located in the 3'UTR region of the gene, have been associated with variations in tenderness in several studies (e.g., Nonneman, Kappes, & Koohmaraie, 1999). However, among the different combinations of these polymorphisms (haplotypes), only two have been associated with improved tenderness. Therefore, specific polymorphisms do not have a consistent effect, which suggests that the known polymorphisms are not the functional variations but are linked to it. The use of *CAST* polymorphisms for assessing animals that may produce tender meat has been patent protected, but it should be noted that no QTL for tenderness have been reported on chromosome 7 in cattle, in the genomic region containing the *CAST* gene.

A QTL associated with pork tenderness has been reported on pig chromosome 2, close to the CAST locus (Malek, Dekkers, Lee, Baas, & Prusa, 2001). Several polymorphisms have been reported in the porcine *CAST* gene (Rothschild, Ciobanu, & Daniel, 2004), some of which affected the protein structure and had a large effect on tenderness. Lysyl oxidase, an enzyme involved in cross-linking collagen fibers, is encoded by a gene (*LOX*), which is located close to *CAST*. Two alleles at the *LOX* locus have been associated with variation in meat tenderness in cattle. It is therefore difficult to separate the activities of these two neighboring genes to determine which is responsible for the observed variation in meat texture.

Intramuscular Fat and Marbling

The second major factor involved in the appreciation of meat quality, defined by sensory panel testing, is intramuscular fat, which affects both flavor and juiciness. The level of intramuscular fat may vary because of the number of fat cells, or because of variations in fat synthesis. Two genes located on chromosome 14 potentially affect both of these factors. The first Thyroglobulin (*TG*) is proteolytically cleaved to thyroid hormones involved in the regulation of adipocyte development. This gene was initially implicated in fat associated traits as it is close to a microsatellite marker, *CSSM66*, which was associated with carcass fat in QTL mapping studies. Subsequently, polymorphisms within the *TG* gene were directly associated with variation in intramuscular fat (Barendse, 2002b). A second gene located close to *TG* encodes diacylglycerol-O-acyltransferase (*DGAT1*), which is involved in the last stages of fat synthesis and has been identified as the gene underlying a QTL for milk fat synthesis (see below). The polymorphism of the *DGAT1* gene, responsible for the Ala232Lys variation in the protein, is associated with variations in milk fat and also seems to affect intramuscular fat (Thaller et al., 2003). A QTL for beef marbling has been reported on bovine chromosome 14, where both the *TG* gene and the *DGAT1* gene are located, but both the genes, *TG* and *DGAT1*, seem to have independent effects, because the alleles of the two genes are not in linkage disequilibrium.

QTL Mapping

Most of the traits associated with variations in meat quality are likely to be complex in-so-far as several genes will to contribute to the observed variation. Each of these genes will be responsible for a different proportion if the variation, some being responsible for a large amount of the variation, some with only a minor effect. For some traits, it may be possible to postulate candidate genes that contribute to the variation in these *quantitative traits*, but even with a good knowledge of the physiology of the trait, other genes not obviously part of relevant biochemical pathways may contribute to the variation and so would not be considered as candidates. Conversely, many apparently obvious candidate genes will not contribute significantly to the observed variation, e.g., because there are no functional or regulatory variations within the gene. Therefore, for complex quantitative traits, it may be better to start with no prior assumptions regarding candidate genes, and to use a genetic mapping approach. This approach first identifies the chromosomal location of the quantitative trait locus (QTL) and then uses this information as the starting point for identifying the gene and the functional variation within that gene – the *QTN*. The genetic, or linkage, mapping approach has two requirements: families which are segregating for the trait of interest and DNA markers to track the inheritance of chromosomal regions segregating in the families. The DNA marker information is then correlated with measurements characterizing the variations in the trait, and statistical methods are used to localize the trait genes to broad chromosomal regions.

Families and Data for QTL Mapping

Families segregating traits of interest, in which the traits have been recorded, are a primary requirement for locating the genes controlling the traits. Unfortunately, the range of traits that are routinely recorded in commercial populations is very limited, and through necessity focused on simple traits, such as growth rates and milk yields. An additional consideration is that traits are often sex specific and while selection is applied most strongly in the males, because of artificial insemination (AI), the trait is often expressed in the females, e.g., fertility. In the case of meat quality, accurate measurements of the traits can only be measured post-slaughter and thus cannot be made in the animals intended for breeding.

The structure of the commercial dairy population includes many large half-sib cows produced by AI from a limited number of elite bulls. This population structure is particularly appropriate for mapping QTL. The genetic contribution of bulls to milk production traits can be determined with high accuracy by measuring the phenotypes of their daughters. Bulls with high breeding value are then used extensively through AI to improve the dairy cow population. The sons of bulls with high breeding values are in turn used as AI sires to produce a large number of daughters. Georges et al. (1995) used the US Holstein population to map several QTL involved in milk yield. However, the way bulls are selected in the beef industry is somewhat different. Beef is produced from a large number of breeds, which have not been under the same intense selection as the dairy breeds, and therefore the beef population structure is different. In some countries, there is some systematic recording of beef production related traits in the live animal, such as growth, fat, and conformation traits in order to select the better bulls for breeding. This data provides some opportunities for mapping QTL for simple beef production traits. However, up to now, there has been little direct recording of meat quality traits. Thus, the majority of information published on QTL controlling beef quality traits comes from specifically bred "resource" herds in which the animals used and mating schemes are carefully controlled. These resource herds are usually managed under standardized protocols, which provide the opportunity to record the more difficult traits, which would be impossible using commercial herds (MacNeil, Miller, & Grosz, 2002). Several of the studies investigating meat quality have used resource herds created by cross-breeding between Bos taurus x Bos. indicus types of cattle. There is known to be very large differences in meat quality traits, particularly toughness between these two cattle types. QTL for several beef associated traits have been localized using these extreme cross-bred herds, e.g., for intra-muscular fat or marbling, muscle mass, meat texture, etc. (e.g. Stone, Keele, Shackelford, Kappes, & Koohmaraie, 1999; Casas et al., 2003). However, the value of these QTL in pure-bred populations has yet to be demonstrated.

Genetic Maps

To carry out QTL studies the second requirement is genetic markers to track the inheritance of chromosomal regions. Thus, to be useful, the location of the markers in the genome and their relationship to each other (genetic distances) must be known. The location, or map position, allows markers to be selected that covered the whole genome, or for fine mapping to be concentrated in targeted regions. The chromosomal location of the genetic markers is also used to place the QTL on chromosomes. Over the past decade, genetic and physical maps have been developed for the genomes of all the major domestic species. Two types of genome maps exist: genetic and physical maps. Genetic maps are created by determining the linkage between markers, estimated from their inheritance in families (Snelling et al., 2005; Barendse et al., 1997; Georges et al., 1995; Bishop et al., 1994; Archibald et al., 1995; Rohrer et al., 1996). Recombination will occur between markers that are far apart on chromosomes more frequently than those that are close together. Thus, the recombination frequency between markers can be used to calculate the genetic distance between them. The genetic maps of the livestock species were initially composed predominantly of "anonymous" markers such as microsatellite loci, but more recently genes and expressed sequence tags (ESTs) have been added to the maps. Genetic maps of the cow (Ihara et al., 2004) and other species now contain around 4000 markers. Genetic maps, however, have limited utility for identifying positional candidate genes near QTL; as in general, there are a few markers within genes included within the linkage maps. The resolution of the available genetic maps is also fairly low, as the order and genetic distance between markers can only be determined when there are recombination events between them. For closely linked markers, the probability of a recombination between them is low, and so to determine their order, a large number of individuals would have to be genotyped.

To overcome the deficiencies of the genetic maps three types of physical maps have been constructed. The first examines the chromosomes directly, by *in situ* hybridization of gene probes to chromosomes and direct visualization of the location of the probe by microscopy (e.g., Hayes et al., 1995; Solinas-Toldo et al., 1995). The second type of physical map is created by observing the retention of chromosome fragments in hybrid cells created by fusing together cells of the target species and immortalized cells of another species, typically hamster. Such somatic hybrid cells retain only a fraction of the chromosomes coming from the target species, and therefore the frequency of co-retention of two marker loci can be used to estimate how close together they are on the genome. A refinement of somatic cell hybrid mapping is the radiation hybrid map. These maps are constructed based on the probability of a radiation induced break between the loci in the genome of the target species. To create a radiation hybrid map, cells from the target species are irradiated before fusion to the immortalized cells. As with somatic hybrid cells, the presence or absence of markers in a series of the hybrid cells can then used to determine how close they are in the genome (Itoh et al., 2005; Jann et al., 2006; Everts-van der Wind et al., 2005). High resolution radiation hybrid maps are achieved by increasing the radiation dose, and hence causing a greater fragmentation of the target genome. The third type of physical map is composed of an ordered set of DNA clones built from large fragment DNA libraries. Currently, the most widely used large fragment cloning vector is the bacterial artificial chromosome (BAC, Zhu et al., 1999). The recovery of DNA from particular genomic regions can be achieved rapidly by selecting the appropriate BAC clone. Using sequence information obtained from clones such as BAC end sequences, clone-based maps can also be used to enhance connections between annotated genome sequences of different species (Gregory et al., 2002). The drawback of physical maps compared with genetic maps is that they lack the polymorphic markers that are needed to refine the locations of the QTL. Therefore, to locate the QTL on a chromosomal region, then fine map that region and ultimately identify the gene(s) controlling the trait, it is often necessary to use the information from both physical and genetic maps.

QTL Discovery by Linkage Mapping

The basic principle underpinning genetic mapping studies and to identify QTL is linkage mapping. This concept is very simple, and is that the likelihood of a recombination event between locations on a chromosome is proportional to the distance they are apart. Therefore, alleles at loci that are on different chromosomes or which are far apart on the same chromosome will be randomly assorted in the successive generations, whereas the closer together genes are, on a chromosome, the more likely they are to be inherited together. By examining the patterns of inheritance of markers and phenotypes, the markers close to the genes controlling the trait are identified as those that show the same, or a very similar, pattern of inheritance. Hence, from knowing the position of the markers in relation to each other in the genetic map, the position of the genes controlling the trait being studied can be deduced. However, the first problem is that each QTL explains only a small fraction of the phenotypic variation, and hence the correlation between the phenotype and the genotype is low. The second problem is that not all of the variation seen in the phenotype is genetic, and environmental factors can have a large impact on the observed variation. The ease with which a particular QTL can be detected is proportional to the amount of the variation in the trait that it explains. A QTL that explains a large proportion of the phenotypic variance, in the extreme case a Mendelian gene which explains all the genetically controlled phenotypic variation, will be easier to detect than one that explains only a small proportion of the variation. The amount of variation in a trait, and the accuracy with which the trait can be measured, is also very important. For traits that show a large variation in relation to the accuracy of measurement, detection of QTL will be easier than a QTL with a smaller effect. Meat quality traits are in general poorly defined, and hence not measured with great precision. Sensory panels, for example, are subjective, and so the definition of a trait by different people will vary. The use of objective measures as surrogates for human taste panels, such as mechanical shear force to define toughness, are poorly correlated with the appreciation of toughness while eating. With such poorly defined traits, only QTL with major effect, i.e., accounting for a large proportion of the phenotypic variation, will be identified with confidence. To identify a QTL with a reasonable certainty, a large number of individuals will be required to make statistically robust associations between the markers and the traits. As the size of a study increases, and in particular by increasing the number of loci (or number of traits) tested, the likelihood of detecting a chance association between marker and phenotypes increases. It is then necessary to set high thresholds for accepting the associations detected, which risks rejecting some real associations.

The methods suggested to detect QTLs in livestock populations were initially developed to use either two or three generation pedigrees to analyze linkage between a single marker and a QTL (Neimann-Sorensen & Robertson, 1961; Weller, Kashi, & Soller, 1990). The drawback of these methods is that they use information from a single marker at a time. If information from multiple markers is used simultaneously, the confidence with which QTL can be detected is increased. Interval mapping methods, which use information from all markers within a defined genetic interval, have been developed and successfully used to identify QTL using half-sib families (Georges et al., 1995; Knott, Elsen, & Haley, 1996). A web-based software package "QTL Express" (Seaton, Haley, Knott, Kearsey, & Visscher, 2002) implements a multi-marker linear regression method (Knott et al., 1996). This software is publicly available and has been used to analyze the data from many QTL studies in livestock. Alleles at different, independent QTL potentially interact and may influence the same trait, with positive or negative effects, depending on the alleles present at each locus. By considering the interactions of multiple chromosomal regions simultaneously in the analyses, it is possible to increase the power and the precision of QTL mapping in out-bred populations (de Koning et al., 2001).

The results of QTL studies should only be taken as preliminary evidence for the role of a particular genomic region in the control of a trait. Before using QTL information, either in selection programs, or as the starting point to identify the trait gene, it is wise to obtain independent data that supports the existence of the QTL and the size of its' effect. Supporting evidence may include the identification of the QTL in independent studies. Examples of QTL studies, their confirmation, and identifying trait genes from the QTL are discussed below.

Intramuscular Fat

Intramuscular fat, seen as marbling fat in meat, is important for eating quality, as it affects flavor, juiciness, and possibly toughness. The accumulation of subcutaneous fat and intramuscular fat are affected by management, and selection for one is highly correlated with the other. While intramuscular fat is desirable, subcutaneous fat is largely unwanted and is trimmed and discarded when meat is dressed. It would be advantageous for producers if animals could be selected that maintained the accumulation of intramuscular but had a reduced accumulation subcutaneous fat.

A QTL for marbling as an indicator of the amount of intramuscular fat was identified by Casas et al. (1998) on bovine chromosome 2. The QTL effects were initially attributed to myostatin as a candidate gene in the region. Four further studies also identified a marbling QTL on chromosome 2 (Stone et al., 1999; Casas, Keele, Shackelford, Koohmaraie, & Stone, 2004; Schimpf et al., 2000; MacNeil & Grosz, 2002), which provided some evidence to confirm the first study; however, the QTL peak in each study seemed to be at a different chromosomal position. This uncertainty is because each study used different sets of markers, and genetic maps, in addition to different breeds and measurements of the trait. Thus, it is not certain if the same or different QTL is being detected in each case. QTL with an effect on muscle marbling have be reported on other chromosomes, with evidence for the QTL provided by more than one study; fat associated QTL have been detected on chromosome 3 in three independent studies (Casas et al., 2004; Stone et al., 1999; Casas et al., 2003) and on chromosome 27 in two separate studies (Casas et al., 2000; Casas et al., 2003), although again the positions of the QTL differed between each study. Other QTL affecting fat traits have been reported on chromosome 5 (Stone et al., 1999), 8, 10 (Casas et al., 2001), 9, 14, 23 (Casas et al., 2003) 16 (Casas et al., 2004), 17 (Casas et al., 2000), and 29 (MacNeil & Grosz, 2002) have not yet been detected in other studies.

Tenderness

Tenderness is a primary factor influencing the consumers' reaction to meat. However, tenderness is difficult to define objectively. The appreciation of tenderness when eating is not explained by the force required to cut through a piece of meat, but is affected by the way the muscle fibers breakdown and the release of juice and flavor while chewing. Tenderness has been measured experimentally by three different approaches: either as shear force required for artificial jaws to pass through a piece of meat, the most common mechanical shear method for defining tenderness is Warner-Bratzler Shear Force (WBSF), by taste panel analysis, and by determining the myofibriler fragmentation index (MFI), which is related to the way that meat fibers are broken down. Several QTLs have been reported for WBSF including QTL regions on chromosomes 5, 9, 15, 20 (Keele, Shackelford, Kappes, Koohmaraie, & Stone, 1999, Casas et al., 2001, Casas et al., 2003, Casas et al., 2004), but none of these have been confirmed in independent studies of the same measurement. Gutiérrez-Gil et al. (2007), identified a suggestive QTL for myofibriler fragmentation index (MFI) on chromosome 15, but did not identify a shear force QTL at the same position. Several independent studies have identified a locus on bovine chromosome 29 with effect on tenderness (Casas et al., 2003; Casas et al., 2001; Casas et al., 2000). The calpain 1 (*CAPN1*) gene that codes for a calcium dependent protease involved in meat tenderization post-mortem (see above) is located within this QTL region and is a strong candidate gene for this QTL for meat tenderness.

Unexpected Discoveries

Globally, consumer requirements and trends in preferences of meat quality vary. Currently, the European market demands meat that is low in fat, while the Asian market requires meat with a high fat content, particularly intramuscular fat. To address local markets selection has resulted in breeds with very different meat characteristics in different global regions. In Europe, cattle breeds, such as the Belgian Blue, Charolais, Limousin, etc., have been selected for rapid, lean growth and good feed conversion efficiency. Traditional European breeds that tended to produce meat with high levels of fat, such as the Hereford and Angus have recently undergone selection to change the breed characteristics and reduce the levels of fat in their meat. In contrast, e.g., in Japan, selection of the Wagu cattle has been strongly focused on developing cattle that produce exceptionally high levels of intra-muscular fat, which is sold at high value in the home market. Similar divergent selection criteria have been applied in pig production. In response to consumer demand, pig breeds used extensively in Europe have been selected for lean growth, particularly in the last 20 years, whereas Meishan pigs from China, have large quantities of fat.

Studies designed to identify QTL controlling fertility and carcass composition in pigs have been carried out using specifically designed cross-bred resource populations from founder breeds with different phenotypes, these include crosses between wild boar, commercial landraces, and local breeds. QTL for carcass and fertility traits have been identified in studies on such extreme breed crosses (e.g., Rattin et al., 2000; Nagamine, Haley, Sewalem, & Visscher, 2003). A QTL that is of particular interest was identified on pig chromosome 7 in a cross between the lean Large White and the fat Meishan pig breeds. This QTL had a particularly large effect, explaining about 30% of the difference in back fat thickness between the two breeds (de Koning et al., 1999). The surprising finding, however, was that the allele associated with lean growth originated from the phenotypically fat Meishan breed.

This example demonstrates that the most beneficial allele for a particular trait may not be found in population showing a desirable phenotype for the trait. This may be because of founder effects, i.e., that the most favorable allele was not present in the individuals initially used to create the population, or breed, or that the most favorable gene may have been lost over time, e.g., because of genetic drift. Alternatively, the favorable allele may have a deleterious effect on another trait and so the benefit is either not seen or the association with undesirable characteristics has placed the gene under negative selection.

Finding the Trait Genes Starting from a QTL

Using a linkage mapping approach for localizing a QTL will define the chromosomal location through flanking DNA markers. These *linked* markers can be used to enhance selection programs by identifying animals that carry the favorable allele at the QTL, which can then be used for breeding. This process is called marker-assisted selection (MAS, Kashi, Hallerman, & Soller, 1990). However, these markers are likely to be at a significant genetic distance from the gene controlling the trait; therefore recombination is likely to occur between the marker and the trait gene. As a result, at a population level, the alleles present at the marker loci will not predict the alleles present at the trait locus. Therefore, before the linked markers can be use for MAS, it is first necessary to determine the *phase* of the markers, i.e., which alleles at each of the marker loci are linked to the favorable or unfavorable alleles at the trait gene. Determining the phase of markers has to be done within a family by recording the phenotype of individuals in the family and relating this information to the genotype at the linked markers. The phase of allele at the flanking markers and the trait gene may not be the same in different families. Furthermore, recombination can occur at each generation and so the phase of allele at the marker loci and the trait gene has to be frequently reconfirmed. For these reasons, marker assisted selection is not particularly efficient if implemented simply using flanking markers. In contrast, identifying the trait gene, or better, knowing the functional polymorphism that is responsible for the variation in the trait provides markers that can be used directly in the population, without first having to determine their phase. Using information on the gene in selection programs is called gene assisted selection (GAS) and is more effective for enhancing selection than linked markers. Nevertheless, the first step for identifying the trait gene is currently a linkage mapping approach.

There are now a large number of QTLs identified for production traits in livestock (e.g., see http://bovineqtl.tamu.edu/ and http://www.animalgenome.org/QTLdb/). However, so far, few trait genes, and specifically the functional mutations within these genes, have been identified. Identification of the trait genes starting from the chromosomal location is not an easy task. Genetic mapping studies generally localize QTL at low-resolution. Typically studies using microsatellite markers have placed the QTL within a 20–40 centi-Morgan (cM) interval, which roughly equates to 20–40 Mb DNA or a quarter of a chromosome. A QTL region of this size could contain 200 or more genes. Thus, it is necessary to either refine the map position before trying to identify the specific gene that controls the trait, or use other information to select genes within the region for which there is good evidence that they have an effect on the trait.

Linkage-mapping relies on recombination to determine the order of the markers in relation to the trait gene(s) on the chromosome. The unrecombined region of a chromosome that is inherited together with the trait gene is in *linkage disequilibrium* with it. In order to fine map a QTL, it is necessary to examine additional meioses which provide the opportunity for addition recombination events within the QTL region closer to trait gene, and hence reduce the region flanking the gene that is in linkage disequilibrium (LD). In practice, a large number of individuals are required to find those with recombination occurring within the region in LD flanking a QTL. Therefore, fine mapping a QTL region requires a large number of individuals that have been recorded for a trait, within families that are segregating for the QTL. Once the QTL region has been fine mapped, two approaches can then be adopted for identifying the trait gene. The most popular and successful approach so far has been a refinement of the candidate gene approach described earlier, using both the information on a biological function of the gene and its' chromosomal position to select a "positional candidate" gene. The identification of positional candidate genes is assisted using comparative genomics to examine the equivalent chromosomal regions across species, together with information on known functions of the genes in controlling phenotypes, e.g., from comparative studies in mice. In the absence of a candidate gene, or when the candidate genes identified prove not to include the trait gene, it is then necessary to adopt alternative strategies. To date this has required the cloning of the QTL region followed by sequencing in order to expand information available on the genes and variations present within the QTL region. Sequences of individuals showing differences in their phenotype is then compared to identify animals in which recombination has occurred within the QTL region to reveal specific genetic variations that are associated with differences observed in the trait. Depending on the positions of recombinations, the position of the causative mutation may be refined, or possibly a specific functional polymorphism identified that is responsible for the observed phenotypic variation.

Identification of the trait genes and functional variations

There are now several examples of studies that have identified trait genes using both position candidate gene and positional cloning approaches. Illustrative examples of both approaches and a combination of these approaches are discussed below.

Myostatin

Double muscling is a distinctive phenotype that has been recognized by breeders of several European beef breeds. It is characterized by pronounced muscular development, resulting from both muscular hypertrophy and hyperplasia, and is associated with reduced intra-muscular fat (Ménissier, 1982). Hence, double muscled animals produce carcasses that have an increased yield of the expensive cuts of meat, and are exceptionally lean. The most extreme form of double muscling is found in the Belgian Blue breed where the trait behaves as if it is controlled by a single major gene. To localize the gene responsible, a research population of cattle was created by crossing double muscled Belgian Blue cattle to a non-carrier breed. Using this population and a conventional linkage mapping approach, the double muscling gene was localized to a region on bovine chromosome 2 (Charlier et al., 1995). Examination of the genes within this region identified no strong candidate genes for the trait. A collagen gene that was a possible candidate was ruled out, and so a fine mapping and positional cloning strategy was started. However, at the same time work in mice on the transforming growth factor (TGFβ) family of genes identified a new member of the gene family, GDF8, that had an effect on muscle development. Transgenic mice in which expression of GFF8 gene was "knocked-out" developed hyper-muscularity similar to the double muscling phenotype in cattle (McPherron, Lawler, & Lee, 1997). The GDF-8 gene product was found to be a negative regulator of muscle growth and was therefore called myostatin. Myostatin acts in the developmental pathway which regulates the differentiation of progenitor cells into muscle and fat tissues. In the absence of myostatin, the cells preferentially develop into muscle instead of fat. Belgian Blue cattle that showed the double muscling phenotype were found to have an 11 bp deletion within the coding region of the GDF8 gene (Grobet et al., 1997; Kambadur, Sharma, Smith, & Bass, 1997). Cattle of other breeds with the double muscled phenotype were subsequently found to have mutations within the coding region of their myostatin gene, e.g., the Piedmontase breed has a single base mutation within the coding region of the myostatin gene (McPherron et al., 1997). Individuals of the Marchigiana beef breed that carry a single nucleotide polymorphism in the 5' promoter region of the gene have a muscularity index 25% higher than individuals without the variation (Crisà, Marchitelli, Savarese, & Valentini, 2003).

Interestingly, the same variation as found in the Marchigiana had also been identified in pigs, where it is associated with higher average daily gain, as well as improved muscling (Jiang, Li, Du, & Wu, 2002). Muscular hypertrophy in Texel sheep is also associated with a variation in the GDF8 gene. In this case, a polymorphism in the 3'UTR in some individuals in this breed creates a target site for microRNAs that destabilize the myostatin gene product. The result is a reduction in myostatin and hence the muscular hypertrophy seen in some Texel sheep (Clop et al., 2006).

Calypyge

In sheep excessive muscular development was observed in the progeny of a single ram, Solid Gold, in the USA in the early 1990s. The trait showed an unusual mode of inheritance, with the expression of the phenotype dependent on whether the "Calypyge" allele was inherited from the sire or the dam, a phenomenon known as imprinting. Linkage mapping studies localized the gene controlling the *CLPG* to sheep chromosome 18 (Fahrenkrug et al., 2000; Berghmans et al., 2001). The region containing the QTL was then sequenced and the polymorphisms identified used to refine the locations of recombination events and reduce the candidate interval to approximately 400 Kb. This interval contained a good candidate gene, *DLK1*. The region of the human chromosome 14 and mouse chromosome 12 that correspond to this region of sheep chromosome 18 have been intensively studied because they show parental imprinting, i.e., allelic expression is dependent on the parent of origin (Schmidt, Matteson, Jones, Xiao-Juan, & Tilghman, 2000). In fact, two genes within this chromosomal region both in mouse and human are imprinted, with paternal allele of *DLK1*, but the maternal allele of *MEG3 is* expressed. This region was fully sequenced from sheep carrying the *CLPG* mutation, and an A to G polymorphism was identified in an inter-genic regulatory region between *MEG3* and *DLK1* that segregated with the trait. The mutation seems to affect a site that regulates the epigenetic modifications involved in the imprinting of the *DLK1-GTL2* region (Freking et al., 2002). While the increase in muscularity associated with *CLPG* is a positive feature which was initially selected for, subsequently the trait was associated with poor meat quality.

Insulin-Like Growth Factor (IGF2)

Resource populations, in which genetically divergent founders with distinct phenotypes are crossed, have been used extensively as a tool to identify QTL, particularly QTL involved in variations in carcass traits. A QTL on chromosome 2 in pigs controlling variation in muscle traits and fat depth was mapped, initially in several resource populations, some created by crossing European with Asian pig breeds. Using the combined data from several different studies, in a meta-analysis, the position of this QTL was refined to a small region including the insulin-like growth factor 2 (*IGF2*) gene (Nezer et al., 1999). Sequencing across the *IGF2* locus revealed 258 polymorphisms in and around the gene (van Laere et al., 2003). These polymorphisms could be assembled into two haplotypes, one defining a chromosomal region of European origin and one from the Chinese breeds. Examination of these haplotype blocks within individual animals in the crossbred resource populations identified several individuals with recombinations between the two haplotypes that further refined the chromosomal location of the causative variation and identified a single SNP, a G to A transition within intron 3 that appeared to be the causative mutation, or "quantitative trait nucleotide" (QTN). Interestingly, intron 3, is not within the region of the gene coding for the *IGF2* protein. One mechanism for the regulation of gene expression through imprinting involves the methylation of cytosine bases to inactivate the gene. The IGF2 QTN occurs in a region of the DNA that is imprinted as a result of cytosine methylation. The expression pattern of *IGF2* is known to show imprinting, and the specific SNP seems to be within a binding site for a protein that regulates gene expression.

Both the *IGF2* and the Callipyge example discussed above (Freking et al., 2002), are traits associated with muscle composition that are controlled by genetic variations outside the protein coding region of the genes involved, and within regions that show imprinting.

Hints from, Fat in Bovine Milk (DGAT1)

Many studies performed on dairy cattle using commercially collected data have identified QTL for milk yield and composition (e.g., Georges et al., 1995). One of these QTL, associated with variations in milk fat content, was mapped to chromosome 14 then fine mapped using new markers produced from a sequencing contiguous set of BAC clones spanning the QTL region (Coppieters et al., 1998). This region of bovine chromosome 14 is equivalent to a region of human chromosome 8 which contain the same set of genes, i.e., shares conserved synteny. Studies in mice have shown that a gene acyiCoA:diacylglycerol acyltransferase 1 (*DGAT1*) with this region on human chromosome 8 affects milk synthesis and mice in which there is no expression of the gene fail to establish lactation (Casas et al., 1998). This *DGAT1* gene was found within a contig of BAC clones spanning the QTL milk region on bovine chromosome 14 which made it a strong candidate gene for the milk fat QTL. Sequencing the *DGAT1* gene from bulls carrying different alleles at the QTL revealed a number of polymorphisms within the gene. These polymorphisms included one in exon 12 of the gene that resulted in an amino acid change (alanine to lysine K232A) in the mature protein (Grisart, Farnir, & Karim, 2002). Extensive studies of Holstein cattle have shown that cows with the lysine allele have consistently higher milk fat than those with the alanine allele and that the lysine allele is associated with decreased protein content and milk yield.

Knowledge that *DGAT1* is involved in the biochemistry of fat synthesis, and the discovery that variations in this gene present in cattle had an effect on milk fat, prompted studies to investigate the possible effect of this gene on muscle fat. A study of the K232A substitution polymorphism within *DGAT 1* in a crossbred Holstein vs Charolais experimental herd suggested that the lysine allele was associated with increased enzyme activity and higher lipid content of different tissues, including muscle (Thaller et al., 2003).

Applications

The information now available on the association between specific variation in genes and traits such as the examples given here can, and in some cases, are being implemented in selection programs to improve quality, quantity, and efficiency of meat production. The advances in genomic information and the technology available for genomic studies in livestock species should make the discovery of the trait genes and causative variations within the genes more rapid in the future. In addition, the new technologies that are now available for large volume genomic studies, open the possibility to apply novel strategies to enhancing selection programs, using whole genome assisted selection, rather than applying information locus by locus. However, to achieve this it is necessary to have markers at sufficiently high density to identify linkage disequilibrium block in the population.

New Opportunities

Initially, studies to identify the genes controlling commercially important traits used either a candidate gene or gene mapping approach, as discussed above. These strategies were appropriate to the knowledge and technology available at that time. However, information on the role of a gene could also come from other sources, e.g., the level of expression of a gene, or by extension the variation in expression of groups of genes involved in particular metabolic pathways. The study of gene expression has traditionally been carried out gene by gene. The first methods used gel electrophoresis and hybridization of radioactive probes to reveal the expression and crudely quantify the levels of particular RNAs (Northern blotting). More recently, techniques using the polymerase chain reaction (reverse transcribed PCR: RT PCR), and derivatives of this technique (e.g., TaqMan. Applied Biosystems Inc), have facilitated a more accurate quantification of the amounts of a particular RNA molecule present in a sample. Nevertheless, the use of these techniques follows from the identification of a specific gene, or at least a small number of genes that are believed to play an important role in controlling variations in the target trait. To identify candidate genes, when little prior information is available, it would be necessary to examine many genes, preferably simultaneously, instead of individuals genes.

Several approaches have been developed to examine the expression profiles of several genes, starting from samples obtained from appropriate tissues taken from individuals with divergent phenotypes in the trait of interest. These approaches begin with no prior hypothesis regarding the genes or gene products that play a role in controlling the observed variation. Differential Display-Reverse Transcribed-PCR (DDRT-PCR) uses a "quasi-random" set of oligonucleotide primers and reverse transcriptase to create cDNA fragments representing a large proportion of the expressed genes in a sample. These are then compared with cDNA fragments created from other samples. The fragments are separated by polyacrylamide gel electrophoresis and side-by-side comparisons allow the identification of transcripts that are differentially expressed between the two samples (Liang & Pardee, 1992). This approach has been applied to examine the variations expression associated with muscle fat by comparing the expression in muscle from Charolais vs Holstien breeds which are phenotypically very different for muscle properties (Dorroch et al., 2001). The DDRT-PCR approach identified 277 differentially genes between the breeds. Sequencing followed by mapping of these differentially transcribed genes to assign them to chromosomal regions (Goldammer et al., 2002) provided a number of candidate genes for muscle related traits. Comparing the genome location with the QTL mapping data for intramuscular fat identified one cluster of differentially expressed genes located on bovine chromosome 3 within a QTL for marbling traits. Other studies have used this technique to ask more focused questions, e.g., to associate variation of gene expression with variations in hormone levels (Bellmann, Wegner, Teuscher, Schneider, & Ender, 2004).

The DDRT-PCR allows differences in expression patterns to be observed from PCR fragments produced from expressed genes. However, identifying the gene from which the fragments originate then requires cloning and sequencing of the DNA fragment. A major technological advance that has opened the way for more extensive studies of gene expression patterns, e.g., between cells from animals with different phenotypes or following particular treatments, has been the development of expression-arrays. By creating cDNA copies of RNA coding for particular genes and binding this cDNA to a solid matrix in an organized array, so that the location of each RNA type is known, the expression of that RNA species, e.g., from different cell types can be investigated by hybridizing RNA from cells to the array. The first arrays, macroarrays, were created by spotting the target cDNA onto a nylon matrix and hybridizing with RNA labeled with radioactive probes. Such arrays could investigate the expression of several hundreds to thousands of genes of simultaneously, but quantifying small difference in the expression from the radioactive signal was not very accurate. More recently, microarrays have been developed by printing at high density many thousands of either cDNAs coding for specific genes, or oligonucleotide probes representing fragments of the cDNA sequences, onto glass slides or other matrices. These arrays are then probed using fluorescently labeled probes which can more precisely measure the radioactively labeled probes. In addition, the labeling of the samples to be compared with two different colored fluors, allows the samples to be hybridized simultaneously to the same array and differences in expression directly compared.

Initial studies used cross species array probes, from humans and mice, as insufficient information was available from the livestock species. A dual color fluorescence approach, using a human array, for example, was used to compare expression patterns in muscle samples from cattle selected for high and low growth potential. From the 1300 human gene probes on the array, 34 genes were identified with different levels of expression between the genetic types (Casser-Malek et al., 2003). Many of the genes identified with differential expression were associated with muscle structure (e.g., titin) or cell regulation (e.g., thyroid hormone receptor). Data from studies using cross species probes can indicate genes with differential expression patterns, but should be treated with caution as the genes identified may not be the equivalents between species, especially when the genes are members of gene families.

Currently, sophisticated technologies are being used to create arrays with hundreds of thousands of "gene features" which can be used to investigate the relative expression of genes with high precision. Using this array technology, the expression of a very large number of genes can be compared between samples, e.g., of tissues from animals with different phenotypes or in different physiological states. With the increase in genomic information available for livestock in recent years, arrays are now available with probes from the target species. Many arrays have been developed for specific tissues, or biochemical pathways, and arrays are now available for cattle and pigs that contain probes for a major part of the expressed genome. Using a microarray approach to investigate fat deposition in Japanese Black (JB) vs Holstein cattle many genes expressed at a higher level in the JB were found that are associated with unsaturated fatty acid synthesis, fat deposition, and the thyroid hormone pathway (Wang et al., 2005). These array data are consistent with the increased monounsaturated fatty acids observed in beef from Japanese Black Cattle.

The microarray approach has also been used to compare gene transcription profiles between pig breeds with different muscle characteristics (Lin & Hsu, 2005). This study identified several genes involved in, e.g., muscle structure (myosin light and heavy chains, and troponin) and energy metabolic enzymes (electron-transferring flavo-protein dehydrogenase, NADH dehydrogenase, malate dehydrogenase, and ATP synthases) associated with lean growth. However, the selection of the tissue for gene profiling studies requires careful consideration. Liver is an organ involved in energy metabolism; therefore, variations in liver function may influence carcass size and quality. Studies carried out on liver samples comparing expression in German Landrace, which has a relatively high fat carcass compared with the Pietrain that is very lean (Ponsuksili, Murani, Walz, Schwerin, & Wimmers, 2007), showed that animals with fatter carcasses have higher expression of genes associated with lipid metabolism pathways (FASN, ACSS2, ACACA) while lean growth was associated with expression of genes for cell growth and/or maintenance, protein syntheses, and cell proliferation pathways (PPARD, POU1F1, IGF2R).

eQTL

The information obtained from mapping and expression studies is different but complimentary. Genes containing polymorphisms identified from mapping studies are not necessarily those that will be differentially expressed, and conversely genes with differential expression will not necessarily contain genetic variations. The polymorphic gene may, e.g., be a receptor that regulates another gene which is involved in the development of the phenotype. Specific polymorphisms in the receptor may not affect the expression of the receptor gene itself, but may be within the ligand binding site, and hence genes which are regulated by the receptor may show differences in expression. Such a case is referred to as trans-acting effects: i.e., when the location of genetic polymorphism and gene with affected expression are at different locations. Alternatively, a variation within a gene may directly affect the expression of that gene, e.g., variation is in a promoter region necessary for expression. Such variations act in "cis", i.e., the differentially expressed gene, and the gene with the genetic variation are at the same location.

In order to locate the gene regulating the variations in expression, an "expression QTL" (eQTL) approach is used. If the expression data is considered as a phenotype, the location of the genes responsible for the variation can be mapped using a conventional genetic mapping approach. If the eQTL(s) maps to the location of the gene(s) with differential expression, i.e., are in *cis*, it is likely that the differential expression is the result of a polymorphism within that gene. If not, a trans-eQTL, implies that the gene with differential expression is being regulated by another gene at a different location on the genome. If the expression eQTL and QTL data for the trait under investigation are combined, together they may suggest the gene(s) and regulatory pathways important controlling variations in the trait. To date, there have been no eQTL studies completed for livestock species, partly because obtaining the expression data is expensive, and partly because the designs of such experiments are not simple. For meat associated traits suitable tissues should be obtained from animals within families that differ in the target trait. However, the choice of tissue is not always obvious, as a trait displayed in one tissue may be the result of a variation in the activity in another. As discussed above, variation in muscle composition may arise from the muscle, liver, or adipose tissues, or indeed, from another tissue, e.g., pituitary.

Proteomic Analysis

The study of gene regulation at the level of the transcript (RNA) may be misleading as amount of RNA coding for a specific protein may not correlate very accurately with the concentration of that protein in the cell, or e.g. in the case of a hormone in the serum. There are many mechanisms in the cell that can alter the rate of translation of an RNA molecule into a protein. Therefore, it may be better to examine the level of a protein directly, instead of inferring the level of a protein from the amount of RNA that codes for it. Additionally, the protein may be post-translationally modified, either by processing of a precursor, or chemically e.g., by addition of sialic-acid or phosphate, before the protein is functional. By taking a proteomic approach it is possible to consider the levels of a protein and also post-translational modifications, and so relate more accurately the level of active protein with phenotypic effect.

The basic technique for proteomic analysis has not changed significantly over the past 20 years and involves analysis of proteins by two-dimensional gel electrophoresis, where the proteins are separated using two different physical properties, typically electrical charge, using iso-electric focusing electrophoresis, and then by size, on denaturing SDS gel electrophoresis. The patterns produced from the different samples are then compared and where differences observed, the protein extracted from the gel for analysis. The protein is then identified by analyzing the fragments produced following digestion, typically, by trypsin. The resent advances to improve proteomic analysis have not been in the basic technique, but with improvements in the reproducibility of the electrophoresis and the image analysis to compare the patterns. Analysis of tryptic digests is now carried out by mass spectroscopy and large libraries with information on the tryptic digests of many proteins are available to interpret the data (http://www.expasy.ch/tools/). To date, there have been few reports of proteomic studies in livestock associated with meat quality traits. One example investigated the proteins expressed in muscle from double muscled and normal cattle. The data identified that muscles from animals with the 11 bp deletion in the myostatin gene have a protein pattern consistent with an increase in fast-twitch glycolytic fiber number, which is inhibited when myostatin is expressed normally (Bouley et al., 2005).

Genome Sequence

The ultimate map of the genome of any species is the genome sequence. Following the publication of human genome sequence (Lander et al., 2001), the sequencing capacity that had been assembled was deployed to sequence the genomes of other species. A project to sequence the bovine genome started in 2003 and the first draft sequence with a threefold coverage of the genome was made publicly available in November 2004. Continued work has now produced a full draft sequence constructed from a sixfold genome shotgun (random) sequencing of DNA from an inbred Hereford cow combined with a 1.1 fold sequence coverage from a minimum tiling path of BAC clones from a of a Hereford bull. The assembly of the shotgun sequence was assisted using a composite genome map of the bovine genome that incorporated all the available genome mapping data (Snelling et al., 2007). A project to sequence the pig genome is currently underway and is using a well ordered genome-wide set of ordered BAC clones to create the whole genome sequence.

One of the important outputs of the human genome sequencing project has been a large number of SNP markers, that are being used in genetic studies. Currently, more than 4 million human SNPs have been validated (International HapMap Consortium, 2007, http://www.genome.gov/10001688). The cattle genome sequencing project identified over 2 million putative SNPs, but many of these are artefacts from sequencing errors. Additional genome sequencing from genetically divergent breeds in the bovine Hapmap project identified 30K SNPs, that have been tested and confirmed across many breeds. A similar approach to SNP discovery has been adopted for pig.

Genotyping Technologies

SNP polymorphisms have advantages over other marker types, in-so-far as they can be detected by methods other than electrophoresis, which is slow and difficult to automate. Following the discovery of many hundreds of thousands of SNPs from the human sequencing project, automated assays have been developed using, e.g., fluorescence or Mass-spectroscopy, to genotype SNPs. It is now possible to rapidly genotype hundreds to thousands of individuals for tens of thousands of SNP markers in a few hours.

A panel of about 50,000 SNPs was recently created for the bovine genome by a consortium of researchers in the USA. The power of this panel of markers was recently demonstrated using *DGAT1* (see above) as an example. Using this marker panel with Holstein cattle families, a QTL with an exceptionally high LOD score and a 3 Kb 90% confidence interval was identified centred on the *DGAT1* gene (Schnabel et al., 2008). This panel is now commercially available (Illumina Inc, San Diego USA). This high density of SNP markers will allow genes to be identified directly by the association of regions of DNA that are inherited in *linkage disequilibrium* with the trait gene. To date, the relatively low density of markers (typically around 150–200) used in QTL mapping studies has required using families.

Application of the Data in Breeding

As discussed above, breed improvement, up to now, has been achieved through phenotypic selection focused on easily measured traits. Over the last four decades, the approaches to selection have been refined and trait measurements, made on the individual, have been replaced by calculated "breeding values" that make use of all the available information on the genetic merit of the individual, including information from relatives, parents, progeny, and siblings. However, many of the economically important traits, and certainly those involved in variation in meat quality are difficult to measure and are "quantitative" in nature. The phenotypic variations in these traits were originally thought to result from the interactions between the many genes, each having a small effect on the phenotype - the infinitesimal model (see review by Flint & Mott, 2001). If this were the case, it was thought that identifying the genes controlling a quantitative trait would be impossible. Fortunately, as demonstrated by the QTL examples given above, for at least some economically important traits it seems that, although there may be many genes involved in controlling the variation in the trait, there are usually a small number of *major* genes that control a reasonable proportion of the observed variation. Information on a few genes can be readily incorporated in to selection programs, by genotyping individuals and adding the information into the calculation of the breeding value currently estimated from phenotypic measurements.

The use of genetic markers to improve estimated breeding values was suggested over 15 years ago (Fernando & Grossman, 1989), but in general the use of marker to increase the accuracy of selection has been restricted by a lack of knowledge of QTLs with large effects. There are now several examples where the genes controlling important meat production traits have been identified, as discussed above, and a few pilot programs are including this information among the selection criteria. However, for most traits the information available is a QTL position based on loosely linked markers, and not the gene, or variation within the gene. The QTL information is specific to the study population in so far as different alleles at the marker loci can be associated with favorable or unfavorable alleles in different populations, thus linkage phase between a marker and QTL had to be established for every family in which the markers are used. The inclusion of neighboring marker information, to create haplotypes containing the QTL, increases the confidence for correctly identifying genomic regions containing favorable alleles compared with using markers individually. Using a high density of markers to identify haplotypes spanning relatively short genetic distance regions that are identical by descent (IBD) and which are conserved at the population level allows recombination events close to the trait gene to be recognized. This information can identify possible errors in correctly assigning alleles at the trait locus. And, as discussed before, knowing the trait gene and variations in that gene allows the information to be used with certainly.

The rate of genetic improvement that can be achieved in selection for meat quality traits, is dependent on the amount of the variation that is genetically controlled. The improvement that is possible using markers in the selection program is proportional to the amount of variation that is explained by the genes included in the selection criteria. For both beef and pork, meat quality traits between 10 and 30% of the variation is genetically controlled (e.g., Burrow, Moore, Johnston, Barendse, & Bindon, 2001), but individual genes may explain only a few percent of this variation. However, some traits that have a well defined biological basis, and which affect specific aspects of meat quality, have a much higher genetic contribution, e.g., the size and number of fibers in particular muscles, which will affect lean muscle development (Rehfeldt, Fiedler, Dietl, & Ender, 2000). Indeed the myostatin gene, which is associated with double muscling in several breeds of cattle, has been shown in mice to have a major influence in regulating muscle fiber size, type, and number (Rehfeldt et al., 2006). However, although the mutation in the myostatin gene has been shown to control a major part of the double muscling phenotype in the Belgian Blue breed, in other breeds, e.g., the South Devon, the same mutation has a more limited effect (Wiener, Smith, Lewis, Woolliams, & Williams, 2002). Therefore, even when the trait gene is known and the effect has been characterized in one population, care should be exercised in extrapolating information for use in another. It is likely that even for a gene that is responsible for a large proportion of the genetic variation the affect on the phenotype may be dependent on interactions with other genes (epistasis).

Eventually, sufficient information will be accumulated to define the biochemical pathways that control particular traits and phenotypes. It will then be possible to select for improvement on several criteria and multiple genetic loci, each of which are involved in the development of the desired phenotype. To identify these pathways, QTL mapping and individual trait gene identification is just the first step. Several approaches will be required to improve the factors involved in regulating meat quality parameters. One route to identifying particular biochemical or developmental pathways that are involved in the meat quality traits, will be to examine the expression patterns of genes and identify those that are co-regulated during particular developmental processes and are associated with, e.g., specific nutritional status or with particular phenotypes. This information along with the genetic information can be used to optimize the selection strategies.

Genome Selection

The use of genome wide marker information has been proposed by several authors (Meuwissen, Hayes, & Goddar, 2001; Gianola et al., 2004), to select the best animals based on their whole genome, rather than on one, or a few markers. The genome selection approach is used to estimate the genetic value of an individual based on all the available genetic marker information, rather than detection of quantitative trait loci. The main challenge with this approach, not withstanding the collection of suitable data, is to develop a functional statistical model that simultaneously relates phenotypes to SNP genotypes taking into account additive genetic effects between different loci and other nuisance effects, such as sex or age of an individual, environment etc. Standard quantitative genetics theory gives a mechanistic basis to the mixed-effects linear model, treated either from classical (Sorensen & Kennedy, 1983; Henderson, 1984) or from Bayesian (Gianola & Fernando, 1986) perspectives. With the availability of a 50,000 SNP genotyping panel for cattle, it has been possible to assess the feasibility of the genome selection approach. Using the data from Holstein cattle, van Tassel et al. (2008) demonstrated the value of a hybrid approach in which genome-wide marker data combined with the phenotype data to calculate a genome assisted breeding value (GAEBV). GAEBV calculated using data from bulls in selection programs between 1995 and 2000 was used to test the accuracy of predictions compared with the standard EBV by examining the actual performance estimated for bulls progeny tested in 2000–2003. For milk yield, protein, fat content and SCC, the GAEBV were considerably more accurate than the standard EBVs.

Conclusions

Use of genetic markers will facilitate more effective genetic selection for improvement in production associated traits in livestock. In the first instance, before the genes controlling the traits under selection have been identified and characterized, marker-assisted selection based on markers that are linked to the gene controlling variation in the traits can be applied. However, this has to be done within families where the allelic associations between markers and traits (phase) had been determined. Once identified, the functional allelic variation within the trait gene can be used to select more effectively for the best alleles at particular loci. With the discovery of tens of thousands of markers genome-wide and the development of techniques to genotype large numbers of markers efficiently, it has become possible to use genomic selection methods. Using a sufficiently high density of markers, it is possible to use those in markers in strong disequilibrium with trait genes to select for chromosomal regions with the best alleles for a trait, even before the trait gene is identified. This is similar to marker assisted selection, but instead of selecting for individual loci, the technique is applied across the whole genome.

There are several advantages of using markers in selection programs, rather than relying on phenotype-based selection. These include a more rapid and accurate prediction of the phenotype and hence earlier selection of breeding stock. Using the genome selection approach, the estimates of breeding values are increased allowing for a more accurate selection and increased confidence in the choice of the best animals. Such selection methods will also help to optimize improvements in traits controlled by loci with pleiotropic effects, where current selection suggests that progress in one trait may have a negative impact on another important trait. Knowledge of the genetic effects of each region of the genome may suggest ways of improving apparently confounded traits simultaneously.

Although there has been rapid advances in knowledge of the genome sequence, in techniques for assaying variations and in the statistical methods for using this information effectively in selection programs, the major barrier to implementing the techniques remains the lack of populations in which phenotypic traits are systematically recorded. Such information is essential to estimate the proportion of the phenotypic variation that is under environmental and genetic control. Phenotypic information is also necessary to calculate the breeding value of each chromosomal segment for a given trait. Genes with a major effect on a phenotype in one population may be associated with little phenotypic variation in another, depending on the genetic background. It is therefore important to collect phenotypic information from different genetic types, and in different environments. Further information on gene interactions may come from gene expression studies. Gene expression microarrays have now been produced for the majority of livestock species, allowing the expression patterns of many thousands of genes to be assayed simultaneously. Building up information on patterns of gene expression in different tissues and species could reveal co-regulated physiological pathways that are currently unknown. This data will add to the information derived from the analysis of the genome and contribute to our understanding of variations in meat quality and other traits.

The increasing availability of genomic sequence from a wide range of species will allow comparison of coding and non-coding regions across species, to identify functional domains and regulatory sequences. When this information is combined with gene expression studies it will be possible to identify those regions of the genome that regulate expression of metabolic pathways involved responsible for complex traits. In some, possibly many, cases the genetic polymorphism responsible for phenotypic variation in traits may not be obvious if it lies in a regulatory region rather than within the coding part of a gene, as found for fat deposition controlled by *IGF2* in pigs or the *Callipyge* muscling phenotype in sheep. And, as with these examples, the expression of these genes may not be simple, with the expression of specific functional allele dependent on the parent of origin (imprinting). Improved knowledge of genetic and epigenetic effects and interactions between genes (epistasis and pleioptropy) and genes with the environment will allow information to be effectively incorporated into selection and management strategies. These ever-increasing refinements will provide breeders with better tools to rapidly respond to changing market demands for meat products with different qualities.

References

Anderson, L., & Georges, M. (2004). Domestic animal genomic: deciphering the genetics of complex traits. *Nature Reviews Genetics, 5,* 202–212.

Archibald, A. L., Haley, C. S., Brown, J. F., Couperwhite, S., McQueen, H. A., Nicholson, D., et al. (1995). The PiGMaP consortium linkage map of the pig (Sus scrofa). *Mamm Genome, 6,* 157–175.

Barb, C. R., Hausman, J. H., & Hoseknechtm, K. L. (2001). Biology of leptin in the pig. *Domestic Animal Endocrinology, 21,* 297–317.

Barendse, W. (2002a). DNA markers for meat tenderness. *Patent WO02064820.*

Barendse, W. (2002b). Assessing lipid metabolism. *Patent WO9923248.*

Barendse, W., Bunch, R. J., & Harrison, B. E. (2005). The leptin C73T missense mutation is not associated with marbling and fatness traits in a large gene mapping experiment in Australian cattle. *Animal Genetics, 36,* 71–93.

Barendse, W., Vaiman, D., Kemp, S., Sugimoto, Y., Armitage, S., Williams, J. L., et al. (1997). A medium density genetic linkage map of the bovine genome. *Mammalian Genome, 8,* 21–28.

Bellmann, O., Wegner, J., Teuscher, F., Schneider, F., & Ender, K. (2004). Muscle characteristics and corresponding hormone concentrations in different types of cattle. *Livestock Production Science, 85,* 45–57.

Berghmans, S., Segers, K., Shay, T., Georges, M., Cockett, N., & Charlier, C. (2001). Breakpoint mapping positions the callipyge gene within a 285 kilobase chromosome segment containing the GTL-2 gene. *Mammalian Genome, 12,* 183–185.

Bishop, M. D., Kappes, S. M., Keele, J. W., Stone, R. T., Sunden, S. L. F, Hawkins, G. A., et al. (1994). A genetic linkage map for cattle. *Genetics, 136,* 619–639.

Blott, S. C., Williams, J. L., & Haley, C. S. (1999). Discriminating among between cattle breeds using genetic markers. *Heredity, 6,* 613–619.

Bouley, J., Meunier, B., Chambon, C., DeSmet, S., Hocquette, J. F., & Picard, B. (2005). Proteomic analysis of bovine skeletal muscle hypertrophy. *Proteomics, 5,* 490–500.

Buchanan, F. C., Fitzsimmons, C. J., Van Kessel, A. G., Thue, T. D., Winkelman-Sim, D. C., & Schmutz, S. M. (2002). Association of a missense mutation in the bovine leptin gene with carcass fat content and leptin mRNA levels. *Genetic Selection Evolution, 34,* 105–116.

Buchanan, F. C., Thue, T. D., Yu, P., & Winkelman-Sim, D. C. (2005). Single nucleotide polymorphisms in the corticotrophin-releasing hormone and pro-opiomelanocortin genes are associated with growth and carcass yield in beef cattle. *Animal Genetics, 36,* 127–131.

Burrow, H. M., Moore, S. S., Johnston, D. J., Barendse, W., & Bindon, B. M. (2001). *Australian Journal of Experimental Agriculture, 41,* 893–919.

Casas, E., Keele, J. W., Shackelford, S. D., Koohmaraie, M., & Stone, R. T. (2004). Identification of quantitative trait loci for growth and carcass composition in cattle. *Animal Genetics, 35,* 2–6.
Casas, E., Shackelford, S. D., Keele, J. W., Koohmaraie, M., Smith, T. P. L., & Stone, R. T. (2003). Detection of quantitative trait loci for growth and carcass composition in cattle. *Journal of Animal Science,81,* 2976–2983.
Casas, E., Shackelford, S. D., Keele, J. W., Stone, R. T., Kappes, S. M., & Koohmaraie, M. (2000). Quantitative trait loci affecting growth and carcass composition of cattle segregating alternate forms of myostatin. *Journal of Animal Science, 78,* 560–569.
Casas, E., Stone, R. T., Keele, J. W., Shackelford, S. D., Kappes, S. M., & Koohmaraie, M. (2001). A comprehensive search for quantitative trait loci affecting growth and carcass composition of cattle segregating alternative forms of the myostatin gene. *Journal of Animal Science, 79,* 854–860.
Casas, S., Smith, S. J., Zheng, Y.-W., Myers, H. M., Lear, S. R., Sande, E.,et al.. (1998). Identification of a gene encoding an acyl CoA:diacylglycerol acyltransferase, a key enzyme in triacylglycerol synthesis. *Proceedings of the National Academy of Sciences of the United States of America, 95,* 13018–13023.
Casser-Malek, I., Sundre, K., Listrat, A., Ueda, Y., Jurie, C., Briand, Y., et al. (2003). Integrated approach combining genetics genomics and muscle biology to manage beef quality. *British Society of Animal Science* York.
Charlier, C., Coppieters, W., Farnir, F., Grobet, L., Leroy, P. L., Michaux, C.,et al. (1995). The mh gene causing double-muscling in cattle maps to bovine Chromosome 2. *Mammalian Genome, 6,* 788–792.
Chowdhary, B. P., Fronicke, L., Gustavsson, I., & Scherthan, H, (1996). Comparative analysis of the cattle and human genomes: detection of ZOO-FISH and gene mapping-based chromosomal homologies. *Mammalian Genome, 7,* 297–302.
Clop, A., Marcq, F., Takeda, H., Pirottin, D., Tordoir, X., Bibe, B., et al. (2006). A mutation creating a potential illegitimate microRNA target site in the myostatin gene affects muscularity in sheep. *Nature Genetics, 38,* 813–818.
Coppieters, W., Riquet, J., Arranz, J.-J., Berzi, P., Cambisano, N., Grisart, B.,et al. (1998). A QTL with major effect on milk yield and composition maps to bovine Chromosome 14. *Mammalian Genome, 9,* 540–544.
Crisà, A., Marchitelli, C., Savarese, M. C., & Valentini, A. (2003). Sequence analysis of myostatin promoter in cattle. *Cytogenetics Genome Research, 102,* 48–52.
D'Andrea, M., Fidotti, M., & Pilla, F. (2005). Differences in MC4R mRNA levels between Casertana and large white pig breeds. *Italian Journal of Animal Science, 4* (Suppl. 2), 94–96.
de Koning, D. J., Janss, L. L. G., Rattink, A. P., van Oers, P. A. M., de Vries, B. J., Groenen, M. A. M.,et al. (1999). Detection of quantitative trait loci for backfat thickness and intramuscular fat content in pigs (sus scrofa). *Genetics, 152,* 1679–1690.
de Koning, D. J., Schulman, N. F., Elo, K., Moisio, S., Kinos, R., et al. (2001). Mapping of multiple quantitative trait loci by simple regression in half-sib designs. *Journal of Animal Science, 79,* 616–622.
Dekkers, J. C. M. (2004). Commercial application of marker- and gene-assisted selection in livestock: Strategies and lessons. *Journal of Animal Science, 82* (E. Suppl.), E313–E328.
Dorroch, U., Goldammer, T., Brunner, R. M., Kata, S. R., Kühn, C.,Womack, J. E., et al. (2001). Isolation and characterization of hepatic and intestinal expressed sequence tags potentially involved in trait differentiation between cows of different metabolic type. *Mammalian Genome, 12,* 528–537.
Everts-van der Wind, A., Larkin, D. M., Green, C. A., Elliott, J. S., Olmstead, C. A., Chiu, R., Schein, J. E., Marra, M. A., Womack, J. E. & Lewin, H. A. (2005). A high resolution whole-genome cattle-human comparative map reveals details of mammalian chromosome evolution. *Proceedings of the National Academy of Sciences USA, 102,* 18526–18531.
Fahrenkrug, S. C., Freking, B. A., Rexroad III, C. A., Leymaster, K. A., Kappes, S. M., & Smith, T. P. L. (2000). Comparative mapping of the CLPG locus. *Mammalian Genome, 11,* 871–876.

Fernando, R. L., & Grossman, M. (1989). Marker-assisted selection using best linear unbiased prediction. *Genetics, Selection, Evolution, 21*, 467–477.
Flint, J., & Mott, R. (2001). Finding the molecular basis of quantitative traits: Successes and pitfalls. *Nature Reviews Genetics, 2*, 437–445.
Freking, B. A., Murphy, S. K., Wylie, A. A., Rhodes, S. J., Keele, J. W., Leymaster, et al. (2002). Identification of the single base change causing the callipyge muscular hypertrophy phenotype, the only known example of polar over dominance in mammals. *Genome Research, 12*, 1496–1506.
Fujii, J., Otsu, K., Zorzato, F., De Leon, S., Khanna, V. K., Weiler, J. E., et al. (1991). Identification of a mutation in the porcine ryanodine receptor that is associated with malignant hypertemia. *Science, 253*, 448–451.
Georges, M., Lathrop, M., Hilbert, P., Marcotte, A., Schwers, A., Swillens, S.,et al. 1990. On the use of DNA fingerprints for linkage studies in cattle. *Genomics, 6*, 461–474.
Georges, M., Nielsen, D., Mackinnon, M., Mishra, A., Okimoto, R., Pasquino, A. T.et al. (1995). Mapping quantitative trait loci controlling milk production in dairy cattle by exploiting progeny testing. *Genetics, 139*, 907–920.
Gianola, D., & Fernando, R. L. (1986). *Journal Animal Science, 63*, 217–244.
Gianola, D., Ødegård, J., Heringstad, B., Klemetsdal, G., Sorensen, D., Madsen, P.,et al. (2004). Mixture model for inferring susceptibility to mastitis in dairy cattle: A procedure for likelihood-based inference. *Genetics, Selection, Evolution. 36*, 3–27.
Gilbert, R. O., Rebhun, C. A., Kim, C. A., Kehrli, M. E. Jr., Shuster, D. E., & Achermann, M. R. (1993). Clinical manifestation of leukocyte adhesion deficiency in cattle: 14 cases (1977–1991). *Journal of American Veterinary Medical Association, 202*, 445–449.
Goldammer, T., Dorroch, U., Brunner, R. M., Kata, S. R., Womack, J. E., & Schwerin, M. (2002). Identification and chromosome assignment of 23 genes expressed in meat and dairy cattle. *Chromosome Research, 10*, 411–418.
Gregory, S. G., Sekhon, M., Schein, J., Zhao, S., Osoegawa, K., Scott, C. E., et al. (2002). A physical map of the mouse genome. *Nature, 418*, 743–750.
Grisart, B., Farnir, F., Karim, L., Cambisano, N., Kim, J.-J., Kvasz, A., et al. (2004). Genetic and functional confirmation of the causality of the DGAT1 K232A quantitative trait nucleotide in affecting milk yield and composition. *Proceedings of the National Academy of Sciences of the United States of America, 101*, 2398–2403.
Grobet, L., Martin, L. J. R., Poncelet, D., Pirottin, D., Brouwers, B., Riquet, J.,et al. (1997). A deletion in the bovine myostatin gene causes the double-muscled phenotype in cattle. *Nature Genetics, 17*, 71–74.
Gutiérrez-Gil, B., Wiener, P., Nute, G. R., Gill, J. L., Wood, J. D. & Williams, J. L., (2007). Detection of Quantitative Trait Loci for Meat Quality Traits in Cattle. *Animal Genetics, 39*, 51–61.
Hayes, H. (1995). Chromosome painting with human chromosome-specific DNA libraries reveals the extent and distribution of conserved segments in bovine chromosomes. *Cytogenetics and Cell Genetics, 71*, 168–174.
Henderson, C. R. (1984). *Applications of linear models in animal breeding*. Ontario, ON, Canada: University of Guelph.
Huston, R. D., Cameron, N. D., & Rance, K. A. (2004). A melanocortin-4 receptor (MC4R) polymorphism is associated with performance traits in divergently selected large white pig populations. *Animal Genetics, 35*, 386–390.
Ihara, N., Takasuga, A., Mizoshita, K., Takeda, H., Sugimoto, M., Mizoguchi, Y.,et al. (2004). A comprehensive genetic map of the cattle genome based on 3802 microsatellites. *Genome Research, 14*, 1987–1998.
International Chicken Genome Sequencing Consortium. (2004). Sequence and comparative analysis of the chicken genome provide unique perspectives on vertebrate evolution, *Nature, 432*, 695–716.
International HapMap Consortium. (2007). A second generation human haplotype map of over 3.1 million SNPs. *Nature, 449*, 851–861.

Itoh, T., Watanabe, T., Ihara, N., Mariani, P., Beattie, C. W., Sugimoto, Y., et al. (2005). A comprehensive radiation hybrid map of the bovine genome comprising 5593 loci. *Genomics, 85,* 413–424.

Jann, O. C., Aerts, J., Jones, M., Hastings, N., Law, A., McKay, S., et al. (2006). A second generation radiation hybrid map to aid the assembly of the bovine genome sequence. *BMC Genomics, 7,* 283.

Jeffreys, A. J., Wilson, V., & Thein, S. L. (1985). Hypervariable 'minisatellite' regions in human DNA. *Nature, 314,* 67–73.

Jiang, Y. L., Li, N., Du, L. X., & Wu, C. X. (2002). Relationship of T–>A mutation in the promoter region of myostatin gene with growth traits in swine. *Yi Chuan Xue Bao, 29,* 413–416.

Jiang, Z.-H., & Gibson, J. P. (1999). Genetics polymorphism in the leptin gene and their association with fatness in four pig breeds. *Mammalian Genome, 10,* 191–193.

Kambadur, R., Sharma, M., Smith, T. P. L., & Bass, J. J. (1997). Mutations in myostatin (GDF-8) in double muscled Belgian Blue and Piedmontese cattle. *Genome Research, 7,* 910–915.

Kappes, S. S., Keele, J. W., Stone, R. T., McGraw, R. A., Sonstegard, T. S., Smith, T. P.,et al. (1997). A second-generation linkage map of the bovine genome. *Genome Research, 7,* 235–249.

Kashi, Y., Hallerman, E., & Soller, M. (1990). Marker-assisted selection of candidate bulls for progeny testing programs. *Animal Production, 51,* 63–74.

Keele, J. W., Shackelford, S. D., Kappes, S. M., Koohmaraie, M., & Stone, R. T. (1999). A region on bovine chromosome 15 influences beef longissimus tenderness in steers. *Journal of Animal Science, 77,* 1364–1371.

Kennes, Y. M., Murphy, B. D., Pothier, F., & Palin, M.-F. (2001). Characterization of swine leptin (LEP) polymorphisms and their association with production traits. *Animal Genetics, 32,* 215–218.

Kim, K. S., Larsen, N., Short, T., Plastow, G., & Rothschild, M. F. (2000). A missense variant of porcine melanocortin-4 receptor (MC4R) gene is associated with fatness, growth, and feed intake traits. *Mammalian Genome, 11,* 131–135.

Kim, K. S., Reecy, J. M., Hsu, W. H., Anderson, L. L., & Rothschild. (2004). Functional and phylogenetic analyses of a melanocortin-4 receptor mutation in domestic pigs. *Domestic Animal Endocrinology, 26,* 75–86.

Knott, S. A., Elsen, J. M., & Haley, C. S. (1996). Methods for multiple-marker mapping of quantitative trait loci in half-sib populations. *Theoretical Applied Genetics, 93,* 71–80.

Koohmaraie, M., Killefer, J., Bishop, M. D., Shackelford, S. D., Wheeler, T. L., & Arbona, J. R. (1995). Calpastatin-based method for predicting meat tenderness. In A. Ouali, D. Demeyer, & F. Smulders (Eds.), *Expression of tissue proteinases and regulation of protein degradation as related to meat quality* (pp. 395–410). Utrecht, The Netherlands: ECCEAMST.

Lagonigro, R., Wiener, P., Pilla, F., Woolliams, J. A., & Williams, J. L. (2003). A mutation in coding region of the bovine leptin gene associated with feed intake. *Animal Genetics, 34,* 371–374.

Lander, E. S., Linton, L. M., Birren, B., Nusbaum, C., Zody, M. C., Baldwin, J., et al. (2001). Initial sequencing and analysis of the human genome. *Nature, 40,* 9860–9921.

Liang, P., & Pardee, A. B. (1992). Differential display of eukaryotic messenger RNA by means of the polymerase chain reaction. *Science, 257,* 967–971.

Liefers, S. C., Veerkamp, R. F., Te Pas, M. F., Chilliard, Y., & Van der Lende, T. (2005). Genetics and physiology of leptin in periparturient dairy cows. *Domestic Animal Endocrinology, 29,* 227–238.

Lin, C. S., & Hsu, C. W. (2005). Differentially transcribed genes in skeletal muscle of Duroc and Taoyuan pigs. *Journal of Animal Science, 83,* 2075–2086.

MacLennan, D. H., Zorzato, F., Fujii, J., Otsu, K., Phillips, M., Lai, F. A.,et al. (1989). Cloning and localization of the human calcium release channel (ryanodine receptor) gene to the proximal long arm (cen-q13.2) of human chromosome 19. (Abstract) *American Journal of Human Genetics, 45* (Suppl.), A205.

MacNeil, M. D., & Grosz, M. D. (2002). Genome-wide scans for QTL affecting carcass traits in Hereford x composite double backcross populations. *Journal of Animal Science, 80*, 2316–2324.

MacNeil, M. D., Miller, R. K., & Grosz, M. D. (2003). Genome-wide scan for quantitative traits loci affecting palatability traits of beef. *Plant and Animal Genomes XI Conference*, San Diego, USA.

Malek, M., Dekkers, J. C. M., Lee, H. K., Baas, T. J., Prusa, K., Huff-Lonergan, E., et al. (2001). A molecular genome scan analysis to identify chromosomal regions influencing economic traits in the pig. II. Meat and muscle composition. *Mammalian Genome, 12*, 637–645.

McCarthy, L. C. (1996). Whole genome radiation hybrid mapping. *Trend in Genetics, 12*, 491–493.

McPherron, A. C., Lawler, A. M., & Lee, S.-J. (1997). Regulation of skeletal muscle mass in mice by a new TGF-b superfamily member. *Nature, 387*, 83–90.

Ménissier, F. 1982. General survey of the effect of double muscling on cattle performance. In J. W. B. King & F. Ménissier (Eds.), M*uscle hypertrophy of genetic origin and its use to improve beef production* (pp. 437–449). London: Martinus Nijhoff Publishers.

Meuwissen, T. H. E. (1998). Optimizing pure line breeding strategies utilizing reproductive technologies. *Journal of Dairy Science, 81* (Suppl. 2), 47–54.

Meuwissen, T. H. E., Hayes, B. J., & Goddar, M. E. (2001). Prediction of total genetic value using genome-wide dense marker maps. *Genetics, 157*, 1819–1829.

Nagamine, Y., Haley, C. S., Sewalem, A., & Visscher, P. M. (2003). Quantitative trait loci variation for growth and obesity between and within lines of pigs (sus scrofa). *Genetics, 164*, 629–635.

Neimann-Sorensen, A., & Robertson, A. (1961). The association between blood groups and several production characteristics in three Danish cattle breeds. *Acta Agricultural Scandinavia, 11*, 163–196.

Nezer, C., Moreau, L., Brouwers, B., Coppieters, A., Detilleux, J., Hanset, R.,et al. (1999). An imprinted QTL with major effect on muscle mass and fat deposition maps to the IGF2 locus in pigs. *Nature Genetics, 21*, 155–156.

Nonneman, D., Kappes, S. M., & Koohmaraie, M. (1999). Rapid communication: A polymorphic microsatellite in the promoter region of the bovine calpastatin gene. *Journal of Animal Science, 77*, 3114–3115.

Page, B. T., Casas, E., Heaton, M. P., Cullen, N. G., Hyndman, D. L., Morris, C. A., et al. (2002). Evaluation of single-nucleotide polymorphisms in CAPN1 for association with meat tenderness in cattle. *Journal of Animal Science, 80*, 3077–3085.

Ponsuksili, S., Murani, E., Walz, C., Schwerin, M., & Wimmers, K. (2007). Pre- and postnatal hepatic gene expression profiles of two pig breeds differing in body composition: Insight into pathways of metabolic regulation. *Physiol Genomics, 29*, 267–279.

Rattink, A. P., De Koning, D. J., Faivre, M., Harlizius, B., van Arendonk, J. A. M., & Groenen, A. M. (2000). Fine mapping and imprinting analysis for fatness trait QTLs in pigs. *Mammalian Genome, 11*, 656–661.

Rehfeldt, C., Fiedler, I., Dietl, G., & Ender, K. (2000). Myogenesis and postnatal skeletal muscle cell growth as influenced by selection. *Livestock Production Science, 66*, 177–188.

Rehfeldt, C., Ott, G., Gerrard, D. E., Varga, L., Schlote, W., Williams, J. L., Renne, U. & Bünger L. (2006). Effects of the Compact mutant myostatin allele Mstn (Cmpt-dl1Abc) introgressed into a high growth mouse line on skeletal muscle cellularity. *Journal of Muscle Research and Cell Motility, 26*, 103–112.

Rohrer, G. A., Alexander, L. J., Hu, Z., Smith, T. P., Keel, J. W., & Beattie, C. W. (1996). A comprehensive map of the porcine genome. *Genome Research, 6*, 371–391.

Rothschild, M., Ciobanu, F., & Daniel, C. (2004). Novel calpastatin (CAST) alleles. United States Patent Application 20040048267.

Schimpf, R. J., Winkelman-Sim, D. C., Buchanan, F. C., Aalhus, J. L., Plante, Y., & Schmutz, S. M. (2000). QTL for marbling maps to cattle chromosome 2. *27th International Conference on Animal Genetics*, Minneapolis, USA.

Schmidt, J. V., Matteson, P. G., Jones, B. K., Xiao-Juan, G., & Tilghman, S. M. (2000). The Dlk1 and Gtl2 genes are linked and reciprocally imprinted. *Genes Development, 14,* 1997–2002.

Schnabell, R. D., Van Tassell, C. O. P., Matukumalli, L. K., Sonstegard, T. S., Smith, T. P., Moore, S. S., et al. Application of the BovineSNP50 assay for QTL mapping and prediction of genetic merit in holstein cattle. *Plant & Animal Genomes XVI Conference* (p. 521).

Seaton, G., Haley, C. S., Knott, S. A., Kearsey, M., & Visscher, P. M. (2002). QTL Express: User-friendly software to map quantitative trait loci in outbred populations. *Bioinformatics, 18,* 339–340.

Shackelford, S. D., Koohmaraie, M., Cundiff, L. V., Gregory, K. E., Rohrer, G. A., & Savell, J. W. (1994). Heritabilities and phenotypic and genetic correlations for bovine postrigor calpastatin activity, intramuscular fat content, Warner-Bratzler shear force, retail product yield, and growth rate. *Journal of Animal Science, 72,* 857–863.

Smith, T. P. L., Casas, E., Rexroad III, C. E., Kappes, S. M., & Keele, J. W. (2000). Bovine CAPN1 maps to a region of BTA29 containing a quantitative trait locus for meat tenderness. *Journal of Animal Science, 78,* 2589–2594.

Snelling, W. M., Casas, E., Stone, R. T., Keele, J. W., Harhay, G. P., Bennett, G. L., et al. (2005). Linkage mapping bovine EST-based SNP. *BMC Genomics, 6,* 74–78.

Snelling, W. M., Chiu, R., Schein, J. E., & The International Bovine BAC Mapping Consortium. (2007). A physical map of the bovine genome. *Genome Biology, 8,* R165 doi:10.1 186/gb-2007-8-8-r165.

Solinas-Toldo, S., Lengauer, C., & Fries, R. (1995). Comparative genome map of human and cattle. *Genomics, 27,* 489–596.

Sorensen, D. A., & Kennedy, B. W. (1983). Estimation of response to selection using least-squares and mixed model methodology. *Journal of Animal Science, 58,* 1097–1106.

Stone, R. T., Keele, J. W., Shackelford, S. D., Kappes, S. M., & Koohmaraie, M. (1999). A primary screen of the bovine genome for quantitative trait loci affecting carcass and growth traits. *Journal of Animal Science, 77,* 1379–1384.

Thaller, G., Kühn, C., Winter, A., Ewald, G., Bellmann, O., Wegner, J., et al. (2003). DGAT1, a new positional and functional candidate gene for intramuscular fat deposition in cattle. *Animal Genetics, 34,* 354–357.

Van Laere, S.-A., Nguyen, M., Braunschweig, M., Nezer, C., Collette, C., Moreau, L., et al. (2003). A regulatory mutation in IGF2 causes a major QTL effect on muscle growth in the pig. *Nature, 425,* 832–836.

Wang, Y. H., Byrne, K. A., Reverter, A., Harper, G. S., Taniguchi, M., McWilliam, S. M., et al. (2005). Transcriptional profiling of skeletal muscle tissue from two breeds of cattle. *Mammalian Genome. 16,* 201–210.

Weller, J. I., Kashi, Y., & Soller, M. (1990). Power of daughter and granddaughter designs for determining linkage between marker loci and quantitative trait loci in dairy cattle. *Journal of Dairy Science, 73,* 2525–2537.

Wheeler, T. L., Cundiff, L. V., Shackelford, S. D., & Koohmaraie, M. (2004). Characterization of biological types of cattle (Cycle VI): Carcass, yield, and longissimus palatability traits. *Journal of Animal Science, 82,* 1177–1189.

Wiener, P., Smith, J. A., Lewis, A. M., Woolliams, J. A., & Williams, J. L. (2002). Muscle-related traits in cattle: The role of the myostatin gene in the South Devon breed. *Genetic Selection and Evolution, 34,* 221–232.

Yeo, G. S., Lank, E. J., Farooqi, I. S., Keogh, J., Challis, B. G., & O'Rahilly, S. (2003). Mutations in human melanocortin-4 receptor gene associated with severe familial obesity disrupts receptor function through multiple molecular mechanism. *Human Molecular Genetics, 12,* 561–574.

Zhang, Y., Proenca, R., Maffel, M., Barone, M., Leopold, L., & Friedman, J. M. (1994). Positional cloning of the mouse obese gene and its human homologue. *Nature, 372,* 425–432.

Zhu, B., Smith, J., Tracey, S., Konfortov, B., Welzel, K., Schalkwyk, L., et al. (1999). A five fold coverage BAC library: Production, characterisation and distribution. *Mammalian Genome, 10,* 706–709.

Chapter 3
DNA-Based Traceability of Meat

G.H. Shackell and K.G. Dodds

Introduction

Definitions of meat traceability are as varied as the people who write them. In essence, they have been encapsulated by McKean (2001) who defined traceability of meat as "the ability to maintain a credible custody of identification for animals or animal products through various steps within the food chain from the farm to the retailer".

The principles of meat traceability adhere very closely to those of "chain of custody" in forensic science. Chain of custody is a legal term that refers to the ability to guarantee the identity and integrity of a specimen by documenting its chronological history, from acquisition through analysis to storage, for future reference and/or disposal (Tomlinson, Elliott-Smith, & Radosta, 2006). By referring to custody, the definition of traceability implies the same level of confidence in all the identification and record keeping from when an animal is born, while it is being raised, during processing and its transfer, as meat, into the marketplace and ultimately on to the consumer's plate.

When traceability follows a non-empirical pathway, surety demands an independent verification system to confirm (or refute) the claims against producers, manufacturers, wholesalers, or retailers (Arana, Soret, Lasa, & Alfonso, 2002). DNA-based traceability systems offer that unequivocal verification tool.

From "Gate to Plate"

Throughout the pathway that meat follows before it reaches the consumer (see Fig. 3.1), there are a variety of risk factors that can mitigate against the value of the product. Furthermore, there are a number of stakeholders involved, with each group having a vested interest at different stages of the pathway. For this reason, traceability is often viewed as a "blame-shifting" initiative, whereby stakeholders

---

G.H. Shackell
AgResearch Invermay, Private Bag 50 034, Mosgiel 9053, New Zealand

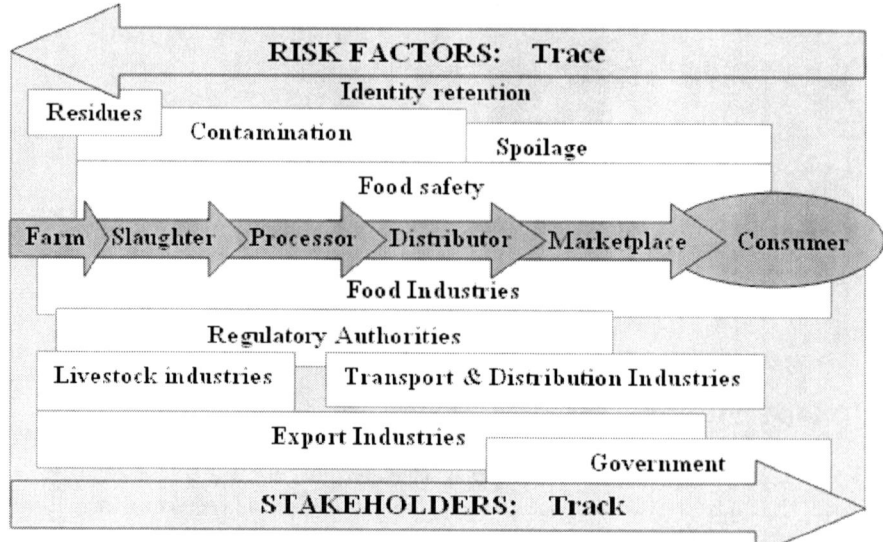

Fig. 3.1 The progression of a meat product from farm to plate and the relationship with risk factors and stakeholders

each seek to transfer blame, and therefore cost, to another (any other?) party. All participants in the continuum hope that the traceability system will absolve them from culpability in any scenario that compromises the product during its progress. Since there is no way to know in advance which samples and data may be involved in a traceability issue, it is imperative that records are both complete and possess integrity. A secure pathway is mandatory for an irrefutable reporting of the outcome.

The Traceability Continuum

The Farm

The first requirement for traceability occurs prior to an animal reaching slaughter age. A traceability system must be able to record feed additives that the animal may have ingested, any chemicals applied to the animal for physiological reasons or animal health treatments, and to adequately address changes of ownership during the period of growing the animal out to the slaughter weight. The level of sophistication required is a function of the farming system. Australia and New Zealand, for example, have pastoral based meat industries where the majority of sheep, and most beef cattle, are raised on the property that they are born on. In North America, feedlotting is common. Feedlot cattle are usually sourced from many farms, and all finished together on the same property. In Europe, farm animals are regularly traded and may change ownership (and location) several times before they enter the food chain.

The Slaughterhouse

The first parts of the carcass to be removed and discarded are the head, the skin, and the viscera. Therefore, any physical identification cues are lost in relation to the meat. Carcasses may be re-identified as they move through the slaughter plant by the use of barcodes or radio frequency identification devices (RFIDs) attached to the body, and/or the hardware associated with each carcass. In order to maintain a tracking/traceability chain, the original mark and the slaughterhouse identification must be linked in such a way as to be unambiguous.

The Processor

Meat may be further processed on premises associated with the slaughterhouse, or the carcasses may be shipped to a third-party processor. The animal is dismantled during processing for distribution either as primal cuts, which are further reduced prior to sale, or as pre-packaged individual cuts. The procedure is usually non-linear, so meat from different animals becomes jumbled during processing and packaging. For the process to allow full traceability, a label or identifier must be correctly attached to each individual piece of meat as it is created, and remain with that piece of meat at all times. In a system designed to meet USDA regulations, requiring that the carcasses and their parts can be identified "as being derived from a particular [live] animal", Heaton et al. (2005), found that 9.5% of liver samples did not match the animal they purportedly came from.

The majority of ground meat products is made from off-cuts or trimmed meat. Processing times and costs may increase considerably if each animal or each piece of trim meat is identified individually throughout the boning process.

The Distributor

Even if it has been possible to keep all of the components of a carcass together during slaughter and processing, it becomes much harder to create a verifiable trail once the meat is placed with a distributor. It is highly likely that a distributor will supply to multiple retailers, and that this will be done on the basis of inventory requirements, with no reference to the origin of the meat.

The Marketplace

In the marketplace, it is likely that large retail chains may in fact further distribute meat within their own organization, thereby adding to the complication of traceability. The marketplace is also where substitution and/or fraudulent packaging (or re-packaging) is most likely to occur.

The Consumer

When a meat product reaches the consumer, the integrity of any traceability trail is almost invariably lost. It is highly unlikely that the consumer will retain the packaging from a meat cut for traceability purposes. However, in the restaurant trade, where meat is often purchased as primal cuts, a record of the supplier and any identification of the meat are more likely to be kept. In some Japanese restaurants, the information is made available for those diners who want to check the origin of the steak they are eating (Ozawa, Lopez-Villalobos, & Blair, 2005).

Risk

Identity Retention

The greatest risk in any system where the objective is to provide traceability is identity retention. The most common animal identification system is printed tags (which can be changed), usually inserted in the ear. Less common, but still widely used, is branding the hide of the animal by burning an identification mark onto the skin using heat or ultra-cold temperatures. Even less commonly used are tattoos. These are generally on the ear, but could be on the lip, or in hair/wool by free skin such as the inside of the shoulder. Advances in electronics has resulted in the increasing use of RFIDs, which can be embedded in ear tags, implanted under the skin or encased in a protective layer and inserted into the digestive tract of the animal. More recently in live animals, biometric technologies have also been developed, including scanning retinal and/or nose patterns, which have a similar ability to differentiate individual animals as do fingerprints in the human.

Each of these methods offers a way of tracing animals through on-farm record keeping, changes of ownership, and during transport. Whichever method is used, most records are stored by transcribing the information into a computer file and/or onto a hard copy or PDA. Other than with RFID, an observer must read the tag or brand and then write the number(s) down. This provides potential error points, where the integrity of the traceability trail is only as reliable as the reading and transcription of the data.

Food Safety

Food safety and quality, along with concomitant environmental and animal welfare issues, are increasingly important in the world marketplace. The meat industry globally has responded to a series of crises such as Bovine Spongiform Encephalopathy (Smith, Cousens, Huillard D'Aignaux, Ward, & Will, 2004), labeling scandals (Castaldo, 2003), chemical contamination scares (e.g., Dioxin, Verbeke, 2001; ElAmin, 2006) and microbiological poisoning (Evans, Lane, Frost, & Nylen, 1998; Anonymous, 2002; Cagney et al., 2004). These issues, plus an increasing public awareness of animal ethics (Hobbs, Hobbs, Isaac, & Kerr, 2002) and negativity toward genetically modified food (Miraglia et al., 2004), have been cited as the drivers of increased consumer concern about the quality, origin, and integrity of meat at all levels of the food chain (Arana et al., 2002; Hoban, 1997; Aumaître, 1999).

Stakeholders

Bernués, Olaizola, and Corcoran (2003) found that consumers can be placed in demographic groups, based on age, family, disposable income, and meat consumption patterns, which have widely differing requirements for traceability. Interestingly, in another consumer survey, Latouche, Rainelli, and Vermersch (1998) found that 25–30% of the respondents who were unwilling to pay more for meat with a traceability guarantee held that view because they believed that the cost should not be borne by the consumer. Verbeke, Frewer, Scholderer, and De Brabander (2007) suggested that some consumer behavior toward food safety and risk information was sometimes irrational because they did not necessarily equate behavior with risk information. Despite this, retailers, wholesalers, and in many cases legislators, insist that the call for traceability is consumer driven. However, it is highly likely that the individual consumer desire for traceability varies with socio-economic status (Dickinson & Bailey, 2002; Hobbs, Bailey, Dickinson, & Haghiri, 2005).

Traceability is critical for the successful identification and recall of contaminated products from the market, and as a support for the quality assurance processes (Caporale, Giovannini, Di Francesco, & Calistri, 2001). Microbial transfer can actually occur during the slaughter process (Buncic, McKinstry, Reid, & Anil, 2002).

Some producers are interested in traceability for the protection and even value-enhancement of their brands, especially in niche markets or during a food crisis (Arana et al., 2002; Cavani and Petracci, 2004).

The industries that transport, process, and distribute animals and/or meat products have an obligation to maintain records that will enhance traceability. In the interests of food safety, it is now becoming increasingly common to have real-time monitoring of location and conditions during transport.

Considerable effort has resulted in meat traceability schemes, which are now required, by regulation, in some regions such as Japan (Clemens, 2003) and the EU (Cavani & Petracci, 2004), as governments have reacted to the growing public concern. The Japanese system, for example, is available at point-of-sale (Ozawa et al., 2005). In the Netherlands, a slaughterhouse may not receive a pig of unknown origin (Madec, Geers, Vesseur, Kjeldsen, & Blaha, 2001).

What is DNA?

DNA is the blueprint that determines how species, breeds, strains, and ultimately individual animals differ from each other. Within an individual, the DNA of all tissues is always the same. The only way two individuals can have the same DNA code is if they are identical twins (or clones) or in some cases when they come from inbred lines or crosses between inbred lines. Therefore, DNA provides an in-built proof of identity in a meat traceability system.

DNA is made up of four molecules (adenine, thiamine, guanine, and cytosine) called nucleotides. The way nucleotides are arranged is a sequence, and the complete sequence of nucleotides for an individual is its genome. A specific region within the genomic DNA is referred to as a locus. Within that locus there are sub-sets of sequences called alleles. An allele may be comprised of many apparently random nucleotides, a group of (usually di, tri, or tetra) repeated nucleotides or just a single nucleotide. The underlying precept is that each locus has its own set of alleles.

Alleles are usually inherited by the Mendelian genetic principles; one allele at each locus is inherited from each parent. If both parents contribute the same allele, the individual is said to be homozygous at that locus and if each parent contributes a different allele the individual is heterozygous.

The pattern of allele differences at a set of multiple loci can distinguish between individuals and the frequency with which the components of those differences occur, can distinguish between breeds or species. These differences are referred to as DNA markers, and it is the variation between the markers that underpins DNA-based traceability technologies.

DNA Markers

The first genetic markers were for simple phenotype variants such as eye color, or protein polymorphisms such as the hemoglobin blood groups. DNA markers have an advantage over protein markers in that they are not degraded by cooking and they are present in all cells in the body. The advent of DNA technologies, in particular, the polymerase chain reaction (PCR) technique, which amplifies small amounts of DNA into analyzable quantities, has allowed the rapid development of informative (variable) genetic markers.

Multi-locus Markers

Minisatellite or VNTR (variable number tandem repeat) markers: Minisatellites, discovered by Jeffreys, Wilson, and Thein (1985), were the first markers sufficiently informative that each individual had a unique genotype. They were mostly dominant markers with only one allele able to be identified. A few, highly informative, single loci minisatellites identified in livestock (Georges et al., 1990) have been found useful. However, with the advent of microsatellites (see below) they have fallen out of favor because they are technically demanding and require a large quantity of DNA for analysis.

RAPD (random amplified polymorphic DNA fragments) markers: RAPD markers were among the first PCR-based markers to be used (Williams, Kubelik, Livak, Rafalski, & Tingey, 1990). They are very easy to generate and require only small quantities of DNA. They also have disadvantages. Their appearance (and disappearance) is very sensitive to slight changes in PCR conditions, so they are not easily reproducible. More seriously however, they are not locus specific. The bands generated by a particular primer in one pedigree may not bear any relation to bands generated by the same primer in a second pedigree.

AFLP (amplified fragment length polymorphisms) markers: AFLPs are now the multi-locus markers of choice (Vos et al., 1995). Genomic DNA is cut with restriction enzymes and linkers are attached to each end of the fragments. Subsets of the fragments are then selectively amplified from the mixture of genomic restriction fragments and separated according to size. Bands that are present in some individuals but absent in others can be used as genetic markers.

Single Locus Markers

Restriction Fragment Length Polymorphisms: RFLPs were the first DNA markers developed and they pre-date the development of PCR methods. The RFLP methodology utilizes restriction enzyme digestion of the genomic DNA, its separation by size using agarose gel electrophoresis and detection and analysis of the DNA sequence by a technique called Southern blotting (Sambrook, Fritsch, & Maniatis, 1989). RFLP markers detect the presence or absence of a restriction site, including insertions and deletions of additional sequences, provided it contains an additional restriction site or changes the band size seen on the Southern blot.

Microsatellites: Microsatellites, like minisatellites, are multi-allelic tandem repeats. However, they are single locus, co-dominant, and spread throughout the genome. They require only small amounts of template DNA and are relatively easy to find and characterize.

A microsatellite is a simple sequence that is repeated 10–50 times. Virtually all of the microsatellites that have been found for livestock have the sequence (AC/GT) as the repeat unit as it is the most abundant type within livestock genomes and therefore much easier to find and characterize. Allelic variation of microsatellites is due to different numbers of simple sequence repeats, which alters the size of the DNA fragment and allows them to be separated, usually by electrophoresis on a DNA sequencing gel.

Single Nucleotide Polymorphisms: The variation in sequence at a particular position is called a single nucleotide polymorphism (SNP). It has been estimated that an SNP occurs about once in every kilo base in humans (Cooper, Smith, & Cooke, 1985). In other mammals, such as livestock, the figure is likely to be similar giving potentially two to three million SNPs. SNPs represent by far the richest source of genetic variation available for traceability applications. A disadvantage is that SNPs are almost always bi-allelic. Therefore, individually they are much less informative than the multi-allelic markers and differentiation between individuals requires many more SNPs to be analyzed than highly informative multi-allelic markers. For this reason, as they were first becoming popular, Vignal, Milan, SanCristobal, and Eggen (2002) referred to the increase in use of SNPs as an "apparent step backward". The development of new technologies, such as "DNA chips" containing high density arrays of DNA (Chee et al., 1996), allows many SNPs to be genotyped at once, thus off-setting the disadvantage of requiring large numbers of such markers by significantly decreasing the unit cost per marker.

DNA as a Traceability Tool

Traceability systems usually revolve around devices that are attached to an animal or a piece of meat. Because DNA is unique to an individual it is a useful tool for proof of identity and/or audit in meat traceability systems (Shackell, 2005).

DNA is already used for meat traceability (Meghen et al., 1998; Cunningham & Meghen, 2001; Shackell, Tate, & Anderson, 2001), and it can also be used for brand protection (Castaldo, 2003), fraud detection (Arana et al., 2002; Vázquez et al., 2004; Pancaldi et al., 2006), and detecting contamination with non-label species (Calvo, Osta, & Zaragoza, 2002) and pathogens (Schroeder et al., 2003; Cagney et al., 2004). Meat processing systems can be forward audited quite simply by tracking individuals and their components, during the progress through the plant. However, DNA is the only way to unequivocally reverse audit a meat production plant.

The simple, schematic supply chain shown in Fig. 3.2, illustrates the traceability/audit complications between the farm and the consumer. For example, meat sold in Restaurants 1 and 3 is sourced from two processors, which between them are supplied by four farms, one of which (Farm 4) supplies to both the processors. Similarly, although Retailers 1 and 2 only source their meat from one processor, each of those processors is supplied by more than one farm, and again Farm 4 supplies to both of those processors. Only the meat sold at Restaurant 4 can be intuitively traced back to a single farm, as this restaurant sources meat from only one processor, who in turn sources from only one farm (#5).

DNA Sampling

It is not possible to simply take a DNA sample from a piece of meat, or for that matter any biological product, and be able to identify the individual that it came from. It may well be possible to deduce some of the characteristics of the individual, but the identity can only be established if there is a reference sample to compare with. Therefore, any DNA-based traceability system, by definition, requires that reference samples are available from any animal for which traceability might be

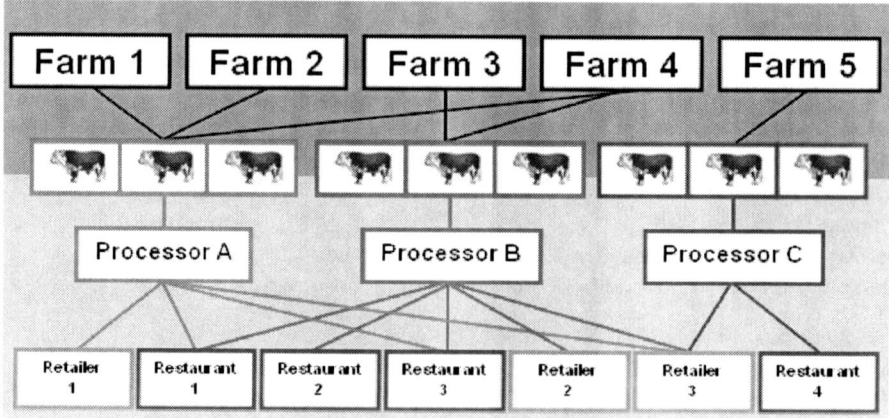

Fig. 3.2 Schematic supply chain of pasture grazed beef

required, and that these reference samples are stored in a way that they can be easily accessed. Once the shelf life of the product has been exceeded, the reference samples can be discarded.

Reference samples can be blood, semen, meat/tissue, tail hair follicles, or even saliva or milk. When a DNA sample has been taken from an animal as a reference sample, it can later be used to verify that animal at any time simply by comparing it to the test sample. This makes the animal fully traceable. For example, the reference DNA sample can be used to ensure that meat supposedly from a specific animal is in fact from that animal only and has not been substituted.

Although the primary focus of this chapter is meat, DNA traceability is not limited to meat from domesticated animals. Among others, the seafood and aquaculture industries are increasingly using DNA traceability (Håstein, Hill, Berthe, & Lightner, 2001; Hayes, Sonesson, & Gjerde, 2005), especially to detect species fraud (Aranishi, Okimoto, & Izumi, 2005; Maldini, Nonnis Marzano, Fortes, Papa, & Gandolfi, 2006).

Using DNA Markers to Make Inferences About the Origin of a Sample

When a DNA sample is obtained from a test item, traceability addresses the need to learn something about its relationship to (one or more) reference items. The test and reference samples are assayed for a set of genetic markers (usually microsatellites or SNPs). The simplest relationship to test is that the samples are from the same individual i.e., a test for identity. Genetic marker results can also be used to make inferences about more distant types of relationships, for example, parent–offspring, full or half-sibling, geographic region of origin, breed, or species. Identity and parental relationships lend themselves to exclusion tests – the hypothesized relationship is excluded if the samples do not share at least two or one, respectively, alleles at every locus. Other types of relationships cannot be tested in this way, for example, full-sibs may share any number of alleles (including zero) at a locus. In these cases we are interested in how likely such results would be.

Notation and Assumptions

For a particular marker, let p_i be the frequency of the ith allele, A_i, $i = 1, 2, \ldots, v$.

Let $S_t = \sum_{i=1}^{v} p_i^t$

Unless stated otherwise, it is assumed that the markers are in Hardy–Weinberg equilibrium (Falconer & Mackay, 1996) in the source population, and that the allele frequencies are known without error for this population. Further assumptions are that the markers are autosomal (nuclear but not X-linked) and have alleles that are transmitted in a Mendelian fashion, that there is no mutation and that genotypes are determined without error. Different markers are assumed to be in linkage equilibrium and segregating independently. A final assumption is that individuals and mates are sampled at random from the population.

Identity

Probability of Identity

Identity testing using DNA markers is an extremely powerful technique that has led to its application in forensics and traceability. The genotype results for an individual are often called its DNA "profile" or "fingerprint". A test and a reference sample are compared, and if their profiles are not the same, the samples are supposed to be derived from different individuals (assuming no errors are made in the testing process). This is a powerful technique because, with even a few markers, the probability that two unrelated individuals share the same genotypes is extremely small.

The power of this test is often quantified using probability statements. There are two situations. The first is before the test sample has been genotyped, and we want to know the usefulness of the marker set. The probability that two individuals, chosen at random from the population, have the same profile is known as the (average) probability of identity (PI) and is (e.g., Taberlet & Luikart, 1999)

$$\text{PI} = \sum_{i=1}^{v} p_i^4 + \sum_i \sum_{j>i} (2 p_i p_j)^2$$
$$= 2 S_2^2 - S_4$$

(see Dodds, Tate, McEwan, & Crawford, 1996). For multiple markers, the PI is the product of the individual marker PIs (assuming the markers are independent, i.e., are in linkage equilibrium). The probability that the two individuals do not share the same profile is known as the identity exclusion. Some values for 1-PI (i.e., identity exclusions) are shown in Figs. 3.3–3.5.

Once a test individual has been profiled, we would like to know what the probability is of that particular profile being seen in another individual. This is often referred to as the match probability (for unrelated individuals). The match probability for each marker is the population frequency of that genotype:

$$P(A_i A_i) = p_i^2$$
$$P(A_i A_j) = 2 p_i p_j, \quad i \neq j$$

Once again, the product of these for each marker is the match probability of the profile. Sometimes both of the probabilities described above are referred to as match probabilities or identity probabilities.

These calculations make a number of assumptions, but refinements are available which allow some of these to be relaxed. The reference samples may contain individuals which are more related than randomly chosen individuals from the

3 DNA-Based Traceability of Meat

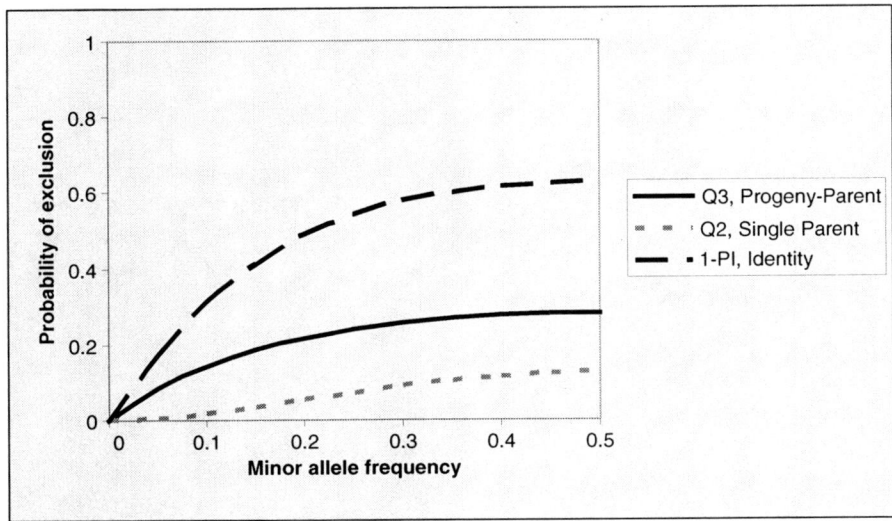

Fig. 3.3 Exclusion probabilities for a single locus with two alleles

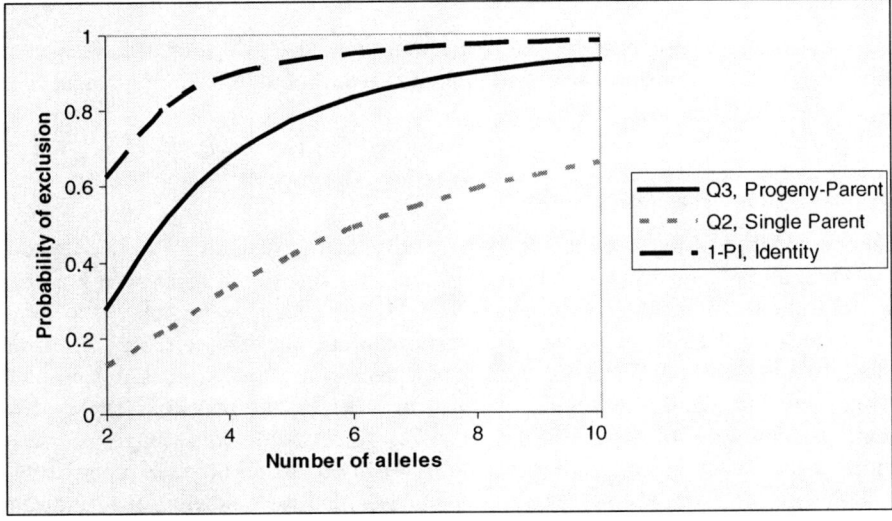

Fig. 3.4 Exclusion probabilities for a single locus with multiple alleles with equal frequencies

population. Animals processed at the same location and on the same date, are likely to contain cohorts including half- and full-siblings. The probability of identity can be modified to account for expected relationships, or expected population substructure (Ayres & Overall, 2004). Similar adjustments are available for match probabilities (Evett & Weir, 1998). Vázquez et al. (2004) used a simulation procedure to accommodate their population structure consisting of two breeds and the cross between them. The assumption of non-independent loci is not considered a serious

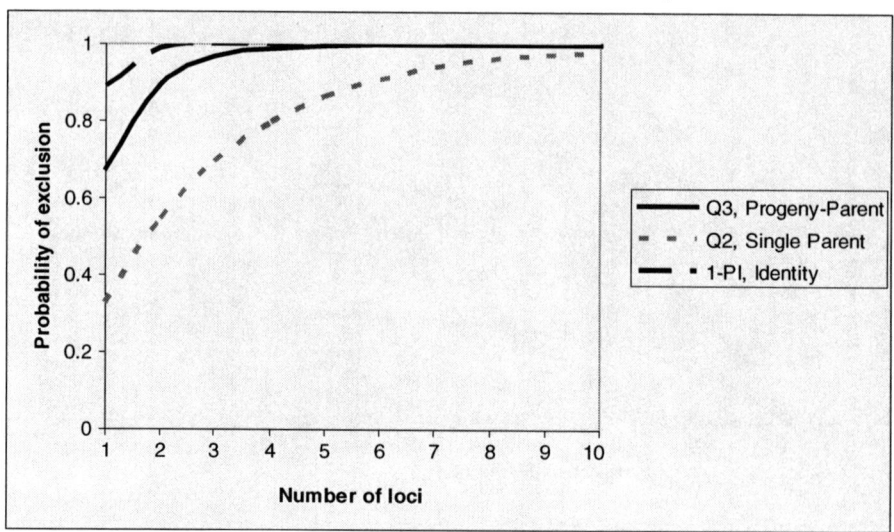

Fig. 3.5 Exclusion probabilities for several loci, all with four equally frequent alleles

problem, as it is usually only very close loci that show strong dependence. Buckleton and Triggs (2006) consider the calculation of match probabilities for half- and full-sibs with linked loci.

Application to Meat Traceability

The traceability of meat products using identity testing (Cunningham & Meghen, 2001) requires that a reference sample is collected and stored from each possible source animal or carcass (Shackell et al., 2001). These samples are labeled to show the identity of the source animal and are stored until after the saleable life of the meat from that animal. When the identity of a piece of meat is required, it is profiled along with the possible reference samples in order to find a match. The system requires some level of paper-based traceability, for example, knowledge of one or more of the source batch, processor and processing date and time for both the test sample and the reference samples. In some cases, the meat labeling may indicate the individual animal that the meat is supposed to have come from (Cunningham & Meghen, 2001; Shackell et al., 2001; Vázquez et al., 2004). Only those reference samples that correspond to the test sample are profiled.

There are three possible outcomes from this process. The test sample could match zero, one or more than one reference sample. If only one reference sample matches, this is assumed to be the same individual (assumes that all possible sources are in the reference set). If there are several individuals matching, then a more powerful marker set is required to exclude all but the correct individual. In some cases, the non-excluded individuals might be found to be highly related (e.g., half- or full-sibs) and from the same farm, which may be sufficient information. If there are no matching individuals then some sort of error has occured, such as a genotyping error or batch identification error, and this will require further investigation to correct the error, if possible. Failure to match a test sample to a group of reference samples can identify tampering or errors in internal traceability systems (Shackell et al., 2001).

Marker Set Requirements

In general, the DNA marker system used should be powerful enough so that matching a test to a reference sample from an individual other than its correct source is rare. Vázquez et al. (2004) suggest that a PI of 10^{-3} is adequate when trying to match a test sample with a single reference sample, and they used a three-marker multiplex to achieve this value. Cunningham and Meghen (2001) recommend that the probability of an error (an incorrect match in any of the reference samples) should be kept below 10^{-5}. Orrú, Napolitano, Catillo, and Moioli (2006) show that the best marker set may differ according to the breed. The marker set requirements will increase with increasing reference set size. The probability that there are no incorrect matches to any of the k reference individuals is approximately $1-(1-PI)^k$. This approximation assumes that exclusions are independent, which is not true, but still gives a reasonable idea of the marker set requirements. Some approximate marker set requirement examples are shown in Table 3.1, for multi-allelic markers (such as microsatellites) and for bi-allelic markers (such as SNPs). It can be seen that each ten-fold increase in the size of the reference set requires approximately one extra multi-allelic marker or two to three extra bi-allelic markers (with the given allele frequencies).

Complications

The process outlined is somewhat idealized, and a number of complications may arise. As mentioned above, genotyping errors are a possibility (Taberlet & Luikart, 1999). Providing a reasonable number of markers have been assayed, an allowance for these errors could be made by allowing one (or possibly a few if

Table 3.1 Approximate number of markers required to reduce the probability that there are no matches to any reference individuals to below 10^{-5}

Reference set size	Marker type	
	Four alleles, each with frequency 0.25 (Microsatellites)	Two alleles, with frequencies 0.4 and 0.6 (SNPs)
1	6	13
10	7	15
100	8	17
1000	9	20
10,000	10	22
100,000	11	25
1000,000	12	27

very many markers are assayed) mismatching genotype in the profiles. Depending on the application, the test sample and reference individuals with few mismatching genotypes may have their profiles rechecked or re-assayed. A possible strategy is to duplicate the genotyping of all samples as a matter of course, but this doubles the genotyping costs. Another approach would be to allow for genotyping errors in a likelihood model. The reference sample with the highest likelihood is declared the source individual, provided no other samples have a similar likelihood. Similar approaches have been used in parentage testing applications (Sancristobal & Chevalet, 1997; Marshall, Slate, Kruuk, & Pemberton, 1998).

There can also be errors associated with the labeling of samples and products. In such cases, it might be possible to find a match by widening the group of reference samples tested. This would be successful, for example, if the error was to use a label from the previous batch.

There may be reasons (apart from genotyping error) for why two samples from the same individual do not match, although these cases are rare. Sometimes, individuals are chimeric (e.g., arising from the fusion of two early embryos) or mosaic (e.g., resulting from somatic cell mutations during early development), with differing genotypes in different parts of their body. The most common livestock example of chimerism is when cattle litter mates share blood in utero, resulting in non-fertile freemartin females in the case of mixed-sex twins. This can be seen by finding more than two alleles for an individual (with multi-allelic markers).

There may also be reasons why two different individuals have matching profiles (even with highly powerful marker sets). The individuals may have the same genotypes, because they are monozygous twins, or because they are clones. Again, such cases are rare with production livestock, but may occur more frequently in the breeding stock generated by using in vitro technologies. Currently, many countries have a voluntary moratorium on allowing cloned animals or their offspring to be used for human consumption. Most are waiting for an FDA decision (Rudenko & Matheson, 2007).

Parentage

Traceability by Parent Matching

One approach to DNA traceability is to match the test sample to its family (Håstein et al., 2001). This could be achieved by collecting reference samples of parents (male, female, or both). With this approach, it is only possible to trace back to the set of progeny of the parent(s). For example, if a sire is known to have progeny only on one farm (with no stock movements to other farms), then a match to that sire would narrow the source of the meat to animals from that farm. The main advantage of the approach is that possibly many fewer reference samples are required (although they would need to be kept until after the retail life of the meat from any of their progeny). This approach might be particularly useful for species that are small in size but have large half- or full-sibling families, such as in aquaculture.

Parentage Exclusion Probabilities

The power of DNA markers to test for parentage has usually been presented as an exclusion probability. For example, a paternity exclusion probability is the probability that a randomly chosen unrelated male from the population has a genotype incompatible with the paternity for the given genotypes of the individual and its dam. Parentage exclusion probabilities are given in, for example, Dodds et al. (1996). Those that are most likely to be of relevance to traceability are the parent-offspring exclusion (i.e., without knowledge of the other parent's genotype):

$$Q_2 = 1 - 4S_2 + 4S_3 - 3S_4 + 2S_2^2$$

and the parent-pair (offspring) exclusion probability:

$$Q_3 = 1 + 4S_4 - 4S_5 - 3S_6 - 8S_2^2 + 2S_3^2 + 8S_3S_2$$

For multiple independent markers inclusion probabilities are found by multiplying together the inclusion probabilities for each marker. Exclusion probabilities are one minus the inclusion probability. Some values of these exclusion probabilities are shown in Figs. 3.3–3.5.

Relatedness

It is also possible to make inferences about relationships other than identity and parentage. In general, it is more difficult to make definitive statements, as it is not possible to use exclusion-based methods. Some types of relationship that may be of use in traceability are an individual with its grandparent(s), full-sib(s) or half-sib(s). If these relationships can be inferred, then this might help narrow the source of a meat product.

Relationships are inferred by estimating relatedness (r), which is twice the probability that random alleles from each of the individuals are identical by descent (IBD). Some types of relationships with the same or similar values of r can be distinguished by considering the kinship coefficients, i.e., the probability of sharing 0, 1, or 2 alleles that are IBD (k_0, k_1, and k_2), where $r = k_1/2 + k_2$. Table 3.2 shows these values for some common relationships.

There are several methods for estimating relatedness or kinship coefficients. These methods have been reviewed by Blouin (2003) and Weir, Anderson, & Hepler (2006). The estimators tend to have a high sampling variance (Van De Casteele, Galbusera, & Matthysen, 2001), and many markers are required to provide acceptable standard errors (Blouin, 2003; Weir et al., 2006). An alternative approach is to conduct hypothesis tests using likelihood ratios (Göring & Ott, 1997), but these also require many markers, especially if there is a need to distinguish between reasonably close relationships such as half- and full-sibs. These might be used, for example, when trying to match an individual to a reference set thought to contain

Table 3.2 Values of relatedness (r) and kinship coefficients (k_0, k_1, and k_2), for some common relationships, assuming non-inbred individuals

Relation	r	k_0	k_1	k_2
Self	1	0	0	1
Parent–offspring	0.5	0	1	0
Full-sibs	0.5	0.25	0.5	0.25
Half-sibs	0.25	0.5	0.5	0
Grandparent–grandchild	0.25	0.5	0.5	0
Cousin	0.125	0.75	0.25	0

a full-sib. If the reference sets contain whole families of known half- or full-sibs, then the information on these groups can be used to reconstruct the likely genotypes of the parent(s), after which the exclusion-based methods of the previous section could be used. There are also methods to reconstruct family groupings, if these are unknown prior to genotyping (Thomas & Hill, 2002).

Breed, Species or Brand Protection

A number of techniques are available for making an inference about the origin of an individual with respect to one or more groups, by using genetic markers. To be able to assign an individual to a group, these groups need to be genetically differentiated. For example, the groups may represent different geographical regions, with individuals within a region more closely related than between regions. Other applications involve groups that relate to different species, different breeds, or a branded product that is genetically distinct from the other sectors of that breed or species.

The simplest method for assigning or excluding an individual from a particular group is when the group has one or more "private" alleles i.e., alleles that are seen only in that group, and for which the group is fixed. One such application of this strategy is to use genes influencing coat color, as these have been used to select breed types for many years so that many breeds have a specific combination of alleles. For example, Maudet and Taberlet (2002) propose using alleles at the MC1R gene (the extension locus) to detect the presence of Holstein milk in cheese of purportedly non-Holstein origin.

Methods that make a probability-based inference about the origin of an individual are known as assignment methods (Manel, Gaggiotti, & Waples, 2005). These have mainly been developed for applications in population genetics, but all of these methods can also be used for traceability applications. The main uses in traceability are the following:

- Assigning a test sample to a reference group from a set of known possible reference groups (*assignment test*)
- Assigning a test sample to a reference group or rejecting the hypothesis that it comes from that group (*exclusion-based assignment test*)
- Determining genetic groupings from a population sample (*clustering*)

The main difference between the first two applications is that in the first, all possible reference groups are known (and characterized for the markers being used) while in the second application only the group of prime interest is characterized. An example of the first is when meat is claimed to come from an animal of a particular breed, but it could be from one of several breeds. An example of the second is when meat is claimed to be of a particular brand, but if not it could come from any other group within the species. The third application (finding genetic groupings) might be undertaken to help create a genetically distinct brand.

Assignment Test

An assignment test endeavors to classify an unknown into its group of origin, using DNA marker information (Manel et al., 2005). The methods rely on all possible source groups being known and having genotype information (for the same markers) (i.e., they are reference groups). They generally assume that the reference groups are randomly sampled and are in Hardy–Weinberg and linkage equilibrium. In its simplest form, the assignment test calculates the likelihood of the test sample coming from each of the reference groups (i.e., the probability of the test genotypes given the allele frequencies for the reference group). The test sample is then assigned to the reference group that yields the highest likelihood (Manel et al., 2005). An alternative method is to calculate a genetic distance between the test sample and the reference groups, and assign the closest reference group. This method relaxes the assumptions of Hardy–Weinberg and linkage equilibrium (Cornuet, Piry, Luikart, Estoup, & Solignac, 1999).

One difficulty with the method is that if the test sample contains an allele not seen in a particular reference sample, then the likelihood will be zero. This is unsatisfactory as it may be due to insufficient sampling of the reference population. A number of solutions have been suggested (Ciampolini et al., 2006). Another difficulty is that the test may be assigned to one reference group, but its likelihood for another group is similar, that is, an assignment is made even though there is little discrimination between two groups. This is less of a problem in population genetics, where the tests were developed, than for traceability applications. In population genetics, it is likely that many individuals are being assigned to populations and an error in one is not critical. In traceability applications, it is usually a single individual that is the focus of the investigation. A possible solution is to use an exclusion-based test (see below) as a follow up, or to establish a statistical confidence level using simulation procedures, similar to those used in parentage studies by Marshall et al. (1998).

Exclusion-Based Assignment Test

An exclusion test can be constructed using either likelihood or distance-based approaches. A large number of individuals are simulated for a particular reference population, by drawing alleles from the population (using estimated frequencies) at random. The assignment test value is calculated for each simulated individual giving a distribution of values for members from that population, which allows a confidence interval to be constructed (Cornuet, 1999). A modification of this procedure allows for linkage disequilibrium (created, e.g., by population admixture) by sampling each set of an individual's alleles from a randomly chosen individual (rather than from many different individuals) in the reference population (Paetkau, Slade, Burden, & Estoup, 2004).

Clustering

A number of methods exist for genetic clustering, for example, phylogenetic methods. These assume that there is no interchange of genetic material between groups, and a number of methods without this assumption have been developed, including the one used by STRUCTURE (Pritchard, Stephens, & Donnelly, 2000). This method allows individuals to be partially assigned to several groups.

Application to Food Safety

DNA can also be used for traceability related to food safety issues. In this case, it is the extrinsic DNA of pathogens and microbes that is tested, as opposed to the intrinsic DNA of the individuals that are the source of the meat or meat products.

As living organisms, the same principles of DNA apply. In fact, because of the very short generation interval, mutations can quickly identify specific strains and populations. In some instances, specific geographic locations develop "signature" microbes, which can be used to isolate the origin of contamination. DNA technologies can, much more rapidly than the traditional culture methods, identify the presence and the species of microbes (Kim et al., 2007), and/or strains of microbes (O'Hanlon, Catarame, Blair, McDowell, & Duffy, 2005; Villani, Russo, Blaiotta, Moschetti, & Ercolini, 2005).

Application to Audit

By definition, a paper/electronic based traceability system can only be forward audited. If traceability through the supply chain is to an animal level, the only way to effectively audit a paper trail, is to follow an animal right through the chain, from the Farm in a forward direction via the Processor to the Retailer or Restaurant. What an auditor cannot do is enter the paper/e-trail pathway and work in reverse. It is possible to follow the trail, but the integrity of any record keeping cannot be guaranteed. DNA identification technologies offer the potential to effectively identify meat from individual animals, by providing the ability to augment and verify a traceability paper trail (Meghen et al., 1998; Cunningham & Meghen, 2001; Shackell et al., 2001; Tate, 2001), or to detect fraud (Arana et al., 2002; Vázquez et al., 2004). For example, DNA reference samples can be used as a random audit simply by taking a sample from the retailer's shelf and testing it against the reference sample(s) indicated by packing information.

If a sample at the retailer's shelf can be reliably linked to an individual animal, the sample only has to be tested against one reference sample to find the match. If it can be linked to a farm, the reference samples can be limited to the animals supplied to the processor on a specific day.

DNA Traceability of Meat in Use

All the meat industry DNA-based traceability systems follow similar pathways, for example, AgResearch, New Zealand (**easiTrace**^{TM}, see Fig. 3.6); Eurofins, Belgium (**Eurofins-TAG**©); Catapult Genetics, Australia (**SureTrak**®); Identigen, Ireland (**TraceBack**^{TM}), and were generally developed using microsatellites, although some are now moving to SNP technology. Prior to (or at the time of) slaughter, a blood or tissue reference sample is taken from each animal, or carcass, and held in storage. When a carcass enters the cutting room, each primal or packaged cut is identified in such a way that either the individual animal or the cohort that went through the slaughter facility at the same time can be identified. If a trace is required, a sample of the meat and/or the identifying packaging information, are conveyed to the facility that holds the DNA samples. The group of stored DNA samples that is likely to contain the sample from the carcass involved is selected, and DNA profiles

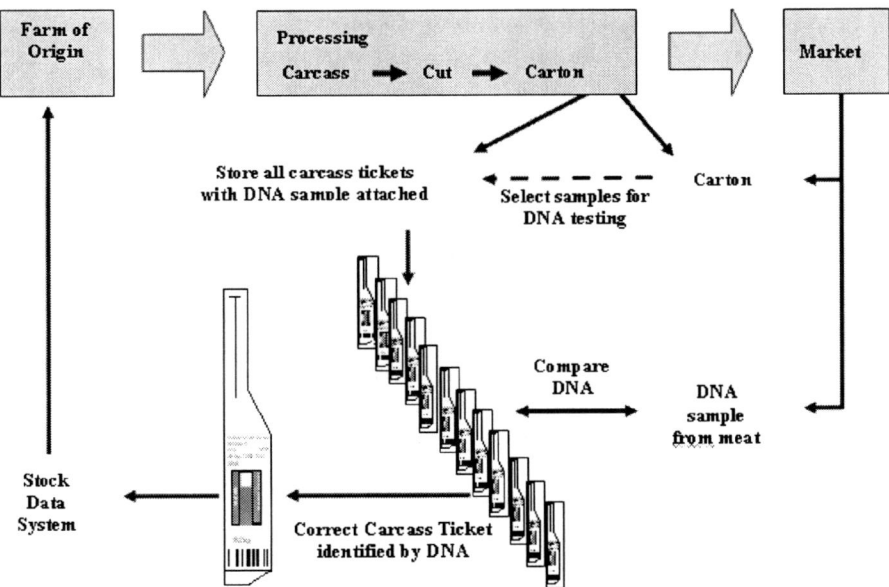

Fig. 3.6 A generic pathway for meat traceability from Shackell et al., 2001

are used to unequivocally match the sample to the carcass or animal that it originated from. For any DNA-based system to be effective an auditable production trail through the processing facility is essential and the processor must follow standard operating procedures to avoid contamination among reference samples (Shackell et al., 2001). Provided these conditions are met, a standardized and validated analytical procedure will match the DNA profiles of a carcass at slaughter to a meat sample in a packing carton.

There are other systems (e.g., Maple leaf Foods in Canada and Kyodo International Inc. in Japan) that use mitochondrial SNP technologies. In these cases, samples are taken from the parent stock and the Genotypes stored in databases. Test samples are analyzed and the resultant genotypes are compared to the database information to determine the farm of origin.

The philosophy of traceability and audit is to reduce the reference samples required to match a test sample to the smallest number possible. This increases the cost-effectiveness of the testing.

Compound Meat Products

Compound or blended meat products are not easily traceable since the origin of the source material is difficult to establish (McKean, 2001), and ground beef and processed products are generally exempt from traceability legislation (e.g., *Japan*, Clemens, 2003; *European Union*, Regulation (EC) No 1760/2000).

Individual DNA profiles show one or a maximum of two alleles at each DNA marker (see Fig. 3.7). Because mixed samples contain many individuals, and each has its own unique alleles, mixed samples have a distinctive DNA "profile", whereby many of the potential alleles at each marker are present. Therefore, while conventional DNA traceability schemes work well for tracing primal and/or sub-primal meat cuts, they have not been so readily adopted for compound meat products, which use ground meat from an unknown number of (usually) unidentified animals. For DNA traceability to be applicable to sausages and ground beef, methodologies must be developed to allow genotyping of DNA mixtures.

Determining the number of contributors to a mixture of randomly selected animals is not feasible when the mixture contains DNA from more than five or six individuals (Dodds & Shackell, 2004). However, if the product can be separated into its constituents, the principle of identifying individuals by their DNA profile can be applied (Shackell & Mathias, unpublished). Furthermore, given the nature of most mixed products, it is unlikely that the trim meat from a set of animals will go to two different manufacturers. Therefore, if at least some of the contributors to mixed products are known, it is not necessary to identify every individual in the mixture to know that the mixture is in fact the one being tested (Vetharaniam, Shackell & Upsdell, unpublished).

By using microsatellite markers as a tool for traceability of ground beef product, it is possible to match anonymous ground beef samples to the correct batch of

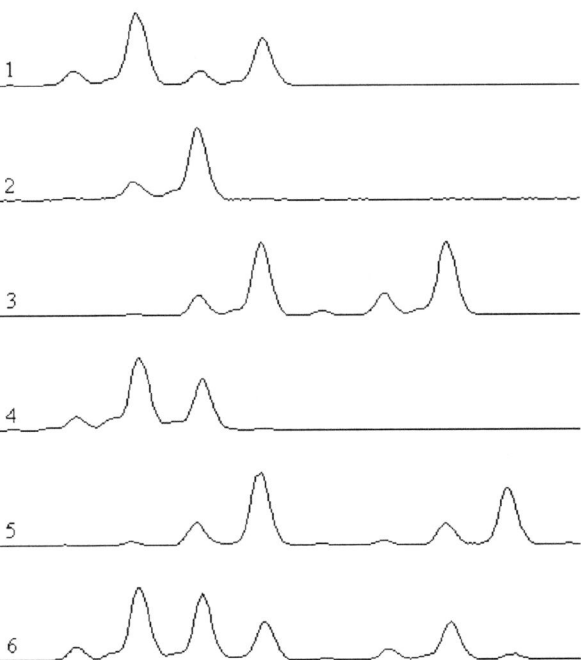

Fig. 3.7 Examples of profiles at a locus where the alleles have fragment lengths between 176 and 192 base pairs. Rows 1–5 show genotypes for one homozygous (row 2) and four heterozygous (rows 1, 3, 4, and 5) contributors. Row 6 shows the genotype of a mixture in which all of the alleles are present

manufacture with a high success rate (Shackell, Mathias, Cave, & Dodds, 2005). Reference samples are taken from each production batch, and their multi-allelic characteristics for a specific microsatellite marker set are determined. The data are analyzed using canonical discriminant functions – linear combinations of the original variables (relative allele intensities) are formed which provide maximum separation between the groups. The ability of the discriminant functions to separate the groups is assessed by re-substitution of the original variables back into the discriminant function and by using cross-validation. In the cross-validation process, each observation was in turn omitted from the data-set. The discriminant functions are then computed from all other observations, and the model classifies the omitted observation, which is then compared with its true group.

Species and Strain Differentiation

DNA technologies can detect meat from different species (Matsunaga et al., 1999; Saez, Sanz, & Toldrá, 2004), from closely-related subspecies (Matsunaga et al., 1998), and are readily adaptable to determine contamination or substitution of compound meat products, by non-labeled species (Colombo, Marchisio, Pizzini, & Cantoni, 2002; Di Pinto, Forte, Conversano, & Tantillo, 2005).

Recently, there have been new and powerful computational methods developed for determining an admixture in a population, or in an individual using clustering algorithm, based upon Bayesian statistics (Pritchard et al., 2000).

DNA Traceability as Part of a Value Chain

DNA traceability offers much more than merely linking a piece of meat to an animal. Figure 3.8, is a schematic representation of a meat industry pyramid. The upper two portions of the pyramid are "behind the farm gate" and the non-shaded section occurs after the animals leave the farm and become meat. DNA can link each part of the production pyramid. By using DNA information, benefits can be gained at each value point. The application of DNA traceability principles enables each part to feed information to the sectors above and below it, as well as for maintaining the integrity of the information.

Breeding

Breeders generate the animals that are the base stock for production systems; the sires, and dams, whose genetic characteristics are the basis for the production animals. By DNA testing for known genetic mutations or markers that confer or are associated with desirable production characteristics (Quantitative Trait Loci, Haley, 1995), the breeder can make informed decisions for breeding programs. By knowing the inheritance patterns of specific mutations that code for valuable traits, the breeder can ensure that those traits are passed to the rams bred for the next tier in the pyramid. The principles of parentage testing maintain the integrity of breeding programs (Israel & Weller, 2000) and the brands can also be protected (Che Man, Aida, Raha, & Son, 2007; García et al., 2006).

Meat Quality

By definition, meat quality (tenderness, yield, fat color, composition, and deposition) can only be identified post-mortem, unless they are under genetic control and DNA mutations or markers can be identified. As high-throughput technologies become available, the identification of the genetics of meat quality will increase rapidly. Already, there are commercial DNA marker tests available

3 DNA-Based Traceability of Meat

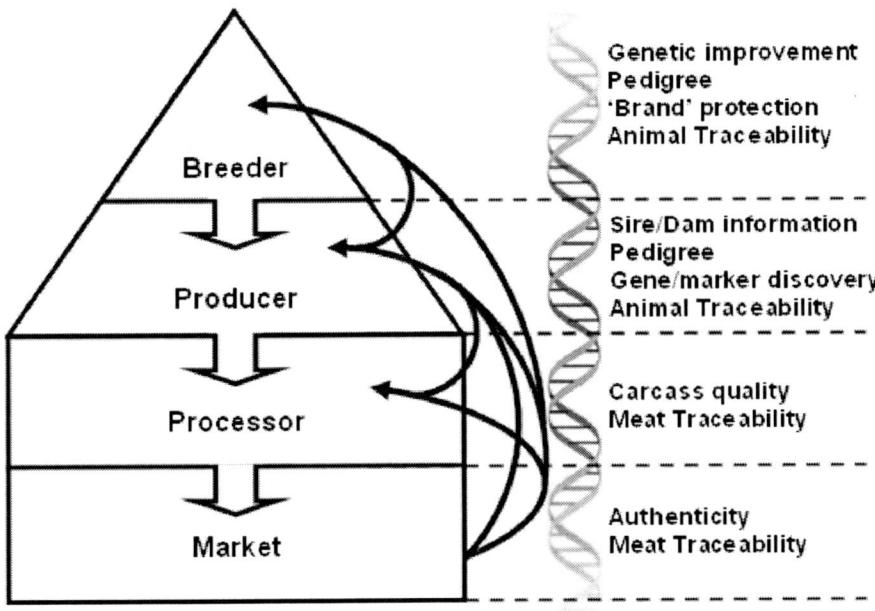

Fig. 3.8 The contribution of the DNA loop to the value chain, when used as a traceability tool during the production of meat

for feed efficiency, tenderness, marbling in beef cattle (e.g., Catapult Genetics, Australia), increased muscling on the leg and rump and in the loin of sheep (e.g., Catapult Genetics, New Zealand), breed assignment and the presence of *Bos indicus* genes in meat products in cattle (e.g., Identigen, Ireland).

Authenticity

In the market, one of the most difficult things to prove is authenticity. Products that demand a premium price are especially at risk of fraud and "passing off". By applying the DNA technologies it is possible to determine breed (Blott, Williams, & Haley, 1999; Ciampolini, Leveziel, Mazzanti, Grohs, & Cianci, 2000) and species (Janssen, Hägele, Buntjer, & Lenstra, 1998), which can in turn be used to detect fraud, for example, beef containing non-European *Bos indicus* DNA labeled as "Irish" beef (Castaldo, 2003). DNA can also detect non-labeled contamination, for example, pork in Italian horse meat sausages (Di Pinto et al., 2005) and in beef and pâté products (Calvo et al., 2002). Verification of the product sold to specification can also benefit from DNA analysis, for example, pork contamination of halal meat products (Che Man et al., 2007). Many regional products are produced from specific breeds or strains and DNA can detect fraudulent substitution or breeding, for example, Iberian ham (García et al., 2006).

Summary

DNA offers an in-built traceability verification tool that cannot be removed from a piece of meat and is not destroyed by cooking. The only criterion is as follows: for traceability to be possible, reference samples are required. For an entire meat industry, DNA offers much more than a simple traceability option by:

- providing information that allows genetic improvement of production animals
- validating that improvement by verifying of pedigree
- verifying the origin of meat
- protecting the brand of niche market meats
- identifying and enabling selection for meat quality
- detecting fraud

References

Anonymous. (2002). Source of deadly US listeria outbreak unknown. Retrieved October 21, 2002, from http://www.foodnavigator.com.
Arana, A., Soret, B., Lasa, I., & Alfonso, L. (2002). Meat traceability using DNA markers: Application to the beef industry. *Meat Science, 61*, 367–373.
Aranishi, F., Okimoto, T., & Izumi, S. (2005). Identification of gadoid species (Pisces, Gadidae) by PCR-RFLP analysis. *Journal of Applied Genetics, 46*(1), 69–73.
Aumaître, A. (1999). Quality and safety of animal products. *Livestock Production Science, 59*(2–3), 113–124.
Ayres, K. L., & Overall, A. D. J. (2004). api-calc 1.0: A computer program for calculating the average probability of identity allowing for substructure, inbreeding and the presence of close relatives. *Molecular Ecology Notes, 4*(2), 315–318.
Bernués, A., Olaizola, A., & Corcoran, K. (2003). Labelling information demanded by European consumers and relationships with purchasing motives, quality and safety of meat. *Meat Science, 65*(3), 1095–1106.
Blott, S. C., Williams, J. L., & Haley, C. S. (1999). Discriminating among cattle breeds using genetic markers. *Heredity, 82*(Part 6), 613–619.
Blouin, M. S. (2003). DNA-based methods for pedigree reconstruction and kinship analysis in natural populations. *Trends in Ecology & Evolution, 18*(10), 503–511.
Buckleton, J., & Triggs, C. (2006). The effect of linkage on the calculation of DNA match probabilities for siblings and half siblings. *Forensic Science International, 160*(2–3), 193–199.
Buncic, S., McKinstry, J., Reid, C. A., & Anil, M. H. (2002). Spread of microbial contamination associated with penetrative captive bolt stunning of food animals. *Food Control, 13*(6–7), 425–430.
Cagney, C., Crowley, H., Duffy, G., Sheridan, J. J., O'Brien, S., Carney, E., et al. (2004). Prevalence and numbers of *Escherichia coli* O157:H7 in minced beef and beef burgers from butcher shops and supermarkets in the Republic of Ireland. *Food Microbiology, 21*(2), 203–212.
Calvo, J. H., Osta, R., & Zaragoza, P. (2002). Quantitative PCR detection of pork in raw and heated ground beef and pâté. *Journal of Agricultural and Food Chemistry, 50*(19), 5265–5267.

Caporale, V., Giovannini, A., Di Francesco, C., & Calistri, P. (2001). Importance of the traceability of animals and animal products in epidemiology. *OIE Revue Scientifique et Technique, 20*(2), 372–378.
Castaldo, D. (2003). Brazilian beef labeled Irish. Retrieved September 26, 2003, from MeatNews.com.
Cavani, C., & Petracci, M. (2004). Rabbit meat processing and traceability. *Proceedings of the 8th World Rabbit Congress*, 1318–1336.
Che Man, Y. B., Aida, A. A., Raha, A. R., & Son, R. (2007). Identification of pork derivatives in food products by species-specific polymerase chain reaction (PCR) for halal verification. *Food Control, 18*(7), 885–889.
Chee, M., Yang, R., Hubbell, E., Berno, A., Huang, X. C., Stern, D., et al. (1996). Accessing genetic information with high-density DNA arrays. *Science, 274*(5287), 610–614.
Ciampolini, R., Cetica, V., Ciani, E., Mazzanti, E., Fosella, X., Marroni, F., et al. (2006). Statistical analysis of individual assignment tests among four cattle breeds using fifteen STR loci. *Journal of Animal Science, 84*(1), 11–19.
Ciampolini, R., Leveziel, H., Mazzanti, E., Grohs, C., & Cianci, D. (2000). Genomic identification of the breed of an individual or its tissue. *Meat Science, 54*(1), 35–40.
Clemens, R. (2003). Meat traceability in Japan. *Iowa Ag Review, 9*(4), 4–5.
Colombo, F., Marchisio, E., Pizzini, A., & Cantoni, C. (2002). Identification of the goose species (Anser anser) in Italian "Mortara" salami by DNA sequencing and a Polymerase Chain Reaction with an original primer pair. *Meat Science, 61*(3), 291–294.
Cooper, D. N., Smith, B. A., & Cooke, H. J. (1985). An estimate of unique DNA sequence heterozygosity in the human genome. *Human Genetics, 69*(3), 201–205.
Cornuet, J. M., Piry, S., Luikart, G., Estoup, A., & Solignac, M. (1999). New methods employing multilocus genotypes to select or exclude populations as origins of individuals. *Genetics, 153*(4), 1989–2000.
Cunningham, E. P., & Meghen, C. M. (2001). Biological identification systems: genetic markers. *Scientific and Technical Review of the Office International des Épizooties, 20*(2), 491–499.
Di Pinto, A., Forte, V. T., Conversano, M. C., & Tantillo, G. M. (2005). Duplex polymerase chain reaction for detection of pork meat in horse meat fresh sausages from Italian retail sources. *Food Control, 16*(5), 391–394.
Dickinson, D. L., & Bailey, D. (2002). Meat traceability: Are U.S. consumers willing to pay for it? *Journal of Agricultural and Resource Economics, 27*(2), 348–364.
Dodds, K. G., & Shackell, G. H. (2004). *The number of alleles seen in a mixture. Abstract MTU42.* Paper presented at the XXIInd International Biometric Conference, Cairns, Australia.
Dodds, K. G., Tate, M. L., McEwan, J. C., & Crawford, A. M. (1996). Exclusion probabilities for pedigree testing farm animals. *Theoretical and Applied Genetics, 92*, 966–975.
ElAmin, A. (2006). Belgium, Netherlands meat sectors face dioxin crisis. Retrieved January 31, 2006, from http://www.foodproductiondaily.com.
Evans, M. R., Lane, W., Frost, J. A., & Nylen, G. (1998). A campylobacter outbreak associated with stir-fried food. *Epidemiology and Infection, 121*(2), 275–279.
Evett, I. W., & Weir, B. S. (1998). *Interpreting DNA evidence—statistical genetics for forensic scientists.* Sunderland, MA: Sinauer Associates.
Falconer, D. S., & Mackay, T. F. C. (1996). *Introduction to Quantitative Genetics* (4th ed.). Harlow, Essex: Addison Wesley Longman.
García, D., Martínez, A., Dunner, S., Vega-Pla, J. L., Fernández, C., Delgado, J. V., & Cañón, J. (2006). Estimation of the genetic admixture composition of Iberian dry-cured ham samples using DNA multilocus genotypes. *Meat Science, 72*(3), 560–566.
Georges, M., Lathrop, M., Hilbert, P., Marcotte, A., Schwers, A., Swillens, S., et al. (1990). On the use of DNA fingerprints for linkage studies in cattle. *Genomics, 6*(3), 461–474.
Göring, H., & Ott, J. (1997). Relationship estimation in affected sib pair analysis of late-onset diseases. *European Journal of Human Genetics, 5*(2), 69–77.
Haley, C. S. (1995). Livestock QTLs – bringing home the bacon. *Trends in Genetics, 11*(12), 488–492.

Håstein, T., Hill, B. J., Berthe, F., & Lightner, D. V. (2001). Traceability of aquatic animals. *Scientific and Technical Review of the Office International des Épizooties*, 20(2), 564–583.
Hayes, B., Sonesson, A. K., & Gjerde, B. (2005). Evaluation of three strategies using DNA markers for traceability in aquaculture species. *Aquaculture*, 250(1–2), 70.
Heaton, M. P., Keen, J. E., Clawson, M. L., Harhay, G. P., Bauer, N., Shultz, C., et al. (2005). Use of bovine single nucleotide polymorphism markers to verify sample tracking in beef processing. *Journal of the American Veterinary Medical Association*, 226(8), 1311–1314.
Hoban, T. J. (1997). Consumer acceptance of biotechnology: An international perspective. *Nature Biotechnology*, 15(3), 232–234.
Hobbs, A. L., Hobbs, J. E., Isaac, G. E., & Kerr, W. A. (2002). Ethics, domestic food policy and trade law: Assessing the EU animal welfare proposal to the WTO. *Food Policy*, 27(5–6), 437–454.
Hobbs, J. E., Bailey, D., Dickinson, D. L., & Haghiri, M. (2005). Traceability in the Canadian red meat sector: Do consumers care? *Canadian Journal of Agricultural Economics/Revue canadienne d'agroeconomie*, 53(1), 47–65.
Israel, C., & Weller, J. I. (2000). Effect of misidentification on genetic gain and estimation of breeding value in dairy cattle populations. *Journal of Dairy Science*, 83(1), 181–187.
Janssen, F. W., Hägele, G. H., Buntjer, J. B., & Lenstra, J. A. (1998). Species identification in meat by using PCR-generated satellite probes. *Journal of Industrial Microbiology and Biotechnology*, 21(3), 115–120.
Jeffreys, A. J., Wilson, V., & Thein, S. L. (1985). Individual-specific 'fingerprints' of human DNA. *Nature*, 316(6023), 76–79.
Kim, J. S., Lee, G. G., Park, J. S., Jung, Y. H., Kwak, H. S., Kim, S. B., et al. (2007). A novel multiplex PCR assay for rapid and simultaneous detection of five pathogenic bacteria: *Escheria coli* O157:H7, *Salmonella*, *Staphylococcus aureus*, *Listeria monocytogenes* and *Vibrio parahaemolyticus*. *Journal of Food Protection*, 70(7), 1656–1662.
Latouche, K., Rainelli, P., & Vermersch, D. (1998). Food safety issues and the BSE scare: Some lessons from the French case. *Food Policy*, 23(5), 347–356.
Madec, F., Geers, R., Vesseur, P., Kjeldsen, N., & Blaha, T. (2001). Traceability in the pig production chain. *OIE Revue Scientifique et Technique*, 20(2), 523–537.
Maldini, M., Nonnis Marzano, F., Fortes, G. G., Papa, R., & Gandolfi, G. (2006). Fish and seafood traceability based on AFLP markers: Elaboration of a species database. *Aquaculture*, 261(2), 487–494.
Manel, S., Gaggiotti, O. E., & Waples, R. S. (2005). Assignment methods: matching biological questions with appropriate techniques. *Trends in Ecology & Evolution*, 20(3), 136–142.
Marshall, T. C., Slate, J., Kruuk, L. E. B., & Pemberton, J. M. (1998). Statistical confidence for likelihood-based paternity inference in natural populations. *Molecular Ecology*, 7, 639–655.
Matsunaga, T., Chikuni, K., Tanabe, R., Muroya, S., Nakai, H., Shibata, K., et al. (1998). Determination of mitochondrial cytochrome B gene sequence for red deer (*Cervus Elaphus*) and the differentiation of closely related deer meats. *Meat Science*, 49(4), 379–385.
Matsunaga, T., Chikuni, K., Tanabe, R., Muroya, S., Shibata, K., Yamada, J., & Shinmura, Y. (1999). A quick and simple method for the identification of meat species and meat products by PCR assay. *Meat Science*, 51(2), 143–148.
Maudet, C., & Taberlet, P. (2002). Holstein's milk detection in cheeses inferred from melanocortin receptor 1 (MC1R) gene polymorphism. *Journal of Dairy Science*, 85(4), 707–715.
McKean, J. D. (2001). The importance of traceability for public health and consumer protection. *OIE Revue Scientifique et Technique*, 20(2), 363–371.
Meghen, C. N., Scott, C. S., Bradley, D. G., MacHugh, D. E., Loftus, R. T., & Cunningham, E. P. (1998). DNA based traceability techniques for the beef industry. *Animal Genetics*, 29S1, 48–49.
Miraglia, M., Berdal, K. G., Brera, C., Corbisier, P., Holst-Jensen, A., Kok, E. J., et al. (2004). Detection and traceability of genetically modified organisms in the food production chain. *Food and Chemical Toxicology*, 42(7), 1157–1180.
O'Hanlon, K. A., Catarame, T. M. G., Blair, I. S., McDowell, D. A., & Duffy, G. (2005). Comparison of a real-time PCR and an IMS/culture method to detect Escherichia coli O26 and O111 in minced beef in the Republic of Ireland. *Food Microbiology*, 22(6), 553–560.

Orrú, L., Napolitano, F., Catillo, G., & Moioli, B. (2006). Meat molecular traceability: How to choose the best set of microsatellites? *Meat Science, 72*(2), 312–317.

Ozawa, T., Lopez-Villalobos, N., & Blair, H. T. (2005). An update on beef traceability regulations in Japan. *Proceedings of the New Zealand Society of Animal Production, 65,* 80–84.

Paetkau, D., Slade, R., Burden, M., & Estoup, A. (2004). Genetic assignment methods for the direct, real-time estimation of migration rate: a simulation-based exploration of accuracy and power. *Molecular Ecology, 13*(1), 55–65.

Pancaldi, M., Carboni, E., Paganelli, A., Righini, G., Salvi, A., Fontanesi, L., et al. (2006). Use of a natural tracer combined with the analysis of its DNA to guarantee the authenticity of agro-food products: The Authentifood system applied to typical dry-cured hams. *Industrie Alimentari, 45*(463), 1147–1155.

Pritchard, J. K., Stephens, M., & Donnelly, P. (2000). Inference of population structure using multilocus genotype data. *Genetics, 155*(2), 945–959.

Regulation (EC) No 1760/2000 of the European Parliament and of the Council of July 17, 2000 establishing a system for the identification and registration of bovine animals regarding the labelling of beef products and repealing Council Regulation (EC), No. 820/97.

Rudenko, L., & Matheson, J. C. (2007). The US FDA and animal cloning: Risk and regulatory approach. *Theriogenology, 67*(1), 198–206.

Saez, R., Sanz, Y., & Toldrá, F. (2004). PCR-based fingerprinting techniques for rapid detection of animal species in meat products. *Meat Science, 66*(3), 659–665.

Sambrook, J., Fritsch, E. F., & Maniatis, T. (1989). *Molecular cloning: a Laboratory Manual* (2nd ed.). Cold Spring Harbor, New York: Cold Spring Harbor Laboratory Press.

Sancristobal, M., & Chevalet, C. (1997). Error tolerant parent identification from a finite set of individuals. *Genetical Research, 70*(1), 53–62.

Schroeder, C. M., White, D. G., Ge, B., Zhang, Y., McDermott, P. F., Ayers, S., Zhao, S., & Meng, J. (2003). Isolation of antimicrobial-resistant Escherichia coli from retail meats purchased in Greater Washington, DC, USA. *International Journal of Food Microbiology, 85*(1–2), 197–202.

Shackell, G. H. (2005). Traceability systems in the meat industry. *Proceedings of the New Zealand Society of Animal Production, 65,* 97–101.

Shackell, G. H., Mathias, H. C., Cave, V. M., & Dodds, K. G. (2005). Evaluation of microsatellites as a potential tool for product tracing of ground beef mixtures. *Meat Science, 70*(2), 337–345.

Shackell, G. H., Tate, M. L., & Anderson, R. M. (2001). Installing a DNA-based traceability system in the meat industry. *Proceedings of the Association for the Advancement of Animal Breeding and Genetics, 14,* 533–536.

Smith, P. G., Cousens, S. N., Huillard D'Aignaux, J. N., Ward, H. J. T., & Will, R. G. (2004). The Epidemiology of Variant Creutzfeldt-Jakob Disease. *Current Topics in Microbiology and Immunology, 284,* 161–191.

Taberlet, P., & Luikart, G. (1999). Non-invasive genetic sampling and individual identification. *Biological Journal of the Linnean Society, 68,* 41–55.

Tate, M. L. (2001). Traceability of meat products – application of DNA technology. *Proceedings of the New Zealand Grasslands Association, 63,* 255–257.

Thomas, S. C., & Hill, W. G. (2002). Sibship reconstruction in hierarchical population structures using Markov chain Monte Carlo techniques. *Genetical Research, 79*(03), 227–234.

Tomlinson, J. J., Elliott-Smith, W., & Radosta, T. (2006). Laboratory information management system chain of custody: Reliability and security. *Journal of Automated Methods and Management in Chemistry,* Article id 74907 4 pages doi: 10.1155 JAMMC/2006/74907.

Van De Casteele, T., Galbusera, P., & Matthysen, E. (2001). A comparison of microsatellite-based pairwise relatedness estimators. *Molecular Ecology, 10*(6), 1539–1549.

Vázquez, J. F., Pérez, T., Ureña, F., Gudín, E., Albornoz, J., & Domínguez, A. (2004). Practical Application of DNA Fingerprinting To Trace Beef. *Journal of Food Protection, 67,* 972–979.

Verbeke, W. (2001). Beliefs, attitude and behaviour towards fresh meat revisited after the Belgian dioxin crisis. *Food Quality and Preference, 12*(8), 489–498.

Verbeke, W., Frewer, L. J., Scholderer, J., & De Brabander, H. F. (2007). Why consumers behave as they do with respect to food safety and risk information. *Analytica Chimica Acta, 586*(1–2 SPEC. ISS.), 2–7.

Vignal, A., Milan, D., SanCristobal, M., & Eggen, A. (2002). A review on SNP and other types of molecular markers and their use in animal genetics. *Genetics Selection Evolution, 34*(3), 275–305.

Villani, F., Russo, F., Blaiotta, G., Moschetti, G., & Ercolini, D. (2005). Presence and characterisation of verotoxin producing E. coli in fresh Italian pork sausages, and preparation and use of an antibiotic-resistant strain for challenge studies. *Meat Science, 70*(1), 181–188.

Vos, P., Hogers, R., Bleeker, M., Reijans, M., Van de Lee, T., Hornes, M., et al. (1995). AFLP: A new technique for DNA fingerprinting. *Nucleic Acids Research, 23*(21), 4407–4414.

Weir, B. S., Anderson, A. D., & Hepler, A. B. (2006). Genetic relatedness analysis: modern data and new challenges. *Nature Reviews Genetics, 7*(10), 771–780.

Williams, J. G. K., Kubelik, A. R., Livak, K. J., Rafalski, J. A., & Tingey, S. V. (1990). DNA polymorphisms amplified by arbitrary primers are useful as genetic markers. *Nucleic Acids Research, 18*(22), 6531–6535.

Useful Websites:

http://www.aic.gov.au/publications
http://www.catapultgenetics.com
http://archives.foodsafetynetwork.ca
http://www.epa.gov
http://www.eurofins.com
http://www.fda.gov
http://www.foodnavigator.com
http://www.foodproductiondaily.com
http://www.geneticsolutions.com.au
http://www.hypor.com
http://www.identigen.com
http://www.kyodo-inc.co.jp
http://www.mapleleaf.com
http://www.species-dna-detection.com
http://www.toxlab.co.uk

Part II
Biotechnology of Starter Cultures for Meat Fermentation

Chapter 4
Molecular Methods for Identification of Microorganisms in Traditional Meat Products

Luca Cocolin, Paola Dolci, and Kalliopi Rantsiou

Introduction

Traditional fermentations are those that have been used for centuries and even pre-date written historical records. Fermentation processes have been developed to upgrade plant and animal materials, to yield a more acceptable food, to add flavor, to prevent the growth of pathogenic and spoilage microorganisms, and to preserve food without refrigeration (Hesseltine & Wang, 1980). Among fermented foods, sausages are the meat products with a longer history and tradition. It is often assumed that sausages were invented by the Sumerians, in what is Iraq today, around 3000 BC. Chinese sausage *làcháng*, which consisted of goat and lamb meat, was first mentioned in 589 BC. Homer, the poet of The Ancient Greece, mentioned a kind of blood sausage in the Odyssey (book 20, verse 25), and Epicharmus (ca. 550 BC–ca. 460 BC) wrote a comedy entitled *"The Sausage"*. Evidence suggests that sausages were already popular both among the ancient Greeks and Romans (Lücke, 1974).

Today, there is a great variety and diversity of fermented sausages in the market world-wide, as a consequence of different formulations used in their production. The ingredients, such as raw materials and spices, together with manufacturing practices and fermentation techniques are the main factors that lead to the production of fermented sausages with specific organoleptic profiles and physico-chemical characteristics. However, a general distinction can be made between fermented products that are produced in the Northern countries and the ones manufactured in areas with temperate climate, mainly the Mediterranean countries. Northern products have a pH below 5, while Mediterranean products have a pH of 5.3–6.2 and are highly desiccated (Talon, Leroy, & Lebert, 2007).

The fermentation of sausages is a microbial process that has been investigated since 1960 (Lerche & Reuter, 1960; Lücke, 1974; Reuter, 1972), and these studies highlighted that the main microorganisms responsible for the transformation are lactic acid bacteria (LAB, *Lactobacillus* spp.) and coagulase-negative cocci

---

L. Cocolin
Dipartimento di Valorizzazione e Protezione delle Risorse Agroforestali, University of Turin, Faculty of Agriculture, via Leonardo da Vinci 44, 10095 Grugliasco – Turin, Italy

(CNC, *Staphylococcus* and *Kocuria* spp.). Moreover, in some fermented sausages, especially those produced in France, Italy, and Spain, the characteristics of the final products are influenced by the activity of molds and yeasts that are developing on the surface of the product (Lücke, 2000).

Microbial dynamics during fermentation have a tremendous impact on the sensory properties of the product. Sausages with lower pH are characterized by a short ripening time and have more lactobacilli from the early stages of fermentation. An acid flavor predominates in the product and commercialization takes place after less than two weeks of ripening. Long ripening times allow for more diverse microbial activities. LAB are mainly responsible for acidification. They are able to reduce the pH of the sausages by production of lactic acid from carbohydrates (Hammes, Bantleon, & Min, 1990; Hammes & Knauf, 1994). Moreover, they influence the sensory characteristics of the fermented sausages by the production of small amounts of acetic acid, ethanol, acetoin, pyruvic acid, and carbon dioxide (Bacus, 1986; Demeyer, 1982), and they are able to initiate the production of aromatic substances thanks to the proteolytic activity of muscle sarcoplasmatic proteins (Fadda et al., 1999a, 1999b). Apart from LAB, also CNC and yeasts contribute to the final characteristics, due to their capability to produce higher levels of volatile compounds with low sensory thresholds (Demeyer, Verplaetse, & Gistelink, 1986). CNC also have a fundamental role in the development and stability of the red color through the formation of nitrosomyoglobin by nitrate reductase activity, possibly involved in the limitation of the lipid oxidation, as well (Talon, Walter, Chartier, Barriere, & Montel, 1999). Furthermore, CNC are able to produce proteolytic and lipolytic enzymes responsible for the release of low molecular weight compounds, such as peptides, amino acids, aldehydes, amines, and free fatty acids, that are influencing the aromatic profile of the final product (Demeyer et al., 1986; Schleifer, 1986).

In the last 10 years, the approach to study microbial biodiversity has changed dramatically. With the advancement of molecular biology and the invention of the polymerase chain reaction (PCR), a new range of techniques were developed that can help in the understanding of the microbial complexity in natural ecosystems. As a consequence, the traditional microbiological techniques, based on plating, isolation, and biochemical identification, are now supported by new methods that rely on the analysis of the nucleic acids for detection, identification, and characterization of the microorganisms. In this context, several groups of food scientists started to apply molecular methods to various food fermentations. The aim of this chapter is to report on the molecular methods used so far to identify, characterize, and profile microbial diversity during the fermentation of sausages.

Approaches Used to Study the Microbial Diversity

The scientific community recognizes that the use of methods that are relying on the cultivation of the microorganisms (culture-dependent techniques), do not properly profile the microbial diversity present in a specific ecosystem (Hugenholtz, Goebel,

4 Identification of Microorganisms in Traditional Meat Products

& Pace, 1998). As a matter of fact, populations that are numerically limited or microorganisms that are stressed or in a sub-lethal state cannot be recovered, thereby they are eliminated from consideration. Moreover, viable but not culturable (VNC) cells that are not able to form colonies on agar plates but possess metabolic activity will not be picked up by culture-dependent methods. Methods that do not depend on cultivation (culture-independent techniques) have attracted the attention of many scientists in different domains of investigation, spanning from the environmental microbiology to food fermentations. Generally, they are based on the analysis of the DNA and RNA that is extracted from the sample directly without any kind of cultivation. The nucleic acids are then amplified by PCR and subjected either to cloning and sequencing or to profiling techniques, such as denaturing gradient gel electrophoresis (DGGE) or temperature gradient gel electrophoresis (TGGE), or single strand conformation polymorphism (SSCP). An alternative, culture-independent approach that does not rely on extraction of nucleic acids from the sample matrix is the fluorescence in situ hybridization (FISH). In this case, not only it is possible to identify the microorganisms by using specific probes, but it also permits the localization of species within the sample being investigated. In Fig. 4.1, the culture-dependent and independent methods are summarized.

Finally, the extensive development of molecular methods resulted in new tools that could be used for molecular characterization of the strains isolated by culture-dependent methods. Randomly amplified polymorphic DNA (RAPD)-PCR,

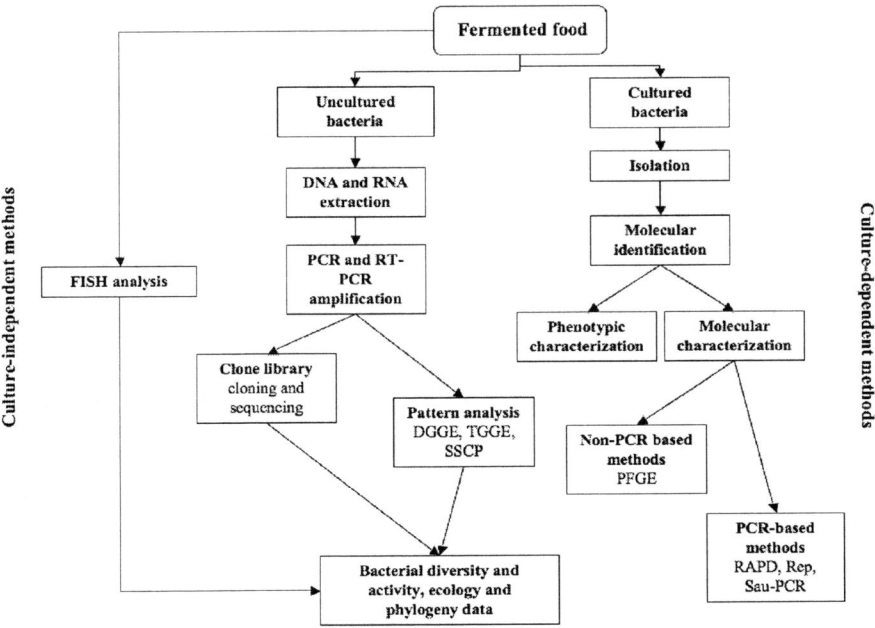

Fig. 4.1 Culture-dependent and -independent methods used to study the microbial ecology of fermented foods (Modified from Pontes et al., 2007).

repetitive bacterial DNA elements (Rep)-PCR, and enterobacterial repetitive intergenic consensus (ERIC)-PCR are common practices in almost all the laboratories that deal with microbial ecology and diversity. At present, it is widely accepted that different methods for the identification and characterization of microbial species in a specific ecosystem are necessary. The combination of different methods to study the microbial ecology of a system is defined as a "polyphasic approach" (Pontes, Lima-Bitterncourt, Chartone-Souza, & Amaral Nascimiento, 2007).

Molecular Methods Used to Investigate the Diversity of Microorganisms in Fermented Sausages

As mentioned above, an important distinction that has to be made is between culture-dependent and independent methods. While the culture-dependent methods are commonly used to identify and molecularly characterize microbial isolates, the culture-independent methods are used to profile directly the microbial populations during fermentation of sausages (Rantsiou & Cocolin, 2006).

Culture-Dependent Techniques

Strain Identification

The identification of strains isolated from sausage fermentation is most commonly achieved by targeting the ribosomal RNA operon (rRNA). It possesses some requisites and attributes that are very important for microbial identification. Apart from being present in all microbial species, it is characterized by the presence of conserved and variable regions, that become very important for differentiation purposes, and it functions as an evolutionary clock, giving the possibility to draw conclusions regarding evolution and phylogeny. In the case of the bacteria, the 16S rRNA gene is the most used target for molecular applications (Collins et al., 1991), while the D1-D2 loop of the 26S RNA gene became a generally accepted target for yeast identification (Kurtzman & Robnett, 1997). The process of identification of microbial strains by targeting the rRNA usually consists of a first step in which the specific gene (16S or 26S rRNA), or portion of it, is amplified by PCR that subsequently is subjected to DNA sequencing. By alignment of the sequence in the GeneBank, using available software programs online (Blast, Altschul et al., 1997), it is possible to achieve the identification. This experimental tactic is not applicable for ecology studies because of the high number of isolates; it is not feasible and economically convenient to sequence hundreds of strains. For this reason, researchers developed species-specific PCR and ribosomal RNA probes as fastest ways to obtain species identification. Ribosomal RNA probes (Nissen & Dainty, 1995), species-specific PCR primers (Berthier & Ehrlich, 1998; Blaiotta, Pennacchia, Parente, & Villani, 2003; Morot-Bizot, Talon, & Leroy-Setrin, 2003; Rossi, Tofalo, Torriani, & Suzzi, 2001), and multiplex PCR (Corbiere Morot-Bizot, Talon, & Leroy, 2004) have been used in the field of fermented sausages for the identification of LAB and CNC isolated during fermentation.

Alternatively, the PCR-sequencing methodology is coupled with techniques that are able to differentiate strains based on fingerprinting profiles, allowing grouping of the isolates and reducing the number of strains requiring sequencing. The methods used to group strains and reduce the number of isolates to sequence are once more based on an amplification step. The differences in the strains to identify can be detected either by exploiting primers that are annealing in various regions of the genome of the strain to identify, thereby producing a band pattern that is representative for a species, or by detecting differences in the DNA sequences specifically amplified by PCR. The first group of methods is mainly represented by RAPD-PCR and Rep-PCR. RAPD-PCR is characterized by an amplification process, using a single random primer that will anneal in several portions of the genomic DNA and give amplification products when the forward and reverse annealing sites are allowing extension by the DNA polymerase (Caetono-Anollés, 1993; Power, 1996). The main drawback of this technique is its low reproducibility. The annealing step takes place at very low temperature (37–40°C), and for this reason, the same strain subjected to RAPD-PCR in different moments can produce different profiles. While it is possible to reach a good intra-laboratory reproducibility, if all the experimental steps are optimized and standardized, it is practically impossible to be able to compare the results obtained in different laboratories. Concerning Rep-PCR, the primers used are specific to the repetitive elements spread around the bacterial genomes (Gilson, Clement, Brutlag, & Hofnung, 1984), and for this reason the profiles obtained are highly specific for a species and they are highly reproducible, as well. RAPD-PCR and Rep-PCR are methods that were first applied for identification purposes (Andrighetto, Zampese, & Lombardi, 2001; Berthier & Ehrlich, 1999; Gevers, Huys, & Swings, 2001; Rebecchi, Crivori, Sarra, & Cocconcelli, 1998), and only recently they have been used for molecular characterization of LAB and CNC (Cocolin et al., 2004; Comi et al., 2005; Iacumin, Comi, Cantoni, & Cocolin, 2006a; Rantsiou, Drosinos, Gialitaki, Urso, et al., 2005, Rantsiou, Iacumin, Cantoni, Comi, & Cocolin, 2005; Rantsiou, Drosinos, Gialitaki, Metaxopoulos, et al., 2006; Urso, Comi, & Cocolin, 2006), as well as yeasts (Cocolin, Urso, Rantsiou, Cantoni, & Comi, 2006a) isolated from fermented sausages.

The techniques that are allowing strain grouping based on differences in a DNA sequence and that which have been applied in the field of fermented sausages are the restriction fragment length polymorphism (RFLP) analysis of the 16S rRNA gene (Lee et al., 2004; Sanz, Hernandez, Ferrus, & Hernandez, 1998) and the denaturing/temperature gradient gel electrophoresis (D/TGGE) (Blaiotta, Pennacchia, Ercolini, Moschetti, & Villani, 2003; Cocolin, Manzano, Cantoni, & Comi, 2000; Cocolin, Manzano, Aggio, Cantoni, & Comi, 2001; Cocolin, Manzano, Cantoni, & Comi, 2001; Ercolini, Moschetti, Blaiotta, & Coppola, 2001a). With the RFLP approach, the differences in the DNA sequence are identified by using a restriction endonuclease that is cutting the DNA in specific restriction sites, giving a specific pattern for a given species. The D/TGGE methods are detecting differences in the sequences by analyzing their denaturation behavior. More specifically, as the

DNA molecule encounters an appropriate denaturant concentration, a sequence-dependent partial denaturation of the double strand occurs. The change in the DNA conformation determines a reduction of the migration rate of the molecule, and thereby DNA with different nucleotide sequences will show differential electrophoretic patterns. This approach was first used in environmental ecosystems (Muyzer, de Waal, & Uitterlinden, 1993), and at the beginning of the 21st century an impressive number of papers applying D/TGGE in the field of food microbiology have been published (for review see Ercolini, 2004).

Strain Characterization

Strains isolated during the fermentation of sausages can be also subjected to molecular characterization, in order to understand the dynamics and diversity within the same species. Usually, these approaches are applied to isolates that have previously been identified with other molecular techniques (see above) and intra-species differences are highlighted. Only the strains belonging to the same species can be subjected to this kind of characterization. Also in this case, the methods that are used are divided into methods that are exploiting a PCR amplification and non-PCR methods. In the case of PCR-based methods, once more the RAPD-PCR and Rep-PCR are well-established techniques to look into the molecular diversity within selected species of bacteria or yeasts. RAPD-PCR has been applied to examine the diversity of *Lactobacillus sakei* in naturally fermented Italian sausages, in order to understand the RAPD-types that were able to take over the fermentation (Urso et al., 2006), and to differentiate *Debaryomyces hansenii* strains from the sausages fermented at low temperature (Cocolin et al., 2006a). Rep-PCR was used in order to characterize *Staphylococcus xylosus* strains isolated from three fermentation processes in Northern Italy (Iacumin et al., 2006a). In addition, in this study, a new technique, named Sau-PCR, developed for the characterization of dairy-related microorganims (Corich, Mattiazzi, Soldati, Carraro, & Giacomini, 2005), was applied for the first time to differentiate the strains of meat origin. This method is based on DNA digestion with the restriction endonuclease *Sau*3A, followed by amplification, using one primer that contains the sequence of the restriction site of the enzyme, and thereby is able to anneal to the ends of the fragments obtained. The result is a pattern of amplified bands that represents the restriction fragments obtained by digestion with the specific endonuclease. Using the same rational, several enzymes can be used in order to obtain a more precise molecular characterization.

Regarding the molecular methods that do not exploit an amplification step, the only one used in the field of fermented meats is represented by the pulsed-field gel electrophoresis (PFGE). Currently, this technique is considered the "gold standard" for the characterization of strains, since it is very precise, reproducible, and reliable. In the field of food borne pathogens, a database of PFGE profiles was created (PulseNet, http://www.cdc.gov/pulsenet/), thereby allowing epidemiologic analysis of food borne disease outbreaks. This is possible because the results obtained by PFGE are comparable between different laboratories thanks to the reproducibility of the method. PFGE is an electrophoretic method that allows separation of the chromosomes for the yeasts and of macro-restriction fragments obtained from the genomic DNA for bacteria (Tenover et al., 1995). The first step is the extraction of the DNA from the microorganism that is embedded in agarose gel and subjected to lysis treatments, in order to digest the cell wall. Once the DNA has been freed, it can be subjected directly to PFGE or it is digested with a restriction endonuclease that generally has the characteristic to be rare-cutting. In this way, a relatively small set of restriction fragments is obtained that are simply resolved in an agarose gel. The electrophoresis is carried out in specific chambers, in which the electric field is changing orientation. In this way, the DNA molecules are separated more efficiently than in conventional electrophoresis. The result obtained is a band pattern that is specific to the strain that was subjected to the PFGE analysis. Despite the fact that PFGE has been applied several times to characterize strains of dairy-origin, it was applied rarely to LAB and CNC isolated from fermented sausages. PFGE was used to monitor *S. xylosus* starter cultures, both in Southern Italy (Di Maria, Basso, Santoro, Grazia, & Coppola, 2002) and in two French processing plants (Corbiere Morot-Bizot, Leroy, & Talon, 2007), and to study the diversity of *Lactobacillus* strains isolated from fermented sausages to be used as potential probiotics (Pennacchia, Vaughan, & Villani, 2006).

Culture-Independent Techniques

As mentioned above, the culture-independent methods are able to profile the microbial populations in complex microbial ecosystems without any cultivation. The strategy based on which these methods are working is the analysis of the nucleic acids extracted directly from the matrix. Once the DNA and RNA are available, they can be subjected to several types of analysis that can be preceded by a PCR step or not. The culture-independent method that has been more extensively applied to sausage fermentation is the DGGE. As already reported above, the method is able to differentiate DNA molecules based on their denaturation behaviors. When the method is used for microbial population profiling, the PCR is carried out with universal primers, able to prime amplification for all the microbes present in the sample. After this step, a complex mixture of DNA molecules will be obtained, that can be differentiated if separated in gels with denaturant gradients. Every single band that is visible in D/TGGE gels represents a component of the micro-biota. The more bands are visible, the more complex is the ecosystem. By using these methods, it is possible not only to profile the microbial populations, but also to follow their dynamics during time. However, these methods are not quantitative. DGGE analysis has been applied mainly to Italian fermented sausages (Cocolin, Manzano, Cantoni, et al., 2001; Rantsiou, Urso, et al., 2005; Silvestri et al., 2007; Villani et al., 2007), but studies on the fermentation dynamics of Argentinean sausages are available as well (Fontana, Cocconcelli, & Vignolo, 2005; Fontana, Vignolo, & Cocconcelli, 2005).

The use of species-specific primers in the PCR amplification, analyzing the DNA extracted directly from the sample, can be considered to be another culture-independent method. This approach was so far used to identify LAB and CNC species in Spanish fermented sausages (Aymerich, Martin, Garriga, & Hugas, 2003). Moreover, with the last technological improvements that allowed the PCR to become a quantitative method, direct enumeration of technologically important species during fermentation of sausages can be achieved. This was described for the first time by Martin, Jofré, Garriga, Pla, and Aymerich, (2006), who optimized a quantitative PCR (qPCR) protocol for the rapid quantitative detection of *Lb. sakei* in fermented sausages.

Lastly, a promising culture-independent method that, unfortunately, has not yet been efficiently exploited to study the microbial diversity in fermented sausages, is the FISH. In this technique a set of specific probes are used to target different microorganisms directly in the sample. The probes are labeled with different fluorophores, thereby allowing the detection of several species simultaneously. Since the probes are generally designed on the ribosomal RNA, only alive cells are detected by FISH (Bottari, Ercolini, Gatti, & Neviani, 2006). One of the most fascinating features of FISH is the possibility to localize the microorganisms directly into the food matrix. This has been applied to dairy products (Ercolini, Hill, & Dodd, 2003), but never to fermented sausages. The only application of FISH to fermented and fresh sausages to profile the microbial populations in food products has been described by Cocolin et al. (2007).

The Microbial Ecology of Fermented Sausages as Determined by Culture-Dependent Methods

The experimental approach used by researchers in order to describe the microbial ecology of fermented sausages, using culture-dependent methods, is summarized in Fig. 4.2, and the list of studies published using these techniques for the identification and characterization of isolates are reported in Table 4.1. After homogenization, the sample is analyzed by plating onto a specific media to allow the selective growth of the microorganisms of interest. After incubation, counts and random isolations are performed. The isolates are subsequently subjected to molecular identification and characterization.

Molecular Methods for the Identification of Isolated Strains

In the nineties, molecular techniques for the identification of microorganisms isolated from fermented meat products, started to place side by side or to substitute biochemical assays. Although the two different approaches, in most of the cases, arrived at similar results, molecular techniques immediately showed a higher level of reproducibility, automatism, and fastness.

4 Identification of Microorganisms in Traditional Meat Products

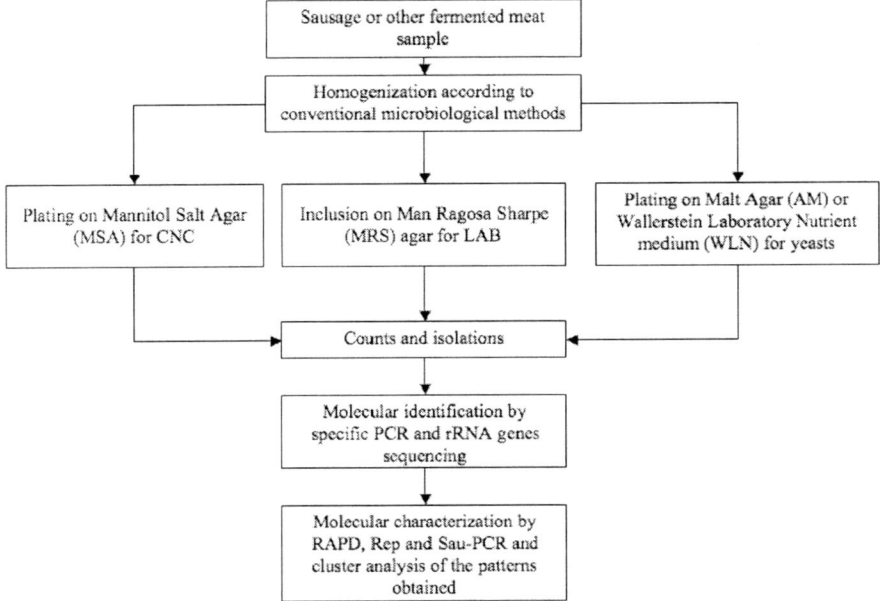

Fig. 4.2 Flowchart describing the application of culture-dependent methods for the identification and characterization of microbial species isolated from fermented sausages.

Specific Hybridization Probes

Although, in later papers, 16S rRNA became the gene more often targeted for bacterial identification (Aymerich et al., 2006; Blaiotta et al., 2004; Gory, Millet, Godon, & Montel, 1999; Rantsiou, Drosinos, Gialitaki, Metaxopoulos, et al., 2006), in 1991, Hertel et al. designed specific probes based on differences in the vicinity of the 5'-terminus of the 23S rRNA gene, to identify the meat lactobacilli *Lb. curvatus*, *Lb. sakei*, and *Lb. pentosus*. The specificity of the probes was checked by dot blot and colony hybridization. Oligonucleotide probes hybridized only with the rRNA gene targets of strains belonging to the corresponding species *Lb. curvatus* and *Lb. sakei*, closely related genetically but not easy to differentiate phenotypically. On the contrary, the probe directed towards the 23S rRNA gene of *Lb. pentosus* reacted positively with *Lb. plantarum*, too. These probes, however, were never used to identify isolates from fermented sausages.

Species-Specific PCR

A large quantity of papers, regarding the design of PCR species-specific primers, for the rapid and reliable identification of closely related bacterial species, were published from 1998, when Berthier and Ehrlich designed primers complementary to lactobacilli species-specific sequences in the 16S/23S rRNA gene spacer region. They were able to distinguish strains belonging to two groups of closely related

Table 4.1 Culture-dependent methods used for the identification and characterization of strains isolated from fermented sausages

Method	Type of analysis	Target group	Type of product	Type of study	Reference
Species-specific PCR	Identification	CNC	Traditional Slovak fermented sausages	Final product	Simonova et al., 2006
		LAB	Traditional French dry sausages	Throughout the fermentation	Ammor et al., 2005
Species-specific PCR and sequencing of 16S rRNA gene	Identification	LAB	Low acid Spanish fermented sausages	Final product	Aymerich et al., 2006
			Fermented Greek sausages		Rantsiou, Drosinos, Gialitaki, Metaxopoulos, et al., 2006
Multiplex PCR	Identification	CNC	Traditional French fermented dry sausages	Throughout the fermentation	Corbiere Morot-Bizot et al., 2006
PCR-TGGE	Identification	LAB	Naturally fermented Italian sausages	Final product	Cocolin, Manzano, et al., 2000
PCR-DGGE and sequencing of 16S rRNA gene	Identification	CNC	Naturally fermented Italian sausages	Throughout the fermentation	Cocolin, Manzano, Aggio, et al., 2001
		LAB			Cocolin, Manzano, Cantoni, et al., 2001; Comi et al., 2005
PCR-DGGE, sequencing of 16S rRNA gene and species-specific PCR	Identification	LAB	Naturally fermented sausages from Greece, Hungary and Italy	Throughout the fermentation	Rantsiou, Drosinos, Gialitaki, Urso, et al., 2005
			Naturally fermented Italian sausages		Urso et al., 2006
		CNC	Naturally fermented Italian sausages		Iacumin et al., 2006b

4 Identification of Microorganisms in Traditional Meat Products 101

Table 4.1 (continued)

Method	Type of analysis	Target group	Type of product	Type of study	Reference
PCR-DGGE and sequencing of 26S rRNA gene	Identification	Yeasts	Naturally fermented Italian sausages	Throughout the fermentation	Cocolin et al., 2006a
ISR-PCR and DGGE	Identification	CNC	Traditional Southern Italian fermented sausages	Final product	Blaiotta, Pennacchia, Ercolini, et al., 2003
ISR-PCR	Identification	CNC	Artisanal Italian dry sausages	Throughout the fermentation	Rossi et al., 2001
ISR-PCR and species-specific PCR	Identification	CNC	Slightly fermented Spanish sausages	Final product	Martin, Garriga, et al., 2006
RFLP	Characterization	LAB	Dry-cured sausages	Final product	Sanz et al., 1998
	Identification	LAB and CNC	Italian dry sausages	Final product	Rebecchi et al., 1998
	Identification and characterization	LAB	Traditional fermented sausages	Final product	Andrighetto et al., 2001
RAPD-PCR	Characterization	LAB CNC *Bacillus* sp. Yeasts	Typical Southern Italian sausages (Salame di Senise)	Manufacturing and ripening	Baruzzi et al., 2006
		CNC	Artisanal Italian dry sausages		Rossi et al., 2001
		LAB	Artisanal Argentine fermented dry sausages		Fontana, Cocconcelli, & Vignolo, 2005
		CNC			
		Yeasts	Fermented Italian sausages	Throughout the fermentation	Cocolin et al., 2006a
		LAB	Naturally fermented Italian sausages		Comi et al., 2005

Table 4.1 (continued)

Method	Type of analysis	Target group	Type of product	Type of study	Reference
RAPD-PCR	Characterization	LAB	Naturally fermented Italian sausages	Throughout the fermentation	Urso et al., 2006
			Naturally fermented sausages from Greece, Hungary and Italy		Rantsiou, Drosinos, Gialitaki, Urso, et al., 2005
			Fermented Greek sausages	Final product	Rantsiou, Drosinos, Gialitaki, Metaxopoulos, et al., 2006
RAPD-PCR and plasmid profiling	Characterization	LAB	Low acid Spanish fermented sausages	Final product	Aymerich et al., 2006
		CNC	Slightly fermented Spanish sausages		Martin, Garriga, et al., 2006
RAPD-PCR and rep-PCR	Characterization	LAB	Traditional Italian fresh sausages	Throughout the fermentation	Cocolin et al., 2004
RAPD-PCR, Sau-PCR and rep-PCR	Characterization	CNC	Naturally fermented Italian sausages	Throughout the fermentation	Iacumin et al., 2006a
rep-PCR	Characterization	CNC	Traditional Italian fresh sausages	Throughout the fermentation	Rantsiou, Iacumin, et al., 2005
PFGE and RAPD-PCR	Characterization	CNC	Fermented Italian sausages (Soppressata)	Throughout the fermentation	Di Maria et al., 2002
PFGE and multiplex PCR	Characterization and identification	LAB	Traditional Italian dry fermented sausages	Final product	Pennacchia et al., 2006
PFGE	Characterization	CNC	Traditional French dry fermented sausages	Throughout the fermentation	Corbiere Morot-Bizot et al., 2006

Lactobacillus species: one composed of *Lb. curvatus*, *Lb. graminis*, and *Lb. sakei* and the other of *Lb. paraplantarum*, *Lb. pentosus*, and *Lb. plantarum*. The next year the same authors (Berthier & Ehrlich, 1999) developed pairs of PCR primers that could be used to specifically detect *Lb. curvatus* and *Lb. sakei*. They evaluated the phenotypic and genotypic diversity among 165 isolates, previously assigned to *Lb. curvatus* and *Lb. sakei* species, by comparing the results of biochemical assays and RAPD data. This allowed them to consider the presence of species-specific RAPD bands that were cloned and sequenced. On the basis of differences in these sequences, species-specific primers were designed. Likewise, a pair of primers specific for *S. xylosus* were designed by Morot-Bizot et al. (2003) on a RAPD product of 539 bp, shared by *S. xylosus* strains from international collections. The validity of the method was confirmed by comparing the results obtained with the designed primers and by PFGE.

Since among the CNC species, *S. xylosus*, *S. equorum*, and *S. carnosus* are frequently involved in the fermentation of several typologies of meat products, a number of papers regarding the design of primers specific for these microorganisms were published. Blaiotta, Pennacchia, Parente, et al. (2003) used xylulokinase and a 60 kDa heat-shock protein coding genes as targets for the design of *S. xylosus* specific primers. They successfully evaluated the specificity of two sets of primers on 27 references strains of the Deutsche Sammlung von Mikroorganismen and Zellkulturen GmbH (DSM) (Germany) collection. Later on, Blaiotta, Ercolini, Mauriello, Salzano, & Villani, (2004) developed primers specific for *S. equorum* that targeted the manganese-dependent superoxide dismutase (*sodA*) gene and were successful in differentiating *S. xylosus* and *S. equorum* that are otherwise difficult to separate by biochemical or physiological traits. The reliability of the method was tested on a total of 112 strains, representing 26 different species of the genus *Staphylococcus*, 3 species of the genus *Kocuria*, and different strains of the *Macrococcus caseolyticus* species. Similarly, Blaiotta, Casaburi, and Villani (2005) designed two different sets of primers, targeting the *sodA* genes of *S. carnosus* and *S. simulans* and were able to differentiate the two species.

Simonova et al. (2006) employed primers previously described by other authors (Aymerich et al., 2003) to detect *S. xylosus* and *S. carnosus* strains in different types of Slovak traditional sausages, for the selection of potential starter cultures to be used in sausage processing. As in many Italian and Spanish fermented meat products (Blaiotta et al., 2004; Coppola, Iorizzo, Saotta, Sorrentino, & Grazia, 1997; Garcia-Varona, Santos, Jaime, & Rovira, 2000), *S. xylosus* was the dominating CNC (63.6%), while *S. carnosus* showed 10.7% frequency.

By using species-specific primers, Ammor et al. (2005) carried out a survey on LAB isolated in a small-scale facility producing traditional dry sausage in France. While *Enterococcus faecium*, *Vagococcus carniphilus*, and *Lactococcus garvieae* were the predominant species isolated from processing equipment and raw material, *Lb. sakei* became the predominant species during the fermentation of the product. *Lb. sakei* strains prevailed (43.3%) also among homo-fermentative mesophilic rods found by Greco, Mazzette, De Santis, Corona, & Cosseddu (2005) in the production and ripening of a typical dry fermented sausage from Sardinia (Italy). The authors used species-specific primers to detect also *Lb. plantarum* and *Lb. curvatus* strains, which were isolated from the product in the percentages of 16.6 and 13.3%, respectively.

The biodiversity of lactobacilli was also evaluated in low acid Spanish fermented sausages and in naturally fermented Greek sausages, produced in three different processing plants, by Aymerich et al. (2006) and Rantsiou, Drosinos, Gialitaki, Metaxopoulos, et al. (2006), respectively. In both cases, isolates that could not be identified by species-specific PCR were identified by the partial sequencing of the 16S rRNA gene that allowed overcoming the limit of species-specific PCR method. In fact, in microbial ecology studies, it is unlikely to reach the identification of all isolates by species-specific PCR, because of the lack of primers or the presence of unexpected species. *Lb. sakei* was the predominant species (74%) in the Spanish fermented sausages (Aymerich et al., 2006), followed by *Lb. curvatus* (21.2%) and *Leuconostoc mesenteroides* (4.8%). The results obtained from Rantsiou, Drosinos, Gialitaki, Metaxopoulos, et al. (2006) highlighted that the main populations involved in the fermentations analyzed, belonged to the species *Lb. sakei, Lb. plantarum,* and *Lb. curvatus,* differently predominating depending on the plants considered.

Corbiere Morot-Bizot et al. (2004) developed a multiplex PCR for the identification of bacteria belonging to the *Staphylococcus* genus and in particular to the species *S. xylosus, S. saprophyticus, S. epidermidis,* and *S. aureus.* Later, the same authors (Corbiere Morot-Bizot et al., 2007) used this approach to investigate the diversity of the staphylococcal community occurring in both the environment and meat products of a small French unit, manufacturing traditional dry fermented sausages. The staphylococcal flora of the environment and meat products of this small unit was dominated by *S. equorum* and *S. succinus* species, rarely described in meat products and never in the environment. They represented the 49% and 33% of isolates, respectively. Other staphylococci belonged to the species *S. saprophyticus, S. xylosus, S. carnosus, S. simulans,* and *S. warneri.*

PCR-DGGE

In the last years, the single or combined use of sets of species-specific primers, started to be used with or be substituted by new, versatile approaches such as D/TGGE. The first paper, applying uniquely this technique for the identification of microorganisms involved in meat fermentations, was published by Cocolin, Manzano, et al. (2000). This work describes the use of PCR-TGGE to identify *Lactobacillus* spp. strains from naturally fermented Italian sausages. This method, based on the amplification of a small fragment from the 16S rRNA gene, followed by TGGE, allowed the differentiation of the strains of *Lb. sakei, Lb. curvatus, Lb. alimentarius, Lb. casei, Lb. plantarum,* and *Lb. brevis*.

The same authors (Cocolin, Manzano, Aggio, et al., 2001) used CNC control strains from international collections to optimize a PCR-DGGE method for their differentiation that was subsequently used to identify, based on co-migration with control strains, 90 isolates from natural fermented Italian sausages. The results showed that the strains of *S. xylosus, S. cohnii, S. intermedius, S. carnosus, K. varians,* and *K. kristinae* could be reliably identified on the basis of the DGGE mobility of the

16S rRNA gene V3 region; once again, S. xylosus was the main bacterium involved in fermented sausage production, representing, from the tenth day of ripening, the only CNC species isolated.

In the next years, PCR-DGGE, more than PCR-TGGE, found wide applications in studies of microbial ecology. These techniques showed the possibility to process high number of isolates in a shorter time; in fact, as already mentioned in the previous sections, these methods allowed grouping of strains and reduced the number of isolates to sequence.

Using these methods, Cocolin, Manzano, Cantoni, et al., (2001) monitored LAB dynamics in fermented sausages, at 3, 10, 20, 30, and 45 days of ripening, prepared in a local meat factory in Italy by traditional techniques. One-hundred and ninety two LAB strains were identified by PCR with the primers targeting the V1 region of the 16S rRNA gene and DGGE. Only two DGGE profiles were detected, belonging to Lb. sakei and Lb. curvatus. In the first stages of fermentation, Lb. sakei was the main LAB present, and only at the end of maturation Lb. curvatus became the predominant microorganism.

The same approach was applied to study the LAB and CNC ecology of fresh sausages stored at 4°C at days 0, 3, 6, and 10 (Cocolin et al., 2004; Rantsiou, Iacumin, et al., 2005). Strains with the same DGGE profile were grouped, and representatives of each group were amplified with primers targeting the V1–V3 region of the 16S rRNA gene and sequenced. A total of 80 LAB were analyzed, among which 69 isolates were identified as Lb. sakei. To study the Staphylococcus spp. ecology, PCR-DGGE was coupled with S. xylosus-specific primers because of the very similar DGGE migration patterns shown by strains belonging to S. xylosus and S. haemolyticus. Primers based on the gehM gene, coding for the lipase of S. xylosus, gave a definitive and sure identification of this species. Almost 50% of isolates were recognized as S. xylosus. Interesting population dynamics were observed, characterized by a succession of S. pasteuri and S. warneri at day 0, with S. xylosus at 3 days and S. equorum at 6 and 10 days.

Comi et al. (2005), studied three productions of a traditional fermented sausage of Northeast Italy. One-hundred and fifty LAB were isolated and identified by means of PCR-DGGE, as described above (Cocolin, Manzano, Cantoni, et al., 2001). The only species that were isolated in all three fermentations were Lb. curvatus and Lb. sakei. A low number of strains belonged to other species, namely E. pseudoavium, L. lactis, Lb. brevis, Lc. mesenteroides, Lb. plantarum, Lb. paraplantarum/pentosus, Lc. citreum and W. paramesenteroides/hellenica.

Rantsiou, Drosinos, Gialitaki, Urso, et al. (2005), using both species-specific PCR and PCR-DGGE, defined the geographic distribution of LAB populations responsible for the fermentation and maturation of naturally fermented sausages produced in three European countries: Greece, Hungary, and Italy. Three-hundred and fifty eight strains belonging to the LAB group were identified. Three species were common to all the three countries studied and were also the most numerous; Lb. curvatus, Lb. plantarum, and Lb. sakei. These three species were identified by species-specific PCR. Isolates that did not belong to any of three species above were identified by PCR-DGGE grouping and 16S rRNA gene sequencing.

Country specific-species: *E. faecium/durans*, *Lb. alimentarius*, *Lb. casei/paracasei*, were only found in sausages from Greece, *E. pseudoavium*, *L. lactis* subsp. *lactis* and *Lb. paraplantarum/pentosus* were only found in sausages from Italy, while *Lc. mesenteroides* and *W. viridescens* were specific to the sausages from Hungary. *Lb. paraplantarum* was isolated from both Italian and Greek products, and *Lb. paraplantarum/plantarum* was from both Greek and Hungarian sausages. *Lc. citreum* and *W. paramesenteroides/hellenica* were found in Italian and Hungarian products. A similar approach, coupling species-specific PCR with PCR-DGGE, was described by Urso et al. (2006), who investigated the ecology of the LAB microflora of three naturally fermented sausages produced in Northeast Italy. Four-hundred and sixty five strains were identified. The only species that were isolated in all three fermentations were *Lb. sakei* and *Lb. curvatus* (353 and 67 isolates, respectively). A low number of other LAB species were found as well, *Lb. plantarum*, *Lb. casei*, *Lb. paraplantarum*, *L. lactis*, *L. garvieae*, *Lc. mesenteroides*, *Lc. carnosum*, *Weissella hellenica*, and *W. paramesenteroides*.

Likewise, Iacumin, Comi, Cantoni, and Cocolin (2006b) used Six couples of species-specific primers (for *S. xylosus*, *S. epidermidis*, *S. simulan*, *S. carnosus*, *S. warneri*, and *K. varians*) to identify 617 CNC strains isolated from naturally fermented sausages in three different plants of the Northeast of Italy. Strains that did not give any PCR amplification with the species-specific primers were subjected to PCR-DGGE and sequencing of the V3 region of the 16S rRNA gene, as described by Rantsiou, Iacumin, et al. (2005). The same species of CNC were found in all the three processing plants, but their contribution to the fermentations was different. In two plants, *S. xylosus* was the main species involved in the fermentation process, while in the third the maturation was carried out equally by three species: *S. xylosus*, *S. warneri*, and *S. pasteuri*.

Cocolin, Urso, Rantsiou, Cantoni, & Comi (2006b) evaluated the ability of a commercial starter culture to perform a 28 day sausage fermentation, combining species-specific PCR and DGGE analysis. As was declared on the starter culture label, LAB isolated from the starter belonged to *Lb. plantarum* species. CNC strains of *S. xylosus*, together with *S. carnosus*, were also isolated and identified in disagreement with what was declared from the factory producing the starter.

Recently, PCR-DGGE was also used to investigate the yeast populations in Italian fermented sausages (Cocolin et al., 2006a). Yeast isolates were subjected to PCR-DGGE according to Cocolin, Bisson, & Mills (2000), and strains giving identical migration patterns were grouped together; at least two representatives were amplified with the primers NL1 and NL4 (Kurtzman, & Robnett, 1998) and sequenced. The work highlighted the dominance of *D. hansenii*, which was already present, as the major species, at the beginning of the fermentation. This species was usually accompanied by *Candida zeylanoides*, and only at the end of the fermentation, by *Metschnikowia pulcherrima*. At the start of the fermentation, a high biodiversity was observed, as five species could be identified; apart from *D. hansenii*, the species *Pichia triangularis*, *Candida parapsilosis*, *Saccharomyces cerevisiae*, and *Sterigmatomyces elviae* were found.

ISR-PCR

The polymorphism of the inter-gene spacer regions (ISR), was also used in combination with DGGE for the identification of the CNC strains isolated from fermented sausages. Blaiotta, Pennacchia, Ercolini, et al., (2003), applied a polyphasic molecular approach to describe staphylococcal population occurring in traditional fermented sausages of Southern Italy by combining DGGE and ISR-PCR. The results obtained by ISR-PCR and PCR-DGGE analysis allowed a clear diffentiation of the reference strains analyzed, with the exception of the pairs *S. cohnii-S. equorum* and *S. schleiferi-S. carnosus*. The strains that displayed the same ISR-PCR pattern could show a different PCR-DGGE profile and vice versa. Among the isolates, the most abundant species of CNC was *S. xylosus*. The strains of *S. saprophyticus*, *S. lentus*, *S. warneri*, *S. cohnii-S. equorum*, *S. epidermidis*, *S. haemolyticus*, *S. succinus*, *S. vitulus*, *S. pasteuri*, *S. aureus*, *M. caseolyticus*, and *Kocuria* spp. were also isolated.

In a previous research (Rossi et al., 2001), ISR-PCR permitted to successfully identify strains belonging to *S. xylosus*, *S. simulans*, and *S. carnosus* and other closely-related species, and *Kocuria* species, isolated from an Italian artisanal dry sausage. Fifty one CNC strains were genotypically identified by amplification with universal primers (Jensen, Webster, & Strauss, 1993), and once more, *S. xylosus* prevailed over the other CNC.

Recently, 240 CNC strains, isolated from the Spanish slightly fermented sausages, were subjected to ISR-PCR with universal primers (Martin, Garriga, et al., 2006). The results obtained by amplification of the intergenic regions were confirmed by species-specific PCR, as previously described (Aymerich et al., 2003). *S. xylosus* was the predominant species (80.8%), followed by *S. warneri* (8.3%), *S. epidermidis* (5.8%) *S. carnosus* (4.6%), and *K. varians* (0.4%).

Concerning ISR-PCR, discrepancy in the results is often found. While Forsman, Tilsala-Timisijarvi, & Alatossava, (1997), Mendoza, Meugnier, Bes, Etienne, & Freney, (1998), and Villard, Kodjo, Borges, Maurin, & Richard, (2000) reported a high polymorphism, Blaiotta, Pennacchia, Ercolini, et al., (2003), in agreement with Couto, Pereira, Miragaia, Sanches, & de Lancastre, (2001), found a lower discrimination potential of this analysis for staphylococci.

Molecular Methods for the Characterization of Isolated Strains

The growing interest in the studies of microbial ecology of fermented products and in monitoring of specific strains carrying out, spontaneously or after inoculation as starters, fermentation processes, pushed for the availability of methods for subspecies characterization of the isolates.

In 1998, RFLP analysis had been already used by Sanz et al. to characterize *Lb. sakei* strains isolated from dry-cured sausages, and previously identified by DNA-DNA hybridization. RFLP studies, by using the restriction enzymes *EcoRI* and *HindIII*, and, as probes, cDNA from *Escherichia coli* 16S and 23S rRNA genes or from *Lb. sakei* 16S rRNA gene from *Lb. sakei* strains, showed distinct polymorphism levels. *EcoRI*-digested DNA, probed with the cDNA from *E. coli*, highlighted the presence of a unique cluster for the meat *Lb. sakei* isolates tested. When *HindIII*-digested DNA was hybridized with the *E. coli* cDNA probe, strain specific patterns were obtained, showing a higher discrimination power. Considerable strain differentiation was also observed when *EcoRI* and *HindIII* digests were hybridized with *Lb. sakei* 16S rRNA gene probes.

In the following years, RAPD-PCR became the most used technique to study the intra-specific variability even if, initially, it was applied for identification purposes (Andrighetto et al., 2001; Baruzzi, Matarante, Caputo, & Morea, 2006; Berthier & Ehrlich, 1999; Rebecchi et al., 1998). Rebecchi et al. (1998) studied the microbial community in the production of a 2-month-ripened Italian dry sausage. Their results, obtained by RAPD-PCR fingerprints, highlighted that environmental parameters interacted to select a limited number of strains during the fermentation process. The dominant strains belonged to the species *S. xylosus* and *S. sciuri*, among CNC, and *Lb. sakei* and *Lb. plantarum*, among LAB.

RAPD-PCR with primers M13 and D8635 was applied to identify 53 lactobacilli isolates originating from traditional fermented sausages and artisanal meat plants of Veneto region in Italy (Andrighetto et al., 2001). RAPD-PCR assigned most of the isolates to the species *Lb. sakei* and *Lb. curvatus* and, at the same time, an intra-specific variability was detected; in some cases, RAPD subgroups reflected the origin of the isolates; in others, different strains were found in the same environment.

In fermented products, the possibility to detect strains that were able to carry out the fermentation processes became extremely important in the selection of microbial strains to be used as starters. Rossi et al. (2001), compared and combined RAPD profiles, obtained with the primers OPL-01, OPL-02, OPL-05, and Hpy 1, from the *S. xylosus* strains isolated from dry sausages. In this paper, the authors underlined the suitability of the RAPD-PCR analysis to discriminate strains with technologically relevant activities; in fact, strains showing nitrate reduction and amino acid decarboxylase activities clustered separately.

Later, Baruzzi et al. (2006) correlated acidification, proteolytic, lipolytic, and nitrate reduction activities to strains isolated from "Salame di Senise",, a typical sausage produced in the South of Italy. The application of RAPD-PCR to more than 90 isolates made it possible to define 18 bacterial and 2 yeast bio-types, mainly belonging to three species of *Bacillus* (*Bacillus subtilis*, *B. pumilus* and *B. amyloliquefaciens*), three species of *Lactobacillus* (*Lb. curvatus*, *Lb. sakei* and *Lb. casei*), three species of *Staphylococcus* (*S. succinus*, *S saprophyticus*, and *S. equorum*), and *D. hansenii*.

The microbial community responsible for the artisanal fermentation of dry sausages produced in Argentina, was studied by Fontana et al. (2005). The authors carried out RAPD analysis with the primers M13, XD9, RAPD1, and RAPD2, to differentiate and characterize 100 strains of lactobacilli and CNC, identified by 16S rRNA gene sequencing. This approach allowed them to demonstrate that the ripening process of the Argentinean artisanal fermented sausages was driven by a limited number of *Lactobacillus* and *Staphylococcus* strains selected from the environmental micro-biota for the ability to best compete under the prevailing conditions of the ecological niche. The authors concluded that these well adapted strains should have been eventually considered for the selection of starter cultures.

Cocolin et al. (2006b) used an identical approach to follow the development, in fermented sausages, of an inoculated commercial starter. RAPD characterization was carried out on 15 LAB and 15 CNC strains from a starter used in a sausage production and on 70 LAB and 70 CNC strains isolated during the same production. RAPD analysis with primer M13 revealed three *Lb. plantarum* RAPD bio-types, but only one of them was able to conduct the fermentation. *S. xylosus* strains from the starter culture, which clustered mainly with strains isolated at 14 and 28 days of ripening, were able to predominate only in the latter stages of fermentation. The different behavior of *Lb. plantarum* and *S. xylosus* could be explained considering the fact that the starter culture was dissolved in white wine, thus inhibiting the initial development of *S. xylosus*, not able to overcome the ethanol stress.

Recently, some authors evaluated possible correlations between microbial strains clustered on the basis of RAPD profiles and their manufacturing origin or geographical provenience. Comi et al. (2005), in the study of three fermentations of a Northeast Italy traditional fermented sausage, highlighted, by RAPD-PCR with primer M13, that lactobacilli population was distributed in a fermentation-specific way; the authors supposed that strains grouped in fermentation-specific clusters came from the ingredients used in the productions. *Lb. sakei* showed a higher degree of heterogeneity than *Lb. curvatus*.

The same approach was followed by Urso et al. (2006) who reached similar results. They isolated and identified, from three naturally fermented sausages produced in the Friuli-Venezia-Giulia region (Italy), 353 *Lb. sakei* and 67 *Lb. curvatus* strains. RAPD-PCR analysis by using primer M13 detected clusters formed by strains isolated from specific fermentations, confirming that ingredient composition, fermentation, and maturation parameters could play an important role in the selection of populations adapted to a specific environment. However, clusters containing strains isolated from different plants were also observed, underlining a homogeneous population distribution in the three different fermentations.

Rantsiou, Drosinos, Gialitaki, Metaxopoulos, et al., (2006) characterized, by RAPD-PCR with primer M13, *Lb. sakei*, *Lb. plantarum*, and *Lb. curvatus* strains isolated from naturally fermented sausages, produced in three different processing plants in continental Greece. In agreement with the results mentioned above, some strains were plant-specific, whereas others shared a degree of homology independently of the provenience.

In a similar study, *Lb. curvatus*, *Lb. plantarum*, and *Lb. sakei* strains, isolated from naturally fermented sausages produced in three European countries (Greece, Italy, and Hungary), were subjected to RAPD-PCR with primer M13 (Rantsiou, Drosinos, Gialitaki, Urso, et al., 2005). The distribution of the lactobacilli strains reflected, in almost all cases, the provenience of the same. It was thereby possible to distinguish the Italian, Greek, and Hungarian LAB populations.

In order to improve the characterization and differentiation among microbial strains, some authors considered both chromosomal and plasmidic characteristics.

Thus, 250 LAB isolated from low acid Spanish fermented sausages were analyzed (Aymerich et al., 2006) by combining RAPD-PCR and plasmid profiling. One-hundred and forty four different strains could be differentiated, 112 belonging to *Lb. sakei*, 23 to *Lb. curvatus*, and 9 to *Lc. mesenteroides*. The isolates identified as the same strain showed common phenotypic properties, in terms of biogenic amine production and antibiotic susceptibility, although isolates with specific traits could not be grouped in single clusters, indicating a probable horizontal exchange of DNA between bacteria.

Martin, Garriga, et al., (2006) also compared and combined RAPD-PCR and plasmid profiling techniques in the characterization of CNC strains isolated from Spanish slightly fermented sausages. In particular, the authors could differentiate 169 profiles out of the 194 *S. xylosus* isolates, indicating a great genetic intra-specific variability. It should be underlined that, in this study, the plasmid pattern analysis was more discriminatory than RAPD-PCR analysis; in fact, the first detected 140 profiles against the 118 of the latter. However, the number of different profiles obtained in the other CNC species analyzed at the strain level (*S. warneri*, *S. epidermidis*, and *S. carnosus*), was similar with both methods. Moreover, plasmid typing has two inconveniences: strains without plasmid cannot be typed, and many factors could influence the final plasmid patterns. For these reasons this analysis is considered not reliable when used alone.

In 2004, Cocolin et al. compared RAPD-PCR to rep-PCR technique, in order to characterize 69 *Lb. sakei* isolates, obtained from fresh sausages prepared by traditional techniques, and to detect different populations that eventually developed on different days of storage. Rep-PCR had been previously applied, in 2001, by Gevers, Huys, and Swings, to differentiate a wide range of lactobacilli recovered from the different types of fermented dry sausages, and it showed to be a promising tool for rapid and reliable speciation and typing of lactobacilli. In their research, Cocolin and his colleagues found that the *Lb. sakei* isolates were grouped in four RAPD clusters, whereas rep-PCR led to the identification of five different clusters. The authors underlined that most strains isolated at day 10 of ripening were grouped in a unique cluster, likely due to selection during storage at low temperature.

Once more RAPD-PCR and rep-PCR, together with Sau-PCR, were compared in a paper by Iacumin et al. (2006a). The purpose of this study was to characterize the *S. xylosus* strains isolated from naturally fermented sausages, produced in three different processing plants in the Friuli Venezia Giulia region (Italy), and eventually to detect a possible strain differentiation depending on their specific provenience. By applying the rep-PCR method, 10 clusters among a total of 13, were formed solely by strains coming from one plant, and some of these clusters contained a large number of isolates. Similar results were obtained with Sau-PCR, while the RAPD technique produced only 8 clusters, out of the 17, formed by strains coming from one plant. The authors confirmed the theory that, depending on temperature, humidity, and ingredients of specific production plants, there is always a specific selection of microorganisms which influences the characteristics of the final products. In this paper, the only correlation found between technological characterization and molecular typing was that all the strains isolated from one of the plants, grew at a temperature of 10°C. The specific plant was the only one using low temperature maturation, thereby selecting a population of CNC that is able to grow under these environmental conditions.

Due to the efficiency and reliability of the technique, some authors used rep-PCR, exclusively. For example, Rantsiou, Iacumin, et al. (2005) used rep-PCR to characterize the *S. xylosus* strains isolated from fresh sausages, at 0, 3, 6, and 10 days. Molecular characterization and cluster analysis highlighted the presence of six main populations, but no correlation was obtained between the day of isolation and the grouping into clusters. This fact underlined a homogeneous distribution of the *S. xylosus* strains throughout the storage of the fresh sausages.

Lastly, RAPD-PCR analysis with primer M13 was used by Cocolin et al. (2006a) to characterize the *D. hansenii* strains, from Italian fermented sausages. The authors noticed a shift in the *D. hansenii* population from the beginning to the end of sausage maturation. In fact, strains present during the early stages of the fermentation were grouped in clusters that differed from those defined in the final phases of the maturation.

The last technique for molecular characterization that we take into account is PFGE. As reported in the previous sections, nowadays PFGE is considered the "gold standard" for the characterization of microbial strains, since it is very precise, reproducible, and reliable. However, few studies are reported where PFGE is applied to characterize LAB and CNC strains isolated from meat products, probably because it is a labor intensive and time-consuming technique. For this reason, PFGE is commonly applied in combination with other, PCR-based techniques, for strain typing. Di Maria et al. (2002) used RAPD-PCR analysis to monitor a *S. xylosus* starter culture throughout the ripening of "Soppressata molisana", a fermented sausage produced in Southern Italy. RAPD-PCR was successfully employed for the discrimination of the added strains from those naturally present during the ripening of the product. PFGE was then applied to confirm the RAPD results.

Pennacchia et al. (2006), coupled a species-specific multiplex PCR assay with PFGE analysis to identify and characterize 25 potential probiotic *Lactobacillus* strains isolated in a previous study, from fermented sausages.

In a study of the staphylococcal community occurring in the environment and in traditional dry fermented sausages produced in a small French unit, Corbiere Morot-Bizot, Leroy, & Talon, (2006) defined 17 distinct PFGE patterns named pulsotypes. Twelve of the pulsotypes (A to H, J, L, M, N) were found independently of the period of sampling (winter or spring), whereas the pulsotype P was sporadic in winter and the pulsotypes I, K, O, and Q were sporadic in spring. The 201 isolates of *S. equorum*, which dominated both in the environment and in the meat products, together with *S. succinus*, belonged to eight distinct unrelated pulsotypes (A to H), showing a wide diversity among this species. However, the dominant pulsotype C, with 77 clones, was found both in the environment of the small processing unit and in the meat products. The presence of a dominant pulsotype indicated that some strains could be capable for adaptation to food plant environment and processing. Recently, the same authors (Corbiere Morot-Bizot et al., 2007) monitored, by PFGE, the growth and survival of *S. xylosus* and *S. carnosus* starters during sausage manufacture in two French processing plants. The PFGE analysis revealed that all *S. xylosus* and *S. carnosus* strains isolated corresponded to the starter strains inoculated. In particular, *S. xylosus* starter strain dominated the staphylococcal micro-biota, whereas the *S. carnosus* starter strain survived during the process. Since successful implantation of the starter cultures is obviously a pre-requisite for their contribution to sensorial qualities, this work highlighted both the efficiency and ineffectiveness of *S. xylosus* and *S. carnosus*, respectively, as starter cultures. It is noteworthy that neither of them were able to colonize the environment of the two processing plants.

The Microbial Ecology of Fermented Sausages as Determined by Culture-Independent Methods

In recent years, the study of the microbial ecology of fermented foods has been enhanced due to the introduction in this field of direct, culture-independent methods. These methods are based on the extraction of total nucleic acids from any given sample, and description of the microbial groups present in the foods (with identification of their individual members), based on DNA and/or RNA sequences.

The analysis of the nucleic acids can be carried out by hybridization with specific probes, by species-specific PCR or by universal PCR and sequence-based separation and identification of the PCR products. The disadvantage of the species-specific PCR is that there is a limit to the number of species one can detect/identify in a sample. Moreover, one has to know which microorganisms to look for in a sample. Alternatively, with the use of universal primers, theoretically all the species of large groups are amplified. Then, the sequence-based separation is achieved by D/TGGE. D/TGGE was first developed for the study of the microbial ecology in environmental samples (Muyzer & Smalla 1998), but soon found application in food microbiology (Ercolini, 2004).

The main advantages of the direct approaches are: (i) no cultivation takes place and therefore, the bias associated with the use of conventional microbiological media for enumeration/isolation, is negated, (ii) compared to the classic approach used so far in microbiology, that is based on isolation of strains from the food matrix and identification, either by physiological/phenotypical tests or by molecular methods, they are less time-consuming and require less effort, (iii) they allow a parallel description of the populations of different microbial groups. In contrast, these techniques generally require specialized personnel and relatively costly equipment. Furthermore, it has been determined that the detection limit for the most common method used in direct analyzes, that is the D/TGGE, is in order of 10^3–10^4 colony forming units (cfu)/g or ml (Cocolin, Manzano, Cantoni, et al., 2001). As a consequence, microbial groups that are present and active, but their population is lower than 10^3–10^4 cfu/g, will not be taken into consideration.

An interesting contribution to our knowledge of the microbial ecology of fermented products is provided by the application of the PCR-DGGE on the RNA extracted directly from the matrix and after reverse transcription. DNA may persist in any given environment, some times long after a microorganism is dead. In contrast, RNA is rapidly degraded after cell death and as a consequence, the application of RT-PCR-DGGE gives the fingerprint (or profiles) of alive and metabolically active populations. When RT-PCR-DGGE was applied in meat products, the results compared fairly well with those obtained by PCR-DGGE (Cocolin, Manzano, Cantoni, et al., 2001), although in certain cases the RT-PCR-DGGE profiles were richer (Cocolin et al., 2004).

These approaches have been successfully employed in the study of the microflora of fresh and fermented sausages. A scheme of the experimental approach that is followed is presented in Fig. 4.3, while in Table 4.2 the published studies employing

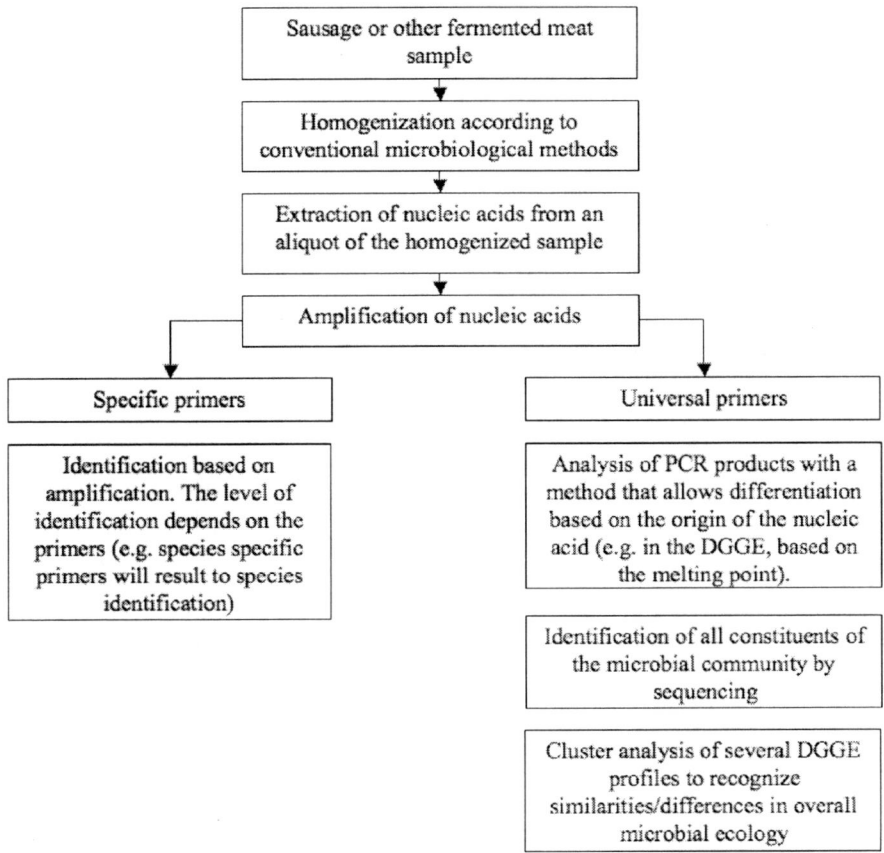

Fig. 4.3 Flowchart showing the experimental approach for the study of the microbial ecology in fermented sausages by using culture-independent methods.

Table 4.2 Meat products and information related to the studies performed with direct, culture-independent approaches

Type of product	Country	Direct Approach	Target group	Type of study	Reference
Fresh sausages	Italy	PCR-DGGE	LAB, CNC and yeasts	During storage	Cocolin et al., 2004
Fermented sausages	Friuli-Venezia-Giulia region, Italy	PCR-DGGE	LAB, CNC and yeasts	Throughout fermentation	Cocolin, Manzano, Cantoni, et al., 2001; Rantsiou et al. 2004; Rantsiou, Urso, et al., 2005
Fermented sausages (Ciauscolo salami)	Marche region, Italy	PCR-DGGE	LAB, CNC and yeasts	Final product	Silvestri et al., 2007
Fermented sausages (Sopressata)	Campania region, Italy	PCR-DGGE	LAB and CNC	Final product	Villani et al., 2007
Low-acid fermented sausages (fuets and chorizos)	Spain	Species-specific PCR	LAB and CNC	Final product	Aymerich et al., 2003
Artisanal dry sausages	Argentina	PCR-DGGE	LAB and CNC	Throughout fermentation	Fontana et al., 2005; Fontana et al. 2005
				Final product	Fontana et al. 2005

culture-independent methods to study the microbial ecology of these products are reported.

The Choice of PCR-DGGE Primers

When applying PCR-DGGE to study complex microbial populations, such as the ones present in food samples, an important parameter, that will influence the results obtained, is the choice of the amplification target gene prior to DGGE.

The target gene has to have two basic characteristics: (i) it should be present in all the members of the microbial group that is under consideration, (ii) it should have conserved regions, where universal primers can be designed, and variable regions, based on which separation is possible. In addition, if the RNA is going to be analyzed, the gene should be characterized by constitutive expression. Genes that fulfill these requirements are those that are involved in important and universal cell functions. Commonly, genes encoding for rRNA fall within this category. In bacteria, various regions of the 16S rRNA coding gene have been used in PCR-DGGE, while in yeasts, the 26S rRNA coding gene is commonly the target.

One important advantage of the 16S and 26S rRNA coding genes is the fact that, for both genes, databases of sequences from a large number of representative species is available. This is important since it allows identification of DGGE bands by sequencing and comparison with the database. A drawback associated with the use of rRNA coding genes is the inherent sequence heterogeneity, within the same species, that is the result of multi-copies of the genes with small differences in the sequence. The multi-copies often result in multi-bands in the DGGE profiles that complicate the analysis.

An alternative gene has been proposed for use in PCR-DGGE. It is the *rpoB* gene, encoding for the β-subunit of the RNA polymerase, which is usually present as a single copy in each genome. The limitation of using the *rpoB* gene is the restricted number of sequences available in the databases that hinders the identification of unknown DGGE bands.

In the DGGE studies performed so far on meat products, several primers have been used, and the relevant information is summarized in Table 4.3. The quality of information produced by PCR-DGGE is dependent on both the number and resolution of the amplicons in denaturing gradient gels (Yu & Morrison, 2004). In fingerprinting the microbial communities of a food product, it is important that the method allows differentiation of individual species that are associated with a specific food. For this reason, the primers used and the conditions employed in PCR-DGGE need to be carefully considered and if necessary, optimized, prior to application in real food samples (Cocolin, Manzano, Cantoni, et al., 2001; Yu & Morrison, 2004).

As shown in Table 4.2, for the profiling of bacterial populations in fresh and fermented sausages, several regions of the 16S rRNA coding genes have been employed.

Table 4.3 Summary of the target genes that have been employed in PCR-DGGE of meat products

Target group	Target region	Primer set	Primer Reference	Application reference
Bacteria	16S rRNA gene, V1	P1/P2	Klijn et al., 1991	Cocolin, Manzano, Cantoni, et al., 2001; Cocolin et al., 2004; Rantsiou, Urso, et al., 2005; Silvestri et al., 2007
	16S rRNA gene, V3	338f/518r	Ampe et al., 1999, Muyzer, de Waal, & Uitterlinden, 1993	Cocolin, Manzano, Cantoni, et al., 2001; Villani et al., 2007
	16S rRNA gene, V9	Ec1055/Ec1392	Ferris, Muyzer, & Ward, 1996	Cocolin, Manzano, Cantoni, et al., 2001
	16S rRNA gene, V2-V3	HDA1/HDA2	Walter et al., 2000	Cocolin, Manzano, Cantoni, et al., 2001
	16S rRNA gene, V6-V8	U968/L1401	Zoetendal, Akkermans, & de Vos, 1998	Cocolin, Manzano, Cantoni, et al., 2001
	16S rRNA gene, V3	P3/P4	Klijn et al., 1991	Cocolin, Manzano, Cantoni, et al., 2001
	16S rRNA gene, V3	V3f/Uni-0515r	Ercolini, Moschetti, Blaiotta, & Coppola, 2001b; Lane, 1991	Fontana et al., 2005; Fontana et al., 2005
	16S rRNA gene, V2-V3	Bact-0124f/Univ-0515r	Lane, 1991	Fontana et al., 2005
	rpoB gene	*rpoB*1698f/*rpoB*2014r	Dahllof, Baillie, & Kjelleberg, 2000	Rantsiou et al., 2004
Yeasts/moulds	26S rRNA gene, D1-D2 loop	NL1-LS2	Cocolin, Bisson, et al., 2000	Cocolin et al., 2004; Rantsiou, Urso, et al., 2005; Cocolin et al., 2006a
	26S rRNA gene	U1-U2	Sandhu et al., 1995	Silvestri et al., 2007

In an early study, conducted by Cocolin, Manzano, Cantoni, et al., (2001) using control strains from culture collections that belonged to bacterial species commonly associated with meat products, it was concluded that region V1 of the 16S rRNA allowed the best differentiation in DGGE. In the DGGE optimization process with the P1/P2 set of primers, it was determined that the 40–60% denaturant gradient allowed differentiation of members of CNC, distinguishing *Kocuria* strains from *Staphylococcus* strains, while the 30–50% denaturant gradient allowed differentiation of all the *Lactobacillus* spp. tested. This set of primers has been subsequently used in DGGE profiling of bacterial populations in different meat products (Cocolin et al., 2004; Rantsiou, Urso, et al., 2005; Silvestri et al., 2007).

Fontana et al. (2005) applied a nested PCR approach in their study of fermented sausages. With primers P0/P4 (Klijn, Weerkamp, & de Vos, 1991), fragments of about 700 bp from the 5' region of the 16S rRNA coding gene were amplified and were used for nested PCR with two different sets of primers targeting the V2–V3 and the V3 regions. They concluded that the best DGGE patterns of the microbial community of fermented sausages were obtained with a 40–60% denaturant gradient for both primer combinations. In a subsequent study by Fontana et al. (2005), three sets of primers were tested, targeting the V1, V2–V3, and V3 regions, and the authors report that the best results from the direct analysis of the microbial community by DGGE were obtained by amplifying the V3 region.

Microbial Ecology of Meat Products Determined by Direct Approaches

Bacterial Ecology by PCR-DGGE

In the study of the microbial ecology of meat products, researchers have followed two approaches: they looked into the evolution of the different microbial populations throughout the production process and until the product was ready for marketing, or they considered the more static picture obtained from the final product. From the studies of the first type, a general observation is that a great diversity, both within the LAB and CNC groups, and complex DGGE profiles characterize the freshly prepared meat mix, just prior to the initiation of the fermentation or storage period, in the case of non-fermented meat products. This biodiversity and the associated complexity in the DGGE profiles, is sharply reduced from the first days of fermentation and just a few species appear to predominate and are commonly detected throughout the rest of the process. On the other hand, when a final product is taken into consideration, then the microorganisms that have managed to persist and most likely were responsible for the transformation process are detected. Of course, in this approach, microorganisms that may play a role at earlier stages but do not survive until the end, cannot be seen.

In the first work published (Cocolin, Manzano, Cantoni, et al., 2001) regarding direct DGGE analysis, traditional fermented sausages produced by local laboratories in the Friuli-Venezia-Giulia region of Italy were analyzed throughout fermentation and ripening that lasted in total for 45 days. In the DNA and RNA-DGGE gels, multiple bands were visible for the first 3 days of fermentation, when different species, most of which were related to *Staphylococcus* spp. were identified. From the tenth day of ripening only the LAB bands were present. The LAB population was characterized by *Lb. sakei* and *Lb. curvatus* throughout the process. *Lb. plantarum* was only detected on the first day in the DNA gel and the band could have been generated from dead cells. *Staphylococcus* species were found only in the meat mixture before the sausages were filled and after 3 days. The only *Staphylococcus* species represented in the DGGE gel after 3 days was *S. xylosus*, which produced a specific band in the gel until the end of fermentation. The corresponding band in the RNA gel was only present at day 0 and day 3 and then disappeared. This could be explained by the large quantity of LAB RNA that restricted amplification of the *S. xylosus* RNA.

In a later study focusing on traditional fermented sausages from the same region of Italy, the goal was to follow the fermentation in three different plants, in which no starters are being used. PCR-DGGE was employed and the bands were identified by sequencing. Furthermore, cluster analysis of the DGGE profiles, obtained from the three fermentations and at different sampling points, was conducted (Rantsiou, Urso, et al., 2005). A general consideration that resulted from this study is that the main differences detected in the ecology, between the three sausages, were not represented by the species of microorganisms identified by band sequencing but by their relative distribution between the fermentations. In all the three fermentations, a stable signal, from the beginning of the period studied was visible for *Lb. curvatus* and *Lb. sakei*, which remained constant throughout the transformation. In one of the fermentation, a band that corresponded to *Lb. paracasei* was present for part of the period, while *L. garviae* was detected in two of the three sausages. No *Lb. plantarum* was detected in these sausages. As seen from the DGGE profiles, important contribution to the microbial ecology was given by the *Staphylococcus* species. *S. equorum* and *S. succinus* were present in all the fermentations, while *S. xylosus* was mainly present in one of them.

When the *rpoB* gene was targeted in PCR-DGGE of the same type of fermented sausages (Rantsiou, Comi, & Cocolin, 2004), significantly simpler profiles, all through the transformation process, were obtained. At the beginning of the fermentation, *S. xylosus* was detected while at the later stages, *Lb. curvatus*, *Lb. sakei*, and *Lb. plantarum* comprised the bacterial population.

In a study conducted by Fontana et al. (2005), PCR-DGGE has been applied to profile the bacterial community of the artisanal Argentinean sausages. Targeting the V3 region of the 16S rRNA coding gene, a highly complex fingerprint was obtained at day 0 of the fermentation that was characterized by the presence of *Lb. plantarum*, *P. acidilactici*, *Lb. sakei*, *Lb. curvatus*, *S. equorum* (identified by co-migration with control strains), and *Corynebacterium variabilis* (identified by sequencing). A more basic fingerprint was obtained at day 5 and day 14 (last day of ripening), with the presence of *Lb. plantarum*, *Lb. sakei*, and *Lb. curvatus*, while *S. saprophyticus* represented the *Staphylococcus* group. In a similar study by the same authors (Fontana et al., 2005), the fermentation of a sausage from the Tucumán region of Argentina was followed during the 14 day production and the same sausage was compared to a ripened sausage from the Córdoba region. A high microbial diversity was observed at day 0 with the presence of *Lb. plantarum*, *Lb. sakei*, *S. saprophyticus*, *C. variabilis* and that with varying intensities were also present in the other two sampling points (day 5 and day 14). When the two sausages were compared, a common band, associated with *Lb. sakei* was detected. Furthermore, the bacterial community of the Tucumán sausage contained *S. saprophyticus* and an uncultured bacterium while in the Córdoba sausage *Brochothrix thermosphacta* was present.

The bacterial ecology of the traditional Italian salami "ciauscolo" was recently surveyed by PCR-DGGE (Silvestri et al., 2007). The V1 region was amplified from 22 samples of 14 artisan and 8 industrial production plants from different geographical areas of the Marche region, in Central Italy. Overall, five LAB species were detected: *Lb. plantarum*, *Lb. curvatus*, *Lb. sakei*, *P. acidilactici*, and *L. lactis*. No *Staphylococcus* bands were observed in the DGGE profiles, either because the populations were below the detection limit or due to the higher number of LAB that may create a masking effect. The most frequently detected species were, once again, *Lb. curvatus* and *Lb. sakei*.

Finally, in a recent study of the ecology of a fermented sausage from Southern Italy, the "soppressata", targeting the V3 and V1 regions, it was possible to identify *S. xylosus*, *S. succinus*, and *S. equorum* among the staphylococci and *Lb. sakei* and *Lb. curvatus* within the lactobacilli (Villani et al., 2007).

A slightly different picture of bacterial ecology was seen when fresh sausages from Italy were considered (Cocolin et al., 2004). These products are highly perishable, since no fermentation takes place. After mixing of the ingredients and casing, they are stored at 4°C. They have a very short shelf life (about 10 days) and are cooked before consumption. Bacterial DGGE profiles were characterized, for both DNA and RNA, by the presence of a constant band throughout the process that was *B. thermosphacta*, whereas the *Lactobacillus* spp., visible at the DNA level, were only marginally present at the RNA level, mainly represented by *Lb. sakei*. *Lc. mesenteroides* was present at both DNA and RNA the last day of sampling. On the fresh mix, at the RNA level, a *Bacillus* sp. band was present, while *Staphylococcus* spp. was also detected during storage of the fresh sausages.

An alternative application of the PCR-DGGE directly in a food matrix was presented by Cocolin et al. (2006b). The authors followed the microbial dynamics of sausage fermentation, with particular interest for *Lb. plantarum* and *S. carnosus* that were inoculated as a starter. PCR-DGGE allowed the detection of a stable band, identified by sequencing as *Lb. curvatus*, that was only marginally detected by the conventional microbiological techniques. This application shows the potential of the method in monitoring, in a culture-independent way, specific microbial groups.

Yeast Ecology by PCR-DGGE

The yeast ecology has been described by targeting the 26S rRNA coding gene for Italian fermented sausages of the Friuli-Venezia-Giulia region (Rantsiou, Urso, et al., 2005; Cocolin et al., 2006a) and of the Marche region (Silvestri et al., 2007) and for fresh sausages (Cocolin et al., 2004).

Overall, the DGGE profiles are less complex than those obtained for the bacteria and oftentimes, they are characteristic of the products. In particular, when a cluster analysis of the yeast DGGE profiles was carried out, from the samples collected during production of three fermented sausages, a fermentation-specific distribution was obtained, with the clusters containing samples from the same fermentation (Rantsiou, Urso, et al., 2005). In this study, the yeast ecology results were characterized by members of the genera *Candida*, *Debaryomyces*, and *Willopsis*. *D. hansenii*, a proteolytic yeast that is associated with fermented sausage production, was dominant in one of the three fermentations but was not seen in the other two productions. *C. krisii* and *Willopsis saturnus* and *C. sake* or *austromarina* characterized, respectively, the other two fermentations. Finally, one band was common in all the three fermentations that after sequencing showed a 86.5% homology to *Mayaca fluviatilis*. This homology score is very low and most likely, the DGGE band corresponds to a yeast species that has not yet been described or for which no 26S rRNA gene sequence is available.

The predominance of *D. hansenii* was also confirmed by a later study in Italian sausages fermented at low temperatures, probably due to its physiological characteristics, and it should therefore be considered well adapted to the specific environment yeast species (Flores, Durà, Marco, & Toldrà, 2004). Throughout the 60 days of fermentation a stable band, at both the DNA and RNA level, identified as *D. hansenii*, was visible in the DGGE gels (Cocolin et al., 2006a).

In the "ciauscolo" salami of the Marche region, the most frequently detected species was *D. hansenii*, while *C. physchrophila* and *Saccharomyces barnettii* were also occasionally found (Silvestri et al., 2007).

D. hansenii was the yeast, characterizing the DNA and RNA profiles of the fresh sausages. It was present during the whole period followed. The only other yeast species detected was *Capronia mansonii*, at day 0 and only in the RNA profile (Cocolin et al., 2004).

Bacterial Ecology by Species-Specific PCR and qPCR

Aymerich et al. (2003) studied the bacterial ecology of 17 commercially available low acid fermented sausages from Spain. Ten of them were fuets (cold-ripened fermented sausages with black pepper) and seven where chorizos (cold-ripened fermented sausages with red pepper). They used 12 sets of primers for 12 different species, six LAB (*Lb. sakei, Lb. curvatus, E. faecium, Lb. plantarum, L. lactis,* and *P. acidilactici*) and six CNC (*S. carnosus, S. warneri, K. varians, S. xylosus, S. simulans,* and *S. epidermidis*). They screened for the 12 species in the fermented sausage homogenate, before and after a 24-hour enrichment period (in MRS for the LAB and in mannitol salt broth [MSB] for the CNC).

Without enrichment, *Lb. sakei* and *Lb. curvatus* were detected in 11.8% of the samples, and *Lb. plantarum* and *S. xylosus* were detected in 17.6%. The low percentages of detection for the species of importance in the fermentation is probably related with a low sensitivity limit of the method, that needed an enrichment step, in order to obtain some information for the species present. However, it should be noted that enrichment can significantly alter the picture of the microbial ecology, since different microbial groups or even species within a microbial group, have different abilities to grow in enrichment media. After the 24h enrichment, it was possible to detect *Lb. sakei* and *S. xylosus* in all the samples, a result that highlights that these two species are predominant in slightly fermented sausages. Regarding other LAB species, *Lb. curvatus* was detected in 71% of the samples (in 80% of the fuets and 57% of the chorizos), while *Lb. plantarum* was present in all of the chorizos and in 50% of the fuets (62% of the samples in total). *P. acidilactici* and *L. lactis* were not detected while *E. faecium* was present in 11.8% of the samples. With regard to the other CNC species, *S. carnosus* was only detected in chorizos (in 14% of them) and *S. epidermidis* in the fuets (in 20% of them). No *S. simulans* or *S. warneri* were found. Overall, the ecology of the low-acid fermented sausages from Spain, as determined after the enrichment, was characterized by the presence of *Lb. sakei* and *S. xylosus*. *Lb. curvatus* and *Lb. plantarum* were also detected but with different frequencies for the two types of sausages. In addition, the presence of *S. epidermidis* and *S. carnosus* characterized the fuets and the chorizos respectively, although they were not detected in all the respective samples.

Recently, a protocol based on qPCR amplification of the 16S-23S rRNA intergenic transcribed region (ITS) of *Lb. sakei*, has been optimized (Martin et al., 2006). The goal of the work was to develop a tool that would allow the quantification of *Lb. sakei*, without cultivation, in the fermented sausages. The protocol was tested in 11 commercial samples of sausages and meat batter; and *Lb. sakei* was detected in the range of 3 to 9 $\log_{10}$ cfu/g. Of the 11 samples, *Lb. sakei* was not detected in three while in a fourth, it was below the quantification limit, which was determined to be in the order of 10^5 cfu/g. The method can be considered as a valid alternative to conventional microbiological testing of fermented sausages that allows rapid and accurate quantification of *Lb. sakei*.

Conclusions

Often at times, when both the culture-dependent and independent methods are applied in parallel to study the microbial ecology of fermented foods, interesting information becomes available. From the experience gained so far in this field, we can conclude that the two approaches complement each other, and their concurrent use highlights important aspects that could otherwise pass un-noticed.

Indisputably, the main breakthrough of the direct methods is that, the inherent limitations of microbial cultivation, related to non-culturable organisms, either due to injury or stress, or due to the selectivity of the media, are circumvented. Furthermore, for the first time, we are able to see in situ which microorganisms are metabolically active and derive information regarding the relative importance of different microbial groups in the transformation process.

The results of the application of molecular approaches to study the microflora of fermented sausages that have been presented here, highlight the unequivocal dominance of *Lb. sakei* and *Lb. curvatus* in these types of products. To a lesser extend, *Lb. plantarum*, in a more product-specific manner, was also shown to be important. The presence and activity of the different *Staphylococcus* members appears to be product-specific as well. Finally, regarding yeasts, *D. hansenii* was detected with a high frequency (Rantsiou & Cocolin, 2006).

Moreover, the use of molecular techniques, able to differentiate strains within the same species, helps in the understanding of the intra-species diversity that may characterize fermented products produced in different countries or in different plants within the same country. Different strains, together with the different ingredients used and the technological processes during production, are responsible for the development of specific sensory profiles. To us, this is an important aspect to be considered in our understanding of the fermentation process, and it needs to be investigated further. Defining and understanding microbial dynamics, as determined by species successions, as well as microbial ecology, as determined by species interactions at each time point and throughout fermentation, is crucial, since these are the parameters that will have a great impact on the organoleptic and sensorial characteristics of the final product.

References

Altschul, S. F., Madden, T. L., Shaffer, A. A., Zhang, J., Zhang, Z., Miller, W., et al. (1997). Gapped BLAST and PSI-BLAST: A new generation of protein database search programs. *Nucleic Acid Research, 25*, 3389–3402.

Ammor, S., Rachman, C., Chaillou, S., Prevost, H., Dousset, X., Zagorec, M., et al. (2005). Phenotypic and genotypic identification of lactic acid bacteria isolated from a small-scale facility producing traditional dry sausages. *Food Microbiology, 22*, 373–382.

Ampe, F., Ben Omar, N., Moizan, C., Wacher, C., & Guyot, J. P. (1999). Polyphasic study of the spatial distribution of microorganisms in Mexican pozol, a fermented maize dough, demonstrates the need for cultivation-independent methods to investigate traditional fermentations. *Applied and Environmental Microbiology, 65*, 5464–5473.

Andrighetto, C., Zampese, L., & Lombardi, A. (2001). RAPD-PCR characterization of lactobacilli isolated from artisanal meat plants and traditional fermented sausages of Veneto region (Italy). *Letter in Applied Microbiology, 33*, 26–30.

Aymerich, T., Martin, B., Garriga, M., & Hugas, M. (2003). Microbial quality and direct PCR identification of lactic acid bacteria and nonpathogenic staphylococci from artisanal low-acid sausages. *Applied and Environmental Microbiology, 69*, 4583–4594.

Aymerich, T., Martin, B., Garrica, M., Vidal-Carou, M. C., Bover-Cid, S., & Hugas, M. (2006). Safety properties and molecular strain typing of lactic acid bacteria from slightly fermented sausages. *Journal of Applied Microbiology, 100*, 40–49.

Bacus, J. N. (1986). Fermented meat and poultry products. In A. M. D. Pearson (Ed.), *Advances in meat and poultry microbiology* (pp. 123–164). London: Macmillan.

Baruzzi, F., Matarante, A., Caputo, L., & Morea, M. (2006). Molecular and physiological characterization of natural microbial communities isolated from a traditional Southern Italian processed sausages. *Meat Science, 72*, 261–269.

Berthier, F., & Ehrlich, S. D. (1998). Rapid species identification within two groups of closely related lactobacilli using PCR primers that target the 16S/23S rRNA spacer region. *FEMS Microbiology Letters, 161*, 97–106.

Berthier, F., & Ehrlich, S. D. (1999). Genetic diversity within *Lactobacillus sakei* and *Lactobacillus curvatus* and design of PCR primers for its detection using randomly amplified polymorphic DNA. *International Journal of Systematic Bacteriology, 49*, 997–1007.

Blaiotta, G., Casaburi, A., & Villani, F. (2005). Identification and differentiation of *Staphylococcus carnosus* and *Staphylococcus simulans* by species-specific PCR assays of *sodA* genes. *Systematic and Applied Microbiology, 28*, 519–526.

Blaiotta, G., Ercolini, D., Mauriello, G., Salzano, G., & Villani, F. (2004). Rapid and reliable identification of *Staphylococcus equorum* by a species-specific PCR assay targeting the *sodA* gene. *Systematic and Applied Microbiology, 27*, 696–702.

Blaiotta, G., Pennacchia, C., Ercolini, D., Moschetti, G., & Villani, F. (2003). Combining denaturing gradient gel electrophoresis of 16S rDNA V3 region and 16S-23S rDNA spacer region polymorphism analyses for the identification of staphylococci from Italian fermented sausages. *Systematic and Applied Microbiology, 26*, 423–433.

Blaiotta, G., Pennacchia, C., Parente, E., & Villani, F. (2003). Design and evaluation of specific PCR primers for rapid and reliable identification of *Staphylococcus xylosus* strains isolated from dry fermented sausages. *Systematic and Applied Microbiology, 26*, 601–610.

Blaiotta, G., Pennacchia, C., Villani, F., Ricciardi, A., Tofalo, R., & Parente, E. (2004). Diversity and dynamics of communities of coagulase-negative staphylococci in traditional fermented sausages. *Journal of Applied Microbiology, 97*, 271–284.

Bottari, B., Ercolini, D., Gatti, M., & Neviani, E. (2006). Application of FISH technology for microbiological analysis: Current state and prospects. *Applied Microbiology and Biotechnology, 73*, 485–494.

Caetono-Anollés, G. (1993). Amplifying DNA with arbitrary oligonucleotide primers. *PCR Methods and Applications, 3*, 85–94.

Cocolin, L., Bisson, L. F., & Mills, D. A. (2000). Direct profiling of the yeast dynamics in wine fermentations. *FEMS Microbiology Letters, 189*, 81–87.

Cocolin, L., Diez, A., Urso, R., Rantsiou, K., Comi, G., Bergmaier, I., et al. (2007). Optimization of conditions for profiling bacterial populations in food by culture-independent methods. *International Journal of Food Microbiology, 20*, 400–409.

Cocolin, L., Manzano, M., Aggio, D., Cantoni, C., & Comi, G. (2001). A novel polymerase chain reaction (PCR)—denaturing gradient gel electrophoresis (DGGE) for the identification of Micrococcaceae strains involved in meat fermentations. Its application to naturally fermented Italian sausages. *Meat Science, 57*, 59–64.

Cocolin, L., Manzano, M., Cantoni, C., & Comi, G. (2000). Development of a rapid method for the identification of *Lactobacillus* spp. isolated from naturally fermented Italian sausage using a polymerase chain reaction—temperature gradient gel electrophoresis. *Letters in Applied Microbiology, 30*, 126–129.

Cocolin, L., Manzano, M., Cantoni, C., & Comi, G. (2001). Denaturing gradient gel electrophoresis analysis of the 16S rRNA gene V1 region to monitor dynamic changes in the bacterial population during fermentation of Italian sausages. *Applied and Environmental Microbiology, 67*, 5113–5121.

Cocolin, L., Rantsiou, K., Iacumin, L., Urso, R., Cantoni, C., & Comi, G. (2004). Study of the ecology of fresh sausages and characterization of populations of lactic acid bacteria by molecular methods. *Applied and Environmental Microbiology, 70*, 1883–1894.

Cocolin, L., Urso, R., Rantsiou, K., Cantoni, C., & Comi, G. (2006a). Dynamics and characterization of yeasts during natural fermentation of Italian sausages. *FEMS Yeast Research, 6*, 692–701.

Cocolin, L., Urso, R., Rantsiou, K., Cantoni, C., & Comi, G. (2006b). Multiphasic approach to study the bacterial ecology of fermented sausages inoculated with a commercial starter culture. *Applied and Environmental Microbiology, 72*, 942–945.

Collins, M. D., Rodriguez, U., Ash, C., Aguirre, M., Farrow, J. E., Martinezmurcia, A., et al. (1991). Phylogenetic analysis of the genus *Lactobacillus* and related lactic acid bacteria as determined by reverse-transcriptase sequencing of 16S ribosomal-RNA. *FEMS Microbiology Letters, 77*, 5–12.

Comi, G., Urso, R., Iacumin, L., Rantsiou, K., Cattaneo, P., Cantoni, C., et al. (2005). Characterization of naturally fermented sausages produced in the North East of Italy. *Meat Science, 69*, 381–392.
Coppola, R., Iorizzo, M., Saotta, R., Sorrentino, E., & Grazia, L. (1997). Characterization of micrococci and staphylococci isolated from soppressata molisana, a Southern Italy fermented sausage. *Food Microbiology, 14*, 47–53.
Corbiere Morot-Bizot, S., Leroy, S., & Talon, R. (2006). Staphylococcal community of a small unit manufacturing traditional dry fermented sausages. *International Journal of Food Microbiology, 108*, 210–217.
Corbiere Morot-Bizot, S., Leroy, S., & Talon, R. (2007). Monitoring of staphylococcal starters in two French processing plants manufacturing dry fermented sausages. *Journal of Applied Microbiology, 102*, 238–244.
Corbiere Morot-Bizot, S., Talon, R., & Leroy, S. (2004). Development of a multiplex PCR for the identification of *Staphylococcus* genus and four staphylococcal species isolated from food. *Journal of Applied Microbiology, 97*, 1087–1094.
Corich, V., Mattiazzi, A., Soldati, E., Carraro, A., & Giacomini, A. (2005). Sau-PCR, a novel amplification technique for genetic fingerprinting of microorganisms. *Applied and Environmental Microbiology, 71*, 6401–6406.
Couto, I., Pereira, S., Miragaia, M., Sanches, I. S., & de Lancastre, H. (2001). Identification of clinical staphylococcal isolates from humans by internal transcribed spacer PCR. Journal of Clinical Microbiology, 39, 3099–3103.
Dahllof, I., Baillie, H., & Kjelleberg, S. (2000). rpoB-based microbial community analysis avoids limitations inherent in 16S rRNA gene intraspecies heterogeneity. Applied and Environmental Microbiology, 66, 3376–3380.
Demeyer, D. I. (1982). Stoichiometry of dry sausage fermentation. *Antonie Leeuwenhoek, 48*, 414–416.
Demeyer, D. I., Verplaetse, A., & Gistelink, M. (1986). Fermentation of meat: An integrated process. *Belgian Journal of Food Chemistry and Biotechnology*, 41, 131–140.
Di Maria, S., Basso, A. L., Santoro, E., Grazia, L., & Coppola, R. (2002). Monitoring of *Staphylococcus xylosus* DSM 20266 added as starter during fermentation and ripening of soppressata molisana, a typical Italian sausage. *Journal of Applied Microbiology, 92*, 158–164.
Ercolini, D. (2004). PCR-DGGE fingerprinting: Novel strategies for detection of microbes in food. *Journal of Microbiological Methods, 56*, 297–314.
Ercolini, D., Hill, P. J., & Dodd, E. R. (2003). Bacterial community structure and location in Stilton cheese. *Applied and Environmental Microbiology, 69*, 3540–548.
Ercolini, D., Moschetti, G., Blaiotta, G., & Coppola, S. (2001a). Behavior of variable V3 region from 16S rDNA of important lactic acid bacteria in denaturing gradient gel electrophoresis. *Current Microbiology, 42*, 199–202.
Ercolini, D., Moschetti, G., Blaiotta, G., & Coppola, S. (2001b). The potential of a polyphasic PCR-DGGE approach in evaluating microbial diversity of natural whey cultures from water-buffalo Mozzarella cheese production: Bias of "culture dependent" and "culture independent" approaches. *Systematic and Applied Microbiology, 24*, 610–617.
Fadda, S., Sanz, Y., Vignolo, G., Aristoy, M.-C., Oliver, G., & Toldrà, F. (1999a). Hydrolysis of pork muscle sarcoplasmatic proteins by *Lactobacillus curvatus* and *Lactobacillus sake*. *Applied and Environmental Microbiology, 65*, 578–584.
Fadda, S., Sanz, Y., Vignolo, G., Aristoy, M.-C., Oliver, G., & Toldrà, F. (1999b). Characterization of muscle sarcoplasmatic and myofibrillar protein hydrolysis caused by *Lactobacillus plantarum*. *Applied and Environmental Microbiology, 65*, 3540–3546.
Ferris, M. J., Muyzer, G., & Ward, D. M. (1996). Denaturing gradient gel electrophoresis profiles of 16S rRNA-defined populations inhabiting a hot spring microbial mat community. *Applied and Environmental Microbiology, 62*, 340–346.
Flores, M., Durà, M.-A., Marco, A., & Toldrà, F. (2004). Effect of *Debaryomyces* spp. on aroma formation and sensory quality of dry-fermented sausages. *Meat Science, 68*, 439–446.

Fontana, C., Cocconcelli, P. S., & Vignolo, G. (2005). Monitoring the bacterial population dynamics during fermentation of artisanal Argentinean sausages. *International Journal of Food Microbiology, 103*, 131–142.

Fontana, C., Vignolo, G., & Cocconcelli, P. S. (2005). PCR-DGGE analysis for the identification of microbial populations from Argentinean dry fermented sausages. *Journal of Microbiological Methods, 63*, 254–263.

Forsman, P., Tilsala-Timisijarvi, A., & Alatossava, T. (1997). Identification of the staphylococcal and streptococcal causes of the bovine mastitis using 16S–23S rRNA spacer regions. *Microbiology, 143*, 3491–3500.

Garcia-Varona, M., Santos, E. M., Jaime, I., & Rovira, J. (2000). Characterization of *Micrococcaceae* isolated from different varieties of chorizo. *International Journal of Food Microbiology, 54*, 189–195.

Gevers, D., Huys, G., & Swings, J. (2001). Applicability of rep-PCR fingerprinting for identification of *Lactobacillus* species. *FEMS Microbiology Letters, 205*, 31–36.

Gilson, E., Clement, J. M., Brutlag, D., & Hofnung, M. (1984). A family of dispersed repetitive extragenic palindromic DNA sequences in *E. coli*. *EMBO Journal, 3*, 1417–1421.

Gory, L., Millet, L., Godon, J. J., & Montel, M. C. (1999). Identification of *Staphylococcus carnosus* and *Staphylococcus warneri* isolated from meat by florescent *in situ* hybridization with 16S-targeted oligonucleotide probes. *Systematic and Applied Microbiology, 22*, 225–228.

Greco, M., Mazzette, R., De Santis, E. P. L., Corona, A., & Cosseddu, A. M. (2005). Evolution and identification of lactic acid bacteria isolated during the ripening of Sardinian sausages. *Meat Science, 69*, 733–739.

Hammes, W. P., Bantleon, A., & Min, S. (1990). Lactic acid bacteria in meat fermentation. *FEMS Microbiology Reviews, 87*, 165–174.

Hammes, W. P., & Knauf, H. J. (1994). Starters in processing of meat products. *Meat Science, 36*, 155–168.

Hertel, C., Ludwig, W., Obst, M., Vogel, R. F., Hammes, W. P., & Schleifer, K. H. (1991). 23S rRNA-targeted oligonucleotide probes for the rapid identification of meat lactobacilli. *Systematic and Applied Microbiology, 14*, 173–177.

Hesseltine, C. W., & Wang, H. L. (1980). The importance of traditional fermented foods. *Bioscience, 30*, 402–404.

Hugenholtz, P., Goebel, B. M., & Pace, N. R. (1998). Impact of culture-independent studies on the emerging phylogenetic view of bacterial diversity. *Journal of Bacteriology, 180*, 4765–4774.

Iacumin, L., Comi, G., Cantoni, C., & Cocolin, L. (2006a). Molecular and technological characterization of *Staphylococcus xylosus* isolated from Italian naturally fermented sausages by RAPD, Rep-PCR and Sau-PCR analysis. *Meat Science, 74*, 281–288.

Iacumin, L., Comi, G., Cantoni, C., & Cocolin, L. (2006b). Ecology and dynamics of coagulase-negative cocci isolated from naturally fermented Italian sausages. *Systematic and Applied Microbiology, 29*, 480–486.

Jensen, M. A., Webster, J. A., & Strauss, N. (1993). Rapid identification of bacteria on the basis of polymerase chain reaction-amplified ribosomal DNA spacer region polymorphisms. *Applied and Environmental Microbiology, 59*, 945–952.

Klijn, N., Weerkamp, A. H., & de Vos, W. M. (1991). Identification of mesophilic lactic acid bacteria by using polymerase chain reaction-amplified variable regions of 16S rRNA and specific DNA probes. *Applied and Environmental Microbiology, 57*, 3390–3393.

Kurtzman, C. P., & Robnett, C. J. (1997). Identification of clinically important ascomycetous yeast based on nucleotide divergence in the 5' and of the large-subunit (26S) ribosomal DNA gene. *Journal of Clinical Microbiology, 35*, 1216–1223.

Kurtzman, C. P., & Robnett, C. J. (1998). Identification and phylogeny of ascomycetous yeasts from analysis of nuclear large subunit (26S) ribosomal DNA partial sequence. *Antonie van Leeuwenhoek 73*, 331–371.

Lane, D. J. (1991). 16S/23S sequencing. In E. Stackebrandt, & M. Goodfellow (Eds.), *Nucleic acid techniques in bacterial systematics* (pp. 115–175). Chichester: John Wiley & Sons.

Lee, J., Jang, J., Kim, B., Kim, J., Jeong, G., & Han, H. (2004). Identification of *Lactobacillus sakei* and *Lactobacillus curvatus* by multiplex PCR-based restriction enzyme analysis. *Journal of Microbiological Methods*, 59, 1–6.

Lerche, M., & Reuter, G. (1960). A contribution to the method of isolation and differentation of aerobic "lactobacilli" (Genus *Lactobacillus* Beijerinck). *Archiv fur Hygiene und Bakteriologie*, *179*, 354–370.

Lücke, F. K. (1974). Fermented sausages. In B. J. B. Wood (Ed.), *Microbiology of fermented foods* (pp. 41–49). London: Applied Science Publishers.

Lücke, F. K. (2000). Utilization of microbes to process and preserve meat. *Meat Science*, 56, 105–115.

Martin, B., Garriga, M., Hugas, M., Bover-Cid, S., Veciana-Nogues, M. T, & Aymerich, T. (2006). Molecular, technological and safety characterization of Gram-positive catalase-positive cocci from slightly fermented sausages. *International Journal of Food Microbiology*, *107*, 148–158.

Martin, B., Jofré, A., Garriga, M., Pla, M., & Aymerich, T. (2006). Rapid quantitative detection of *Lactobacillus sakei* in meat and fermented sausages by real-time PCR. *Applied and Environmental Microbiology*, 72, 6040–6048.

Mendoza, M., Meugnier, H., Bes, M., Etienne, J., & Freney, J. (1998). Identification of *Staphylococcus* species by 16S–23S rDNA intergenic spacer PCR analysis. *International Journal of Systematic Bacteriology*, 48, 1049–105.

Morot-Bizot, S., Talon, R., & Leroy-Setrin, S. (2003). Development of specific primers for a rapid and accurate identification of *Staphylococcus xylosus*, a species used in food fermentation. *Journal of Microbiological Methods*, 55, 279–286.

Muyzer, G., De Waal, E. D., & Uitterlinden, A. G. (1993). Profiling of complex microbial populations by denaturing gradient gel electrophoresis analysis of polymerase chain reaction-amplified genes coding for 16S rRNA. *Applied and Environmental Microbiology*, 59, 695–700.

Muyzer, G., & Smalla, K. (1998). Application of denaturing gradient gel electrophoresis (DGGE) and temperature gradient gel electrophoresis (TGGE) in microbial ecology. *Antonie Van Leeuwenhoek*, 73, 127–141.

Nissen, H., & Dainty, R. (1995). Comparsion of the use of rRNA probes and conventional methods in identifying strains of *Lactobacillus sake* and *L. curvatus* isolated from meat. *International Journal of Food microbiology* 25, 311–315.

Pennacchia, C., Vaughan, E. E., & Villani, F. (2006). Potential probiotic *Lactobacillus* strains from fermented sausages: Further investigations on their probiotic properties. *Meat Science*, 73, 90–101.

Pontes, S. D., Lima-Bitterncourt, I. C., Chartone-Souza, E., & Amaral Nascimiento, A. M. (2007). Molecular approaches: Advantages and artifacts in assessing bacterial diversity. *Journal of Industrial Microbiology and Biotechnology*, 34, 463–473.

Power, E. G. (1996). RAPD typing in microbiology—a technical review. *Journal of Hospital Infection*, 34, 247–265.

Rantsiou, K., & Cocolin, L. (2006). New developments in the study of the microbiota of naturally fermented sausages as determined by molecular methods: A review. *International Journal of Food Microbiology*, *108*, 255–267.

Rantsiou, K., Comi, G., & Cocolin, L. (2004). The *rpoB* gene as a target for PCR-DGGE analysis to follow lactic acid bacteria population dynamics during food fermentations. *Food Microbiology*, 21, 481–487.

Rantsiou, K., Drosinos, E. H., Gialitaki, M., Metaxopoulos, I., Comi, G., & Cocolin, L. (2006). Use of molecular tools to characterize *Lactobacillus* spp. isolated from Greek traditional fermented sausages. *International Journal of Food Microbiology*, *112*, 215–222.

Rantsiou, K., Drosinos, E. H., Gialitaki, M., Urso, R., Krommer, J., Gasparik-Reichardt, J., et al. (2005). Molecular characterization of *Lactobacillus* species isolated from naturally fermented sausages produced in Greece, Hungary and Italy. *Food Microbiology*, 22, 19–28.

Rantsiou, K., Iacumin, L., Cantoni, C., Comi, G., & Cocolin, L. (2005). Ecology and characterization by molecular methods of *Staphylococcus* species isolated from fresh sausages. *International Journal of Food Microbiology*, 97, 277–284.

Rantsiou, K., Urso, R., Iacumin, L., Cantoni, C., Cattaneo, P., Comi, G., et al. (2005). Culture dependent and independent methods to investigate the microbial ecology of Italian fermented sausages. *Applied and Environmental Microbiology, 71*, 1977–1986.

Rebecchi, A., Crivori, S., Sarra, P. G., & Cocconcelli, P. S. (1998). Physiological and molecular techniques for the study of bacterial community development in sausage fermentation. *Journal of Applied Microbiology, 84*, 1043–1049.

Reuter, G. (1972). Experimental ripening of dry sausages using lactobacilli and micrococci starter cultures. *Fleischwirtschaft, 52*, 465–468, 471–473.

Rossi, F., Tofalo, R., Torriani, S., & Suzzi, G. (2001). Identification by 16S-23S rDNA intergenic region amplification, genotypic and phenotypic clustering of *Staphylococcus xylosus* strains from dry sausages. *Journal of Applied Microbiology, 90*, 365–371.

Sandhu, G. S., Kline, B. C., Stockman, L., & Roberts, G. D. (1995). Molecular probes for diagnosis of fungal infections. *Journal of Clinical Microbiology, 33*, 2913–2919.

Sanz, Y., Hernandez, M., Ferrus, M. A., & Hernandez, J. (1998). Characterization of *Lactobacillus sake* isolates from dry-cured sausage by restriction fragment length polymorphism analysis of the 16S rRNA gene. *Journal of Applied Microbiology, 84*, 600–6006.

Schleifer, K. H. (1986). Gram positive cocci. In P. H. A Sneath, N. S. Mair, & Holt, J. G. (Eds.), *Bergey's manual of systematic bacteriology* (Vol. 2, pp. 999–1003). Baltimore: Williams & Wilkins.

Silvestri, G., Santarelli, S., Aquilanti, L., Beccaceci, A., Osimani, A., Tonucci, F., et al. (2007). Investigation of the microbial ecology of Ciauscolo, a traditional Italian salami, by culture-dependent techniques and PCR-DGGE. *Meat Science, 77*, 413–423.

Simonova, M., Strompfova, V., Marcinakova, M., Laukova, A., Vetserlund, S., Moratella, et al. (2006). Characterization of *Staphylococcus xylosus* and *Staphylococcus carnosus* isolated from Slovak meat products. *Meat Science, 73*, 559–564.

Talon, R., Leroy, S., & Lebert, I. (2007). Microbial ecosystems of traditional fermented meat products: The importance of indigenous starters. *Meat Science, 77*, 55–62.

Talon, R., Walter, D., Chartier, S., Barriere, C., & Montel, M. C. (1999). Effect of nitrate and incubation conditions on the production of catalase and nitrate reductase by staphylococci. *International Journal of Food Microbiology, 52*, 47–56.

Tenover, F. C., Arbeit, R. D., Goering, R. V., Mickelsen, P. A., Murray, B. E., Persing, D. H., et al. (1995). Interpreting chromosomal DNA restriction patterns produced by pulsed-field gel electrophoresis: Criteria for bacterial strain typing. *Journal of Clinical Microbiology, 33*, 2233–2239.

Urso, R., Comi, G., & Cocolin, L. (2006). Ecology of lactic acid bacteria in Italian fermented sausages: Isolation, identification and molecular characterization. *Systematic and Applied Microbiology, 29*, 671–680.

Villani, F., Casaburi, A., Pennacchia, C., Filosa, L., Russo, F., & Ercolini, D. (2007). The microbial ecology of the *Soppressata* of Vallo di Diano, a traditional dry fermented sausage from southern Italy, and *in vitro* and *in situ* selection of autochthonous starter cultures. *Applied and Environmental Microbiology, 73*, 5453–5463.

Villard, L., Kodjo, A., Borges, E., Maurin, F., & Richard, Y. (2000). Rybotiping and rapid identification of *Staphylococcus xylosus* by 16S-23S spacer amplification. *FEMS Microbiology Letters, 185*, 83–87.

Walter, J., Tannock, G. W., Tilsala-Timisjarvi, A., Rodtong, S., Loach, D. M., Munro, K., et al. (2000). Detection and identification of gastrointestinal *Lactobacillus* species by using denaturing gradient gel electrophoresis and species-specific PCR primers. *Applied and Environmental Microbiology, 66*, 297–303.

Yu, Z., & Morrison, M. (2004). Comparison of different hypervariable regions of *rrs* genes for use in fingerprinting of microbial communities by PCR-denaturing gradient gel electrophoresis. *Applied and Environmental Microbiology, 70*, 4800–4806.

Zoetendal, E. G., Akkermans, A. D. L., & de Vos, W. M. (1998). Temperature gradient gel electrophoresis analysis of 16S rRNA from human fecal samples reveals stable and host-specific communities of active bacteria. *Applied and Environmental Microbiology, 64*, 3854–3859.

Chapter 5
Characteristics and Applications of Microbial Starters in Meat Fermentations

Pier Sandro Cocconcelli and Cecilia Fontana

Introduction

Fermentation and drying are among the most ancient food preservation techniques used by man. Developed through the years, these processes prolonged the storage time of meats (and meat products) and brought favorable changes to their organoleptic properties, with respect to the original substrate.

The microbiology of fermented sausages is varied and complex. The type of microflora that develops is often closely related to the ripening technique utilized. Several investigations established two groups of microorganisms as being the main organisms responsible for the transformations involved during fermentation and ripening of sausages. The spontaneous fermentation of dry sausages involves the participation of the Lactic Acid Bacteria (LAB), Gram-positive Catalase-positive Cocci (GCC+, mostly *Staphylococcus* and *Kocuria* species), and, less importantly, yeasts and moulds. Most of the commercially available meat starter cultures contain mixtures of LAB and GCC+. These bacterial groups are responsible for the basic microbial reactions that occur simultaneously during fermentation; the decrease of pH values via glycolysis by LAB, the reduction of nitrate, and the development of aroma by GCC+.

Research on the use of starter cultures in meat products began in the USA in the 1940s, inoculating the raw material for fermented sausages using lactobacilli, with the aim to govern and accelerate the fermentation. In the late fifties, the Niinivaara (1955) helped to launch this idea in Europe developing mixed cultures of the *Micrococcus* sp. and *Pediococcus cerevisiae*.

According to the definitions of Hammes, Bantleon, and Min (1990), meat starter cultures are preparations that contain active or dormant microorganisms that develop the desired metabolic activity in the meat. They are, by definition, used to drive the fermentation process in the desired direction, reducing the variability in the quality of the product, limiting the growth of spoilage bacteria by accelerating

---

P.S. Cocconcelli
Istituto di Microbiologia, Centro Ricerche Biotecnologiche, Università Cattolica del Sacro Cuore, via Emilia Parmense 84, 29100 Piacenza-Cremona, Italy

fermentation, and improving the sensory properties of fermented meat products. Appropriate cultures have to be selected according to the specific formulation of the batter and technology of fermentation, since environmental factors will interact to select a limited number of strains that are competitive enough to dominate the process.

In 1994, Buckenhüskes divided the meat starter cultures into two categories. The first generation of starter bacteria, generally derived from cultures for vegetable fermentation, such as *L. plantarum* and pediococci, were mainly selected for their acidification properties. A second generation developed isolating strains of meat origin, such as *L. sakei* and coagulase negative staphylococci (CNS), that harbored phenotypic traits of technological relevance. This second generation is now widely used in the industrial processes of fermented meat production.

Development of a third generation of functional starter cultures with increased diversity, stability, and industrial performance is under way. Using comparative genomics, microarray analysis, transcriptomics, proteomics, and metabolomics, the natural diversity of wild strains that occur in traditional artisan foods is being explored. This approach permits rapid, high-throughput screening of promising wild strains with interesting functional properties that lack negative characteristics, and it will enable the construction of genetically modified starter cultures with a tailored functionality. The most promising bacteria for starter cultures are those that are isolated from the indigenous micro-biota of traditional products, because they are well adapted to the meat environment and are capable of controlling the micro-biota of products.

The aim of this chapter is to provide the most recent information on bacterial inoculants and their potential functionalities, explaining how they (can) improve the quality of fermented meat products.

Starter Cultures for Fermented Meats

Taxonomy

The genus *Lactobacillus* is of utmost importance in meat fermentation and that is why its members are often used as starter cultures in dry fermented sausage production. This genus encompasses more than 100 different species with a large variety of phenotypic, biochemical, and physiological properties (Axelsson, 2004), but only a few of them have been isolated from meat fermentations and used as starter cultures. *L. sakei*, *L. curvatus*, and *L. plantarum*, belonging to the sub-group of facultative hetero-fermentative lactobacilli, are generally used for this purpose. The main energetic metabolism of these bacteria is the dissimilation of sugar to organic acid, by means of the glycolysis and phosphoketolase pathways. When hexoses are the energy source, lactic acid is the major fermentation end product. *L. sakei* is the predominant species in fermented meat products, and its use as a starter culture for sausage production is widespread. Although *L. plantarum* has been identified as part of the meat micro-biota and is used as starter cultures for meat fermentation, this species lacks the specific adaptation to meat environment found in *L. sakei*. *L. plantarum* is a versatile bacterium encountered in a variety of different environments; its genome size, 3.3 Mb (Kleerebezem et al., 2003), is the largest of the *Lactobacillus* genus, reflecting its metabolic and environmental flexibility. Much less information is available on the physiology and genetics of *L. curvatus*, other than for the production of antimicrobial peptides, discussed in Chap. 18.

Pediococci are not often isolated from European fermented sausages but occasionally occur in small percentages (Papamanoli, Tzanetakis, Litopoulou-Tzanetaki, and Kotzekidou, 2003). They are more common in fermented sausages from the United States, where they are deliberately added as starter cultures to accelerate the acidification of the meat batter. Pediococci are Gram-positive, coccus-shaped, LAB, showing the distinctive characteristic of tetrads formation via cell division in two perpendicular directions in a single plane. Pediococci present a homo-fermentative metabolism, where lactic acid is the major metabolic end product (Axelsson, 2004). Phylogenetically, *Pediococcus* species belong to the *L. casei – Pediococcus* sub-cluster of the *Lactobacillus* cluster. The genus consists currently of nine species, but only *P. pentosaceus* is generally used as a starter culture for meat fermentation. The species *P. cerevisiae*, frequently mentioned as a starter culture, has now been reclassified as *P. pentosaceus*. The genome sequencing project of *P. pentosaceus* ATCC 25745 is complete (http://genome.jgi-psf.org/draft_microbes/pedpe/pedpe.info.html).

Micrococcaceae are frequently mentioned as components of meat starter cultures, but this term generally refers to the members of the *Staphylococcus* genus (that belongs to the family *Staphylococcaceae*). *Staphylococcus* was originally grouped with other Gram-positive cocci, such as *Micrococcus* genus, because these two genera often co-habit the same habitats. However, molecular taxonomy has revealed that these genera are phylogenetically separate and distinct. The genus *Staphylococcus* belongs to the *Clostridium* subdivision of Gram-positive bacteria, while *Micrococcus* is part of the *Actinomycetales*.

The genus *Staphylococcus* comprises 42 validly described species and subspecies of Gram-positive, catalase-positive cocci (Ghebremedhin, Layer, König, & König, 2008; Bannerman, 2003; Kwok, & Chow, 2003; Spergser et al., 2003), 10 of which contain subdivisions with subspecies designations (Garrity, Bell, & Lilburn, 2004; Spergser et al., 2003; Place, Hiestand, Gallmann, & Teuber, 2003). The staphylococci present a spherical shape and the cells are often grouped to form clusters. The staphylococci are widespread in nature; their major habitats are skin, skin glands and mucous membranes of mammals and birds. Some species, mainly coagulase-positive such as *S. aureus*, are pathogens responsible for infections and food born intoxications. Staphylococci are facultative anaerobes capable of metabolizing a number of different sugars. Under anaerobic conditions the major end product is lactic acid, but acetate and pyruvate and acetoin are also formed. These organisms show the ability to survive under environmental stress, such as high salt and low temperatures, as encountered during meat fermentation. Moreover, they reduce nitrate to nitrite, a technologically relevant physiological feature. Many CNS such as *Staphylococcus xylosus*, *S. carnosus*, *S. equorum*, and *S. saprophyticus* have been isolated from dry fermented sausages, but other species occur too.

Kocuria varians, formerly classified as *Micrococcus varians*, is a member of *Micrococcaceae* and is used as meat starter cultures for its nitrate reductase ability.

Commercial starter cultures in Europe are generally made up of a balanced mixture of LAB (*Lactobacillus* and *Pediococcus*) and GCC+ (*Staphylococcus* and *Kocuria*). Table 5.1 reports the most frequently used bacterial species and their functional role in meat fermentation.

Genomics

Genetic studies of the main bacterial species *L. sakei*, *S. xylosus*, and *S. carnosus* in the last few years, have provided an understanding of the main metabolic activities

Table 5.1 Starter culture species most frequently used for meat fermentation. * Bacteriocin production is discussed in Chap. 18.

Species	Functional Technological properties for meat fermentation	Quality characteristics
Lactobacillus sakei	Decrease of pH value	Preservation
	Catalase activity	Firmness (consistency)
	Flavor development	Aroma
	Amino acid metabolism	
	Antioxidant properties: catalase and SOD	
	Bacteriocin production*	
Lactobacillus curvatus	Decrease of pH value	Preservation
	Proteolytic activity	Firmness (consistency)
	Antioxidant properties: catalase	Aroma
	Bacteriocin production*	
Lactobacillus plantarum	Decrease of pH value	Preservation
	Antioxidant properties: catalase	Firmness (consistency)
	Bacteriocin production*	
Lactobacillus rhamnosus	Probiotic	
Pediococcus acidilactici	Acidification	Preservation
	Bacteriocin production*	Firmness (consistency)
Pediococcus pentosaceus	Acidification	Preservation
	Bacteriocin production*	Firmness (consistency)
Staphylococcus xylosus	Proteolysis	Color
	Amino acid catabolism	Aroma
	Lipolysis	Preservation
	Antioxidant properties: catalase and SOD	
	Nitrate reduction	
Staphylococcus carnosus	Proteolysis	Color
	Amino acid catabolism	Aroma
	Lipolysis	Preservation
	Antioxidant properties: catalase and SOD	
	Nitrate reduction	
Staphylococcus equorum	Flavor development	Color
	Nitrate reduction	Aroma
		Preservation
Kocuria varians	Nitrate reduction	Color
		Preservation

and stress resistance functions. The maximum information was obtained from the recent analysis of the 1.8 Mb genome sequence from *L. sakei* 23K (Chaillou et al., 2005). This organism has evolved to adapt itself to the meat environment, harboring the genetic function that confers it the ability to grow and survive in the meat environment. *L. sakei* seems to be very well suited to derive energy from other compounds that are more abundant in meat. Its adaptation to the meat, an environment rich in amino acids because of the activity of endogenous proteases, has caused it to lose biosynthetic pathways for amino acid synthesis. *L. sakei* is therefore auxotrophic for all amino acids except aspartic and glutamic acid (Champomier-Verges, Marceau, Mera, & Zagorec, 2002). Amino acid metabolism can provide an alternative energy source for *L. sakei* when glucose is exhausted and this affects the sensory properties of the sausage, as discussed later. The genome shows a particularly well-developed potential for amino acid catabolism, and, in addition, *L. sakei* has the ability to use purine nucleosides for energy production (a unique property among LAB).

The most common CNS species used as starter cultures are *S. carnosus* and *S. xylosus*. The genome sequencing of *S. xylosus* is ongoing and information can be obtained at www.cns.fr/externe/English/Projets/Projet_NN/NN.html. The size of *S. xylosus* strains C2a was estimated to be 2868 kb (Dordet-Frisoni, Talon &. Leroy, 2007) and *S. carnosus* TM300 strain was estimated by PFGE to be 2590 kb (Wagner, Doskar, & Gotz, 1998). The complete genome sequence of these strains enabled the exact size of the chromosome to be determined at 2.56 Mb (Rosenstein, Nerz, Resch, & Götz, 2005).

The chromosome of *S. carnosus* TM300 comprises 2454 ORFs and the annotation of the complete sequence is under way. The first comparative studies showed a high conservation in gene order between *S. carnosus* and other sequenced staphylococci. However, comparison of the gene products revealed that about 20% of them have no staphylococcal homologue (Rosenstein et al., 2005).

Knowledge of the whole chromosome sequence of *S. xylosus* will provide for a better understanding of the physiology of this species. A proteomics approach to study cell-envelope proteins of *S. xylosus* has been developed (Planchon, 2006; Planchon et al., 2007), in which a significant set of cell enveloped proteins can be recovered. When such information is integrated with future analyses of the transcripts, a more integrated and comprehensive knowledge of the mechanism by which *L. sakei* and *S. xylosus* contribute to the fermentation of meat can be obtained, as can how these bacteria interact with each other.

Functional Features

Competitiveness in the Meat Environment

One of the fundamental properties of bacterial starter cultures is their ability to compete with the adventitious micro-biota of meat, to colonize this environment, and to dominate the microbial community of the fermented products. The starter culture must compete with the natural micro-biota of the raw material and will have to undertake the metabolic activities expected, being conditioned by its growth rate and survival in the conditions prevailing in the sausage, that is, an anaerobic atmosphere, rather high salt concentrations, low temperatures, and low pH. Successful colonization of meat products by starter cultures is a pre-requisite for their contribution to sensory qualities of the final product. A summary of the main metabolic pathways used by the meat starter bacteria (*Lactobacillus* and *Staphylococcus*) is given in Fig. 5.1.

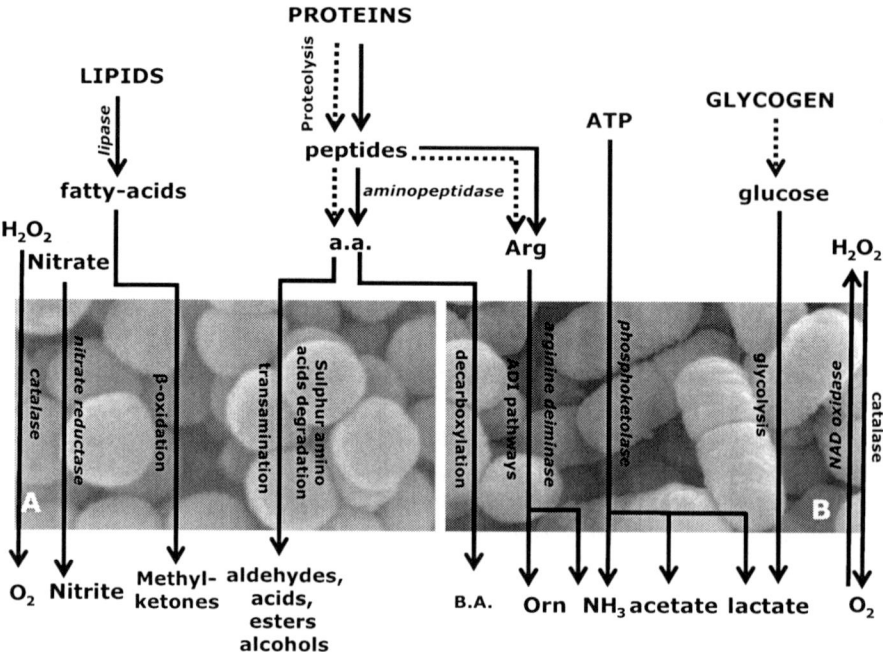

Fig. 5.1 Major metabolic pathways used for the meat starter bacteria, contributing to meat adaptation, color, texture, and flavor development. a) coagulase negative staphylococci. b) Lactobacilli. a.a, amino acids; Arg, arginine; B.A., Biogenic Amines; Orn, ornithine.
Dotted lines indicate the activity of endogenous meat enzymes, continuous lines indicate microbial metabolisms. Names of the main enzymes are indicated in *italics*.
Catabolic pathways (carbohydrates, proteins and lipids), are used for initial microbial growth during fermentation. Glycogen is the main source of glucose. Ribose is released by ATP hydrolysis, and the subsequent metabolism of ribose-derived molecules is used for energy production. When the sugar concentration declines, free amino acids are utilized for microbial growth. Arginine deiminase (ADI) pathways enhance the survival of lactobacilli during the stationary phase. Staphylococci modulate the aroma through the conversion of amino acids (particularly the branched-chain amino acids leucine, isoleucine, and valine), into methyl-branched aldehyde, methyl-branched acids and sulphites, diacetyl, and ethyl ester. The methyl ketones (2-pentanone and 2-heptanone) derive from intermediates of an incomplete ß-oxidation pathway in staphylococci

Two of the most common preservative conditions employed in meat processing are low temperatures and high salt concentrations. *L. sakei* is remarkably well equipped to cope with these conditions. It contains several transporters for cryo- and osmoprotective substances and has more cold stress proteins than other lactobacilli. *L. sakei* has psychotropic and osmo-tolerant properties, being able to grow at low temperatures and in the presence of up to 10% of sodium chloride (NaCl). These physiological features are associated with the presence, in its genome, of a higher number of genes coding for stress response proteins, such as cold shock and osmo-tolerance proteins, than found in other lactobacilli.

The genome contains genes encoding a heme-dependent catalase, a superoxide dismutase, and a NADH oxidase, to cope with reactive oxygen species, and there are several systems to cope with changes in the redox potential. *L. sakei* lacks proteins involved in adhesion to the intestinal mucous, but its genome codes for numerous proteins that may be involved in adhesion to the meat surface (e.g., to collagen), aggregation and biofilm formation. Thus, the bacterium seems well equipped to adhere to and spread on a meat surface (Eijsink & Axelsson, 2005).

Sanz and Toldrá (2002) reported an arginine-specific aminopeptidase activity in *L. sakei* that is important for the release of the free amino acid, since it could be further channeled into the arginine deiminase pathway. The genes encoding the proteins required for arginine catabolism in *L. sakei* are organized in a cluster (Zúriga, Miralles, & Pérez-Martínez, 2002), and their transcription is repressed by glucose and induced by arginine. Arginine, in particular, is an essential amino acid for *L. sakei* and it specifically promotes its growth in meat, being used as an energy source in the absence of glucose (Champomier-Verges et al., 1999). The concentration of free arginine in raw meat is low, although it is relatively abundant in muscle myofibrillar proteins. Moreover, the genome analysis has shown that *L. sakei* harbors a second putative arginine deiminase pathway, containing two peptydil-arginine deaminases, enzymes that can contribute to the metabolism of arginine residues not released from the peptides (Chaillou et al., 2005), thus increasing the competitiveness of *L. sakei* in a meat environment.

Acid Production in Meat Products

Sugars (glucose and occasionally lactose or saccharose) are usually included for the industrial manufacture of fermented meat products, though in Spain, chorizo is traditionally manufactured with little or no added sugar. During fermentation and ripening, LAB converts glucose (their primary energy source) to lactic acid, which is the main component responsible for the pH decrease. This acidification has a preservative effect due to inhibition of pathogenic and spoilage bacteria that are little resistant to low pH, and it contributes to the development of the typical organoleptic characteristics of the fermented sausages (Bover-Cid, Izquierdo-Pulido, & Vidal-Carou, 2001). Although it is well established that fermentable carbohydrates have an influence on flavor, texture and yield of the fermented sausages, carbohydrates for use in dry sausages formulations are generally chosen to ensure an adequate initial drop in meat pH (Bacus, 1984; Lücke, 1985) for preservation reasons, and less importance is given to the product texture. The level of acidification used depends, however, on the desired sensory properties of the product; lactic acid taste is preferred in northern European sausages, but not in typical southern European fermented meats. The pH value during ripening is influenced by the meat fermentation technology, as are the growth dynamics of *Lactobacillus* and *Staphylococcus* (Fig. 5.2). Figure 5.2a shows the growth kinetics of *Lactobacillus* and *Staphylococcus* and the pH during fermentation of an artisanal sausage without addition of starter cultures (Fontana, Cocconcelli, & Vignolo, 2005). A more acidic product can be obtained using a *Lactobacillus* starter culture by adding more carbohydrates to the sausage matrix (Fig. 5.2b), while less acidic products are obtained using a lower concentration of glucose and by adding starter cultures were *Staphylococcus* are predominant (Fig. 5.2c). The increase of pH in the latter stages of fermentation is related to ATP and amino acid metabolism.

Acidification could also be the result of alternative pathways. In *L. sakei*, the presence of genes involved in the energetic catabolism of nucleosides, such as adenosine and inosine, is an example of the adaptation of this organism to the meat environment. Glucose, the favorite carbon source of *L. sakei*, is rapidly consumed in meat, while adenosine and inosine are abundant, reaching twice the concentration of glucose. In addition (as shown in Fig. 5.1), *L. sakei* harbors genes coding for adenosine deaminase, inosine hydrolase, and nucleoside phosphorylase, all of which enable the release of a ribose moiety from the nucleoside (adenosine and inosine) and its subsequent metabolism (Chaillou et al., 2005). Moreover, the presence of methylglyoxal synthase, a novel genetic trait in LAB, has been proposed as a pathway to counteract frequent glucose starvation and modulate metabolism of alternative carbon sources (Chaillou et al., 2005).

Catalase Activity

The metabolism of most LAB, such as the adventitious lactobacilli that contaminate raw meat, could lead to the formation of hydrogen peroxide, a compound that interferes with the sensory properties of meat products, being involved in discoloration of nitroso- heme-pigments and lipid oxidation. Bacterial strains used in meat cultures can produce catalase, an antioxidant enzyme that disproportionates hydrogen peroxide to oxygen and water, preventing the risk of a reduced quality of the fermented meat. Thus, catalase production is considered a relevant technological property of the starter cultures for fermented meat products (Leroy, Verluyten, & De Vuyst, 2006). Production of this antioxidant enzyme is a common trait in aerobic bacteria, such as CNS. The characterization of catalase and superoxide dismutases in *S. carnosus* and *S. xylosus* have been reported. The catalase gene, *kat*A, of *S. xylosus* has been studied in detail (Barriere et al., 2001a; Barriere, Leroy-Sétrin, & Talon, 2001b; Barriere, Bruckner, Centeno, & Talon, 2002). Transcriptional activity of this gene is activated and induced upon entry into the stationary phase, by oxygen and hydrogen peroxide. Moreover, a second gene coding for heme-dependent catalase has been detected in *S. xylosus*.

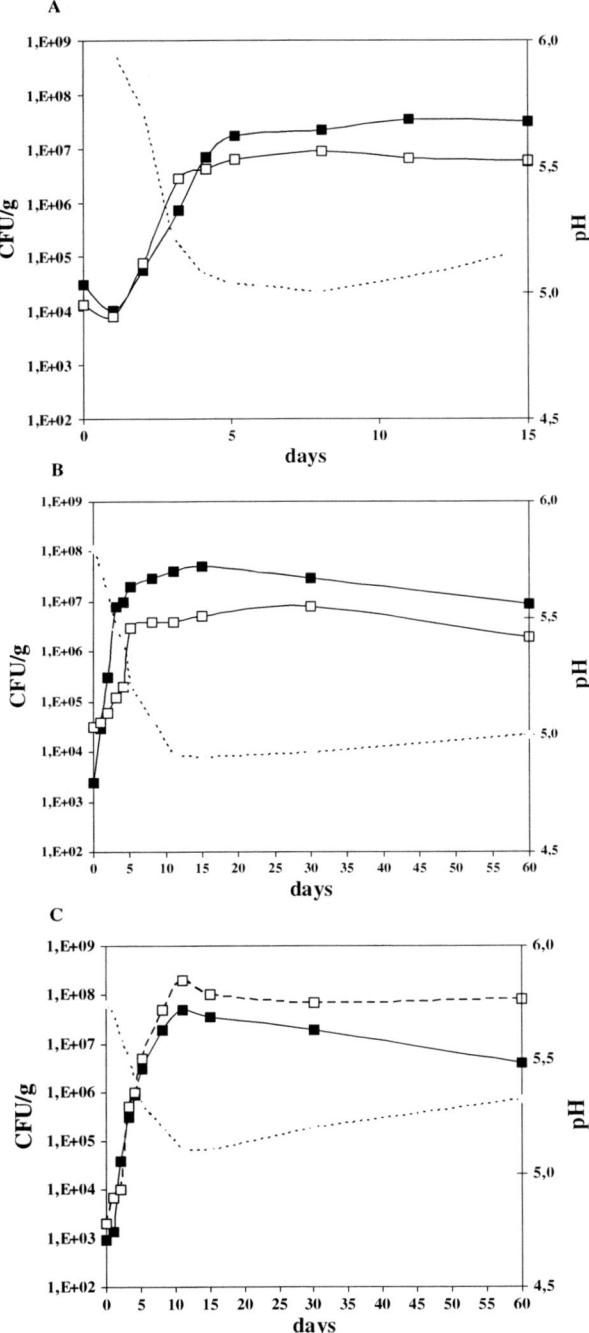

Fig. 5.2 *Lactobacillus* (■) and *Staphylococcus* (□) population growth dynamics and pH decline (- - -) in different kinds of fermentation process. a) without added starter cultures, spontaneous fermentation; b) *Lactobacillus* starter added; c) *Staphylococcus* starter added

Although LAB have long been considered as catalase negative microorganisms, in the last decade, two groups of catalase activity have been reported in the genera *Lactobacillus*, *Pediococcus*, and *Leuconostoc*. The first group is defined as heme catalase and the second group as non-heme Mn-containing catalase. The presence of a heme-dependent catalase has been demonstrated in *L. plantarum* (Igarashi, Kono, & Tanaka, 1996) and *L. sakei* (Noonpakdeea, Pucharoen, Valyasevi, & Panyim, 1996); it can be active in meat products because these substrates contain abundant heme sources (Hertel, Schmidt, Fischer, Oellers, & Hammes, 1998). Moreover, the analysis of the genome of *L. sakei* revealed that this meat organism harbors systems for the protection against reactive oxygen species, such as Mn dependent SOD and heme-dependent catalase (Chaillou et al., 2005).

Nitrate Reduction

Nitrate is added to fermented sausages for its capacity to fix and obtain the typical color of cured products, rather than for its antimicrobial properties. To be effective, the added nitrate must be reduced to nitrite. Besides contributing to flavor, *Staphylocuccus* and *Kocuria* also have a role for their nitrate reductase and antioxidant activities (Talón, Walter, Chartier, Barriere, & Montel, 1999; Talón, Deliere, & Bertrand, 2002). These microorganisms reduce nitrate to nitrite, which is important for the formation of nitrosylmyoglobin, the compound responsible for the characteristic red color of fermented meats. The nitrate reductase activity is widespread in CNS; it has been detected in *S. xylosus*, *S. carnosus*, *S. epidermidis*, *S. equorum*, *S. lentus*, and *S. simulans* (Talón et al., 1999; Mauriello, Casaburi, Blaiotta, & Villani, 2004). In *S. carnosus*, the molecular genetic determinants for nitrogen regulation, the *nre*ABC genes, were identified and shown to link the nitrate reductase operon (*nar*GHJI) and the putative nitrate transporter gene *nar*T. The data provide evidence for a global regulatory system with oxygen as the effector molecule (Fedtke, Kamps, Krismer, & Gotz, 2002).

Flavor Formation

The flavor and aroma of fermented meats is a combination of several elements, such as batter ingredients, manufacture technology, activity of tissue enzymes, proteases and lipases, and microbial metabolism. The bacterial community contributes to flavor development through carbohydrate dissimilation, peptide metabolism, and lipolysis. LAB produce lactic acid and small amounts of acetic acid, ethanol and acetoin, and together with staphylococci are involved in peptide and amino acid metabolism, while only CNS show an effective metabolism of lipids.

Even though microbial proteolytic activity is generally low in the conditions found in fermented sausages (Kenneally, Fransen, Grau, O'Neill, & Arendt, 1999), a minor, strain-dependent activity may still partly contribute to the initial protein breakdown (Molly et al., 1997; Sanz et al., 1999; Fadda et al., 1999a, 1999b; Fadda, Oliver, & Vignolo, 2002). It has been shown that several *Lactobacillus* spp. exhibit proteolytic activity on porcine muscle myofibrillar and sarcoplasmic proteins. Fadda et al. (2001a) reported the contribution of curing conditions to the generation of hydrophilic peptides and free amino acids by the proteolytic activity of *L. curvatus* CRL 705. Moreover, it has been demonstrated that *L. sakei* plays an important role in amino acid generation (Fadda et al., 1999a, 1999b; Sanz et al., 1999).

To ensure the sensory quality of fermented sausages, the contribution of the proteolytic and lipolytic activities of staphylococci is fundamental. The volatiles so far recognized as being produced by staphylococci are primarily amino acid catabolites, pyruvate metabolites, and methylketones from incomplete β-oxidation of fatty acids (Stahnke, Holck, Jensen, Nilsen, & Zanardi, 2002). In particular, *S. xylosus* and *S. carnosus* modulate the aroma through the conversion of amino acids, particularly the branched-chain amino acids (BCAA: leucine, isoleucine, and valine). The BCAA can be degraded into methyl-branched aldehydes, alcohols, and acids by *S. xylosus* and *S. carnosus* (Beck, Hansen, & Lauritsen, 2002; Larrouture, Ardaillon, Pepin, & Montel, 2000; Vergnais, Masson, Montel, Berdague', & Talon, 1998). Furthermore, addition of the *S. carnosus* starter culture has been shown to decrease the maturation time of Italian dried sausages by more than 2 weeks (Stahnke et al., 2002). Olesen, Strunge Meyer, and Stahnke (2004) reported that curing conditions had a considerable influence on the development of volatile compounds in sausages. In addition, major differences were observed in the development of volatile compounds depending on whether *S. xylosus* or *S. carnosus* were used as starter culture.

Bacteriocin and Biopreservation

In recent years, there has been a considerable increase in studies of natural antimicrobial compounds on and in the food produced by LAB, referred to as bio-protective cultures. Bio-protective cultures may act as starter cultures in food fermentation processes, such as dry sausage manufacturing, or they may protect foods from any detrimental organo-leptic changes.

The ability to produce different antimicrobial compounds, such as bacteriocins and/or low-molecular mass antimicrobial compounds may be one of the critical characteristics for effective competitive exclusion. As mentioned above, one of the main roles of meat LAB starter cultures is the rapid production of organics acids; this inhibits the growth of unwanted biota and enhances product safety and shelf-life. The production of bacteriocins, one of the most promising technological features of starter cultures, is discussed in Chap. 18.

Probiotics

Foods which have health benefits beyond their nutritional content (functional foods), and particularly foods containing probiotics, are products that are growing in popularity. Probiotics are available as dietary supplements or they may be incorporated directly into foods. They are live microorganisms that when administered in adequate amounts confer a health benefit on the host (FAO, 2006); they are added to a variety of foods. Recently, attention has been directed to the use of fermented sausages as a food carrier because these products could contain high numbers of viable LAB. To use probiotics as starter cultures for fermented sausages, in addition to the demonstrated probiotic features (FAO, 2006), other properties are demanded.

Although dairy products are the most commonly used food vehicles for delivery of probiotics, the future of dry fermented sausages in this field has been termed "promising" (Incze, 1998). The probiotic culture should be well-adapted to the conditions of the fermented sausage, in order to dominate in the final product, competing with other bacterial populations from meat and from the starter culture. In addition, the culture should not develop off-flavors in the final product.

The potential for dry fermented sausages to serve as a vehicle for probiotic organisms has been comprehensively reviewed by Työppönen, Petäjä, & Mattila-Sandholm (2003). Most of the studies discussed in this review relied on the fermentative abilities of the probiotic organisms used, so the selection of probiotics was limited to organisms that were capable of fermenting carbohydrate in meat.

Various studies have shown that probiotic organisms have poor survival in fermented foods such as yoghurt, fermented milks, and dry fermented sausages (Kailasapathy & Rybka, 1997; Shah 2000; Shah & Ravula, 2000; Lücke, 2000; Erkkilä, Suihko, Eerola, Petaja, & Mattila-Sandholm, et al., 2001). Dry fermented sausages with their low a_w and pH, plus curing salts and competing organisms, would seem to present a challenging environment for the survival of probiotics during processing. Kearney, Upton, and McLoughlin (1990) was the first to report the use of micro-encapsulation in alginate to protect the starter cultures during meat fermentation. Recently, Muthukumarasamy and Holley (2006) used micro-encapsulation technology as a means to protect a recognized probiotic organism (*L. reuteri*), from the harsh environment during sausage processing. Based on their results the authors suggest that micro-encapsulation may be an option for formulation of fermented meat products with viable heath-promoting bacteria. Another approach for selecting bio-protective and probiotic cultures for use in dry fermented sausages involves the isolation of LAB which possesses acid and bile resistance from finished products (Papamanoli et al., 2003; Pennacchia et al., 2004). This approach requires an extensive study of the isolates for other beneficial properties, such as intestinal colonization potential and inhibitory activity against pathogenic bacteria.

Commercial probiotic cultures, such as strains *L. rhamnosus* GG, *L. rhamnosus* LC-705, *L. rhamnosus* E-97800, and *L. plantarum* E-98098, have been tested as functional starter culture strains in northern European sausage fermentation without negatively affecting the technological or sensory properties, with the exception of *L. rhamnosus* LC- 705 (Erkkilä et al., 2001). Klingberg, Axelsson, Naterstad, Elsser, and Budde (2005) identified *L. plantarum* and *L. pentosus* strains, originating from the dominant NSLAB of fermented meat products, as promising candidates of probiotic meat starter cultures, suitable for the manufacture of the Scandinavian-type fermented sausage.

Safety of Selected Bacterial Starter Cultures

Biogenic Amines

The presence in food of biogenic amines (BA), such as cadaverine, putrescine, spermidine, histamine, phenethylamine, agmatine, and tyramine, is of concern for health because their biological effect can lead to toxicological symptoms, such as pseudo-allergic reactions, histaminic intoxication, and interaction with drugs (Shalaby, 1996). Excessive consumption of these amines could cause nervous, gastric, intestinal, and blood pressure problems (Suzzi & Gardini, 2003). Nowadays, increasing attention is given to BA because of the growing number of consumers who are sensitive to them; in such people the action of amine oxidases, the enzymes involved in the detoxification of these substances, is deficient (Suzzi & Gardini, 2003). High levels of BA, especially tyramine, and also histamine and the diamines putrescine and cadaverine, have been described in fermented sausages (Hernández-Jover, Izquierdo-Pulido, Veciana-Noguéz, Mariné-Font, & Vidal-Carou, 1997a, 1997b; Bover-Cid, Hugas, Izquierdo-Pulido, & Vidal-Carou, 2000a; Bover-Cid, Izquierdo-Pulido, Vidal-Carou, 2000b).

The accumulation of BA in foods requires the presence of amino acids precursors, microorganisms with amino acid decarboxylase activity, and favorable conditions for growth and decarboxylation. The large quantities of protein present and the proteolytic activity found during ripening of meat products provide the precursors for later decarboxylase reactions performed by both the starter cultures and wild micro-biota (Suzzi & Gardini, 2003; Komprda et al., 2004). Many LAB from meat and meat products can decarboxylate amino acids (Bover-Cid & Holzapfel, 1999). Many studies have reported a significant correlation between pH and BA contents, the lowest pH generally being characterized by the highest amine levels (Vandekerckove, 1977; Eitenmiller, Koehler, & Reagan, 1978; Halàsz, BaraÂth, Simon-Sarkadi, & Holzapfel, 1994; Bover-Cid, Schoppen, Izquierdo-Pulido, & Vidal-Carou, 1999; Parente et al., 2001), according to the hypothesis that BA production could be a protective mechanism for micro-organisms against acidic environmental conditions.

The final BA contents in fermented sausages depend on the microbial composition of meat used as raw material, but also on the type and activity of the inoculated starter culture. Most strains of *L. curvatus*, one of the main species used as a starter in sausage production, are associated with high BA production (Bover-Cid & Holzapfel, 1999; Pereira, Crespo, & Romao, 2001).

The use of starter cultures with negative decarboxylase activity was shown to prevent the growth of BA producers and lead to end-products nearly free of BA, as long as the raw material was of sufficient quality. Several papers have reported on the ability of a selected starter culture (*L. sakei* CTC494) to greatly reduce BA accumulation in fermented sausages (Bover-Cid et al., 2001; González-Fernández, Santos, Jaime, & Rovira, 2003). This negative-decarboxylate strain, can decrease the pH quickly during the fermentation step and be predominant throughout the process, thus preventing the growth of bacteria that can produce BA.

The introduction of starter strains that possess amine oxidase activity might be a way of further decreasing the amount of BA produced during meat fermentation (Martuscelli, Crudele, Gardini, & Suzzi, 2000; Fadda et al., 2001b; Gardini, Matruscelli, Crudele, Paparella, & Suzzi, 2002; Suzzi & Gardini, 2003).

Antibiotic Resistance

The safety of bacterial strains intentionally added to foods, such as starter cultures used for meat products, is becoming an issue for their application in food. Although meat starter cultures have a long history of apparent safe use, safety concerns can be associated with LAB and, more frequently, with CNS. A risk factor potentially associated with all bacterial groups used as starter cultures for sausage is the presence of acquired genes for antimicrobial resistance.

The food chain has been recognized as one of the main routes for the transmission of antibiotic resistant bacteria between animal and human populations (Witte, 2000). The European Food Safety Authority has recently concluded that bacteria used for or in feed production might pose a risk to human and animal health because of carrying acquired resistance genes (EFSA, 2007). Fermented meats that are not heat-treated before consumption provide a vehicle for such bacteria, and can act as a direct link between the indigenous micro-flora of animals and the human GIT.

Several studies have reported antibiotic resistance in LAB from meats and meat products; a few strains involved in sausage fermentation such as *L. sakei*, *L. curvatus*, and *L. plantarum* have been found to show such resistance (Gevers et al., 2003; Holley & Blaszyk, 1997; Teuber & Perreten, 2000). Some genetic determinants such as chloramphenicol acetyltransferase (*cat*-TC), erythromycin *erm*B, and tetracycline, *tet*L, *tet*M, *tet*S resistance genes have been identified, suggesting that horizontal gene transfer may have occurred (Ahn, Collins-Thompson, Duncan, & Stiles, 1992; Gevers et al., 2003; Lin, Fung, Wu, & Chung, 1996; Tannock et al., 1994; Ammor et al., 2008)

Antimicrobial resistance in CNS has been studied in detail due to its clinical relevance. These bacteria display a high prevalence of antibiotic resistance (Agvald-Ohman, Lund, & Edlun, 2004), and can constitute reservoirs of antibiotic resistance genes that can be transferred to other staphylococci (Wielders et al., 2001). Antibiotic resistant strains were found in food (Gardini, Tofalo, & Suzzi, 2003; Martin et al., 2006) and genes for antimicrobial resistance to tetracycline, *tet*(M), *tet*(K), to erythromycin, *erm*B and *erm*C, and to β-lactams (*blaZ* and mecA), has been detected in the CNS isolated from fermented meat. Moreover, *S. xylosus* strains, isolated from poultry infections, were found to be resistant to streptogramins, harboring the *vatB* and the *vgaB* genes.

An additional concern is that, even in the absence of selective pressure, mobile genetic elements carrying antibiotic resistance can be transferred at a high frequency among the microbial community during sausage fermentation (Vogel, Becke-Schmid, Entgens, Gaier, & Hammes, 1992; Cocconcelli, Cattivelli, & Gazzola, 2003).

Toxigenic Potential

Some members of the CNS group, primarily *S. epidermidis*, are common nosocomial pathogens and the presence of regulatory elements, involved in the control of the virulence factor synthesis, have been recently identified. Remarkably, strains of *S. xylosus* were isolated from patients, who had an underlying disease. The same species has been reported to be involved in infections of poultry (Aarestrup et al., 2000).

Although CNS of food origin have not been found to produce nosocomial infections, some strains that produced enterotoxins have been described (Balaban and Rasooly, 2000). Vernozy-Rozand et al. (1996) reported enterotoxin E as the most frequent enterotoxin found in *S. equorum* and *S. xylosus*, though Martin et al. (2006), reported that the occurrence of staphylococcal enterotoxin genes in CNS from slightly fermented sausages was rare, detecting only *ent*C in *S. epidermidis*. The coagulase positive species, *S. intermedius*, and *S. aureus*, have been more frequently related to staphylococcal intoxication till date. Moreover, *S. intermedius* are opportunistic pathogens in implant infections, due to their ability to form a bio-film on prosthetic materials.

The *S. xylosus* and *S, carnosus* strains, currently used as starter cultures or isolated from fermented meat products, generally lack of toxic genes. Absence of genes coding for staphylococcal enterotoxins or enterotoxin like superantigens is a qualification required for the CNS strains selected for use as starter culture.

Conclusion

The performance of a selected starter culture should be seen in the context of its application, since functionality will depend on the type of sausage, the technology applied, the ripening time, and the ingredients and raw materials used. Besides, the focus on functional characteristics of potential new starter cultures, negative aspects such as antibiotic resistance, virulence genes, and undesirable metabolite formation should not be overlooked. The control of antibiotic resistance, for instance, is a topic of importance because the high load of endogenous bacteria in raw meat material and the inoculation with starters may represent a problem with regard to the spreading of antibiotic resistance.

References

Aarestrup, F. M., Agersø, Y., Ahrens, P., Østergaard Jørgensen, J. C., Madsen, M., & Bogø Jensen, L. (2000). Antimicrobial susceptibility and presence of resistance genes in staphylococci from poultry. *Veterinary Microbiology, 74*, 353–364

Agvald-Ohman, C., Lund, B., & Edlun, C. (2004). Multiresistant coagulase-negative staphylococci disseminate frequently between intubated patient into a multidisciplinary intensive care unit. *Critical Care, 8*, 42–47.

Ahn, C., Collins-Thompson, D., Duncan, C., & Stiles, M. E. (1992). Mobilization and location of the genetic determinant of chloramphenicol resistance from *Lactobacillus plantarum* caTC2R. *Plasmid, 27*, 169–176.

Ammor, M. S., Gueimonde, M., Danielsen, M., Zagorec, M., van Hoek, A. H., de Los Reyes-Gavilán, C. G., et al. (2008). Two different tetracycline resistance mechanisms, plasmid-carried *tet*(L) and chromosomally-encoded transposon-associated *tet*(M), coexist in *Lactobacillus sakei* Rits 9. *Applied Environmental Microbiology* (Epub ahead of print).

Axelsson, L. (2004). Lactic acid bacteria: Classification and physiology. In S. Salminen, A. Ouwehand, & A. Von Wright (Eds.), *Lactic acid bacteria: Microbiology and functional aspects* (3rd ed.). New York: Marcel Dekker, Inc.

Bacus, J. (1984). Utilization of microorganisms in meat processing: A handbook for meat plant operators. Research Studies, Letchworth (UK).

Balaban, N., & Rasooly, A. (2000). Staphylococcal enterotoxins. *International Journal of Food Microbiology, 61*, 1–10.

Bannerman, T. L. (2003). *Staphylococcus, Micrococcus,* and other catalase-positive cocci that grow aerobically. In P. R. Murray, E. J. Baron, J. H. Jorgensen, M. A. Pfaller, R. H. Yolken (Eds.), *Manual of Clinical Microbiology* (Vol. 5, pp. 384–404). Washington: American Society Microbiology.

Barriere, C., Bruckner, R., Centeno, D., & Talon, R. (2002). Characterization of the *kat*A gene encoding a catalase and evidence for at least a second catalase activity in *Staphylococcus xylosus*, bacteria used in food fermentation. *FEMS Microbiology Letters, 216*(2), 277–283.

Barriere, C., Centeno, D., Lebert, A., Leroy-Setrin, S., Berdague', J., & Talon, R. (2001a). Roles of superoxide dismutase and catalase of *Staphylococcus xylosus* in the inhibition of linoleic acid oxidation. *FEMS Microbiology Letters, 201*, 181–185.

Barriere, C., Leroy-Sétrin, S., & Talon, R. (2001b). Characterization of catalase and superoxide dismutase in *Staphylococcus carnosus* 833 strain. *Journal of Applied Microbiology, 91*, 514–519.

Beck, H., Hansen, A., & Lauritsen, F. (2002). Metabolite production and kinetics of branched-chain aldehyde oxidation in *Staphylococcus xylosus*. *Enzyme and Microbial Technology, 31*, 94–101.

Bover-Cid, S., Izquierdo-Pulido, M., & Vidal-Carou, M. (2001). Effectiveness of a *Lactobacillus sakei* starter culture in the reduction of biogenic amine accumulation as a function of the raw material quality. *Journal of Food Protection, 64*, 367–373.

Bover-Cid, S., Hugas, M., Izquierdo-Pulido, M., & Vidal-Carou, M. (2000a). Reduction of biogenic amine formation using a negative amino acid-decarboxylase starter culture for fermentation of "Fuet" sausages. *Journal of Food Protection, 63*, 237–243.

Bover-Cid, S., Izquierdo-Pulido, M., & Vidal-Carou, M. (2000b). Mixed starter cultures to control biogenic amine production in dry fermented sausages. *Journal of Food Protection, 63*, 1556–1562.

Bover-Cid, S., & Holzapfel, W. H. (1999). Improved screening procedure for biogenic amine production by lactic acid bacteria. *International Journal of Food Microbiology, 53*, 33–41.

Bover-Cid, S., Schoppen, S., Izquierdo-Pulido, M., & Vidal-Carou, M. C. (1999). Relationship between biogenic amine contents and the size of dry fermented sausages. *Meat Science 51*, 305–311.

Buckenhüskes, H. J. (1994). Bacterial starter cultures for fermented sausages. *Meat Focus International, 12*, 497–500.

Chaillou, S., Champomier-Verges, M. C., Cornet, M., Crutz-Le Coq, A. M., Dudez, A. M., Martin, V., et al. (2005). The complete genome sequence of the meat-borne lactic acid bacterium *Lactobacillus sakei* 23K. *Nature Biotechnology, 23*, 1527–1533.

Champomier-Verges, M. C., Marceau, A., Mera, T., & Zagorec, M. (2002). The *pep*R gene of *Lactobacillus sakei* is positively regulated by anaerobiosis at the transcriptional level. *Applied and Environmental Microbiology, 68*(8), 3873–3877.

Champomier-Verges, M. C., Zúñiga, M, Morel-Deville, F., Pérez-Martınez, G., Zagorec, M., & Ehrlich, S. D. (1999). Relationships between arginine degradation, pH and survival in *Lactobacillus sakei*. *FEMS Microbiology Letters, 180*, 297–304.

Cocconcelli, P. S., Cattivelli, D., & Gazzola, S. (2003). Gene transfer of vancomycin and tetracycline resistances among *Enterococcus faecalis* during cheese and sausage fermentation. *International Journal of Food Microbiology, 88*, 315–323.

Dordet-Frisoni, E., Talon R., &. Leroy, S. (2007). Physical and genetic map of the *Staphylococcus xylosus* C2a chromosome. *FEMS Microbiology Letters, 266*(2), 184–193.
EFSA The EFSA Journal. (2007). 587, 1–16 © European Food Safety Authority, 2007 Introduction of a Qualified Presumption of Safety (QPS) approach for assessment of selected microorganisms referred to EFSA.
Eijsink, V., & Axelsson, L. (2005). Bacterial lessons in sausages making. *Nature Biotechnology, 23*(12), 1494–1495.
Eitenmiller, R. R., Koehler, P. E., & Reagan, J. O. (1978). Tyramine in fermented sausages: factors affecting formation of tyramine and tyrosine decarboxylase. *Journal of Food Science, 43*, 689–693.
Erkkilä, S., Suihko, M. L., Eerola, S., Petaja, E., & Mattila-Sandholm, T. (2001). Dry sausage fermented by *Lactobacillus rhamnosus* strains. *International Journal of Food Microbiology, 64*, 205–210.
Fadda, S., Oliver, G., & Vignolo, G. (2002). Protein degradation by *Lactobacillus plantarum* and *Lactobacillus casei* in a sausage model system. *Journal of Food Science, 67*, 1179–1183.
Fadda, S., Vignolo, G., Aristoy, M. C., Oliver, G., & Toldrá, F. (2001a). Effect of curing conditions and *Lactobacillus casei* CRL705 on the hydrolysis of meat proteins. *Journal of Applied Microbiology, 91*(3), 478–487.
Fadda, S., Vignolo, G., & Oliver, G. (2001b). Tyramine degradation and tyramine/histamine production by lactic acid bacteria and *Kocuria* strains. *Biotechnology Letters, 23*, 2015–2019.
Fadda, S., Sanz, Y., Vignolo, G., Aristoy, M. C., Oliver, G., & Toldrà, F. (1999a). Hydrolysis of pork muscle sarcoplasmic proteins by *Lactobacillus curvatus* and *Lactobacillus sakei*. *Applied and Environmental Microbiology, 65*, 578–584.
Fadda, S., Sanz, Y., Vignolo, G., Aristoy, M. C., Oliver, G., & Toldra, F. (1999b). Characterization of pork muscle protein hydrolysis caused by *Lactobacillus plantarum*. *Applied and Environmental Microbiology, 65*, 3540–3546.
FAO. (2006). Probiotics in Food. Health and nutritional properties and guidelines for evaluation. FAO food and nutrition paper. 85.
Fedtke, I., Kamps, A., Krismer, B., & Gotz, F. (2002). The nitrate reductase and nitrite reductase operons and the *nar*T gene of *Staphylococcus carnosus* are positively controlled by the novel two-component system NreBC. *Journal of Bacteriology, 184*(23), 6624–6634.
Fontana, C., Cocconcelli, P. S., & Vignolo, G. (2005). Monitoring the bacterial population dynamics during fermentation of artisanal Argentinean sausages. *International Journal of Food Microbiology, 103*, 131–142.
Gardini, F., Tofalo, R., & Suzzi, G. (2003). A survey of antibiotic resistance in *Micrococcaceae* isolated from Italian dry fermented sausages. *Journal of Food Protection, 66*, 937–945.
Gardini, F., Matruscelli, M., Crudele, M. A., Paparella, A., & Suzzi, G. (2002). Use of *Staphylococcus xylosus* as a starter culture in dried sausages: effect on biogenic amine content. *Meat Science, 61*, 275–283.
Garrity, G. M., Bell, J. A. & Lilburn, T. G. (2004). Taxonomic Outline of the Prokaryotes.In Bergey's Manual of Systematic Bacteriology, 2nd edn. Release 5.0, May 2004. Springer-Verlag, NY.
Gevers, D., Masco, L., Baert, L., Huys, G., Debevere, J., & Swings, J. (2003). Prevalence and diversity of tetracycline resistant lactic acid bacteria and their *tet* genes along the process line of fermented dry sausages. *Systematic and Applied Microbiology, 26*(2), 277–283.
Ghebremedhin, B., Layer, F., König, W. & König, B. (2008). Genetic classification and distinguishing of *Staphylococcus* species based on different partial gene sequences: *gap*, 16S rRNA, *hsp*60, *rpo*B, *sod*A, and *tuf* gene. *Journal of Clinical Microbiology* (In press).
González-Fernández, C., Santos, E., Jaime, I., & Rovira, J. (2003). Influence of starter cultures and sugar concentrations on biogenic amine contents in chorizo dry sausage. *Food Microbiology, 20*, 275–284.
Halàsz, A., BaraÂth, A., Simon-Sarkadi, L. & Holzapfel, W. (1994). Biogenic amines and their production by micro-organisms in food. *Trends in Food Science and Technology* 5, 42±48.

Hammes, W. P., Bantleon, A., & Min, S. (1990). Lactic acid bacteria in meat fermentation. *FEMS Microbiology Letters, 87*, 165–173.
Hernández-Jover, T., Izquierdo-Pulido, M., Veciana-Noguéz, M., Mariné-Font, A., & Vidal-Carou, M. (1997a). Biogenic amine and polyamine contents in meat and meat products. *Journal of Agricultural Food Chemistry, 45*, 2098–2102.
Hernández-Jover, T., Izquierdo-Pulido, M., Veciana-Noguéz, M., Mariné-Font, A., & Vidal-Carou, M. (1997b). Effect of starters cultures on biogenic amine formation during sausages production. *Journal of Food Protection, 60*, 825–830.
Hertel, C., Schmidt, G., Fischer, M., Oellers, K., & Hammes, W. P. (1998). Oxygen-dependent regulation of the expression of the catalase gene katA of *Lactobacillus sakei* LTH677 *Applied and Environmental Microbiology, 64*(4), 1359–1365.
Holley, R. A., & Blaszyk, M. (1997). Antibiotic challenge of meat starter cultures and effects upon fermentations. *Food Research International, 30*, 513–522.
Igarashi, T., Kono, Y., & Tanaka, K. (1996). Molecular cloning of manganese catalase from *Lactobacillus plantarum*. *Journal of Biological Chemistry, 271*, 29521–29524.
Incze, K. (1998). Dry fermented sausages. *Meat Science* 49 (Suppl 1) S169–S177.
Kailasapathy, K., & Rybka, S. (1997). *L. acidophilus* and *Bifidobacterium* spp.—their therapeutic potential and survival in yogurt. *The Australian Journal of Dairy Technology, 52*, 28–33.
Kearney, L., Upton, M., & McLoughlin, A. (1990). Meat fermentations with immobilized lactic acid bacteria. *Applied Microbiology and Biotechnology, 33*, 648–651.
Kenneally, M., Fransen, G., Grau, H., O'Neill, E., & Arendt, K. (1999). Effects of environmental conditions on microbial proteolysis in a pork myofibril model system. *Journal of Applied Microbiology, 87*, 794–803.
Kleerebezem, M., Boekhorst, J., van Kranenburg, R., Molenaar, D., Kuipers, O. P., Leer, R., et al. (2003). Complete genome sequence of *Lactobacillus plantarum* WCFS1. *Proceedings of the National Academy of Sciences*, U S A. 100, 1990–1995.
Klingberg, D., Axelsson, L., Naterstad, K., Elsser, D., & Budde, B. (2005). Identification of potential probiotic starter cultures for Scandinavian-type fermented sausages. *International Journal of Food Microbiology, 105*, 419–431.
Komprda, T., Smêlá, D., Pechova', P., Kalhotka, L., Stencl, J., & Klejdus, B. (2004). Effect of starter culture, spice mix and storage time and temperature on biogenic amine content of dry fermented sausages. *Meat Science* 67, 607–616.
Kuipers, P., Buist, G., & Kok, J. (2000). Current strategies for improving food bacteria. *Research in Microbiology, 151*, 815–822.
Kwok, A. Y., & Chow, A. W. (2003). Phylogenetic study of *Staphylococcus* and *Macrococcus* species based on partial *hsp*60 gene sequences. *International Journal of Systematic Evolutionary Microbiology, 53*, 87–92.
Larrouture, C., Ardaillon, V., Pepin, M., & Montel, C. (2000). Ability of meat starter cultures to catabolize leucine and evaluation of the degradation products by using an HPLC method. *Food Microbiology, 17*, 563–570.
Leroy, F., Verluyten, J., & De Vuyst, L. (2006). Functional meat starter cultures for improved sausage fermentation. *International Journal of Food Microbiology, 106*, 270–285.
Lin, C. F., Fung, Z. F., Wu, C. L., & Chung, T. C. (1996). Molecular characterization of a plasmid-borne (pTC82) chloramphenicol resistance determinant (cat-TC) from *Lactobacillus reuteri* G4. *Plasmid, 36*, 116–124.
Lücke, K. (2000). Utilization of microbes to process and preserve meat. *Meat Science, 56*, 105–115.
Lücke, K. (1985). Fermented sausages. In B. J. Wood (Ed.), *Microbiology of fermented foods* (pp. 41–83). London: Elsevier.
Martuscelli, M., Crudele, A., Gardini, F., & Suzzi, G. (2000). Biogenic amine formation and oxidation by *Staphylococcus xylosus* from artisanal fermented sausages. *Letters and Applied Microbiology, 31*, 228–232.
Mauriello, G., Casaburi, A., Blaiotta, G., & Villani, F. (2004). Isolation and technological properties of coagulase negative staphylococci from fermented sausages of Southern Italy. *Meat Science, 67*, 49–158.

Molly, K., Demeyer, D., Johansson, G., Raemaekers, M., Ghistelinck, M., & Geenen, I. (1997). The importance of meat enzymes in ripening and flavour generation in dry fermented sausages. First results of a European project *Food Chemistry, 59*(4), 539–545.

Muthukumarasamy, P., & Holley, R. (2006). Microbiological and sensory quality of dry fermented sausages containing alginate-microencapsulated *Lactobacillus reuteri*. *International Journal of Food Microbiology, 111,* 164–169

Niinivaara, F. (1955). Über den Einfluss von Bacterienreinkulturen auf die Reifung und Umrötung der Rohwurst. In *Acta Agralia Fennica, 84,* 1–128.

Noonpakdeea, W., Pucharoen, K., Valyasevi, R., & Panyim, S. (1996). Molecular cloning, DNA sequencing and expression of catalase gene of *Lactobacillus sake* SR911. *Journal of Molecular Biology and Biotechnology, 4,* 229–235.

Olesen, P., Strunge Meyer, A., & Stahnke, L. (2004). Generation of flavour compounds in fermented sausages—the influence of curing ingredients, *Staphylococcus* starter culture and ripening time. *Meat Science, 66,* 675–687.

Papamanoli, E., Tzanetakis, N., Litopoulou-Tzanetaki, E., & Kotzekidou, P. (2003). Characterization of lactic acid bacteria isolated from a Greek dry fermented sausage in respect of their technological and probiotic properties. *Meat Science, 65,* 859–867.

Parente, E., Martuscelli, M., Gardini, F., Grieco, S., Crudele, M. A., & Suzzi, G. (2001). Evolution of microbial populations and biogenic amine production in dry sausages produced in Southern Italy. *Journal of Applied Microbiology, 90* (6), 882–891.

Pennacchia, C., Ercolini, D., Blaiotta, G., Pepe, O., Mauriello, G., & Villani, F. (2004). Selection of *Lactobacillus* strains from fermented sausages for their potential use as probiotics. *Meat Science, 67,* 309–317.

Pereira, C. I., Barreto Crespo, M. T., & San Romão, M. V. (2001). Evidence for proteolytic activity and biogenic amines production in Lactobacillus curvatus and L. homohiochii. *International Journal of Food Microbiology 68*(3), 211–216.

Place, R., Hiestand, D., Gallmann, H. R., & Teuber, M. (2003). *Staphylococcus equorum* subsp. *linens,* subsp. *nov.,* a starter culture component for surface ripened semi-hard cheese. *Systematic Applied Microbiology, 26,* 30–37.

Planchon, S., Chambon, C., Desvaux, M., Chafsey, I., Leroy, S., Talon, R., et al. (2007). Proteomic analysis of cell envelope proteins from *Staphylococcus xylosus* C2a. *Journal of Proteomics Research,* Submitted.

Planchon, S. (2006). Aptitude de *Staphylococcus carnosus* et *Staphylococcus xylosus* à former des biofilms- Etude d'une souche biofilm positif par une approuche protèomique. These de Docteur de l'Universitè Blaise Pascal, Clemont Ferrand II.

Rosenstein, R., Nerz, C., Resch, A., & Götz, F. (2005). Comparative genome analysis of staphylococcal species. *2nd European Conference on prokaryotic genomes.* Prokagen.

Sanz, Y., & Toldrá, F. (2002). Purification and characterization of an arginine aminopeptidase from *Lactobacillus sakei. Applied and Environmental Microbiology, 68*(4), 1980–1987.

Sanz, Y., Fadda, S., Vignolo, G., Aristoy, C., Oliver, G., & Toldrá, F. (1999). Hydrolysis of muscle myofibrillar proteins by *Lactobacillus curvatus* and *Lactobacillus sakei*. *International Journal of Food Microbiology, 53,* 115–125.

Shah, N. P. (2000). Probiotic bacteria: Selective enumeration and survival in dairy foods. *Journal of Dairy Science, 83*(4), 894–907.

Shah, N. P., & Ravula, R. (2000). Microencapsulation of probiotic bacteria and their survival in frozen fermented dairy desserts. *Australian Journal of Dairy Technology, 55*(3), 139–144.

Shalaby, A. (1996). Significance of biogenic amines to food safety and human health. *Food Research International, 29*(7), 675–690.

Spergser, J., Wieser, M., Taubel, M., Rossello-Mora, R. A., Rosengarten, R., & Busse, H. J. (2003). *Staphylococcus nepalensis* sp. *nov.,* isolated from goats of the Himalayan region. *International Journal of Systematic Evolutionary Microbiology, 53,* 2007–2011.

Stahnke, H., Holck, A., Jensen, A., Nilsen, A., & Zanardi,. E. (2002). Maturity acceleration of Italian dried sausage by *Staphylococcus carnosus*—relationship between maturity and flavor compounds. *Journal of Food Science, 67,* 1914–1921.

Suzzi, G., & Gardini, F. (2003). Biogenic amines in dry fermented sausages: a review. *International Journal of Food Microbiology, 88*, 41–54.

Talón, D., Deliere, E., & Bertrand, X. (2002). Characterization of methicillin-resistant Staphylococcus aureus strains susceptible to tobramycin. *International Journal of Antimicrobial Agents, 20*(3), 174–179.

Talón, R., Walter, D., Chartier, S., Barriere, C., & Montel, M. (1999). Effect of nitrate and incubation conditions on the production of catalase and nitrate reductase by staphylococci. *International Journal of Food Microbiology, 52*, 47–50.

Tannock, G. W., Luchansky, J. B., Miller, L., Connell, H., Thode-Andersen, S., & Mercer, A. (1994). Molecular characterization of a plasmid-borne (pGT633) erythromycin resistance determinant (*erm*GT) from *Lactobacillus reuteri* 100–63. *Plasmid, 31*, 60–71.

Teuber, M., & Perreten, V. (2000). Role of milk and meat products as vehicles for antibiotic-resistant bacteria. *Acta Veterinaria Scandinavica*, (Suppl. 93), discussion 111–7, 75–87.

Työppönen, S., Petäjä, E., & Mattila-Sandholm, T. (2003). Bioprotectives and probiotics for dry sausages. *International Journal of Food Microbiology, 83*, 233–244.

Vandekerckove, P. (1977). Amines in dry fermented sausage: a research note. *Journal of Food Science, 42*, 283–285.

Vergnais, L., Masson, F., Montel, M. C., Berdague', J. L., & Talon, R. (1998). Evaluation of solid-phase microextraction for analysis of volatile metabolites produced by staphylococci. *Journal of Agricultural and Food Chemistry, 46*, 228–234.

Vernozy-Rozand, C., Mazuy, C., Prevost, G., Lapeyre, C., Bes, M., Brun, Y., et al. (1996). Enterotoxin production by coagulase-negative staphylococci isolated from goat's milk and cheese. *International Journal of Food Microbiology, 30*(3), 271–280.

Vogel, R. F., Becke-Schmid, M., Entgens, P., Gaier, W., & Hammes, W. (1992). Plasmid transfer and segregation in *Lactobacillus curvatus* LTH1432 in vitro and during sausage fermentations. *Systematic and Applied Microbiology, 15*, 129–136.

Wagner, E., Doskar, J., & Gotz, F. (1998). Physical and genetic map of the genome of *Staphylococcus carnosus* TM300. *Microbiology, 144*, 509–517.

Wielders, C., Vriens, M., Brisse, S., de Graaf-Miltenburg, L., Troelstra, A., Fleer, A., Schmitz, F., Verhoof, J., & Fluit, A. (2001). Evidence for *in vivo* transfer of *mec*A DNA between strains of *Staphylococcus aureus. Lancet, 357*, 1674–1675.

Witte, W. (2000). Selective pressure by antibiotic use in livestock. *International Journal of Antimicrobial Agents, 16*, S19–S24.

Zúríga, M., Miralles, M., & Pérez- Martínez, G. (2002). The product of *arc*R, the sixth gene of the *arc* peron of *Lactobacillus sakei*, is essential for expression of the arginine deiminase pathway. *Applied Environmental Microbiology, 68*(12), 6051–6058.

Chapter 6
Genetics of Lactic Acid Bacteria

Monique Zagorec, Jamila Anba-Mondoloni, Anne-Marie Crutz-Le Coq, and Marie-Christine Champomier-Vergès

Introduction

Many meat (or fish) products, obtained by the fermentation of meat originating from various animals by the flora that naturally contaminates it, are part of the human diet since millenaries. Historically, the use of bacteria as starters for the fermentation of meat, to produce dry sausages, was thus performed empirically through the endogenous micro-biota, then, by a volunteer addition of starters, often performed by back-slopping, without knowing precisely the microbial species involved. It is only since about 50 years that well defined bacterial cultures have been used as starters for the fermentation of dry sausages. Nowadays, the indigenous micro-biota of fermented meat products is well identified, and the literature is rich of reports on the identification of lactic acid bacteria (LAB) present in many traditional fermented products from various geographical origin, obtained without the addition of commercial starters (See Talon, Leroy, & Lebert, 2007, and references therein). The LAB species that are naturally present in those products and become dominant in the final processing steps essentially belong to *Lactobacillus sakei*, *Lactobacillus curvatus*, and *Lactobacillus plantarum*. These are also the three main species that are sold as starters for the fermentation of dry sausages, essentially in Europe, to which should be added the two other pediococci species *Pediococcus pentosaceus* and *Pediococcus acidilactici* (Hammes & Hertel, 1998).

Since the last 20 years, many microbiologists have investigated the physiology of these LAB, in order to understand the mechanisms by which they contribute to the quality of the final product, and to improve their use. Molecular tools were therefore developed, leading to an increase of the knowledge about their genetics. More recently, the genomes of *L. plantarum* WCFS1 (Kleerebezem, et al, 2003), *L. sakei* 23K (Chaillou, et al., 2005), and *P. pentosaceus* ATCC25745 (Makarova, et al., 2006) were entirely sequenced, giving a general overview on the whole genetic repertoire of those bacteria. However, the description and analysis of all

---

M. Zagorec
Unité Flore Lactique et Environnement Carné, UR309, INRA, Domaine de Vilvert, F-78350 Jouy en Josas, France

the genes composing a genome is a hard and tricky task, and authors usually focus on emblematic characteristics of a species when they aim to describe what are the most important genetic traits encoded by a genome. For instance, the genome of *L. plantarum*, a ubiquitous species found in many fermented products but also present in the gastrointestinal tract of mammals and even described as a putative probiotic, was searched mostly for the functions explaining its ubiquity. The genome of *P. pentosaceus* was compared with that of many other LAB, in order to understand evolution and phylogeny of this vast taxonomic group, and finally the *L. sakei* genome was analyzed in order to understand its adaptation to the meat environment.

In this chapter, we will present an overview of the genetic elements known for these species and we will describe, from the three genomes available, some traits that are required for the development of fermented meat products and that are common to *L. sakei*, *L. plantarum*, and *P. pentosaceus*. We will consider *P. pentosaceus* as belonging to lactobacilli, since its genome analysis confirmed that the pediococci should be included in the lactobacilli genera (Makarova et al., 2006).

Mobile Elements and Plasmids in Meat LAB Starters

These genetic elements can be transferred intra- and inter-species and may largely contribute to bacterial genetic evolution and to the acquisition of new functions that will then influence the ability of the host to adapt or to become more competitive to an ecological niche.

Insertion Sequences and Transposons

Transposable genetic elements are ubiquitous, their presence or absence within a genome can vary between individual strains. Transposable elements have impacts on the host genomes by altering gene expression, providing genomic rearrangements, causing insertional mutations and are sources of phenotypic variation. Moreover, numerous systems of bacterial transferable elements such as plasmids, transposons, integrative and conjugative or mobilizable elements, insertion sequences (IS), and prophages are known to promote horizontal gene transfer events and DNA rearrangements. Many of these elements carry accessory genes that encode various functions such as resistance to antibiotics, virulence, and degradation of xenobiotics (Frost, Leplae, Summers, & Toussaint, 2005). Conjugative elements harbor modules involved in their own conjugative transfer, while mobilizable elements carry a module which includes an origin of transfer and one or two mobilization genes, and require the mating machinery provided by the conjugative elements to be transferred. IS contain genes involved in transposition usually edged by two inverted repeat sequences, that allow insertion and excision processes through site-specific recombination. Transposition, performed by transposons or by bacterial IS may lead to the random mutation of chromosomal genes. Twelve IS are found

6 Genetics of Lactic Acid Bacteria

Table 6.1 Main characteristics of mobile elements in the genomes of LAB meat starters

	Genome size (bp)	Plasmids	Nr of proteins	Nr of prophages	Nr of mobile elements
Lactobacillus sakei 23K	1 884 661	0 (2)*	1886	0 (+1 remnant)	12
Lactobacillus plantarum WCFS1	3 308 274	3	3009	2 (+2 remnant)	15
Pediococcus pentosaceus ATCC25745	1 832 387	0	1757	1	1

* plasmids present in the non cured strain, not yet sequenced.

within the *L. sakei* 23K genome (Chaillou et al., 2005), including five IS*1520*, three IS*Lsa1*, three IS*Lsa2*, and one IS*Lsa3*, belonging to respectively, IS*3*, IS*30*, IS*150*, and IS*4* families (Table 6.1). While only one transposase is present within the genome of *P. pentosaceus*, ATCC 25745 (Makarova et al., 2006), 15 IS are present in *L. plantarum* WCFS1 (Boekhorst et al., 2004). Genetic exchanges between *L. plantarum* and *P. pentosaceus* have been suggested as homologous sucrase genes, sharing 98% nucleic acid identity, were found in the raffinose-sucrose gene clusters that were flanked by IS elements (Naumoff, 2001).

Bacteriophages

Bacteriophages encode genes involved in specific functions like encapsidation, bacterial lysis, recombination, or conjugation that are usually grouped in functional modules. The bacteriophages still cause dramatic damage during fermentation of dairy products.

Despite the increasing number of reports on dairy LAB bacteriophages, information on the meat LAB phages is poorly documented. Very few bacteriophages using *L. sakei*, *P. pentosaceus*, or *L. plantarum* as hosts were isolated from industrial meat fermentation. Basic characteristics of ten virulent phages active on silage-making lactobacilli were investigated (Doi et al., 2003). Morphological properties, host ranges, protein composition, and genome characterization allowed classification into five groups. Morphologically, three phages of group I belonged to the Myoviridae family, while seven other phages of groups II, III, and V belonged to the Siphoviridae family according to the Ackermann classification. The seven phages of groups I, II, and V were active on *L. plantarum* and *L. pentosus*. Phage phiPY4 (group III) infected both *L. casei* and *L. rhamnosus*. Three other *L. plantarum* virulent bacteriophages, phiJL1 (Lu, Altermann, Breidt, Predki, Fleming, & Klaenhammer, 2005; Lu, Breidt, Fleming, Altermann, & Klaenhammer, 2003), phiB2 (Nes, Brendehaug, & von Husby, 1988) and the temperate phage phi g1e (Kakikawa, Yamakawa, et al., 2002; Kakikawa, Yokoi, et al., 2002), have been studied at the morphological, host range, and genome levels. In *L. sakei*

23K, only one remnant prophage is found in the complete genome (Chaillou et al., 2005). One virulent *L. sakei* bacteriophage, PWH2, has been isolated from fermented sausage. Fourteen *L*. sakei strains and five *L. curvatus* strains were tested as potential hosts and only strain *L. sakei* Ls2 was sensitive (Leuschner, Arendt, & Hammes, 1993). These results may explain why no direct proof has ever been shown for a detrimental role of phage attack in meat fermentations.

The PWH2 bacteriophage belongs to the family of Siphoviridae according to the Ackermann classification (Ackermann, 1987). In the genome of *L. plantarum* WCFS1, two entire prophages and two remnants are present (Ventura, et al., 2003). Two prophages were observed in the *P. pentosaceus* ATCC25745 genome (Table 6.1). Little is known about the mechanism of induction or on the effect of their presence in these genomes except for the three prophages of *P. acidilactici*, which can be induced by classical addition of mitomycin C (Caldwell, McMahon, Oberg, & Broadbent, 1999). The new bacteriophages, named pa97, pa40, and pa42 were characterized morphologically and belong to the family Siphoviridae.

Plasmid Content of the Strains

With the rise of molecular biology of the lactobacilli in the late 1980s, their plasmids have become a subject of study, as a means to characterize strains, to search for plasmid-borne functions, or to characterize replicons for potential use as cloning vehicles. Although many plasmids have been isolated, the global plasmid content of lactobacilli has been the object of relatively few and ancient systematic studies (for a review see Wang & Lee, 1997). As other species of lactobacilli, *L. curvatus*, *L. plantarum*, and *L. sakei* have been reported to commonly contain one to about ten plasmids, whatever the (i.e. isolated from/material) vegetable, meat, or fermented origin of the strains (Liu, Kondo, Barnes, & Bartholomew, 1988; Nes, 1984; Ruiz-Barba, Piard, & Jiménez-Diaz, 1991; Vogel, Lohmann, Weller, Hugas, & Hammes, 1991; West & Warner, 1985). *P. pentosaceus* and *P. acidilactici* have also been reported to commonly harbor plasmids (Graham & McKay, 1985; Pérez Pulido et al., 2006). These plasmids are usually circular and their size may range from less than to 2 kb to more than 60 kb. Mega plasmids, larger than 100 kb have been found in *L. plantarum* (Wang & Lee, 1997), although it rather seems to be a specific trait of *L. salivarius* and a few other species of lactobacilli (Li et al., 2007). Linear plasmids among LAB have been described only in *L. salivarius* to date (Li et al., 2007). Some strains have no naturally resident plasmid.

Circular plasmids can be classified according to their mode of replication (rolling-circle replication characterized by an intermediate single-stranded state or theta-type replication) and the need for both host factors and plasmid-borne determinants. The latter comprise at least a segment of DNA called the origin of replication at which replication initiates and may comprise an additional protein called initiator or Rep protein. Usually, newly sequenced plasmids can be assigned to families by sequence homology. A classification of plasmids has been tentatively proposed at the website of the Database of Plasmid Replicon (http://www.essex.ac.uk/bs/staff/osborn/DPR_home.htm).

Several plasmids from the five species mentioned above have been isolated and characterized. Among those, the two main modes of replication, rolling-circle, and theta replication exist. At the time of writing, the Genbank database contains the complete sequence for one plasmid of *L. curvatus*, one plasmid of *L. sakei*, 15 plasmids of *L. plantarum*, two plasmids of *P. pentosaceus*, and two plasmids of *P. acidilactici*. Their characteristics are given in Table 6.2. One obvious conclusion is that plasmids belonging to the same family are distributed across species. A picture of plasmid distribution can also be brought by hybridization experiments, which detects homology relationships. For example, it is known that *L. sakei* and *L. curvatus* strains harbor plasmids related to the pLP1 rolling-circle plasmid, originally isolated from *L. plantarum* (Bringel, Frey, & Hubert, 1989) or that pLC2-type plasmids, originally described in *L. curvatus*, are also detected in *L. sakei* (Vogel et al., 1991).

Another characteristic which was revealed by sequence data is the composite nature of plasmids, made of mosaics of segments or clusters of genes from different origins. Typical examples are given by pMD5057 (Danielsen, 2002) and pRV500 (Alpert, Crutz-Le Coq, Malleret, & Zagorec, 2003).

Plasmid-Encoded Functions

So far, only a few functions, other than those involved in replication or transfer, have been identified on these plasmids. Many cryptic plasmids with no characterized phenotype are known in lactobacilli. Among the functions for which the presence of plasmids correlated with a particular phenotype, we can mention cystein transport in *L. sakei* (Shay, Egan, Wright, & Rogers, 1988), β-galactosidase activity in *L. plantarum* (Mayo, Gonzalez, Arca, & Suarez, 1994), maltose utilization in *L. plantarum* and *Lactobacillus* ssp. isolated from fresh meat (Liu et al., 1988) and utilization of various sugars like raffinose, melibiose, maltose, or sucrose in *P. pentosaceus and P. acidilactici* (Gonzalez & Kunka, 1986; Gonzalez & Kunka, 1987; Halami, Ramesh, & Chandrashekar, 2000). Bacteriocin production, that can be chromosomally encoded, has also been reported to be encoded by plasmids in several species (Giacomini, Squartini, & Nuti, 2000; Gonzalez & Kunka, 1987; Kanatani & Oshimura, 1994; Kantor, Montville, Mett, & Shapira, 1997; Miller, Ray, Steinmetz, Hanekamp, & Ray, 2005; Motlagh, Bukhtiyarova, & Ray, 1994; Osmanagaoglu, Beyatli, & Gunduz, 2000; Simon, Frémaux, Cenatiempo, & Berjeaud, 2002; Skaugen, Abildgaard, & Nes, 1997; Van Reenen, Van Zyl, & Dicks, 2006; Vaughan, Eijsink, & Van Sinderen, 2003).

As seen in Table 6.2, conjugative or mobilizable elements are also commonly encoded by plasmids (Gevers, Huys, & Swings, 2003). They can, in addition, encode antibiotic resistance genes and metabolic functions which can be transferred between strains present in the bacterial population sharing the same ecological

Table 6.2 Main features of sequenced plasmids isolated from meat LAB starters

Isolated from	Name	Size (kb)	Function or genes identified (other than those involved in plasmid maintenance)	Genbank accession	type of replication and replicon family
L. sakei	pRV500	13	Putative type I restriction modification system	NC_004942	theta; pUCL287
L. curvatus	pLC2	2.5	Cryptic	Z14234	"rolling-circle"; pE194 / pWV01
L. plantarum	p256	7.2	Putative prophage protein, putative cold shock protein	NC_006278	theta; without Rep protein
	pWCFS103	36	Arsenical resistance, DNA-damage-inducible gene, plasmid transfer via conjugation	NC_006377	theta; pLS32
	pLKS	2	Phage resistance	AB035265	
	pMD5057	10.9	Tetracycline resistance tet(M) gene	NC_004944	theta; pUCL287
	pLJ42	5.5	Cryptic, mobilization protein	DQ099911	
	pWCFS101	1.9	Cryptic	NC_006375	"rolling-circle"; pC194
	pLP2000	2	Cryptic	NC_003893	pUB110
	pLTK2	2.3	Cryptic	NC_002123	
	pM4	3.3	Cryptic, mobilization protein	NC_009666	
	pLP1	2.1	Cryptic	M31223	
	pWCFS102	2.4	Cryptic, mobilization protein	NC_006376	"rolling-circle"; pE194 / pWV01
	pPB1	3	Cryptic, mobilization protein	NC_006399	
	pLP9000	9.3	Cryptic	NC_003894	"rolling-circle"; unknown
	pLKL	6.8	Phage protein homologs	AB219181	unknown
	pPLA4	8.1	Bacteriocin 423 operon, mobilization protein	AF304384	unknown
P. pentosaceus	pMD136	19.5	Pediocin A production, putative oxyquinone oxydoreductases, putative ABC transporter, mobilization protein	NC_001277	theta; pWV02
	pRS4	3.5	Mobilization protein	AJ968953	"rolling-circle"; pC194 / pUB110
P. acidilactici	pSMB74	9	Pediocin AcH	NC_004832	theta; pUCL287
	pEOC01	11.7	Streptomycin resistance / erythromycin methylase B	DQ220741	theta; pAMβ1

niche. The possibility of the dissemination of antibiotic resistance by meat lactobacilli has been addressed by Gevers, Danielsen, Huys, and Swings, (2003), who showed that tetracycline resistance plasmids could be transferred from *L. sakei* or *L. plantarum* to other Gram-positive bacteria such as *Enterococcus faecalis* and *Lactococcus lactis* by in vitro conjugation (Gevers, Danielsen, et al., 2003; Gevers, Huys, et al., 2003). This suggests that meat lactobacilli might be the reservoir microorganisms for acquired resistance genes that can be spread among other bacteria.

A number of other plasmid-encoded functions were assigned after plasmid sequence determination, such as putative restriction modification systems or putative stress proteins (see Table 6.2). Thus, plasmids may clearly carry a source of genetic diversity and provide an extension of metabolic properties or additional resistance to environmental conditions.

Vectors and Genetic Tools

Geneticists developed various genetic tools in order to construct mutants, either randomly or in targeted genes, to express foreign genes, or to monitor their expression. Such approaches rely on the capability to transfer DNA molecules, therefore efficient transformation protocols have been developed in parallel.

Electroporation, based on transient pore formation in membranes after applying an electrical field, is the method of choice for gene transfer and protocols have been adapted for each species. Depending on the recipient strains used, they give different amount of transformants, the most efficient ones reaching up to 10^4–10^6 transformants per µg of plasmid DNA (Alegre, Rodriguez, & Mesas, 2004; Aukrust & Blom, 1992; Berthier, Zagorec, Champomier-Vergès, Ehrlich, & Morel-Deville, 1996; Luchansky, Muriana, & Klaenhammer, 1988; Rodriguez, Alegre, & Mesas, 2007).

Conjugation is known to naturally occur in lactobacilli but requires conjugative plasmids. It can be made through the use of broad host-range conjugative plasmids isolated from other bacteria such as pAMβ1 or pIP501 (Gonzalez & Kunka, 1983; Langella, Zagorec, Ehrlich, & Morel-Deville, 1996) either alone or as helpers with mobilizable plasmids. Conjugation has proven useful for plasmid transfer into strains that are poorly transformable by electroporation.

Natural competence and transformation were never reported in LAB, although complete genome sequences revealed the presence of genes homologous to bona fide competence genes involved in DNA transport, in other bacterial groups. The question of naturally occurring gene transfer is of particular interest with food-borne organisms. Plasticity of the chromosomes and plasmids has revealed the relative importance of gene transfer especially among bacteria sharing a same biotope. This may then occur by natural, but not yet characterized competence, stress-induced permeation, plasmid mobilization/conjugation, or phage infection. Indeed, conjugative transfers were detected to occur during sausage fermentation with *L. curvatus* (Vogel, Becke-Schmid, Entgens, Gaier, & Hammes, 1992).

Plasmid vectors have been developed for *Lactobacillus* and to a lesser extent for *Pediococcus*, whose molecular biology is still at its beginning. A non-exhaustive list of vectors and methods used in "meat species" is highlighted here. A comprehensive review of the natural plasmids from LAB and derived vectors was recently published by Shareck, Choi, Lee, & Miguez (2004).

Integrative vectors (also called delivery or suicide vectors), that is those that are incapable of self-replication in the recipient host, are used for the construction of chromosomal mutants. The vector part often comes from widely used plasmids of *Escherichia coli* of the colE1-type, like pRV300 (Leloup, Ehrlich, Zagorec, & Morel-Deville, 1997) or pJDC9 (Chen & Morrison, 1988). Chromosomal integration relies on homologous recombination and allows different kinds of mutations to be obtained: (*i*) inactivation of genes after the simple insertion of the recombinant vector, or (*ii*) point mutations (Stentz & Zagorec, 1999) and deletion mutants (Ferain et al., 1996; Leer et al., 1993; Malleret, Lauret, Ehrlich, Morel-Deville, & Zagorec, 1998; Stentz, Loizel, Malleret, & Zagorec, 2000), for which no exogenous DNA persists. The latter strategy of gene replacement usually involves two successive recombination steps; the first one involves selecting for integration of the entire plasmid, the second one consisting of an excision of plasmid sequences between regions duplicated after integration. In the case of high recombination frequency and provided it can be selected or screened for, a double-recombination event may be obtained in one step (Malleret et al., 1998). Conditionally replicative vectors (replicating only in so-called permissive conditions) are considered as the most effective delivery vectors, especially for low transformable strains, because they allow temporal separation of the transformation and integration steps. Some plasmids have been found to be naturally thermo-sensitive at 42°C, and thermo-sensitive vectors have also been developed (see Shareck et al., 2004). However, it should be noted that this character is strongly dependent on the host and that no thermo-sensitive plasmid has been found effective in mesophilic species, such as a number of *L. sakei* strains, which do not grow over 37°C (Gory, Montel, & Zagorec, 2001).

Replicative vectors can be used to easily bring new functions in the cell or to express genes at a high level (expression vectors). Inducible promoters have been used for over-expression (Sørvig, Mathiesen, Naterstad, Eijsink, & Axelsson, 2005) or to modulate functions (Stentz et al., 2000).

Reporter genes are also interesting tools in the molecular biologist's arsenal and comprise *lacZ* (Stentz et al., 2000), *gusA* (Hertel, Schmidt, Fischer, Oellers, & Hammes, 1998) and the Green Fluorescent Protein (GFP) gene (Gory et al., 2001), the activity of the latter being detected by fluorescence, thereby allowing to follow the strains in complex environments. It should be noted that the use of *lacZ* may be hampered by the presence of β-galactosidase activity often existing in a number of lactobacilli.

In a view of food-grade applications, plasmids should not contain DNA originating out of LAB species and in particular should not harbor genes for resistance to antibiotics. In this context, appropriate selective cassettes have been developed; nisin immunity conferred by *nisI* has been used in *L. plantarum* (Takala & Saris, 2002), and selection based on specific metabolic properties could be used provided the encoding genes can complement a deficient strain as described in Bron, et al., (2002); and in Takala, Saris, and Tynkkynen, (2003).

Random mutagenesis via transposition is still to be developed in meat starter species, this approach also requires thermo-sensitive delivery vectors. Genetic tools have been developed in LAB using IS, such as IS946 or IS*AS*1, in combination with non-replicative vectors to efficiently perform site-directed or random mutagenesis by transposition (Maguin, Prévost, Ehrlich, & Gruss, 1996; Romero & Klaenhammer, 1992). No system based on the IS elements present in their genome have yet been developed for *L. sakei*, *P. pentosaceus*, and *L. plantarum*. Attempts for obtaining random transposition were performed in *L. plantarum* with variable success. Whereas Tn*917* delivered by pTV1Ts essentially targeted the naturally harbored plasmids (Cosby, Axelsson, & Dobrogosz, 1989), pGhost9:IS*S*1 was successfully used at 42°C to obtain a set of random transposition mutants (Gury, Barthelmebs, & Cavin, 2004).

Chromosomal Elements

In addition to the genetic elements mentioned above, that largely participate in the genetic repertoire and biodiversity of bacteria, the chromosomally encoded functions represent the large majority of the genetic elements that characterize a bacterial species and its properties.

The chromosomes of *L. plantarum* WCFS1, *L. sakei* 23K, and *P. pentosaceus* ATCC27745 encode respectively 3009, 1886, and 1757 coding sequences (Table 6.1), showing a large difference between those species. Besides housekeeping functions, the genetic repertoire of those bacteria shows specific features that are common to the three species, and absent from other closely related bacteria (lactobacilli) which do not usually develop in meat products. One can thus hypothesize that these specific features are important for the adaptation to the fermented meat environment. They mostly allow bacteria to grow in meat as a substrate, and to resist the stressing environment that the steps of meat fermentation represent, for instance the presence of high amount of salt, redox potential variations, or temperature changes during the process.

Functions Involved in Growth and Fermentation

The main role of LAB starters during meat fermentation is to degrade sugars to produce mainly lactic acid. Some traditional sausages can be manufactured without the addition of any sugar. In this case, LAB performs fermentation through the utilization of the carbon sources that are naturally present in meat, mainly ribose and glucose. We previously noticed that in *L. sakei*, the gene cluster involved in ribose utilization had a characteristic organization (Stentz & Zagorec, 1999). Indeed, ribose transport was not performed by an ABC transporter as usually observed in other bacteria. Instead of the *rbsABCD* genes encoding the ribose specific ABC transporter, we found a gene, named *rbsU* that was similar to glucose transporters mediating sugar transport by a facilitated diffusion mechanism. Interestingly, a similar situation is observed in the genomes of *L. plantarum* and *P. pentosaceus*, but also in the genome of *Enterococcus faecalis* V583, a bacterial species that can contaminate meat products. Such a situation is not encountered in other LAB species, the genome of which has been sequenced. It is therefore plausible that such a property, shared by meat-borne bacteria confers them an advantage to compete for carbon sources, useful for energy production and growth. As a comparison, the gene syntheny observed around the *ldhL* gene of *L. sakei*, encoding the unique lactate dehydrogenase in this bacterium, responsible for L-lactate production, is conserved in *L. plantarum* and *P. pentosaceus* but also in other LAB genomes such as those of *L. casei* ATCC334 and *L. brevis* ATCC367, which are not used for meat fermentation. Concerning the capability to ferment sugars, it was previously noticed that *L. plantarum* has the widest range of sugar utilization due to the presence of many sugar transport systems, a characteristic trait of this species that is certainly linked to the many ecosystems it can colonize (Kleerebezem et al., 2003).

Meat is an iron rich medium, especially regarding complexed iron like haemoglobin or myoglobin. Iron metabolism seems to be emblematic of meat LAB species. More particularly, the ability to transport complex iron seems to be restricted to meat lactobacilli; only two other LAB species (*L. casei* and *Lactococcus lactis*) possess such an equipment (Table 6.3). *L. plantarum*, is also the only LAB species possessing a ironIII ABC transporter (AfuA). Moreover, iron dependent transcriptional regulators (Fur family) are also well represented in these meat LAB species. In fact, three are present in the *L. sakei* genome and two in both *L. plantarum* and *P. pentosaceus*. It seems thus likely that the ability to use iron compounds and especially iron complexes could be considered as a kind of meat signature for meat LAB species.

Functions Involved in Resistance to the Stressing Environment of Fermented Meat Products

Addition of NaCl to fermented meat products, as well as the drying process induce a high osmolarity environment which the bacterial should resist in order to survive in the processed meat. We previously noticed in the genome of *L. sakei* 23K the presence of three gene clusters encoding putative ABC transporters that may be involved in the uptake of osmoprotectants, such as glycine betaine, carnitine, or choline (Chaillou et al., 2005). Interestingly, *Listeria monocytogenes*, a pathogenic Gram-positive bacterium that can contaminate meat products shares the same equipment. However, among those three systems, only the cluster *lsa*0616-0619 was also conserved in *L. plantarum* WCFS1 (corresponding to *lpl*1607-1610) and *P. pentosaceus* ATCC25745 (genes *pepe*1655-1652). The two other osmoprotectant transporters of *L. sakei*, encoded by *lsa*1694-1695 and *lsa*1869-1870 may result from a duplication

Table 6.3 Comparison of redox equipment in *L. sakei* 23K, *L. plantarum* WCFS1, and *P. pentosaceus* 25,745

	L. sakei	L. plantarum	P. pentosaceus	Other LAB
NADH oxidase	2	6	1	L. casei, L. brevis, L. reuteri, L. sanfransiscensis
NADH peroxidase	1	1	1	L. casei, L. salivarius, L. brevis, L. acidophilus, L. gasseri
SOD	1	none	none	L. casei
Thiol peroxidase	1	1	1	L. reuteri, L. brevis, L. salivarius, L acidophilus, L. jonhsonii
NADH, dye-type peroxidases	1	1	none	L. casei, L. brevis L. salivarius, L. reuteri
Hydroperoxide resistance	1	none	none	Lc. lactis
Catalase	1	1	none	L. brevis
CytB5	1	1	none	L. acidophilus, L. salivarius
CytP450	1	none	none	None in LAB
Glutaredoxine,	1	1	1	L. casei, L. salivarius L. brevis, L. delbrueckii
Glutathione reductase	1	4	1	L. casei, L. salivarius L. brevis, L. gasseri, L. reuteri
Glutathion synthase	1	2	none	L. brevis, L. reuteri, L. salivarius, L. casei
Thioredoxines	4	4	3	L. acidophilus, L. brevis, L. casei, L. delbrueckii, L. johnsonii
Thioredoxine reductases	3	2	2	L. acidophilus, L. brevis, L. casei, L. delbruecki, L. johnsonii
Ferric Uptake Regulators	3	2	2	L. brevis, L. salivarius L. casei
Iron uptake Complex iron (Fhu)	1	1	none	Lc. lactis, L.brevis
Iron III	none	1	none	None in LAB
IronII/IIIMn	1	1	none	L. casei
IronII/Mn	3	4	2	Lc. Lactis, L. johnsonnii, L. casei, L. brevis, Lc. lactis, L. brevis, L. salivarius, L. delbrueckii
p-coumaric acid decarboxylase	1	1	1	Lactobacillus species
Aromatic hydrocarbon decarboxy-lases	2	1	1	L. reuteri, L. brevis

as both *L. plantarum* and *P. pentosaceus* had only one additional system (*lpl*0367-0368, and *pepe*0242-0243, respectively) that are equally similar to both the systems of *L. sakei*. Thus, *L. sakei* may be better adapted to high osmolarity than the two other LAB starters.

Meat storage and processing are characterized by numerous changes in oxygen levels, including the use of reducing agents such as nitrate and nitrite in fermented meat products. All the bacterial species present in the meat ecosystem thus have to cope with this main stressing environmental factor in order to grow and survive in fermented meat products. Protection against oxidative damage implies both defense systems against Reactive Oxygen Species (ROS) and repair systems to restore protein functionality or DNA damage. Iron can also enhance oxygen toxicity due to a possible arising of chemical reactions and the consecutive generation of ROS by the Fenton reactions. The emblematic meat species *L. sakei* has previously been reported as a bacterium with complete genetic equipment dedicated to defense against oxidative stress (Chaillou et al., 2005). These features have been searched in the two other meat LAB species for which whole genome sequence was available (Table 6.3). This genome comparison revealed some common features together with species specific traits that could be considered specific for meat LAB starters.

Analysis of the genome content of *L. plantarum* WCFS1, *L. sakei* 23K, and *P. pentosaceus* ATCC25745 shows that the three species harbor a genetic equipment dedicated to protection against oxidative damages. However, this equipment seems to be more similar between *L. plantarum* and *L. sakei*, *P. pentosaceus* being devoid of the main functions related to peroxide and hydroperoxide resistance. *L. sakei* harbors the most complete equipment for fighting against oxygen toxicity, including, in particular, one superoxide dismutase, a hydroperoxide resistance protein and an uncharacterized NADH oxidase that is unique in LAB. *L. plantarum* is clearly characterized by a high redundancy in glutathion reductases and glutathion synthase, a compound involved in protection against oxidative stress. It appears then that *P. pentosaceus* might be less resistant, than the two other species, to oxidative damage conditions encountered during meat fermentation.

Aromatic compounds can arise from exposure to spices and smoke, that can be used as meat bio-preservation methods, in addition to the fermentation process. Detoxification of such compounds can be achieved by a *p*-coumaric acid decarboxylase (PdcA) and aromatic hydrocarbon decarboxylases. PdcA homologs are largely found in *Lactobacillus* species. Homologs of the aromatic hydrocarbon decarboxylases are present in meat LAB and less represented in other LAB (only *L. brevis* and *L. reuteri*). Remarkably, *L. sakei* is unique among LAB, since it is the only species possessing a specific aromatic hydrocarbon hydroxylase for which no homologous protein is found in other LAB. Moreover, a gene encoding a putative cytochrome P450, whose function is yet unknown, is present only in *L. sakei*.

Conclusion

Genetic studies on LAB emerged a few decades ago, as a mean for microbiologists to better understand the mechanisms of food fermentation, in order to control and improve it or to select the better starter strains. Most of the literature was, for a long time, almost exclusively dedicated to the dairy LAB *Lc. lactis*. However, although with a certain delay, the genetics of other species, including that of meat LAB starters, gained the interest of the scientific community. The study of the plasmidic genetic elements was relatively poorly developed, mainly aimed as a tool for species characterization or for studies on systematics or classification. The search for bacteriocins, to improve meat safety, led to the isolation of several plasmids but, unfortunately, only the genes responsible for bacteriocin production were characterized, plasmids by themselves being poorly considered. The global conclusion of the genetic knowledge gained from the study of plasmids in LAB in general is that those genetic elements, as other mobile elements, are largely shared and exchanged between strains, species, and genera.

In the 1990s, molecular biologists started to develop tools and methods to genetically modify strains, as a basis to gain knowledge on the metabolism of LAB, and with the dream to control and orientate metabolic fluxes for improving fermentation processes. However, the use of genetically modified organisms in food, especially in fermented food that are eaten raw, i. e., with living microorganisms, such as fermented meat is not acceptable for the consumer. The use of genetics for LAB thus stayed just as a tool for laboratory studies to generate information.

In the mean time, genomics emerged and an increasing number of bacterial genomes were sequenced and analyzed each year during the last decade. After the sequencing of bacterial species, which were considered as models by the microbiologists, many pathogenic species were also sequenced, and finally, the genomes of most of the LAB used for food fermentation have been sequenced in the last years. The genome sequence of *L. plantarum* was published in 2003, that of *L. sakei* in 2005, and that of *P. pentosaceus* in 2006. Consequently, the delay traditionally observed with the knowledge of meat LAB compared to dairy LAB stopped.

Whole genome analysis revealed specific features of the various species. The comparison between several species and genera of LAB confirmed the status of *P. pentosaceus* as belonging to the *Lactobacillus* genus (Makarova et al., 2006). A careful analysis of the genome features common to meat LAB has still to be performed. However, some characteristic traits have already been detected: (*i*) all specifically show a metabolic capability oriented to the meat environment, which is not present in other LAB. This is exemplified by the use of ribose or iron, two nutrients present in meat and absent or substituted by other components in other materials that are also commonly fermented such as milk or vegetables; (*ii*) all show, with some species differences, the ability to resist the stressing environment that characterizes meat fermentation, such as oxidative or salt stresses or detoxification of molecules issued from spices or smoking.

References

Ackermann, J. W. (1987). Bacteriophage taxonomy in 1987. *Microbiology Sciences, 4*, 214–218.
Alegre, M. T., Rodriguez, M. C., & Mesas, J. M. (2004). Transformation of *Lactobacillus plantarum* by electroporation with in vitro modified plasmid DNA. *FEMS Microbioogy Letters, 241*, 73–77.

Alpert, C. A., Crutz-Le Coq, A.-M., Malleret, C., & Zagorec, M. (2003). Characterization of a theta-type plasmid from *Lactobacillus sakei*: A potential basis for low-copy vectors in lactobacilli. *Applied and Environmental Microbiology, 69*, 5574–5584.

Aukrust, T., & Blom, H. (1992). Transformation of *Lactobacillus* strains used in meat and vegetable fermentation. *Food Research International, 25*, 253–261.

Berthier, F., Zagorec, M., Champomier-Vergès, M., Ehrlich, S. D., & Morel-Deville, F. (1996). High-frequency transformation of *Lactobacillus sake* by electroporation. *Microbiology, 142*, 1273–1279.

Boekhorst, J., Siezen, R. J., Zwahlen, M. C., Vilanova, D., Pridmore, R. D., Mercenier, A., et al. (2004). The complete genomes of *Lactobacillus plantarum* and *Lactobacillus johnsonii* reveal extensive differences in chromosome organization and gene content. *Microbiology, 150*, 3601–3611.

Bringel, F., Frey, L., & Hubert, J. C. (1989). Characterization, cloning, curing, and distribution in lactic acid bacteria of pLP1, a plasmid from *Lactobacillus plantarum* CCM1904 and its use in shuttle vector construction. *Plasmid, 22*, 193–202.

Bron, P. A., Benchimol, M. G., Lambert, J., Palumbo, E., Deghorain, M., Delcour, J., et al. (2002). Use of the *alr* gene as a food-grade selection marker in lactic acid bacteria. *Applied and Environmental Microbiology, 68*, 5663–5670.

Caldwell, S. L., McMahon, D. J., Oberg, C. J., & Broadbent, J. R. (1999). Induction and characterization of *Pediococcus acidilactici* temperate bacteriophage. *Systematic and Applied Microbiology, 22*, 514–519.

Chaillou, S., Champomier-Vergès, M. C., Cornet, M., Crutz Le Coq, A.-M., Dudez, A.-M., et al. (2005). Complete genome sequence of the meat-borne lactic acid bacterium *Lactobacillus sakei* 23K. *Nature Biotechnology, 23*, 1527–1533.

Chen, J. D., & Morrison, D. A. (1988). Construction and properties of a new insertion vector, pJDC9, that is protected by transcriptional terminators and useful for cloning of DNA from *Streptococcus pneumoniae*. *Gene, 64*, 155–164.

Cosby, W. M., Axelsson, L. T., & Dobrogosz, W. J. (1989). Tn917 transposition in *Lactobacillus plantarum* using the highly temperature-sensitive plasmid pTV1Ts as a vector. *Plasmid, 22*, 236–243.

Danielsen, M. (2002). Characterization of the tetracycline resistance plasmid pMD5057 from *Lactobacillus plantarum* 5057 reveals a composite structure. *Plasmid, 48*, 98–103.

Doi, K., Zhang, Y., Nishizaki, Y., Umeda, A., Ohmomo, S., & Ogata, S. (2003). A comparative study and phage typing of silage-making *Lactobacillus* bacteriophages. *Journal of Bioscience and Bioengineering, 95*, 518–525.

Ferain, T., Hobbs, J. N., Jr., Richardson, J., Bernard, N., Garmyn, D., Hols, P., et al. (1996). Knockout of the two *ldh* genes has a major impact on peptidoglycan precursor synthesis in *Lactobacillus plantarum*. *Journal of Bacteriology, 178*, 5431–5437.

Frost, L. S., Leplae, R., Summers, A. O., & Toussaint, A. (2005). Mobile genetic elements: The agents of open source evolution. *Nature Reviews Microbiology, 3*, 722–732.

Gevers, D., Danielsen, M., Huys, G., & Swings, J. (2003). Molecular characterization of *tet*(M) genes in *Lactobacillus* isolates from different types of fermented dry sausage. *Applied and Environmental Microbiology, 69*, 1270–1275

Gevers, D., Huys, G., & Swings, J. (2003). In vitro conjugal transfer of tetracycline resistance from *Lactobacillus* isolates to other Gram-positive bacteria. *FEMS Microbiology Letters, 225*, 125–130.

Giacomini, A., Squartini, A., & Nuti, M. P. (2000). Nucleotide sequence and analysis of plasmid pMD136 from *Pediococcus pentosaceus* FBB61 (ATCC43200) involved in pediocin A production. *Plasmid, 43*, 111–122.

Gonzalez, C. F., & Kunka, B. S. (1983). Plasmid transfer in *Pediococcus* spp.: Intergeneric and intrageneric transfer of pIP501. *Applied and Environmental Microbiology, 46*, 81–89.

Gonzalez, C. F., & Kunka, B. S. (1986). Evidence for plasmid linkage of raffinose utilization and associated alpha-galactosidase and sucrose hydrolase activity in *Pediococcus pentosaceus*. *Applied and Environmental Microbiology, 51*, 105–109.

Gonzalez, C. F., & Kunka, B. S. (1987). Plasmid-associated bacteriocin production and sucrose fermentation in *Pediococcus acidilactici*. *Applied and Environmental Microbiology, 53,* 2534–2538.

Gory, L., Montel, M. C., & Zagorec, M. (2001) Use of Green Fluorescent Protein to monitor *Lactobacillus sakei* in fermented meat products. *FEMS Microbiology Letters, 194,* 127–133.

Graham, D. C., & McKay, L. L. (1985). Plasmid DNA in strains of *Pediococcus cerevisiae* and *Pediococcus pentosaceus*. *Applied and Environmental Microbiology, 50,* 532–534.

Gury, J., Barthelmebs, L., & Cavin, J. F. (2004). Random transposon mutagenesis of *Lactobacillus plantarum* by using the pGh9:IS*S1* vector to clone genes involved in the regulation of phenolic acid metabolism. *Archives in Microbiology, 182,* 337–345.

Halami, P. M., Ramesh, A., & Chandrashekar, A. (2000). Megaplasmid encoding novel sugar utilizing phenotypes, pediocin production and immunity in *Pediococcus acidilactici* C20. *Food Microbiology, 17,* 475–483.

Hammes, W. P., & Hertel, C. (1998). New development in meat starter culture. *Meat Science, 49,* S125–S128.

Hertel, C., Schmidt, G., Fischer, M., Oellers, K., & Hammes, W. P. (1998). Oxygen-dependent regulation of the expression of the catalase gene *katA* of *Lactobacillus sakei* LTH677. *Applied and Environmental Microbiology, 64,* 1359–1365.

Kakikawa, M., Yamakawa, A., Yokoi, K. J., Nakamura, S., Taketo, A., & Kodaira K. (2002) Characterization of the major tail protein gpP encoded by *Lactobacillus plantarum* phage phi gle. *Journal of Biochemistry, Molecular Biology, and Biophysics, 6,* 185–191.

Kakikawa, M., Yokoi, K. J., Kimoto, H., Nakano, M., Kawasaki, K., Taketo, A., et al. (2002). Molecular analysis of the lysis protein Lys encoded by *Lactobacillus plantarum* phage phi g1e. *Gene, 299,* 227–234.

Kanatani, K., & Oshimura, M. (1994). Plasmid-associated bacteriocin production by a *Lactobacillus plantarum* strain. *Bioscience Biotechnology Biochemistry, 58,* 2084–2086.

Kantor, A., Montville, T. J., Mett, A., & Shapira, R. (1997). Molecular characterization of the replicon of the *Pediococcus pentosaceus* 43200 pediocin A plasmid pMD136. *FEMS Microbiology Letters, 151,* 237–244.

Kleerebezem, M., Boekhorst, J., van Kranenburg, R., Molenaar, D., Kuipers, O. P., Leer, R., et al. (2003). Complete genome sequence of *Lactobacillus plantarum* WCFS1. *Proceedings of the National Academy of Science USA, 100,* 1990–1995.

Langella, P., Zagorec, M., Ehrlich, S. D., & Morel-Deville, F. (1996). Intergeneric and intrageneric conjugal transfer of plasmids pAMβ1, pIL205 and pIP501 in *Lactobacillus sake*. *FEMS Microbiology Letters, 139,* 51–56.

Leer, R. J., Christiaens, H., Verstraete, W., Peters, L., Posno, M., & Pouwels, P. H. (1993). Gene disruption in *Lactobacillus plantarum* strain 80 by site-specific recombination: isolation of a mutant strain deficient in conjugated bile salt hydrolase activity. *Molecular and General Genetics, 239,* 269–272.

Leloup, L., Ehrlich, S. D., Zagorec, M., & Morel-Deville, F. (1997). Single cross-over integration in the *Lactobacillus sake* chromosome and insertional inactivation of the *ptsI* and *lacL* genes. *Applied and Environmental Microbiology, 63,* 2127–2133.

Leuschner, R. G. K., Arendt, E. K., & Hammes, W. P. (1993). Characterization of a virulent *Lactobacillus sake* phage PWH2. *Applied Microbiology and Biotechnology, 39,* 617–621.

Li, Y., Canchaya, C., Fang, F., Raftis, E., Ryan, K. A., van Pijkeren, J. P., et al. (2007). Distribution of megaplasmids in *Lactobacillus salivarius* and other lactobacilli. *Journal of Bacteriology, 189,* 6128–6139.

Liu, M. L., Kondo, J. K., Barnes, M. B., & Bartholomew, D. T. (1988). Plasmid-linked maltose utilization in *Lactobacillus* ssp. *Biochimie, 70,* 351–355.

Lu, Z., Altermann, E., Breidt, F., Predki, P., Fleming, H. P., & Klaenhammer, T. R. (2005). Sequence analysis of the *Lactobacillus plantarum* bacteriophage PhiJL-1. *Gene, 348,* 45–54.

Lu, Z., Breidt, F. Jr., Fleming, H. P., Altermann, E., & Klaenhammer, T. R. (2003). Isolation and characterization of a *Lactobacillus plantarum* bacteriophage, phiJL-1, from a cucumber fermentation. *International Journal of Food Microbiology, 84,* 225–235.

Luchansky, J. B., Muriana, P. M., & Klaenhammer, T. R. (1988). Application of electroporation for transfer of plasmid DNA to *Lactobacillus, Lactococcus, Leuconostoc, Listeria, Pediococcus, Bacillus, Staphylococcus, Enterococcus* and *Propionibacterium. Molecular Microbiology, 2*, 637–646.

Maguin, E., Prévost, H., Ehrlich, S. D., & Gruss, A. (1996). Efficient insertional mutagenesis in lactococci and other Gram-positive bacteria. *Journal of Bacteriology, 178*, 931–935.

Makarova, K., Slesarev, A., Wolf, Y., Sorokin, A., Mirkin, B., Koonin, E., et al. (2006). Comparative genomics of the lactic acid bacteria. *Proceedings of the National Academy of Science USA, 103*, 15611–15616.

Malleret, C., Lauret, R., Ehrlich, S. D., Morel-Deville, F., & Zagorec, M. (1998). Disruption of the sole *ldhL* gene in *Lactobacillus sakei* prevents the production of both L- and D-lactate. *Microbiology, 144*, 3327–3333.

Mayo, B., Gonzalez, B., Arca, P., & Suarez, J. E. (1994). Cloning and expression of the plasmid encoded beta-D-galactosidase gene from a *Lactobacillus plantarum* strain of dairy origin. *FEMS Microbiology Letters, 122*, 145–151.

Miller, K. W., Ray, P., Steinmetz, T., Hanekamp, T., & Ray, B. (2005). Gene organization and sequences of pediocin AcH/PA-1 production operons in *Pediococcus* and *Lactobacillus* plasmids. *Letters in Applied Microbiology, 40*, 56–62.

Motlagh, A., Bukhtiyarova, M., & Ray, B. (1994). Complete nucleotide sequence of pSMB74, a plasmid encoding the production of pediocin AcH in *Pediococcus acidilactici. Letters in Applied Microbiology, 18*, 305–312.

Naumoff, D. G. (2001). Beta-fructosidase superfamily: Homology with some alpha-L-arabinases and beta-D-xylosidases. *Proteins, 42*, 66–76.

Nes, I.. F. (1984). Plasmid profiles of ten strains of *Lactobacillus plantarum. FEMS Microbiology Letters, 21*, 359–361.

Nes, I. F., Brendehaug, J., & von Husby, K. O. (1988). Characterization of the bacteriophage B2 of *Lactobacillus plantarum* ATCC 8014. *Biochimie, 70*, 423–427.

Osmanagaoglu, O., Beyatli, Y., & Gunduz, U. (2000). Cloning and expression of a plasmid-linked pediocin determinant trait of *Pediococcus acidilactici* F. *Journal of Basic Microbiology, 40*, 41–49.

Pérez Pulido, R., Abriouel, H., Ben Omar, N., Lucas López, R., Martínez Canamero, M., & Gálvez, A. (2006). Plasmid profile patterns and properties of pediococci isolated from caper fermentations. *Journal of Food Protection, 69*, 1178–1182.

Rodríguez, M. C., Alegre, M. T., & Mesas, J. M. (2007). Optimization of technical conditions for the transformation of *Pediococcus acidilactici* P60 by electroporation. *Plasmid, 58*, 44–50.

Romero, D. A., & Klaenhammer, T. R. (1992). IS946-mediated integration of heterologous DNA into the genome of *Lactococcus lactis* subsp. *lactis. Applied and Environmental Microbiology, 58*, 699–702.

Ruiz-Barba, J. L., Piard, J. C., & Jiménez-Díaz, R. (1991). Plasmid profiles and curing of plasmids in *Lactobacillus plantarum* strains isolated from green olive fermentations. *Journal of Applied Bacteriology, 71*, 417–421.

Shareck, J., Choi, Y., Lee, B., & Miguez, C. B. (2004). Cloning vectors based on cryptic plasmids isolated from lactic acid bacteria: Their characteristics and potential applications in biotechnology. *Critical Reviews in Biotechnology, 24*, 155–208.

Shay, B. J., Egan, A. F., Wright, M., & Rogers, P. J. (1988). Cystein metabolism in an isolate of *Lactobacillus sake*: Plasmid composition and cystein transport. *FEMS Microbiology Letters, 56*, 183–188.

Simon, L., Frémaux, C., Cenatiempo, Y., & Berjeaud, J.-M. (2002). Sakacin G, a new type of antilisterial bacteriocin. *Applied and Environmental Microbiology, 68*, 6416–6420.

Skaugen, M., Abildgaard, C. I., & Nes, I. F. (1997). Organization and expression of a gene cluster involved in the biosynthesis of the lantibiotic lactocin S. *Molecular and General Genetics, 253*, 674–686.

Sørvig, E., Mathiesen, G., Naterstad, K., Eijsink, V. G., & Axelsson, L. (2005). High-level, inducible gene expression in *Lactobacillus sakei* and *Lactobacillus plantarum* using versatile expression vectors. *Microbiology, 151*, 2439–2449.
Stentz, R., Loizel, C., Malleret, C., & Zagorec, M. (2000). Development of genetic tools for *Lactobacillus sakei*: Disruption of the β-galactosidase gene and use of *lacZ* as a reporter gene to study regulation of the putative copper ATPase, AtkB. *Applied and Environmental Microbiology, 66*, 4272–4278.
Stentz, R., & Zagorec, M. (1999). Ribose utilization in *Lactobacillus sakei*: Analysis of the regulation of the *rbs* operon and putative involvement of a new transporter. *Journal of Molecular Microbiology and Biotechnology, 1*, 165–173.
Takala, T. M., & Saris, P. E. (2002). A food-grade cloning vector for lactic acid bacteria based on the nisin immunity gene *nisI*. *Applied Microbiology and Biotechnology, 59*, 467–471.
Takala, T. M., Saris, P. E., & Tynkkynen, S. S. (2003). Food-grade host/vector expression system for *Lactobacillus casei* based on complementation of plasmid-associated phospho-beta-galactosidase gene *lacG*. *Applied Microbiology and Biotechnology, 60*, 564–570.
Talon, R., Leroy, S., & Lebert, I. (2007). Microbial ecosystems of traditional fermented meat products: the importance of indigenous starters. *Meat Science, 77*, 55–62.
Van Reenen, C. A., Van Zyl, W. H., & Dicks, L. M. (2006). Expression of the immunity protein of plantaricin 423, produced by *Lactobacillus plantarum* 423, and analysis of the plasmid encoding the bacteriocin. *Applied and Environmental Microbiology, 72*, 7644–7651.
Vaughan, A., Eijsink, V. G., & Van Sinderen, D. (2003). Functional characterization of a composite bacteriocin locus from malt isolate *Lactobacillus sakei* 5. *Applied and Environmental Microbiology, 69*, 7194–7203.
Ventura, M., Canchaya, C., Kleerebezem, M., de Vos, W. M., Siezen, R. J., & Brüssow, H. (2003). The prophage sequences of *Lactobacillus plantarum* strain WCFS1. *Virology, 316*, 245–255.
Vogel, R. F., Becke-Schmid, M., Entgens, P., Gaier, W., & Hammes, W. P. (1992). Plasmid transfer and segregation in *Lactobacillus curvatus* LTH1432 *in vitro* and during sausage fermentations. *Systematics and Applied Microbiology, 15*, 129–136.
Vogel, R. F., Lohmann, M., Weller, A. N., Hugas, M., & Hammes, W. P. (1991). Structural similarity and distribution of small cryptic plasmids of *Lactobacillus curvatus* and *Lactobacillus sake*. *FEMS Microbiology Letters, 68*, 183–190.
Wang, T. T., & Lee, B. H. (1997). Plasmids in *Lactobacillus*. *Critical Reviews in Biotechnology, 17*, 227–272.
West, C. A., & Warner, P. J. (1985). Plasmid profiles and transfer of plasmid-encoded antibiotic resistance in *Lactobacillus plantarum*. *Applied and Environmental Microbiology, 50*, 1319–1321.

Chapter 7
Genetics of Yeasts

Amparo Querol, Mª Teresa Fernández-Espinar, and Carmela Belloch

Introduction

The use of yeasts in biotechnology processes dates back to ancient days. Before 7000 BC, beer was produced in Sumeria. Wine was made in Assyria in 3500 BC, and ancient Rome had over 250 bakeries, which were making leavened bread by 100 BC. And milk has been made into Kefyr and Koumiss in Asia for many centuries (Demain, Phaff, & Kurtzman, 1999). However, the importance of yeast in the food and beverage industries was only realized about 1860, when their role in food manufacturing became evident.

Yeasts are used in many industrial processes, as they grow on a wide range of substrates and can tolerate extreme physico-chemical conditions. Today, the impact of yeasts on food and beverage production extends beyond the original and popular notions of bread, beer, and wine fermentations by *Saccharomyces cerevisiae* (Fleet, 2006; Querol, Belloch, Fernández-Espinar, & Barrio, 2003). In addition to *S. cerevisiae*, *S. bayanus*, and *S. pastorianus*, it is now well established that various species of *Hanseniaspora*, *Candida*, *Pichia*, *Metschnikowia*, *Kluyveromyces*, *Schizosaccharomyces*, and *Issatchenkia* can make a positive contribution in the manufacture of fermented foods, dairy, meats, cereals, coffee, and sauces. In the case of fermentation of meat sausages and maturation of hams, the most predominant and important yeasts are *Debaryomyces hansenii* (and the anamorph *C. famata*), *Yarrowia lipolytica*, and various *Candida* species.

Methods for Yeast Identification

The development of rapid and simple methods for the identification and characterisation of yeast strains is an essential tool for the meat industry. This is obvious, since yeasts in meat products can contribute to beneficial and detrimental aspects, and the

---

D. A. Querol
Instituto de Agroquímica y Tecnología de Alimentos, CSIC, P.O. Box 73, E-46100 Burjassot (Valencia) Spain

identification of individual strains is often required to improve the microbiological quality of these foods.

Classical taxonomy is based predominantly on phenotypic characteristics, such as physiological traits and cell morphology. These approximations have been used in the identification of yeasts isolated from different meat products (Gardini et al., 2001; Núñez, Rodríguez, Córdoba, Bermúdez, & Asensio, 1996). However, nutritional characteristics have been shown to be highly variable as well as mutable, and genetic crosses have linked the characteristics to one or only a few genes, which in some cases could lead to an incorrect classification of a species or a false identification of strains (for a review see Boekhout & Robert, 2003; Fernández-Espinar, Martorell, de Llanos, & Querol, 2006). The conventional methodology for yeast identification requires the evaluation of some 60–90 tests and the process is complex, laborious, and time consuming. Besides, in the case of one of the most relevant yeast isolates in meat products like *Debaryomyces*, many of the results obtained with these tests are variable, thus making correct identification of the *Debaryomyces* species difficult. As an example, serious difficulties were experienced in identifying isolates assigned to *Debaryomyces*, due to the erroneous assimilation of D-xylose, and to a lesser extent, raffinose and L-arabinose (Metaxopoulos, Stavropoulos, Kakouri, & Samelis, 1996). In recent years, rapid kits for yeast identification have been developed to improve the conventional methods. However, these were designed initially for clinical diagnosis and their application is restricted to 40–60 yeast species of medical interest (Deák & Beuchat, 1996). Other methods based on the analysis of total cell proteins and long-chain fatty acids using gas chromatography have also been developed. However, the reproducibility of these techniques is questionable due to the fact that they depend on the physiological state of the yeast cells (Golden, Beuchat, & Hitchcock, 1994).

Recent progress in molecular biology has led to the development of new techniques for yeast identification (for a review see Boekhout & Robert, 2003; Fernández-Espinar et al., 2006). In the next sections we review the new methods that have been developed and their application to the rapid identification of meat yeasts in the industrial practice. The specificity of nucleic acid sequences has prompted the development of several methods for rapid species identification. The comparison of ribosomal RNA (rRNA) and its template ribosomal DNA (rDNA) has been used extensively in recent years, to assess both close and distant relationships among many kinds of organisms. In several hemiascomycetous yeasts, rRNA genes are located in a single genome region composed of 100–150 tandem repeats of a fragment of 9 Kb. The ribosomal genes (5.8S, 18S, and 26S) are grouped in tandem forming transcription units that are repeated in the genome between 100 and 200 times. In each transcription unit, two other regions exist, the internal transcribed spacers (ITS) and the external ones (ETS) that are transcribed but are not processed. In turn, the codifying units are separated by the IGS intergenic spacers, also called NTS. The gene 5S is not included in the previously described transcription unit but is found adjacent to the same repetition unit, in tandem, in the case of yeast. The sequences of the ribosomal genes 5.8S, 18S, and 26S, as well as the spacers ITS and NTS, represent powerful tools to establish phylogenetic relationships and identify species (Kurtzman & Robnett, 1998), due to the conserved sequences found there, as well as their concerted evolution, i.e., the similarity between repeated transcription units is greater within the species than between units belonging to different species, due to mechanisms like unequal crossing-over or genetic conversion (Li, 1997).

Using the information contained in these regions, different methods have been developed to identify yeast species, as we will describe below.

Ribosomal RNA Gene Sequencing

One of these methods is based on the determination and comparison of the nucleotide sequences in these regions. The two most commonly used regions are those corresponding to the domains D1 and D2, located at the 5′ end of the gene 26S (Kurtzman & Robnett, 1998) and the gene 18S (James, Cai, Roberts, & Collins, 1997). The availability of these sequences in databases, especially in the case of the D1/D2 region of gene 26S, make this technique very useful to assign an unknown yeast to a specific species when the percentage of homology of its sequences is over or close to 99% (Kurtzman & Robnett, 1998).

Restriction Fragment Length Polymorphism of rDNA

With an industrial application in mind, other simpler identification methods based on PCR amplification of the regions of the ribosomal DNA and later restriction of the amplified fragment were developed. Guillamón, Sabate, Barrio, Cano, & Querol (1998) proposed a rapid and easy method for routine wine yeast identification, based on RFLPs of the 5.8S rRNA gene and the internal transcribed spacers (ITS1 and 2). Later on its use has been extended to a total of 191 yeast species (de Llanos, Fernández-Espinar, & Querol, 2004; Esteve-zarzoso, Belloch, Uruburu, & Querol, 1999; Fernández-Espinar, Esteve-Zarzoso, Querol, & Barrio, 2000) related with beverages and food including meat yeasts isolates. The amplified fragments and restriction profiles of these species with the enzymes *Hae*III, *Hin*fI, *Cfo*I, and *Dde*I are currently available "online" at the address http://yeast-id.com. The utility of the technique has been proved studying reference strains (Esteve-Zarzoso, Zorman, Belloch, & Querol, 2003; Fernández-Espinar et al., 2000; Ramos, Valente, Hagler, & Leoncini, 1998) and has been applied by numerous authors for yeast isolate identification in different foods and beverages, such as isolates pertaining to the genus *Debaryomyces*, *Yarrowia*, and *Candida* (Andrade, Rodríguez, Sánchez, Aranda, & Córdoba, 2006; Martorell, Fernández-Espinar, & Querol, 2005; Ramos et al., 1998).

PCR-DGGE Denaturing Gradient Gel Electrophoresis

Recently, a genetic fingerprinting technique based on PCR amplification and denaturing gradient gel electrophoresis or DGGE has been introduced into microbial ecology (Muyzer, de Waal, & Uitterlinden, 1993). In PCR-DGGE, DNA fragments of the same length but with different sequences can be separated. The use of DGGE in microbial ecology is still in its infancy, but their future perspectives are promising (Muyzer & Smalla, 1998). Its application to yeast identification in foods and beverages is very recent (for a review see Fernández-Espinar et al., 2006), and only a few studies related with meat products or the yeast ecology of fermented sausages have been performed (Cocolin, Urso, Rantsiou, Cantón, & Comi, 2006; Rantsiou et al., 2005).

Yeast Biodiversity in Meat Products

The most important yeasts in meat products belong to the ascomycetous genera, *Debaryomyces*, *Pichia*, *Yarrowia*, and *Candida*, and the basidiomycetous genera, *Rhodotorula*, *Cryptococcus*, and *Trichosporon*. Fresh meat processing and storage causes the progressive replacement of basidiomycetous by ascomycetous yeasts. A predominance of *Candida* spp. occurs during spoilage of fresh meat, while meat salting, curing and fermentation are selective for *Debaryomyces* (Samelis & Sofos, 2003). Some important physiological and biochemical characteristics of meat yeasts and their interactions result in the selection of certain yeast species in specific meat products, like *Y. lipolytica* in fresh and spoiled poultry (Ismail, Deak, El-Rahman, Yassien, & Beuchat, 2000), and *D. hansenii* in cured dried and fermented meats (Durá, Flores, & Toldrá, 2004a, 2004b; Guerzoni, Lanciotti, & Marchetti, 1993). Consequently, yeasts in meat products have both spoiling and beneficial aspects.

Spoilage Yeasts in Meat

Yeasts have commonly been considered unimportant in meat spoilage because their low initial numbers and their slow growth rates at refrigeration temperatures prevent them from competing effectively with psychrophilic bacteria (Dillon, 1998; Nortje et al., 1990). Most yeasts, however, are more resistant than bacteria to several food related stresses, such as low water activity, low pH, high salinity, and chemical preservatives, while certain species are extremely psychrophilic (Fleet, 1990). Thus, opportunistic yeasts may become important spoilage agents in meat products when bacterial growth is retarded due to the inhibitory effects of the above stress factors (Samelis & Sofos, 2003).

The increasing consumer demand for less processed, more "natural" foods has led the food industries to commercialize new products, in which the use of lower concentration of preservatives, packaging in modified atmospheres, or new formulations occasionally permit the growth of yeasts. Yeast contributes a small, but permanent part of the natural microflora on meat. The ability of some yeast to grow at low temperature, high salt concentration, and under reduced oxygen tension enables them to proliferate in refrigerated, cured and vacuum-packed meat and meat products (Deák & Beuchat, 1996). Meat spoilage caused by yeasts is mainly due to their lipolytic and proteolytic activities, although their action on carbohydrates and associated by-products of bacterial metabolism may also lead to the formation of compounds reducing the sensory quality of meat products, such as organic acids, alcohols, esters, and others (Dillon, 1998).

Numerous studies have been conducted in yeasts isolated from meat and meat products (for a review see Samelis & Sofos, 2003), and the yeasts most frequently isolated on spoiled meat are summarized in Table 7.1 (see Romano, Capece, & Jespersen, 2006; Samelis & Sofos, 2003; Selgas & García, 2007).

Beneficial Aspects of Yeast on Meats: Yeast Starter Cultures

Yeasts can also be considered habitual components of the micro-biota growing on fermented sausages and dry-cured hams, and their origin is mainly related to the environment and the meat used as raw material. In fermented meats, the lactic acid produced by bacteria and the low water activity resulting from the presence of salts or a dehydration process constitute a modified environmental factor that hinders the bacterial growth and favor the development of natural competitors. Thus, yeasts use all the nutrients and energy and grow quickly and easily (Dillon & Board, 1991). Yeasts can grow at pH, water activity, and temperature values usual in fermented sausages (Hammes & Knauf, 1994; Monte, Villanueva, & Domínguez, 1986). The presence of yeasts in fermented raw sausages has been studied less than bacteria and molds.

Sausages and dry-cured hams are traditional meat products obtained after several months of ripening. Proteolysis and lipolysis constitute the main biochemical reactions in the generation of flavor precursors, where the endogenous enzymes play the most important role (Toldrá, 1998). Yeasts and molds traditionally play an important role in sausages fermentation and in dry ham (Martín, Córdoba, Aranda, Córdoba, & Asensio, 2006; Martín, Córdoba, Benito, Aranda, & Asensio, 2003), and their contribution of the fungal population enzymes to proteolysis in minced dry-cured meat products, such as dry-cured sausages, is widely known.

The use of commercially available starters, mainly constituted of lactic acid bacteria and micrococci, may also produce an impoverishment of flavor and aroma and a loss of peculiar organo-leptic characteristics found in naturally fermented sausages. For this reason, in several European countries, artisanal sausages are still preferred by the consumer and are manufactured by relying on an unknown "factory flora" (Samelis, Stavropoulos, Kakouri, & Metaxopoulos, 1994). Lactic acid bacteria, micrococci, and coagulase-negative staphylococci have the most relevant role in the fermentative process and ripening, but also yeasts and molds can be involved as we explained before. Though there are several reports on the yeast populations in various meat products (Fung & Liang, 1990), studies on the yeast biodiversity in sausages are limited. The earliest studies on salami reported that *D. hansenii* was

Table 7.1 The most significant and commonly reported meat yeast species (Deáck & Beuchat, 1996; Martín et al., 2003, 2006; Romano, Capece, & Jespersen, 2006; Samelis & Sofos, 2003)

Fresh meats				Cured fresh and cooked meats		Dried and fermented meats		
Beef	Poultry	Lamb	Pork	Sausages	Frankfurters	Ham	Sausages and salamis	Dry-cured ham
C. albicans	C. catenulata	C. glabrata	C. glabrata	C. catenulata	C. catenulata	C. catenulata	C. gropengiesseri	C. saitoana
C. catenulata	C. intermedia	C. zeylanoides	C. zeylanoides	C. intermedia	C. saitoana	C. saitoana	C. haemulonii	C. zeylanoides
C. diddensiae	C. parapsilosis	Cr. laurentii	Cr. laurentii	C. parapsilosis	C. zeylanoides	C. zeylanoides	C. intermedia	Cr. albidus
C. intermedia	C. rugosa	Cys. infirmominiatum	Cys. infirmominiatum	C. rugosa	Deb. hansenii	Deb. hansenii	C. kruisii	Cr. laurentii
C. norvegica	C. zeylanoides	Deb. hansenii	Deb. hansenii	C. saitoana	Tr. pullulans	Tr. pullulans	C. parapsilosis	Cr. humicolus
C. parapsilosis	Cr. laurentii	Rh. glutinis	Rh. glutinis	C. tropicalis	Y. lipolytica	Y. lipolytica	C. zeylanoides	Deb. hansenii
C. rugosa	P. membranifaciens	Rh. minuta	Rh. minuta	C. zeylanoides			Cit. matritensis	Deb. polymorphus
C. sake	Rh. glutinis	Rh. mucilaginosa	Rh. mucilaginosa	Cr. albidus			Cr. albidus	P. cijerii
C. tropicalis	Rh. minuta	Tr. pullulans	Tr. pullulans	Deb. hansenii			Cr. skinneri	P. holstii
C. versatilis	Sacch. exiguus	Y. lipolytica	Y. lipolytica	Dekkera spp.			Deb. hansenii	P. sydowiorum
C. zeylanoides	Tr. moniliforme			Gal. geotrichum			Deb. maramus	P. guillermondii
Cr. albidus	Tr. pullulans			I. orientalis			Deb. occidentalis	Rh. glutinis
Cr. laurentii				P. anomala			Deb. polymorphus	Y. lipolytica
Cys. infirmominiatum				P. guillermondii			Deb. vanrijiae	
Deb. hansenii				P. membranifaciens			G. geotrichum	
P. burtonii				Rh. minuta			H. uvarum	
P. fermentans				Rh. mucilaginosa			M. pulcherrima	
P. membranifaciens				Sporobolomyces spp.			P. farinosa	
Rh. glutinis				Tr. moniliforme			P. guillermondii	
Rh. minuta				Tr. pullulans			P. philogaea	
Rh. mucilaginosa				Y. lipolytica			Rh. minuta	
Sacch. exiguus							Rh. mucilaginosa	
Tr. moniliforme							Tr. beigelii	
Tr. pullulans							Sp. roseus	
Y. lipolytica							St. halophilus	
							T. delbrueckii	
							Tr. pullulans	
							Y. lipolytica	
							Z. rouxii	

the most commonly isolated yeast. Most recently, several researches confirmed these results, but other yeast genera were found, such as *Candida, Pichia, Rhodotorula, Hansenula* (synonym of *Pichia*), and *Torulopsis* (synonym of *Candida*) (Dalton, Board, & Davenport, 1984; Grazia, Suzzi, Romano, & Giudici, 1989). On this basis, *D. hansenii* was used as a starter with positive effects on the development of a characteristic yeast flavor and stabilization of the reddening reaction. *D. hansenii* strains are presently used in starter preparations to be added to the sausages (Hammes & Knauf, 1994). The yeast *Yarrowia lipolytica*, the perfect form of *Candida lipolytica*, has also frequently been isolated from fresh beef (Dalton et al., 1984; Fung & Liang, 1990) and sausages (Viljoen, Dykes, Collis, & von Holy, 1993). Due to its lipolytic and proteolytic activity, this species can have a high technological potential (Sinigaglia, Lanciotti, & Guerzoni, 1994).

We can conclude that *D. hansenii* is the most frequently isolated yeast in fermented sausages and dry cured ham (Romano, Capece, & Jespersen, 2006; Samelis & Sofos, 2003; Selgas & García, 2007). The contribution of this yeast to the typical aroma of the products is based on their primary and secondary metabolites produced by the activity of lipases and proteinases, which are the key enzymes (Durá et al., 2004a, 2004b). In the next section we will give a detailed description of this yeast species.

Physiological and Genetic Characteristics of *D.hansenii*

According to the last revision, the genus *Debaryomyces* comprises 15 species (Nakase, Suzuki, Phaff, & Kurtzman, 1998). Many representatives can be found in natural habitats such as air, soil, pollen, tree exudates, plants, fruits, insects, and faeces and gut of vertebrates (Barnett, Payne, & Yarrow, 2000). Nine of the *Debaryomyces* species, namely *D. carsonii, D. etchellsii, D. hansenii, D. maramus, D. melissophilus, D. polymorphus, D. pseudopolymorphus, D. robertsiae,* and *D. vanrijiae* have been found in a variety of processed foods, such as fruit juices and soft drinks, wine, beer, sugary products, bakery products, dairy products, and meat or processed meats. The presence of the *Debaryomyces* species in foods usually has no detrimental effects and in some cases is beneficial to the food. However, *D. hansenii* as well as other *Debaryomyces* species may also be responsible for food spoilage, causing off-odors and off-flavors. The best method for identification of this species is the PCR-RFLP of the intergenic spacer region (IGS) region of rDNA (Quiros et al., 2006).

Morpholgy and Physiology

Colonies on agar medium at 25°C are grayish white to yellowish, dull to shiny, and smooth to wrinkle. The two varieties within *D. hansenii* are distinguished by

Panel A.

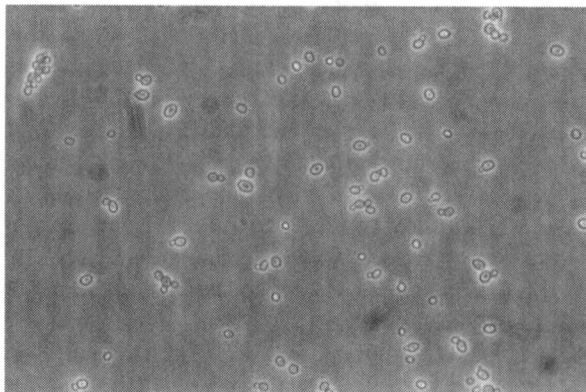

Panel B.

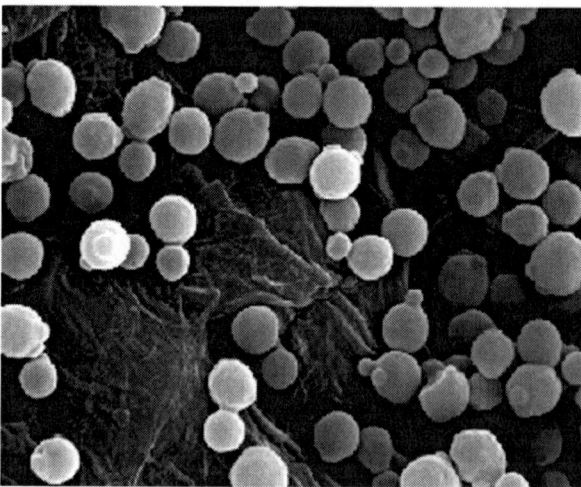

Fig. 7.1 Morphology of *Debaryomyces hansenii* strain PYCC 298. Panel A optic microscopy phase contrast (40X). Panel B, scanning electronic microscopy (JEOL JSM-6300). * Photo panel B, courtesy of Professor José Ramos, Department of Microbiology, University of Córdoba, Spain

maximum growth temperature. *D. hansenii* var. *hansenii* has a maximum growth temperature of 35°C while *D. hansenii* var. *fabryi* has a maximum growth temperature of 40°C and does not growth on cycloheximide. The microscopic morphology following 72 hours incubation at 25°C shows no pseudohyphae, the cells are spheroidal to short-ovoidal and single, in pairs or on short chains (see Fig. 7.1). Pseudo-mycelium is absent, primitive, or occasionally well developed (Nakase et al., 1998).

D. hansenii shows some remarkable biotechnological properties like the ability to grow at 10% NaCl or 5% glucose and assimilates a broad spectrum of carbon substrates. *D. hansenii* has a high chemo-stress tolerance, which means that high concentrations of many substrates can be used in the cultivation. It is characterized by its cryo and halo-tolerance (Prista, Almagro, Loureiro-Dias, & Ramos, 1997; Prista, Loureiro-Dias, Montiel, García, & Ramos, 2005), and its ability to metabolize lactic and citric acids. Such properties allow its development in all types of cheeses (Fleet, 1990) and different fermented meat products (Encinas, López-Díaz, García-López, Otero, & Moreno, 2000) contributing to the ripening process and the generation of aroma precursors by the lipolytic and proteolytic activities (Flores, Durá, Marco, & Toldrá, 2004; Leclercq-Perlat, Corrieu, & Spinnler, 2004; Martín et al., 2003; Petersen, Westall, & Jespersen, 2002).

D. hansenii is generally regarded as non-pathogenic (Swiss Agency for the Environment, Forests and Landscape [SAEFL], 2004); however, clinical isolates are not rare (Nishikawa, Tomomatsu, Sugita, Ikeda, & Shinoda, 1996) and there are reports of bone infection caused by *D. hansenii* (Wong et al., 1982). *D. hansenii* can cause subcutaneous abscesses, osteitis, and keratitis in both immunocompetent and immunocompromised patients (Mattsson, Haemig, & Olsen, 1999) and extrinsic allergic alveolitis (Yamamoto et al., 2002).

Genetic Characteristics

The molecular biology of *D. hansenii* is poorly established. Most strains are haploid, mate very rarely and diploidize transiently by somatogamous autogamy to form asci. Vegetative reproduction is by multilateral budding. Heterogamous conjugation between the cell and its bud precedes ascus formation. The spores are spherical with a warty wall (Nakase et al., 1998). Conjugation between separate cells may also occur, thus making difficult the use of classical genetic techniques.

The genome of *D. hansenii* has been explored in the Genolevures project, and the genome sequence is available in http://cbi.labri.fr/Genolevures/about/GL1_genome.php).

The presence of several linear plasmids designated pDHL1 (8.4 kB), pDHL2 (9.2 kB) and pDHL3 (15.0 kB) [79] and pDH1A and B (Cong, Yarrow, Li, & Fukuhara, 1994) in *D. hansenii* has also been described. The stability of these plasmids seems to depend on the osmotic pressure (Breuer & Harms, 2006).

Chromosomal DNA

Electrophoretic karyotyping of *Debaryomyces* strains showed very divergent chromosomal profiles at both inter-specific (Martorell et al., 2005) and intra-specific levels (Corredor, Davila, Casaregola, & Gaillardin, 2003; Petersen, Westall, & Jespersen, 2002). These studies demonstrated that the chromosomal arrangement in *D. hansenii* strains is heterogeneous, showing variations in both the number and size of the chromosomal bands on a Pulsed field gel electrophoresis (PFGE) gel.

The number of chromosomal bands observed varies from five to ten, but the most common chromosome number was found to be six. However no data about chromosomal profiles in isolates of *D. hansenii* from meat products are available.

Tandem gene duplications are 5–10 times more frequent in *D. hansenii* than in other yeasts, but few duplicated blocks have been detected (Lepingle et al., 2000). Tandem gene duplications refer to head-to-tail repeats of directly adjacent homogeneous genes or gene groups, whereas homologous blocks of genes (up to 250 contiguous genes) on different chromosomes, which duplicated simultaneously as a result of a poly-ploidization event (Friedman & Hughes, 2001) are called duplicated blocks.

Transformation Systems

Several transformation systems have been developed for *Debaryomyces* (for a revision see Breuer & Harms, 2006). The first methods to be tested were designed for *Schwanniomyces occidentalis* cells. However, the halo-tolerance of *Debaryomyces* seemed too difficult for the transformation process, thus no specific and effective system for inserting the genes into the genome of *D. hansenii* was developed.

A transformation system that did effectively transform *D. hansenii* was also developed based on a plasmid that is integrated into the chromosomal DNA. The expression of heterologous genes in *Debaryomyces* has been tested using a bacterial gene coding for a NADPH-dependent acetoacetyl-CoA reductase (*phb*B).

In another study, six heterologous yeast expression vectors were constructed for *D. hansenii*, using heterologous promoters from *S. cerevisiae*. The expression parameters of proteins in *D. hansenii* seemed to be similar to those in *S. cerevisiae*, with transcription being controllable by almost all of the *S. cerevisiae* and *D. hansenii* inducible promoters tested, except for the ADH2 gene promotor for alcohol dehydrogenase 2 from *S. cerevisiae* (Maggi & Govind, 2004).

Therefore, we can conclude that the genetic machinery and operational transformation methods required for efficient expression of the heterologous genes or whole pathways in *D. hansenii* are available. However, further investigation on this subject is recommended if secretory heterologous products are involved.

References

Andrade, M. J., Rodríguez, M., Sánchez, B., Aranda, E., & Córdoba, J. J. (2006). DNA typing methods for differentiation of yeasts related to dry-cured meat products. *International Journal of Food Microbiology, 107*, 48–58.

Barnett, J. A., Payne, R., & Yarrow, D. (2000). In J. A. Barnett, R. W. Payne, & D. Yarrow (Eds.), *Yeasts: Characteristics and identification* (3rd ed.). Cambridge: Cambridge University Press.

Boekhout, T., & Robert, V. (2003). *Yeast in food. Beneficial and detrimental aspects*. Cambridge: CRC Press.

Breuer, U., & Harms, H. (2006). *Debaryomyces hansenii* an extremophilic yeast with biotechnological potential. *Yeast, 23*(6), 415–437.

Cocolin, L., Urso, R., Rantsiou, K., Cantón, C., & Comi, G. (2006). Dynamics and characterization of yeasts during natural fermentation of Italian sausages. *FEMS Yeast Research, 6*, 692–701.

Cong, Y. S., Yarrow, D., Li, Y. Y., & Fukuhara, H. (1994). Linear DNA plasmids from *Pichia etchellsii, Debaryomyces hansenii* and *Wingea robertsiae. Microbiology, 140*, 1327–1335.

Corredor, M., Davila, A. M., Casaregola, S., & Gaillardin, C. (2003). Chromosomal polymorphism in the yeast species *Debaryomyces hansenii. Antonie van Leeuwenhoek, 84*(2), 81–88.

Dalton, H. K., Board, R. G., & Davenport, R. R. (1984). The yeasts of British fresh sausage and minced beef. *Antonie van Leeuwenhoek, 50*, 227–248.

Deák, T., & Beuchat, L. R. (1996). *Food spoilage yeasts*. CRC Boca Ratón, FL: Press.

de Llanos, R., Fernández-Espinar, M. T., & Querol, A. (2004). Identification of species of the genus *Candida* by RFLP analysis of the 5.8S rRNA gene and the two ribosomal internal transcribed spacers. *Antonie Van Leeuwenhoek, 85*, 175–185.

Demain, A. L., Phaff, H. J., & Kurtzman, C. P. (1999). In C. P. Kurtzman & J. W. Fell (Eds.), *The yeasts a taxonomic study*. Amsterdam: Elsevier.

Dillon, V. M. (1998). Yeasts and moulds associated with meat and meat products. In A. Davies & R. Board (Eds.), *The microbiology of meat and poultry* (pp. 85–117). London: Blacie Academic and professional.

Dillon, V. M., & Board, R. G. (1991). Yeast associated with red meats. *Journal of Applied Microbiology, 71*, 93–108.

Durá, M. A., Flores, M., & Toldrá, F. (2004a). Effect of growth phase and dry-cured sausage processing conditions on *Debaryomyces* spp. generation of volatile compounds from branched-chain amino acids. *Food Chemistry, 86*, 391–399.

Durá, M. A., Flores, M., & Toldra, F. (2004b). Effect of *Debaryomyces* spp. on the proteolysis of dry-fermented sausages. *Meat Science, 68*, 319–328.

Encinas, J. P., López-Díaz, T. M., García-López, M. L., Otero, A., & Moreno, B. (2000). Yeast populations on Spanish fermented sausages. *Meat Science, 54*, 203–208.

Esteve-zarzoso, B., Belloch, C., Uruburu, F., & Querol, A. (1999). Identification of yeast by RFLP analysis of the 5.8S rRNA gene and the two ribosomal internal transcribed spacers. *International Journal of Systematic Bacteriology, 49*, 329–337.

Esteve-Zarzoso, B., Zorman, T., Belloch, C., & Querol, A. (2003). Molecular characterisation of the species of the genus *Zygosaccharomyces. Systematic Applied Microbiology, 26*, 404–411.

Fernández-Espinar, M. T., Esteve-Zarzoso, B., Querol, A., & Barrio, E. (2000). RFLP analysis of the ribosomal internal transcribed spacer and the 5.8S rRNA gene region of the genus *Saccharomyces*: A fast method for species identification and the differentiation of flor yeasts. *Antonie Van Leeuwenhoek, 78*, 87–97.

Fernández-Espinar, M. T., Martorell, P., de Llanos, R., & Querol, A. (2006). Molecular methods to identify and characterize yeast in foods and beverages. In A. Querol & G. H. Fleet (Eds.), *Yeast in food and beverages. The yeast handbook* (pp. 1–13). Berlin: Springer.

Fleet, G. H. (1990). Food spoilage yeasts. In F. F. T. Spencer & D. M. Spencer (Eds.), *In yeast technology* (pp. 124–166). Berlin, Heidelberg, Germany: Springer-Verlag.

Fleet, G. H. (2006). The commercial and community significance of yeasts in food and beverage production. In A. Querol & G. H. Fleet (Eds.). *Yeast in Food and beverages. The Yeast Handbook* (pp. 1–13). Berlin: Springer.

Flores, M., Durá, M. A., Marco, A., & Toldrá, F. (2004). Effect of *Debaryomyces spp.* on aroma formation and sensory quality of dry-fermented sausages. *Meat Science, 68*, 439–446.

Friedman, R., & Hughes, A. L. (2001). Gene duplication and the structure of eukaryotic genomes. *Genome Research, 11*, 373–381.

Fung, D. Y. C., & Liang, C. (1990). Critical Review of isolation, detection, and identification of yeasts from meat products. *CRC Critical Reviews in Food Science and Nutrition, 29*, 341–379.

Gardini, F., Suzzi, G., Lombardi, A., Galgano, F., Crudele, M. A., Andrighetto, C., Schirone, M. & Tofalo, R. (2001). A survey of yeasts in tradicional sausages of southern Italy. *FEMS Yeast Research*, 161–167.

Golden, D. A., Beuchat, L. R., & Hitchcock, H. L. (1994). Changes in fatty acid composition of *Zygosaccharomyces rouxii* as influenced by solutes, potassium sorbate and incubation temperature. *International Journal of Food Microbiology, 21*, 293–303.

Grazia, L., Suzzi, G., Romano, P., & Giudici, P. (1989). The yeasts of meat products. *Yeast, 5*, 495–499.

Guerzoni, M. E., Lanciotti, R., & Marchetti, R. (1993). Survey of the physiological properties of the most frequent yeasts associated with commercial chilled foods. *International Journal of Food Microbiology, 17*, 329–341.

Guillamón, J. M., Sabate, J., Barrio, E., Cano, J., & Querol, A. (1998). Rapid identification of wine yeast species based on RFLP analysis of the ribosomal internal transcribed spacer ITS region. *Archives of Microbiology, 169*, 387–392.

Hammes, W. P., & Knauf, H. J. (1994). Starters in the processing of meat products. *Meat Science, 36*, 155–168.

Ismail, S. A., Deak, T., El-Rahman, H. A., Yassien, M. A., & Beuchat, L. R. (2000). Presence and changes in populations of yeasts on raw and processed poultry products stored at refrigeration temperature. *International Journal of Food Microbiology, 62*(1–2), 113–121.

James, S. A., Cai, J., Roberts, I. N., & Collins, M. D. (1997). Phylogenetic analysis of the genus *Saccharomyces* based on 18S rRNA gene sequences: Description of *Saccharomyces kunashirensis* sp. nov. and *Saccharomyces martiniae* sp. nov. *International Journal of Systematic Bacteriology, 47*, 453–460.

Kurtzman, C. P., & Robnett, C. J. (1998). Identification and phylogeny of ascomycetous yeasts from analysis of nuclear large subunit 26S ribosomal DNA partial sequences. *Antonie van Leeuwenhoek, 73*, 331–371.

Leclercq-Perlat, M. N., Corrieu, G., & Spinnler, H. E. (2004). Comparison of volatile compounds produced in model cheese medium deacidified by *Debaryomyces hansenii* and *Kluyveromyces marxianus*. *Journal of Dairy Science, 87*, 1545–1550.

Lepingle, A., Casaregola, S., Neuveglise, C., Bon, E., Nguyen, H., Artiguenave, F., et al. (2000). Genomic exploration of the hemiascomycetous yeasts: 14. *Debaryomyces hansenii* var. *hansenii*. *FEBS Lett, 487*, 82–86.

Li, W. H. (1997). *Molecular evolution*. Sunderland, MA: Sinauer Associates.

Maggi, R. G., & Govind, N. S. (2004). Regulated expression of green fluorescent protein in *Debaryomyces hansenii*. *Journal of Industrial Microbiology and Biotechnology, 31*, 301–310.

Martín, A., Córdoba, J. J., Aranda, E., Córdoba, M. G., & Asensio, M. A. (2006). Contribution of a selected fungal population to the volatile compounds on dry-cured ham. *International Journal of Food Microbiology, 110*, 8–18.

Martín, A., Córdoba, J. J., Benito, M. J., Aranda, E., & Asensio, M. A. (2003). Effect of *Penicillium chrysogenum* and *Debaryomyces hansenii* on the volatile compounds during controlled ripening of pork loins. *International Journal of Food Microbiology, 84*, 327–338.

Martorell, P., Fernández-Espinar, M. T., & Querol, A. (2005). Sequence-based identification of species belonging to the genus *Debaryomyces*. *FEMS Yeast Research, 5*, 1157–1165.

Mattsson, R., Haemig, P. D., & Olsen, B. (1999). Feral pigeons as carriers of *Cryptococcus laurentii*, *Cryptococcus uniguttulatus* and *Debaryomyces hansenii*. *Medical Mycology, 37*, 367–369.

Metaxopoulos, J., Stavropoulos, S., Kakouri, A., & Samelis, J. (1996). Yeasts isolated from traditional Greek dry salami. *Italian Journal of Food Science, 1*, 25–32.

Monte, E., Villanueva, J. R., & Domínguez, A. (1986). Fungal profiles of Spanish country-cured hams. *International Journal of Food Microbiology, 3*, 355–359.

Muyzer, G., de Waal, E. C., & Uitterlinden, A. G. (1993). Profiling of complex microbial populations by denaturing gradient gel electrophoresis analysis of polymerase chain reaction-amplified genes encoding for 16S rRNA. *Applied and Environmental Microbiology, 59*, 695–700.

Muyzer, G., & Smalla, K. (1998). Application of denaturing gradient gel electrophoresis (DGGE) and temperature gradient gel electrophoresis (TGGE) in microbial ecology. *Antonie van Leeuwenhoek, 73*, 127–14.

Nakase, T., Suzuki, M., Phaff, H. J., & Kurtzman, C. P. (1998). *Debaryomyces* Lodder & Kreger-van Rij Nom. Cons. In C. P. Kurtzman & J. W. Fell (Eds.), *The yeasts – A taxonomic study* (pp. 157–173). Amsterdam: Elsevier.

Nishikawa, A., Tomomatsu, H., Sugita, T., Ikeda, R., & Shinoda, T. (1996). Taxonomic position of clinical isolates of *Candida famata*. *Journal of Medical and Veterinary Mycology, 34*, 411–419.

Nortje, G. L., Nel, L., Jordaan, E., Badenhorst, K., Goedhart, E., & Holzapfel, W. H. (1990). The aerobic psychrotrophic populations on meat and meat contact surfaces in a meat production system and on meat stored at chill temperature. *The Journal of Applied Bacteriology, 68*, 335–344.

Núñez, F., Rodríguez, M. M., Córdoba, J. J., Bermúdez, M. E., & Asensio, M. A. (1996). Yeast population during ripening of dry-cured Iberian ham. *International Journal of Food Microbiology, 29*, 271–280.

Petersen, K. M., Westall, S., & Jespersen, L. (2002). Microbial succession of *Debaryomyces hansenii* strains during the production of Danish surfaced-ripened cheeses. *Journal of Dairy Science, 85*, 478–486.

Prista, C., Almagro, A., Loureiro-Dias, M. C., & Ramos, J. (1997). Physiological basis for the high salt tolerance of *Debaryomyces hansenii*. *Applied and Environmental Microbiology, 63*, 4005–4009.

Prista, C., Loureiro-Dias, M. C., Montiel, V., García, R., & Ramos, J. (2005). Mechanisms underlying the halotolerant way of *Debaryomyces hansenii*. *FEMS Yeast Research, 5*, 693–701.

Querol, A., Belloch, C., Fernández-Espinar, M. T., & Barrio, E. (2003). Molecular evolution in yeast of biotechnological interest. *International Microbiology, 6*, 201–205.

Quiros, M., Martorell, P., Valderrama, M. J., Querol, A., Peinado, J. M., & de Silóniz, M. I. (2006). PCR-RFLP analysis of the IGS region of rDNA: A useful tool for the practical discrimination between species of the genus *Debaryomyces*. *Antonie Van Leeuwenhoe, 90*(3), 211–219.

Ramos, J. P., Valente, P., Hagler, A. N., & Leoncini, O. (1998). Restriction analysis of ITS region for characterization of the *Debaryomyces* species. *The Journal of General and Applied Microbiology, 44*, 399–404.

Rantsiou, K., Iacumin, L., Cantoni, C., Cattaneo, P., Comi, G., & Cocolin, L. (2005). *Applied and Environmental Microbiology, 71*, 1977–1986.

Romano, P., Capece, A., & Jespersen, L. (2006). Taxonomic and ecological diversity of food and beverages yeasts. In A. Querol & G. H. Fleet (Eds.), *Yeast in food and beverages. The yeast handbook* (pp. 13–55). Berlin: Springer.

Samelis, J., & Sofos, J. (2003). Yeast in meat and meat products. In T. Boekhout & V. Robert. *Yeasts in food, beneficial and detrimental aspects* (pp. 234–265). Hamburg, Germany Behr's: Verlag Gmbh & Co.

Samelis, J., Stavropoulos, S., Kakouri, A., & Metaxopoulos, J. (1994). Quantification and characterization of microbial population associated with naturally fermented Greek dry salami. *Food Microbiology, 11*, 447–460.

Selgas, M. D., & García, M. L. (2007). Starter cultures: Yeasts. In F. Toldrá, Y. H. Hui, I. Astiasarán, W. K. Nip, J. G. Sebranek, E. T. F. Silveira, L. H. Stahnke, & R. Talon (Eds.), *Handbook of fermented meat and poultry* (pp. 159–169), Ames, Iowa: Blackwell Publishing.

Sinigaglia, M., Lanciotti, R., & Guerzoni, M. E. (1994). Biological and physiological characteristics of *Yarrowia lipolytica* strains in relation to isolation source. *Canadian Journal of Microbiology, 40*, 54–59.

Swiss Agency for the Environment, Forests and Landscape (SAEFL). (2004). *Guideline classifications of organisms: Fungi*. Berne, Switzerland: SAEFL.

Toldrá, F. (1988). Proteolysis and lypolisis in flavour development of dry-cured meat products. *Meat Science, 65*, 935–948.

Viljoen, B. C., Dykes, G. A., Collis, M., & von Holy, A. (1993). Yeasts associated with Vienna sausage packaging. *International Journal of Food Microbiology, 18*, 53–62.

Wong, B., Kiehn, T. E., Edwards, F., Bernard, E. M., Marcove, R. C., de Harven, E., et al. (1982). Bone infection caused by *Debaryomyces hansenii* in a normal host: A case report. *Journal of Clinical Microbiology, 16*, 545–548.

Yamamoto, Y., Osanai, S., Fujiuchi, S., Yamazaki, K., Nakano, H., Ohsaki, Y., et al. (2002). Extrinsic allergic alveolitis induced by the yeast *Debaryomyces hansenii*. *The European Respiratory Journal, 20*, 1351–1353.

Chapter 8
Characteristics and Applications of Molds

Elisabetta Spotti, Elettra Berni, and Cristina Cacchioli

Introduction

Molds on Aged Meat Products

The ripening and maturing techniques used to process meat products quickly lead to a distinctive surface colonization by a great number of mycetes.

In matured products such as dry-cured hams, a typical microbial population can grow on the surface; apart from bacteria such as micrococci and staphylococci, both yeasts and molds can colonize surface layers of dry-cured meats for most of the seasoning time, which usually lasts 12–24 months (Lori, Grisenti, Parolari, & Barbuti, 2005; Martin, Cordoba, Nunez, & Asensio, 2004). In general, yeasts tend to form a film on the whole ham (Comi & Cantoni, 1983), contributing to the development of the final sensory characteristics due to their enzymatic activity (Simoncini, Rotelli, Virgili, & Quintavalla, 2007), while molds tend to develop their mycelium after yeasts have grown. Uncontrolled fungal growth can lead to different detrimental effects: anomalous aspect; changes in technological properties and nutritive value of the product, such as the so-called "phenic acid defect" (Baldini & Spotti, 1995); production of toxic secondary metabolites such as mycotoxins (Pietri, Bertuzzi, Gualla, & Piva, 2006); formation of allergenic substances and spreading of mycosis among workers employed in the meat industry (De Hoog, Guarro, Gené, & Figueras, 2000).

In any case, moulding is generally tolerated in aged products if these molds:

- have an antioxidative effect, contributing to keep the color;
- prevent the surface to become sticky or slimy;
- contribute to lipolysis and proteolysis, concurring at the development of characteristic aromatic compounds at the end of the process (Martín, Cordoba, Aranda, Cordoba, & Asensio, 2006).

E. Spotti
Department of Microbiology, Stazione Sperimentale per l'Industria delle Conserve Alimentari, SSICA (Experimental Station for the Food Preserving Industry), Viale Tanara 31/A, 43100 PARMA, Italy

In ripened products such as salami, surface mycoflora includes both yeasts, which play a leading role in the fermentation process, and filamentous fungi (molds), which are considered fundamental to impart desirable appearance as well as satisfactory technological and sensory traits in some meats (Baldini et al., 2000; Samelis & Sofos, 2003). At the first step of the process, yeasts are the predominant micro-organisms on casings, totalling about 95% of the surface microflora. After the first 2 weeks, molds and yeasts are present in equal amounts. At the end of the ripening process (from 2nd to 8th week for sausages; from 4th to 8th month for larger-sized products) mycoflora is mainly represented by molds, while yeasts may undergo one logarithmic decrement. On these products, molds usually prevail on yeasts; this is due to the progressive reduction of surface water activity in sausages and due to the invasive way molds grow on the surfaces (by apical increase and lateral branching of cells called "hyphae", and by substantial production of conidia, the reproductive units of molds).

In general, molding of ripened products is a favored practice in several European countries. In Italy, Rumania, Bulgaria, France, Hungary, Switzerland, Germany, Spain, Austria, and Belgium selected molds can be used as starter cultures on casings of meat products to improve their quality. This is due to the fact that the mycelium:

- prevents excessive drying, allowing water loss and therefore homogeneous dehydration of the product;
- protects fat portions from oxidation because it metabolizes and consumes peroxides, thus preventing rancidity;
- reduces the O_2 levels on the surface of the product, thus avoiding oxidative processes and improving meat color;
- contributes to enhancing the flavor of the final product (especially when natural casing is used), because it breaks up fats, proteins, and lactic acid, thus favoring an increase in pH;
- makes sausage peeling easier, thanks to the differentiation of the fungal basal hyphae into a sort of root called "rhizoid", which can penetrate the inner part of the mixture.

Parameters Affecting Mold Growth

The physico-chemical parameters recorded in industrial environments are directly connected with microbial growth (Baldini et al., 2000; Battilani et al., 2007)

In particular, surface molding of dry-cured meats (i.e., hams) is influenced by parameters such as levels of environmental contamination by fungal spores and length of exposure of the products to air; RH and T values recorded in plants (air flux is very important for keeping the correct thermohygrometric parameters in each part of the drying chamber as well as to get the correct mass transfer between product and air during the first 5 months of drying); fat and protein content of the exposed muscle portion; presence of autochthonous microrganisms (i.e., yeasts) capable of inhibiting mold growth. Usually, molds growing on these kinds of products tend to be more xerophilic than those which often occur during sausage ripening; this is why *Eurotium* species can be expected to prevail over *Penicillium* ones (Spotti, Berni, & Cacchioli, 2001; Spotti, Mutti, & Campanini, 1989).

Surface molding of ripened meats (i.e., salami) is instead influenced by: chemical composition of the mixtures, with particular focus on the mincing degree of fat portions; the relative humidity of air and the temperature conditions (RH and T) usually applied in industrial plants; drying methods (how surface water activity, a_w, decreases during product ripening); aeration methods (ventilation cycles at high air velocities alternating with cycles in which the product is allowed to rest favoring fungal growth). According to this, it has been demonstrated that the matured sausages of a greater size and with longer ripening times usually show initial growth mainly characterized by hydrophilic molds, while towards the end of the ripening period there is a prevalence of xerotolerant strains. These xerotolerant strains tend to prevail over the species grown for the first time, which often prove to be inhibited by surface a_w reduction below 0.90 (Spotti, Berni, & Cacchioli, 2007).

Environment-Contaminating Mold Species

For a long time, molding was left to contamination by environment-contaminating mold species, mainly belonging to the genera *Penicillium*, *Aspergillus*, and *Eurotium*, which better adapt to the technological conditions to which these products are subjected.

As regards matured products such as dry cured hams, both the prevalence of species belonging to the genus *Eurotium* and the growth of more xerotolerant fungal species belonging to genus *Penicillium*, in association with the former, can be related to peculiar characteristics of adaptation by each fungal strain to surface a_w and to the thermohygrometric conditions applied in the plants; their presence has proved to be minimized by reducing RH down to 80–85% (Battilani et al., 2007; Spotti et al., 1989; Spotti et al., 2001). This is confirmed by most isolations carried out at the SSICA in the last years on exposed muscle portion of hams from different Italian industrial plants (Tables 8.1 and 8.2). As the tables show, the conditions applied in these plants allow a more considerable amount of negative (not contaminated by fungal spores) samples in seasoning rather than in pre-seasoning stages.

Table 8.1 Percentage of hams investigated at the SSICA, on which absence / presence of molds was detected (93 samples investigated)

Amount of molds detected on hams	Pre-seasoning (%)	Seasoning (%)	Total (%)
Absent	45.8	84.4	64.5
Just one species on each ham	23.0	6.9	15.1
Several species on each ham	31.2	8.7	20.4

Note: Percentage of total samples is obtained calculating the average of pre-seasoning and seasoning values.

Table 8.2 Frequency (number of lots investigated and found positive/negative for molds) of the fungal species isolated at SSICA on the exposed muscle portion of hams during the pre-seasoning and seasoning stages from 2000 to 2007

Fungal species	Frequency Pre-ripening	Ripening
Eurotium repens	15	6
E. rubrum	12	4
E. herbariorum	2	0
Penicillium nalgiovense	11	5
P. solitum	11	4
P. griseofulvum	9	1
P. verrucosum / P. nordicum	9	0
P. brevicompactum	4	1
P. expansum	4	0
P. camemberti	3	1
P. aurantiogriseum	3	0
P. citrinum	2	0
P. commune	2	0
P. waksmanii	2	0
P. gladioli	1	1
P. chrysogenum	1	0
P. echinulatum	1	0
P. implicatum	1	0
P. olsonii	1	0
P. viridicatum	1	0
Hypopichia burtonii	0	5
Negative	22	38

In particular, the extent of fungal growth during pre-seasoning has been detected with higher frequency only in the aitchbone area, which often is the wettest part of the ham. In the limited number of cases where a diffused molding had been observed, the prevailing mold was almost always represented by *Eurotium repens*, followed by *E. rubrum*; these molds are two of the most xerophilic species detectable on hams. The presence of the most xerophilic species belonging to the genus *Penicillium*, such as *P. aurantiogriseum, P. camemberti, P. solitum, P. verrucosum, P. nordicum*, and *P. brevicompactum*, has been detected only in a minor percentage of the isolations performed, with prevalence on hams in the pre-seasoning stage, whereas it is only sporadic on those in the seasoning stage (Table 8.1). Sporadicity had also characterized the presence of other species, indicated in Table 8.1, during both pre-seasoning and seasoning stages. Individually taken, they were even less diffused on the product than the previous ones, even if they are recognized as possible contaminants of aged meats. As regards *P. nordicum*, which has been recently differentiated from *P. verrucosum* by molecular techniques (Geisen, Mayer, Karolewiez, & Faerber, 2004) and which was found to be associated with meats and cheeses (Larsen, Svendsen, & Smeedsgaard, 2001) where it is the responsible for Ochratoxin A (OTA) production, it has been detected only occasionally. Despite this, it can be assumed that its growth in the samples found positive for OTA may have occurred before sampling, but that the subsequent growth of different molds and/or the intervention of mites on the ham surface rendered *P. nordicum* undetectable. All this is supported by the fact that fungal species tend to compete each other, one prevailing over the other or undergoing mutual inactivation in the long maturing period, where variations in surface RH occur, that can affect mold growth.

The control of thermohygrometric conditions then results fundamental: this is why in medium- or larger-sized plants equipped with air-conditioning and ventilating systems, thermohygrometric parameters are usually low enough to keep the ham surface dry and to prevent mold growth. In case the above-mentioned parameters reach high values during the first weeks of resting and an homogeneous dehydration of the product is not carried on correctly, unexpected changes should happen and persist in the final product, such as the so-called "potato defect" and the "phenic acid defect". The former has been attributed to some bacteria belonging to the genus *Pseudomonas* (Blanco, Barbieri, Mambriani, Spotti, & Barbuti, 1994) and consists of the development of an odor similar to that of the potatoes; the latter is due to a fungal colonization of the aitchbone area or of the covering fat close to the pigskin by *P. commune* (Baldini & Spotti, 1995; Spotti, Mutti, & Campanini, 1988) and *P. solitum* (Pitt & Hocking, 1997), and it is usually observed when hams with high levels of surface contamination by molds are not correctly processed during pre-resting.

With regard to ripened meats such as salami, the prevalence of xerotolerant strains belonging to the genus *Penicillium* subgenus *Penicillium* over those belonging to more xerophilic genera such as *Aspergillus* or *Eurotium* is due to the easier growth of most *Penicillium* species, which better survive in environments with RH values ranging from 85 to 92% and with temperatures from 10 to 20°C (Table 8.3). Growth of hydrophilic molds can occur on these products, such as: *Scopulariopsis*

Table 8.3 Frequency (number of lots investigated and found positive for molds) of the fungal species isolated at SSICA on ripened salami from 2000 to 2007

Fungal species	Frequency	Fungal species	Frequency
P. nordicum	36	*P. camemberti*	5
P. brevicompactum	32	*P. olsonii*	4
Aspergillus candidus	29	*Eurotium herbariorum*	3
P. nalgiovense	24	*Eurotium repens*	3
P. aurantiogriseum	24	*P. commune*	3
Eurotium rubrum	20	*P. citrinum*	3
P. griseofulvum	20	*Scopulariopsis brumptii*	3
P. solitum	15	*Scopulariopsis flava*	3
P. gladioli	12	*Scopulariopsis brevicaulis*	2
P. waksmanii	12	*Eupenicillium katangense*	2
Scopulariopsis candida	12	*A. versicolor*	2
P. chrysogenum	11	*A. sclerotiorum*	1
P. implicatum	9	*A. niveus*	1
Eupenicillium spp.	6	*P. fellutanum*	1
Aspergillus ochraceus	5	*Talaromyces wortmannii*	1
Eupenicillium cinnamopurpureum	5	*Talaromyces luteus*	1
Mucor spp.	5		

brevicaulis, *S. brumptii*, and *Mucor* spp., which tend to form darkish spots on casings; *S. flava*, whose growth can impart ripened meats either desirable appearance as well as satisfactory technological and sensory traits or reddish undesirable spots, in case it produces fruiting bodies derived from sexual reproduction (Abbott, Lumley, & Sigler, 2002; Spotti et al., 2007); *S. candida*, which produces whitish mycelium and conidia, giving the sausages a good appearance. These species must be removed because the former do not allow homogeneous drying of the product in the first steps of the process and the latter can produce a nasty ammonia odor because of its strong proteolytic activity in the final step of the ripening.

Fungal Starter Cultures in Ripened Meats

In European countries where food industries aim to obtain traditional products, the ripening techniques normally applied in the industrial plants usually allow for the growth of characteristic whitish molds which can also inhibit the multiplication of other molds, especially those which proved to be potentially toxigenic and/or with a mycelium having an undesirable color (Baldini et al., 2000). In particular, recent findings concerning the ability of many *Penicillium* and *Aspergillus* species, frequently found on the surface of dry sausages to produce mycotoxins, have stressed the importance of studies on the possibility of controlling mold growth on raw sausages during ripening by using mold starters.

These starters consist of selected fungal isolates that proved to impart a desirable appearance and good technological and sensory characteristics to the product, proved unable to synthesize antibiotics and/or toxic metabolites such as mycotoxins and do not cause well-known allergies or mycoses. For the reasons mentioned above, they can be used to inoculate fermented sausages by spraying or by immersion in a conidial suspension.

Among the starters, which can be considered both safe and effective, *P. nalgiovense* has been frequently isolated from the surface of ripened meat products as a "domesticated species" from *P. chrysogenum* (wild type) (Pitt & Hocking, 1997); at present it is routinely used as a starter culture in many traditional productions and so it is often present in the environmental air.

Even *P. chrysogenum* has been mentioned as a suitable starter for mold-fermented products (Hammes & Knauf, 1994; Ministero della Sanità, 1995), since some non-toxic strains proved to produce proteases which contributed to the generation of characteristic flavored compounds in dry fermented sausages (Martín et al., 2002; Rodríguez, Núñez, Cordoba, Bermúdez, & Asensio, 1998). However, at present *P. chrysogenum* is not always considered acceptable in the industrial practice because the marketed isolates with a lightly pigmented conidiation may again start producing green conidia, and most of the strains tested may produce antibiotics and toxic substances such as roquefortine C (Pitt & Hocking, 1997).

P. camemberti had also been considered as a starter culture for the meat industry because of its ability to improve the sensory characteristics of dry fermented sausages (Bruna et al., 2003). Nevertheless, all the studied isolates proved to be cyclopiazonic-acid producers, so it has not been taken into account as a commercial starter (Samson & Frisvad, 2004). Other strains belonging to the non-toxigenic *Penicillium* species were also excluded by the industrial practice because they produce darkish-green conidia which may impair the outward appearance of the sausage.

In order to safeguard typical productions, in case both the appearance and aroma imparted to ripened meats by autochthonous molds come up to the expectations of a peculiar quality, molding should be favored by means of preventive and specific studies. These studies should be at first focused on the recognition of the genus and the species to which the most occurring (on that specific meat product) strains belong; then, on the selection of the strain which has been considered safer and more suitable because of its biochemical and physiological characteristics.

At the beginning, the identification of the species will be entrusted to expert mycologists, who are usually able to carry out any morphological, biochemical, and, when possible, molecular assays in order to identify molds. In the industrial practice, the dominance of the above-mentioned autochthonous strains over the environmental contaminating fungal species will be then favored by means of the most appropriate technological interventions. Regular inspections concerning the real presence of the autochthonous starter culture inoculated are part of any Quality Control system, and they must be carried on throughout the ripening process.

In the last years, several experiments have been carried out at the SSICA, making use of two isolates of *P. gladioli*: one from Professor Grazia collection (Grazia, Romano, Bagni, Roggiani, & Guglielmi, 1986), the other from the SSICA mycological collection. At present, both strains, isolated on meat products from Emilia Romagna (Northern Italy), can be considered as "real autochthonous starters". They have proved to be non-mycotoxin producers and impart to ripened meats a desirable appearance, due to a pale-grayish conidiation (Baldini et al., 2005; Grazia et al., 1986; Samson & Frisvad, 2004).

S. flava, for a long time now used as starter culture in some French industries producing dry-smoked sausages (Dragoni, Cantoni, Papa, & Vallone, 1997), has been isolated from salami and used too as "autochthonous starter" in recent experiments carried out at the SSICA. This mold morphologically resembles *S. brevicaulis* but, unlike it, *S. flava* can impart on ripened meats a desirable appearance, due to a grayish-beige conidiation. Only if a high relative humidity is present, *S. flava* must be avoided, because it develops nasty reddish spots and a strong ammonia odor.

Lipolytic and Proteolytic Activity of Molds

In the last years, a lot of studies concerning the lipolytic and proteolytic activity of the molds involved in the ripening process, including *P. nalgiovense, P. chrysogenum*, and *P. camemberti*, were carried out throughout the world (Bruna et al., 2003; Geisen, Lücke, & Kröckel, 1992; Ockerman, Céspedes Sánchez, Ortega Mariscal, & León-Crespo, 2001; Rodriguez et al., 1998).

At the SSICA, a recent experiment (Spotti & Berni, 2005) focused on the biochemical activity of two starters, *P. nalgiovense* and *P. gladioli*, in comparison with that of the environment-contaminating molds grown on ripened products from Northern Italy (*P. griseofulvum*, *P. brevicompactum*, *P. olsonii*, *P. implicatum*, *P. nordicum*, *Talaromyces wortmannii*, *Mucor* sp.). With regard to lipid breakdown, both *P. nalgiovense* and *P. gladioli* proved to have good lipolytic activity; the lipase production of the former was increased by lower temperatures (more marked at 14°C than at 18°C), whereas that of the latter was greater at the higher temperatures. With regard to protein breakdown, *P. nalgiovense* showed a strong proteolytic activity both at 14 and at 18°C, whereas *P. gladioli* did not show this activity at the two tested temperatures. Despite their different metabolisms, *P. gladioli* used as a starter culture in SSICA pilot plants and in industrial practice proved to supply ripened products with good technological and sensory features, similar to *P. nalgiovense*.

Within the framework of a three-year project on Italian traditional salami, the biochemical characteristics of the strains isolated on salami produced in the Nebrodi area (Sicily) were studied by lipolytic and preoteolytic in vitro tests at 10°C, 14°C, and 18°C (Berni, Cacchioli, Castagnetti, Sarra, & Spotti, 2007). Their enzymatic activities have been paired with those of molds isolated from salami produced in the Northern Italy (see Tables 8.4 and 8.5). The results showed that both lipolysis and proteolysis increased with incubation time and temperature. In general, morphologically similar species belonging to *Penicillium* subgenus *Penicillium* such as *P. aurantiogriseum*, *P. solitum*, *P. brevicompactum*, *P. nordicum*, *P. griseofulvum*,

Table 8.4 Lipolytic activity (mm hydrolyzed medium/mm total medium × 100) of the strains isolated on Italian salami, as a function of time and temperature of incubation

Fungal strains tested	10°C			14°C			18°C		
	7 d	14 d	21 d	7 d	14 d	21 d	7 d	14 d	21 d
P. aurantiogriseum N1	0.0	26.3	50.0	17.5	37.5	65.0	25.0	52.5	73.8
P. solitum N1	25.0	68.8	100.0	41.3	71.3	100.0	50.0	80.0	92.5
P. gladioli N1	0.0	22.5	45.0	8.8	35.0	62.5	21.3	43.8	62.5
P. gladioli N2	0.0	25.0	48.8	10.0	37.5	57.5	18.8	47.5	72.5
P. brevicompactum N1	7.5	47.5	67.5	25.0	50.0	77.5	31.3	63.8	87.5
P. nordicum N1	13.8	42.5	62.5	23.8	48.8	100.0	36.3	72.5	100.0
P. griseofulvum N1	5.0	32.5	48.8	13.8	36.3	58.8	23.8	50.0	70.0
P. implicatum N1	0.0	6.3	17.5	0.0	8.8	23.8	0.0	13.8	32.5
P. nordicum N2	8.8	38.8	52.5	20.0	45.0	71.3	27.5	57.5	86.3
P. gladioli SS1	17.5	62.5	90.0	33.8	67.5	100.0	40.0	82.5	100.0
P. nalgiovense PC1	2.5	21.3	42.5	10.0	n.d.	n.d.	13.3	n.d.	n.d.
P. chrysogenum PC1	0.0	17.5	38.8	12.2	n.d.	n.d.	31.1	n.d.	n.d.
P. griseofulvum PC1	0.0	18.8	37.5	12.2	34.6	49.2	13.3	40.2	63.6
P. aurantiogriseum PC1	0.0	17.5	23.8	13.3	26.2	38.1	26.7	50.2	74.8
P. brevicompactum PC1	0.0	20.0	38.8	11.1	27.2	42.3	28.8	56.2	79.9
S. flava PC1	0.0	5.0	11.3	9.6	n.d.	n.d.	15.6	n.d.	n.d.

Note. The capital letter after the strain name indicates where it comes from (N for Nebrodi area; SS for SSICA's collection; PC for Northern Italy); the number after the capital letter identifies the strain tested, in case more than one strain belonging to the same species has been isolated

8 Characteristics and Applications of Molds

Table 8.5 Proteolytic activity (g proteolyzed medium/g total medium × 100) of the strains isolated on Italian salami, as a function of time and temperature of incubation

Strains tested	10° C		14° C		18° C	
	14 d	21 d	14 d	21 d	14 d	21 d
P. aurantiogriseum N1	1.5	11.3	12.5	21.0	17.8	28.6
P. solitum N1	0.0	0.0	3.6	7.5	8.7	17.8
P. gladioli N1	0.0	0.0	7.5	18.4	12.6	23.1
P. gladioli N2	0.0	9.6	8.4	27.5	9.5	21.3
P. brevicompactum N1	2.3	2.3	1.5	23.3	16.5	29.3
P. nordicum N1	0.0	0.0	0.0	5.8	9.3	9.9
P. griseofulvum N1	0.7	12.2	11.4	21.9	18.7	28.8
P. implicatum N1	0.0	0.0	0.0	0.0	0.0	4.0
P. nordicum N2	0.0	0.0	0.0	0.0	0.0	2.9
P. gladioli SS1	0.0	0.0	0.0	0.0	0.0	0.0
P. nalgiovense PC1	4.6	6.7	6.8	15.0	12.9	28.5
P. griseofulvum PC1	0.0	15.6	6.5	21.9	23.6	34.8
P. brevicompactum PC1	2.6	13.7	3.0	13.5	14.8	19.9
P. nordicum PC1	0.0	0.0	4.1	7.9	4.9	13.2
S. flava PC1	0.0	0.0	0.0	0.0	0.0	0.0

Note. The capital letter after the strain name indicates where it comes from (N for Nebrodi area; SS for SSICA's collection; PC for Northern Italy); the number after the capital letter identifies the strain tested, in case more than one strain belonging to the same species has been isolated

and *P. gladioli* proved to have considerable lipolytic activity (≥40%) in the course of time (Table 8.4); this is important since lipolysis is responsible for the formation of aroma compounds typical of salami. On the contrary, the above strains showed greater differences in proteolytic activity (Table 8.5), which proved to be lower than the lipolytic one, but which is anyway less significant, since it is mainly carried out by fermentative bacteria grown in the mince and by proteases native to meat.

Growth and Competition Tests

Several studies were carried out from 1999 to 2007 at the SSICA pilot plants in Parma (Baldini et al., 2000; Baldini et al., 2005; Spotti et al., 2007) to evaluate the growth and the competition ability of some fungal starters available on the market and of some others belonging to different collections against some undesirable species in model systems, reproducing normal maturing conditions on product surface.

Growth

The research works carried out in model systems reproducing traditional Italian aging processes were focused on the growth of different starter strains, in comparison with some of the most occurring environment-contaminating species, as a function of temperature (from 8 to 22°C) and water activity (from 0.78 to 0.92)

Table 8.6 Optimal growth conditions with corresponding values of minimum lag time and growth rate for different fungal strains

Strains tested	Optimum T (°C)	Optimum RH (%)	Minimum lag time (days)	Growth rate (mm/day)
P. camemberti M	17	91	2	1.7
P. camemberti P	12	91	5	1.2
P. gladioli P	19	88	4	1.0
P. gladioli B	17	90	4	1.2
P. nalgiovense	14	90	4	2.0
P. solitum	16	89	5	1.4
P. nordicum[a]	20	88	4	1.3
P. chrysogenum	16	89	3	1.1
A. ochraceus	19	86	1	2.2
E. rubrum	18	90	3	4.23
H. burtonii	14	90	5	1.21

[a]The strain, first identified by morphological methods as *Penicillium verrucosum*, was later named *P. nordicum* according to molecular identification by Professor R. A. Samson in 2004.
Source. Adapted from Spotti, Busolli, & Palmia, 1999.

(Spotti et al., 2007; Spotti & Berni, 2005; Spotti, Busolli, & Palmia, 1999). For each strain, the average growth radial rate (obtained from the linear correlation between colony diameters and growth time during the linear phase) and the average duration of lag-time were determined (see Tables 8.6 and 8.7).

In dry-cured products, *E. rubrum* proves to dominate throughout the seasoning. In fact, it is the most xerophilic species occurring on dry-cured meats such as hams, where it is well favored by the thermohygrometric conditions of the environments and by long-term aging. It also tends to prevail on *H. burtonii*, which often

Table 8.7 Lag time and growth rate at 15°C and RH 88–90% for different fungal strains

Strains tested	Lag time (days)	Growth rate (mm/day)
P. camemberti M	4	1.6
P. camemberti P	5	1.2
P. gladioli P	5	0.7
P. gladioli B	5	1.0
P. nalgiovense	4	2.0
P. nordicum[a]	6	1.0
P. implicatum	5	0.8
P. griseofulvum	5	1.2
P. solitum	5	1.3
P. verrucosum	6	1.0
P. chrysogenum	3	1.0
A. ochraceus	14	1.1
E. rubrum	7	3.8
H. burtonii	5	1.2

[a]The strain, first identified by morphological methods as *Penicillium verrucosum*, was later named *P. nordicum* according to molecular identification by Professor R. A. Samson in 2004.
Source. Adapted from Spotti, Busolli, & Palmia (1999).

derives from the spreadable fat mince ("sugna") after the sixth month of seasoning and which is now beginning to be studied (Simoncini et al., 2007).

On the contrary, the molds most likely to grow during drying in ripened salami are *P. gladioli, P. solitum, P. nordicum*, and *P. chrysogenum*, since they are favored by the short-term ripening and by the thermohygrometric conditions for optimum growth (Table 8.6). However, it is also possible to find *P. camemberti, A. ochraceus* (Mutti, Previdi, Quintavalla, & Spotti, 1988), and *P. griseofulvum* (Spotti & Berni, 2005). Under these conditions, the suitability of initial treatment of the products with starter cultures proves recommendable. It has also been concluded that *P. nalgiovense* and *P. camemberti* are particularly able to proliferate during ripening, on the basis of the temperature for optimum growth as reported in Table 8.6. However, for a better comparison of the various strains, lag time and growth rate values were determined at 15°C keeping the relative humidity at the optimal value for each individual strain (Table 8.7).

Tables 8.6 and 8.7 illustrate different dominance over "undesired molds". In ripened products, the strain for which dominance over the "undesired species" is predictable is *P. nalgiovense*, since it has the highest growth rate under the temperature conditions indicated. Even though *P. chrysogenum* has a shorter lag time (3 days) as compared to *P. nalgiovense* (4 days), its growth rate is half that of *P. nalgiovense* and its mycelium is therefore less invasive. On the contrary, together with *P. camemberti* and *P. gladioli* strains, assuming equal contamination levels, also *P. chrysogenum, P. solitum, P. griseofulvum*, and *P. nordicum* may be present on the same product.

Tables 8.6 and 8.7 also illustrate that RH proves to have a greater influence than temperature for almost all strains; this accounts for the lag time at optimum RH in Table 8.7 not being too different from those in Table 8.6. The only exception is *A. ochraceus*, which is much more sensitive to temperature changes than to humidity, as it is shown by its long lag time at 15°C (14 days), whereas under optimum conditions it is only 1 day.

Recently, the course of growth rate of the fungal species that more frequently occur on the surface of molded meat products during ripening has also been investigated (Spotti & Berni, 2005). Table 8.8 clearly shows how the growth of starter cultures such as *P. nalgiovense* and *P. gladioli* is favored by a higher relative humidity in the first days of ripening.

Competition

According to the techniques tested by Wheeler & Hocking (1993), several competition trials were carried out, where each starter was inoculated in combination with the undesired strain, each at a time (Spotti et al., 1999; Spotti & Berni, 2005; Spotti, Mutti, & Scalari, 1994). The results of these experiments showed that *P. nalgiovense* may be overgrown by *A. ochraceus* during drying ($T \geq 15°C$), whereas *P. nalgiovense* will dominate this undesired mold during ripening ($T \leq 15°C$) as previously referred.

Table 8.8 Trend in growth rate of most frequent fungal species during ripening at 14–15°C as a function of the surface water activity (a_w)

Fungal species	$a_w = 0.85$	$a_w = 0.86$	$a_w = 0.88$	$a_w = 0.90$	$a_w = 0.92$
P. gladioli	→	→	→	→	MG
P. nalgiovense	→	→	→	MG	←
P. chrysogenum	MG	←	←	←	←
P. griseofulvum	MG	←	←	←	←
P. implicatum	MG	←	←	←	←
P. nordicum	MG	←	←	←	←
P. solitum	MG	←	←	←	←
P. camemberti	→	→	MG	nd	nd
A. candidus	→	→	MG	nd	nd

Note. MG indicates maximum growth rate.
→ indicates that growth rate tends to increase if surface water activity increases.
← indicates that growth rate tends to decrease if surface water activity increases.
nd = not determined.
Source. Adapted from Spotti, Mutti, and Scalari (1994) and Spotti and Berni (2005).

On the contrary, *P. nalgiovense* is unable to dominate over *P. chrysogenum*, *P. solitum*, *P. nordicum, and P. griseofulvum* in the case of similar initial contamination levels. During the drying phase, *P. gladioli* is inhibited by *A. ochraceus* and *P. solitum*, whereas during ripening it prevails over *A. ochraceus*, but not over *P. chrysogenum*, *P. solitum*, and *P. nordicum*. *A. ochraceus* may therefore dominate over the starter strains during drying and may grow invasively throughout the product surface even during ripening, if present in equal amounts. The results of the trials also suggested that the starter cultures could coexist with *P. chrysogenum, P. solitum, P. nordicum, P. griseofulvum, P. implicatum* on the product, if their initial levels are the same.

Conclusions

The aim of this chapter is to stress the importance of fungal identification on meat products. At present, morphological techniques are still the most widespread practices all over the world, since they proved to be more accurate than both biochemichal (Biolog MicroStation™, Hayward, CA, USA) and molecular techniques (i.e., Pulsed Field Gel Electrophoresis). The latter two proved to give satisfactory results only if associated with the former.

The identification of the fungal species occurring on the surface of the ripened product, regardless of the good appearance and the high sensory quality of the final product, proves to be fundamental since most spontaneous molds species subjected to consecutives subcultures, as it frequently occurs in the ripening rooms within subsequent productive cycles, proved to rapidly degenerate or to change their morphological appearance and adapt to environmental conditions. This is due to the fact that the genome of the species belonging to the genus *Penicillium* subgenus *Penicillium* appears to be unstable and tends to cause a rapid adaptation of the species to the nutritional niches available (Williams, Pitt, & Hocking, 1985). The above-mentioned variations cannot always be considered beneficial to the final product.

In case it is decided for the growth of a selected fungal strain, the choice must first be addressed to that species which proved not to impair the outward appearance of the sausages, in each step of the ripening process because of its heavy-colored conidiation (this is what frequently happens when *P. solitum*, *P. brevicompactum*, and *P. chrysogenum* grow on surface of meat products), and which proved not to produce toxic metabolites (as for *P. nordicum* or *P. verrucosum*, whose isolates in most cases result to be OTA-producers). Lipolytic and proteolytic activities of molds can be taken into account when selecting a proper starter culture; these two characteristics become important when choosing the strain to inoculate within the same species, in order to obtain the best quality standards.

The poor competitiveness of superficial starter cultures compared with other contaminants requires a low rate of environmental contamination; therefore starter molds should be used at high inoculation levels on sausages to be ripened in order to avoid possible development of undesirable species.

The control on the growth of the selected species and their actual predominance over undesirable molds should be periodically planned on the basis of routine laboratory tests (i.e., isolation and identification of the species employed in the industrial process) in order to avoid the unexpected setting up of environment-contaminating mold species, which are morphologically similar to the selected ones.

References

Abbott, S. P., Lumley, T. C., & Sigler, L. (2002). Use of holomorph characters to delimit *Microascus nidicola* and *M. sopii* sp. nov., with notes on the genus *Pithoascus*. *Mycologia, 92*(2), 362–369.

Baldini, P., Berni, E., Diaferia, C., Follini, A., Palmisano, S., Rossi, A., et al. (2005). Materia prima flora microbica nella produzione dei salumi piacentini. *Rivista di Suinicoltura, 4*, 183–187.

Baldini, P., Cantoni, E., Colla, F., Diaferia, C., Gabba, L., Spotti, E., et al. (2000). Dry sausages ripening: Influence of thermohygrometric conditions on microbiological, chemical and physico-chemical characteristics. *Food Research International, 33*, 161–170.

Baldini, P., & Spotti, E. (1995). Importanza della fase di preriposo sulla probabilità di ritrovare odori estranei (acido fenico) in prosciutti stagionati. *Industria Conserve, 70*, 418–422.

Battilani, P., Pietri, A., Giorni, P., Formenti, S., Bertuzzi, T., Toscani, T., et al. (2007). *Penicillium* populations in dry-cured ham manufacturing plants. *Journal of Food Protection, 70*(4), 975–980.

Berni, E., Cacchioli, C., Castagnetti, G., Sarra, P. G., & Spotti, E. (2007). Microbial surface colonization in Nebrodi salami. *Proceedings of the 6th International Symposium on the Mediterranean Pig*, October 11–13, 2007 – Capo d'Orlando & Messina (Italy)

Blanco, D., Barbieri, G., Mambriani, P., Spotti, E., & Barbuti, S. (1994). Studio sul "difetto di patata" nel prosciutto crudo stagionato. *Industria Conserve, 69*, 230–236.

Bruna, J. M., Hierro, E. M., De la Hoz, L., Mottram, D. S., Fernández, M., & Ordoñez, J. A. (2003). Changes in selected biochemical and sensory parameters as affected by the superficial inoculation of *Penicillium camemberti* on dry fermented sausages. *International Journal of Food Microbiology, 85*, 111–125.

Comi, G., & Cantoni, C. (1983). Yeasts in dry Parma hams. *Industrie Alimentari, 22,* 102–104.
De Hoog, G. S., Guarro, J., Gené, J., & Figueras, M. J. (2000). In G. S. de Hoog, J. Guardo, J. Gené, & M. J. Figueras (Eds.), *Atlas of clinical fungi* (pp. 21–29). Utrecht: CBS.
Dragoni, I., Cantoni, C., Papa, A., & Vallone, L. (1997). In I. Dragoni (Ed.), *Muffe alimenti e micotossicosi*. Milano: UTET, Città*Studi*Edizioni.
Geisen, R., Lücke, F. K., & Kröckel, L. (1992). Starter and protective cultures for meat and meat products. *Fleischwirtschaft, 72*(6), 894–898.
Geisen, R., Mayer, Z., Karolewiez, A., & Faerber, P. (2004). Development of a real time PCR system for detection of *Penicillium nordicum* and for monitoring of ochratoxin A production in foods by targeting the ochratoxin polyketide synthase gene. *Systematic and Applied Microbiology, 27*(4), 501–507.
Grazia, L., Romano, P., Bagni, A., Roggiani, D., & Guglielmi, D. (1986). The role of moulds in the ripening process of salami. *Food Microbiology, 3,* 19–25.
Hammes, W. P., & Knauf, H. J. (1994). Starter in the processing of meat products. *Meat Science, 36,* 155–168.
Larsen, T. O., Svendsen, A., & Smeedsgaard, J. (2001). Biochemical characterization of Ochratoxin A producing strains of the genus *Penicillium*. *Applied and Environmental Microbiology, 67,* 3630–3635.
Lori, D., Grisenti, M. S., Parolari, G., & Barbuti, S. (2005). Microbiology of dry-cured ham. *Industria Conserve, 80,* 23–32.
Martín, A., Asensio, M. A., Bermúdez, M. E., Cordoba, M. G., Aranda, E., & Cordoba, J. J. (2002). Proteolytic activity of *Penicillium chrysogenum* and *Debaryomyces hansenii* during controlled ripening of pork loins. *Meat Science, 62,* 129–137.
Martin, A., Cordoba, J. J., Nunez, F., & Asensio, M. A. (2004). Contribution of a selected fungal population to proteolysis on dry cured ham. *International Journal of Food Microbiology, 94*(1), 55–56.
Martín, A., Cordoba, J. J., Aranda, E., Cordoba, M. G., & Asensio, M. A. (2006). Contribution of a selected fungal population to the volatile compounds on dry-cured ham. *International Journal of Food Microbiology, 110*(1), 8–18.
Ministero della Sanità. (1994). Decreto 28 Dicembre 1994. Gazzetta Ufficiale Repubblica Italiana. Serie generale n. 89 (pp. 4–5), 15.04.1995.
Mutti, P., Previdi, M. P., Quintavalla, S., & Spotti, E. (1988). Toxigenity of mould strains isolated from salami as a function of culture medium. *Industria Conserve, 63,* 142–145.
Ockerman, H., W., Céspedes Sánchez, F. J., Ortega Mariscal, M. A., & León-Crespo, F. (2001). The lipolytic activity of some indigenous Spanish moulds isolated from meat products. *Journal of Muscle Foods, 12,* 275–284.
Pietri, A., Bertuzzi, T., Gualla, A., & Piva, G. (2006). Occurrence of ochratoxin A in raw ham muscles and in pork products from northern Italy. *Italian Journal Food Science, 18*(1), 99–106.
Pitt, J. I., & Hocking, A. D. (1997). *Fungi and Food Spoilage* (2nd ed.). Cambridge: University Press. Blackie Academic & Professional.
Rodríguez, M., Núñez, F., Cordoba, J. J., Bermúdez, M. E., & Asensio, M. A. (1998). Evaluation of proteolytic activity of micro-organisms isolated from dry cured ham. *Journal of Applied Microbiology, 85,* 905–912.
Samelis, J., & Sofos, J. N. (2003). In T. Boekhout, & V. Robert (Eds.) *Yeasts in Foods* (pp. 239–265). Cambridge: Woodhead Publishing Limited.
Samson, R. A., & Frisvad, J. C. (2004). *Penicillium subgenus Penicillium: New taxonomic schemes, mycotoxins and other extrolites*. Uthrecht: CBS.
Simoncini, N., Rotelli, D., Virgili, R., & Quintavalla, S. (2007). Dynamic and characterization of yeasts during ripening of typical Italian dry-cured ham. *Food Microbiology, 24*(6), 577–584.
Spotti, E., & Berni, E. (2005). Variazione della flora microbica superficiale nei salumi oggetto della ricerca. *Oral Presentation of 2d Annual Conference on the Research Project "Salumi piacentini"*. October 25, 2005. Parma, Italy: SSICA.
Spotti, E., Berni, E., & Cacchioli, C. (2007). Le muffe sugli insaccati stagionati. Rassegna e aggiornamento sui progressi della ricerca SSICA *Industria Conserve, 82*(3), 243–250.

Spotti, E., Busolli, C., & Palmia, F. (1999). Growth of mould cultures of primary importance in the meat product industry. *Industria Conserve, 74*, 23–33.

Spotti, E., Chiavaro, E., Lepiani, A., & Colla, F. (2001). Mould and ochratoxin A contamination of pre-ripened and fully ripened hams. *Industria Conserve, 76*, 341–354.

Spotti, E., Mutti, P., & Campanini, M. (1988). Indagine microbiologica sul "difetto dell'acido fenico" sul prosciutto durante la stagionatura. *Industria Conserve, 63*, 343–346.

Spotti, E., Mutti, P., & Campanini, M. (1989). Presenza di muffe sui prosciutti durante la prestagionatura e la stagionatura: Contaminazione degli ambienti e sviluppo sulla porzione muscolare. *Industria Conserve, 64*, 110–113.

Spotti, E., Mutti, P., & Scalari, F. (1994). *P. nalgiovense, P. gladioli, P. candidum* e *A. candidus*: Possibilità d'impiego quali colture starter. *Industria Conserve, 69*, 237–241.

Wheeler, K. A., & Hocking, A. D. (1993). Interactions among xerophilic fungi associated with dried salted fish. *Journal of Applied Bacteriology, 74*, 164–169.

Williams, A. P., Pitt, J. I., & Hocking, A. D. (1985). The closely related species of subgenus Penicillium – A phylogenic exploration. In R. A. Samson, & J. I. Pitt (Eds.), *Advances in Aspergillus and Penicillium Systematics* (pp. 121–128). New York: Plenum Press.

Part III
Biotechnology for Better Quality and Nutritional Properties of Meat Products

Chapter 9
Biotechnology of Flavor Generation in Fermented Meats

Fidel Toldrá

Introduction

Traditionally, meat fermentation was based on the use of natural flora, including the "back-slopping", or addition of a previous successful fermented sausage. However, these practices gave a great variability in the developed flora and affected the safety and quality of the sausages (Toldrá, 2002; Toldrá & Flores, 2007). The natural flora of fermented meat has been studied for many years (Leistner, 1992; Toldrá, 2006a), and more recently, these micro-organisms have been isolated and biochemically identified through molecular methods applied to extracted DNA and RNA (Cocolin, Manzano, Aggio, Cantoni, & Comi, 2001; Cocolin, Manzano, Cantoni, & Comi, 2001; Comi, Urso, Lacumin, Rantsiou, Cattaneo & Cantoni, 2005). Today, most of the fermented sausages are produced by using microbial starters, usually from both *Lactobacilli* and *Micrococcaceae* because this combination ensures rapid acidulation and optimal flavor development (Demeyer & Toldrá, 2004; Toldrá, Nip, & Hui, 2007). The main groups of micro-organisms used as starters are briefly described below.

Main Groups of Microbial Starters

Lactic Acid Bacteria (LAB)

Lactobacilli are the most competitive micro-organisms usually found in fermented meat products and are also essential components of starter cultures. These bacteria are able to metabolize glucose or any other available carbohydrate to generate lactic acid through either homo or heterofermentative pathways. The accumulation of lactic acid produces a pH drop towards acidic values. The homofermentative pathway only produces lactic acid but in the case of heterofermentative pathway, some undesirable secondary products (i.e.,- acetic acid, hydrogen peroxide, acetoin,

---

F. Toldrá
Instituto de Agroquímica y Tecnología de Alimentos (CSIC), PO Box 73, 46100 Burjassot (Valencia), Spain

etc.) may also be produced (Demeyer, 1992). *Lactobacillus sakei* and *Lactobacillus curvatus* grow well at mild temperatures (20–25°C), and this is the main reason why they are mainly used in European sausages (mild fermentation temperatures). On the other hand, *L. plantarum* and *Pediococcus acidilactici* are used in the U.S because they grow better at higher temperatures (30–35°C) (Toldrá, 2007a). LAB have a proteolytic system, with a good number of peptidases, which may partly contribute to the degradation of muscle proteins. Most important proteases found in *L. sakei* are reported in Table 9.1. Further description on the genetics of LAB is given in Chap. 6.

Micrococcacceae

Staphylococcus and *Kocuria* (formerly *Micrococcus*) are the main representatives of this family. These micro-organisms have important enzymes like some proteases and lipases that contribute to flavor generation as will be later described. Other interesting enzymes are nitrate reductase, that reduces nitrate to nitrite and contributes to the formation of nitrosylmyoglobin giving rise to the typical color formation and safety of the sausage and catalase that helps to prevent lipid oxidation and improve flavor formation and color stability (Talon, Walter, Chartier, Barriere, & Montel, 1999).

Yeasts

Debaryomyces hansenii has been found as the predominant yeast in fermented sausages. This yeast grows preferentially in the outer area of the sausage due to its aerobic metabolism. *D. hansenii* has a good proteolytic activity (Bolumar, Sanz, Aristoy, & Toldrá, 2003a, 2003b, 2005) as shown in Table 9.1, and its lipolytic activity is able to generate flavored volatile compounds from branched-chain amino acids (Durá, Flores, & Toldrá, 2004a). Another important contribution to sausage flavor is through its deaminase/deamidase activity on certain free amino acids that generate ammonia as a by-product which rises the pH in the sausage (Durá, Flores, & Toldrá, 2002). Further description on the genetics of yeasts is given in Chap. 7.

Molds

A white coating on the surface of dry fermented sausages is typical in certain areas like the Mediterranean. The most usual molds identified are *Penicillium nalgiovense* and *P. chrysogenum* (Sunesen & Stahnke, 2003). They contribute to the flavor through their proteolytic and lipolytic activity as well as to the ammonia generated by deamination and deamidation. Further extensive description on molds is given in Chap. 8.

9 Biotechnology of Flavor Generation in Fermented Meats

Table 9.1 Main proteases from *L. sakei* and *D. hansenii*

Enzyme	Molecular mass (kDa)	Biochemical similarity	Activation	Preferred substrates	Reference
L. sakei					
Major aminopeptidase	35	Pep L	Ca^{2+}, Sn^{2+}, Mg^{2+}, Ba^{2+}, Mn^{2+}	Leu-peptide, Ala-peptide	Sanz & Toldrá, 1997
Arginine aminopeptidase	60	Pep N like	Reducing agents, salt	Arg-peptide, Lys-peptide	Sanz & Toldrá, 2002
X-prolyl-dipeptidylpeptidase	88	Pep X	-	X-Pro-peptides, Ala-Pro-peptide	Sanz & Toldrá, 2001
Dipeptidase	50	Pep V	-	Met-Ala	Montel, Seronine, Talon, & Hebraud, 1995
Tripeptidase	55	Pep T	-	Ala-Ala-Ala	Sanz, Mulholland & Toldrá, 1998
D. hansenii					
Protease B	430	PrB	-	Sarcoplasmic proteins	Bolumar, 2005
Protease A	55	PrA	-	"	Bolumar, Sanz, Aristoy, & Toldrá, 2008
Prolyl aminopeptidase	370	-	-	Pro-peptide	Bolumar et al., 2003a
Arginyl aminopeptidase	101	ApY	Ca^{2+}, Mg^{2+}, Co^{2+}	Arg-peptide, Lys-peptide	Bolumar et al., 2003b

Table 9.2 Main products, affecting sensory quality, resulting from biochemical reactions by muscle and microbial enzymes

Initial substrate	Type of reaction	Final product	Effect on sensory quality
Carbohydrates	Glycolysis-homofermentative	Lactic acid	Taste
"	Glycolysis-heterofermentative	Lactic acid Diacetyl Acetaldehyde Acetoin Short chain fatty acids Carbon dioxide	Taste and aroma
Proteins	Proteolysis	Peptides Free amino acids	Taste
Amino acids	Degradation reactions	Branched aldehydes Branched alcohols Branched acids	Aroma
"	Deamination	Aldehydes Ketones Acids Ammonia	Aroma
"	Decarboxylation	Amines	-
"	Transamination	Other amino acids	Taste
Lipids	Lipolysis	Free fatty acids	Taste
Free fatty acids	Oxidation	Volatile compounds	Aroma

Flavor Generation in Fermented Meats

The generation of aroma compounds in fermented sausages from the degradation of carbohydrates, proteins, and lipids has been extensively reviewed in recent years (Demeyer & Stahnke, 2002; Ordoñez, Hierro, Bruna & de la Hoz, 1999; Toldrá, Sanz, & Flores, 2001a; Toldrá & Flores, 2007). The main reactions consist of carbohydrate degradation, proteolysis, amino acid degradation reactions (decarboxylation, deamination, transamination), Maillard reactions, Strecker degradation reactions, lipolysis, and lipid oxidation (Table 9.2). Flavor is also determined by other factors; one of the most important is the addition of spices like pepper, paprika, mustard, nutmeg, cloves, orégano, rosemary, thyme, garlic, onion, etc. that may exert a high impact on the aroma of the fermented products (Chi & Wu, 2007; Ordoñez et al., 1999). The smoking, a usual practice in many countries, also imparts a characteristic smoke flavor to the sausages (Ellis, 2001).

The correlation of flavor formation in dry fermented sausages and the types of starter used has been reported by different authors. So, the combination of different LAB and Sthaphylococcus species in dry sausages were reported to produce important differences in the volatile compound profiles and thus had a different impact on flavor (Berdagué, Montel, & Talon, 1993). But different strains of Staphylocci contribute in a different way to the flavor formation. So, after studying 19 strains of *Micrococacceae* in dry fermented sausage models, *S. xylosus* and *S. carnosus* were reported as the most adequate for better sausage aroma (Montel, Reitz, Talon, Berdagué & Rousset-Akrim, 1996). When comparing both the species, *S. carnosus* was reported to give a larger amount of volatile compounds from the degradation of leucine, isoleucine, and valine (Olesen, Meyer, & Stahnke, 2004a). The catabolism of leucine was reported to be reduced by the addition of nitrate for *S. xylosus* and nitrite for *S. carnosus* (Olesen, Stahnke & Talon, 2004b). The maturation time of Italian dry sausages was also reduced by the use of *S. carnosus* as a starter culture (Stahnke, Holck, Jensen, Nilsen, & Zanardi, 2002). Normal inoculation levels of Staphylococci are within the range of 10^5–10^7 cfu/g. Larger inoculation levels of *S. carnosus* are associated to higher amounts of diacetyl, ethanol, and ethyl esters while lower inoculation levels produce more methyl-branched aldehydes and acids, 2-methyl-butanol and sulfides (Tjener, Stahnke, Andersen, & Martinussen, 2004a, 2004b).

Some model systems for the study of microbial action during meat fermentation have been used in recent years. Some of these models have been used for the study of lactic acid generation and proteolysis of LAB (Fadda et al., 1999a, 1999b; Sanz et al., 1999a, 1999b), or proteolysis of *D. Hansenii* (Santos et al., 2001), while others have been developed to examine the formation of volatiles by Staphylococcus strains (Tjener, Stahnke, Andersen, & Martinussen, 2003) or by *D. Hansenii* (Durá, Flores, & Toldrá, 2004b).

The degradation of branched-chain amino acids gives methyl-branched aldehydes and acids which are correlated to sausage aroma (Montel, Masson & Talon, 1998; Stahnke et al., 2002). However, the addition of free amino acids to the sausage has not proven to improve the sensory profile; only a mixture of valine, isoleucine, and leucine has been reported to give a higher amount of volatile compounds and a slight improvement in sensory properties (Herranz, de la Hoz, Hierro, Fernández, & Ordoñez, 2005). Other typical volatile compounds in fermented sausages are pyruvate metabolites and methylketones obtained from the β-oxidation of fatty acids (Montel et al., 1998).

The addition of cell free extracts from *D. hansenii* and *L. sakei*, as an additional source of enzymes, to fermented sausages was reported to promote the generation of volatile compounds from lipid oxidation and carbohydrate fermentation and improve the sensory quality (Bolumar, Sanz, Aristoy, & Toldrá, 2006). Other positive improvements of sensory quality have been obtained with the addition of cell-free extracts of *L. paracasei sbsp. paracasei* (Hagen, Berdagué, Holck, Naes, & Blom, 1996) or molds like *Mucor racemosus* and *P. aurantiogriseum* (Bruna, Fernández, Hierro, Ordoñez, & de la Hoz, 2000).

The pattern of glycolysis, proteolysis, lipolysis, and oxidation reactions are related to the changes in pH, texture, moisture, and a_w, interacting with the type of fermentation as determined by the type of meat and starter culture, fermentation temperature, length of ripening, and sausage diameter (Demeyer & Toldrá, 2004).

The biochemical stoichiometry determines the relative amounts of the substrates (carbohydrates, oxygen, proteins, lipids, etc.), all the intermediate products (free amino acids, peptides, free fatty acids, pyruvic acid, etc.) and the end products (lactic acid, ammonia, volatile compounds, etc.) and how they will affect the sensory quality (see Table 9.1). These reactions are described below.

Glycolysis in Fermented Meat

Carbohydrates serve as substrates for the growth of the microbial starters added to the sausage. Their fermentation produces lactic acid as an end product and its accumulation results in a pH drop. The D(-) or L(+) configuration or a mixture of both will depend on the species of LAB (L and D lactate dehydrogenase, respectively, and the presence of lactate racemase) that is used as a starter (Demeyer & Toldrá, 2004). The intensity of the pH drop depends on the type of LAB used as starter, the type and amount of added carbohydrates, the fermentation temperature (higher in the USA and lower in Europe) and other relevant processing parameters like the amount of salt, time and conditions of ripening, etc. (Demeyer, 1992; Demeyer et al., 2000). If the starters are heterofermentative, some additional end products like acetate, formate, ethanol, and acetoin may be generated and thus affect the sensory quality of the sausage (Demeyer & Stahnke, 2002). For instance, the generation of compounds such as diacetyl, acetoin, or butanediol impart a buttery and yogurt aroma in fermented sausages (Montel et al., 1998). The pH drop towards acid values favors protein coagulation and water release as well as proteolysis and lipolysis by stimulation of cathepsin D and lysosomal acid lipase, respectively (Toldrá, 2004). Furthermore, acidity also contributes to the inhibition of undesirable, pathogens or spoilage bacteria.

The generation of the specific volatile compounds during the carbohydrate fermentation depends on the starter used in the processing of dry fermented sausages. Sugar degradation is an important source of di and tricarboniyls which may react subsequently with amino acids through Strecker reactions (MacLeod & Seyyedain-Ardebili, 1981; Toldrá & Flores, 2007).

Proteolysis in Fermented Meat

Protein breakdown by both muscle and microbial proteases generate a good number of peptides that are then further degraded to small peptides and free amino acids. The relative contribution of each group of enzymes depends on the type of meat and micro-organisms used as starter cultures (Toldrá, 2006c). Studies with antibiotics and other protease inhibitors revealed muscle cathepsin D, which is very active at pH 4.5, as the protease that started the degradation of myosin and actin. Other muscle proteases, like cathepsins B and L, would have a minor role restricted to a few proteins like actin and its degradation products (Molly et al., 1997). The applied temperature, pH values and the duration of the process is also very important for the extent of proteolysis. The highest non-protein nitrogen value has been reported in sausages with pH values below 4.7 (Flores, Marcus, Nieto, & Navarro, 1997). The parent proteins for some of these peptides have been identified as myoglobin, creatin kinase, troponin T, troponin I, and myosin light chain 2 (Hughes et al., 2002). Lactobacilli have shown a good ability to degrade sarcoplasmic proteins and some myofibrillar proteins (Fadda et al., 1999a, 1999b; Sanz et al., 1999a, 1999b).

The addition of cell free extracts from *D. hansenii* to fermented sausages, as an additional source of enzymes, was reported to promote the proteolysis (see Fig. 9.1) (Santos et al., 2001). However, the addition of *Debaryomices* spp. to fermented sausages did not contribute to major proteolysis in comparison to the controls which

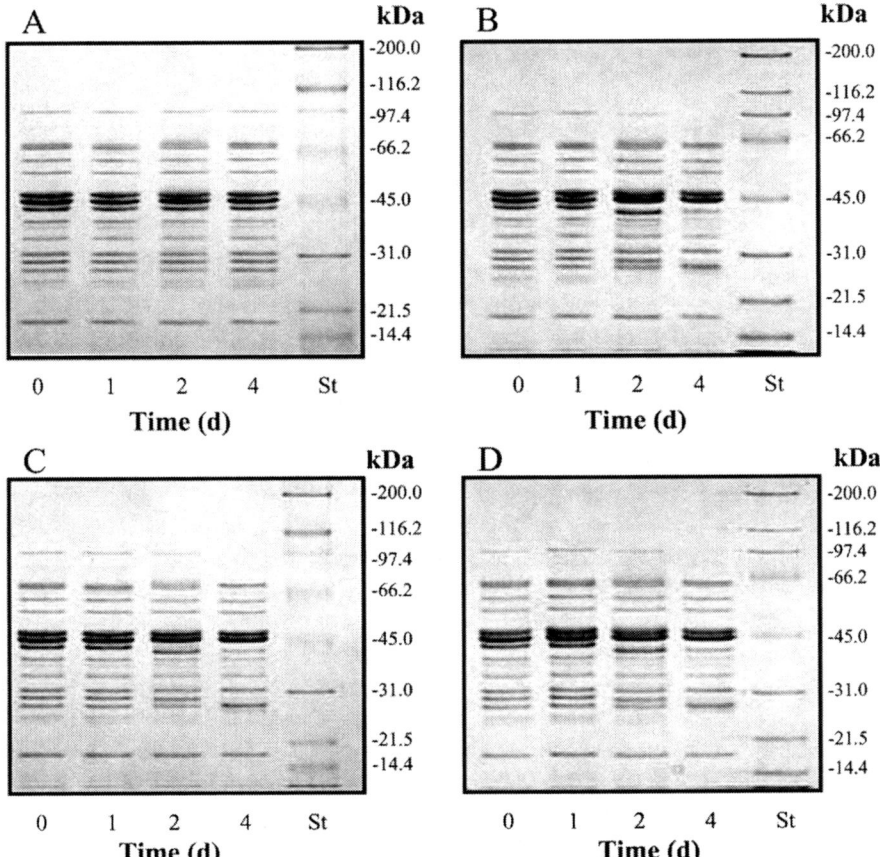

Fig. 9.1 SDS–PAGE of sarcoplasmic protein hydrolysis by *D. hansenii* CECT 12487 during incubation at 0, 1, 2, and 4 days at 27°C. A: Control, B: whole cells, C: cell-free extract and D: whole cells cell-free extract. St lane: Standard proteins. Reproduced from Santos, N. N., Santos-Mendonça, R. C., Sanz, Y., Bolumar, T., Aristoy, M.-C., & Toldrá, F. (2001). Hydrolysis of pork muscle sarcoplasmic proteins by *Debaryomyces hansenii*, *International Journal of Food Microbiology*, 68, 199–206, with permission from Elsevier

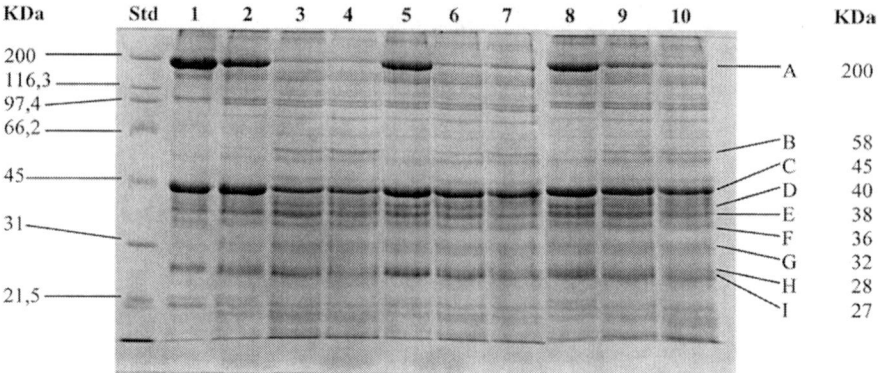

Fig. 9.2 SDS–PAGE of myofibrillar proteins from Control (lanes 2–4), L1 (lanes 5–7) and L2 (lanes 8–10) batches during ripening stages. Std, standards, lanes: (1) Initial (0 days), (2) C-6 days, (3) C-21 days, (4) C-35 days, (5) L1-6 days, (6) L1-21 days, (7) L1-35 days, (8) L2-6 days, (9) L2-21 days, (10) L2-35 days. L1: addition of 5×10^6 cfu/g *Debaryomices* spp. and L2: addition of 15×10^6 *Debaryomices* spp to standard culture (control). Reproduced from Durá, M. A., Flores, M., & Toldrá, F. (2004b) Effect of *Debaryomyces* spp. on the proteolysis of dry-fermented sausages. *Meat Science, 68*, 319–328, with permission from Elsevier

had the same starter culture (*L. sakei, P. pentosaceus, S. xylosus,* and *S. carnosus*) but with no *Debaryomices* inoculation (see Fig. 9.2) (Durá et al., 2004a, 2004b).

The evaluation of proteolysis by lactobacilli using muscle sarcoplasmic and myofibrillar proteins as substrates has been carried out by incubation of these proteins with whole-cells and cell free extracts of different lactobacilli strains (*L. casei* CRL705, *L. curvatus* CECT904, *L. plantarum* CRL681, and *L. sakei* CECT4808), isolated from sausages. Proteolysis was determined by the analysis of proteins, peptides, and free amino acids (Fadda et al., 1999a, 1999b; Sanz et al., 1999a, 1999b). Different profiles for proteins degradation and peptides generation for *L. curvatus* can be observed in Fig. 9.3. The strains of the species *L. plantarum* and *L. casei* provoked the strongest degradation. The observed modifications showed that the activity located at the cell surface could be limiting the intracellular peptidase activity. In relation to the net generation of free amino acids, this was markedly higher in the sarcoplasmic than in myofibrillar protein extracts. In general, it was shown that the incorporation of whole-cells to sarcoplasmic protein extracts caused a reduction in the content of free amino acids, especially glutamine. The addition of only cell-free extracts of *L. casei, L. curvatus,* and *L. sakei* resulted in a net increase in free amino acids, but a decrease was reported when cell-free extracts of *L. plantarum* were incorporated. This indicated that either this strain has lower exoproteolytic activity or higher intracellular metabolic activity for amino acid degradation. The combination of both cell-free extracts and whole-cell suspensions, especially of *L. sakei* and *L. plantarum*, resulted in maximal increases in free amino acids (Sanz, Sentandreu, & Toldrá, 2002). Important protein breakdown patterns have also been observed when using *D. hansenii* and also a combination of *L. curvatus* and *S. xylosus* in traditional Italian fermented sausage (Casaburi et al., 2007). However, when only

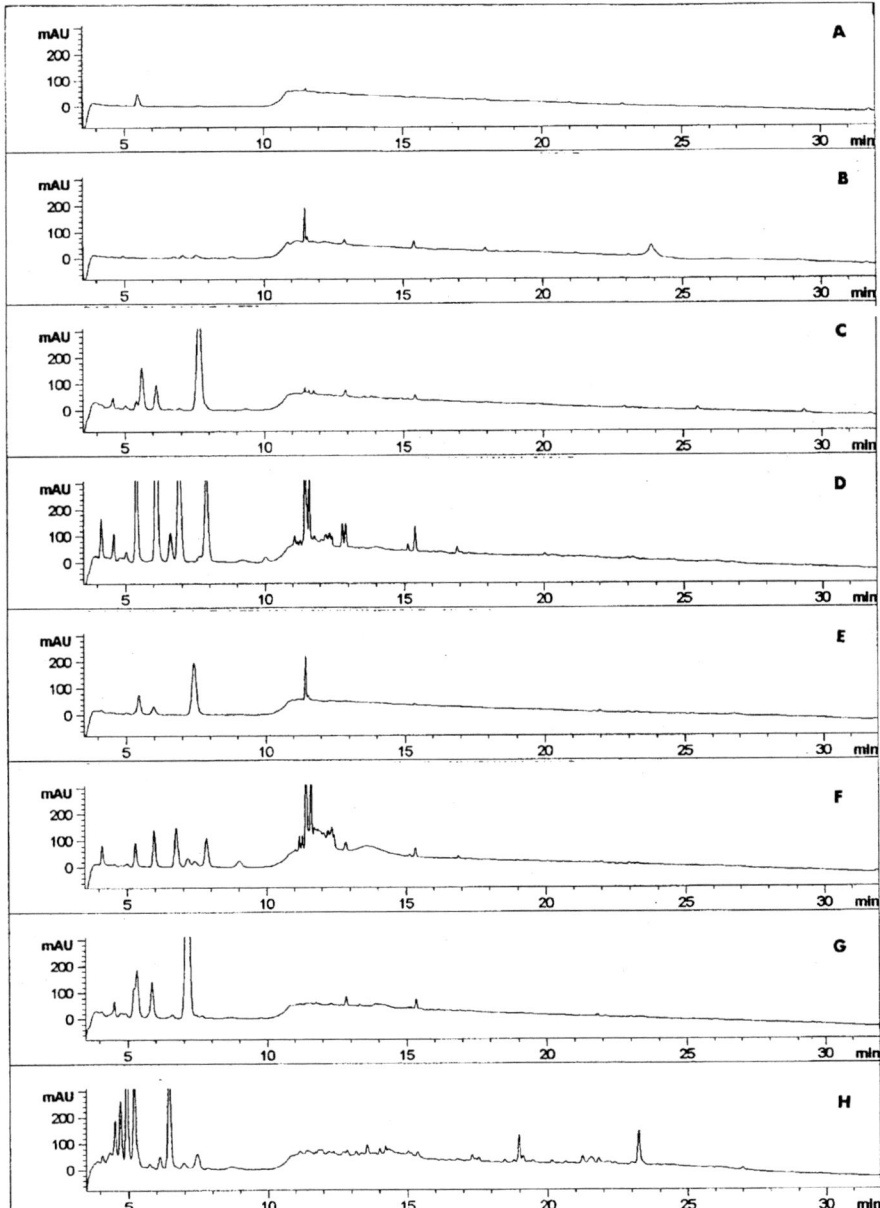

Fig. 9.3 Reverse-phase high performance liquid chromatography (HPLC) patterns of soluble peptides contained in myofibrillar protein extracts treated with *L.curvatus* CECT 904 at 0h (A, C, E, and G) and 96h (B, D, F, and H) of incubation at 378C. Control samples (A and B), samples containing whole-cells (C and D), samples containing CFE (E and F) and samples containing whole-cells plus CFE (G and H). Reproduced from Sanz, Y., Fadda, S., Vignolo, G., Aristoy, M.-C., Oliver, G., & Toldrá, F. (1999). Hydrolysis of muscle myofibrillar proteins by *Lactobacillus curvatus* and *Lactobacillus sake*. *International Journal of Food Microbiology, 53*, 115–125, with permission from Elsevier

S. *xylosus* was used as starter without lactobacilli, the observed proteolysis was identical to that in non-inoculated control sausages, reinforcing the importance of proteolysis contribution by lactobacilli (Casaburi et al., 2008).

Tri- and di-peptides are generated as a result of the action of muscle and microbial tri and dipeptidylpeptidases, respectively. These enzymes are activated or inhibited depending on the pH and the level of salt (Sentandreu & Toldrá, 2001). In fact, the generation of small peptides can be reduced by increasing the salt content, a known inhibitor of muscle peptidases (Sanz et al., 2002; Sanz & Toldrá, 1999; Toldrá, 2004). This is probably an important reason for wide variations (between 2 and 12 fold) in the increase in non-protein-nitrogen. The final step of proteolysis consists of the generation of free amino acids by muscle and microbial aminopeptidases. Muscle alanyl and methionyl aminopeptidases have important activity in sausages and also have a wide scope. Something similar has been reported for the major aminopeptidase from *L. sakei* (Sanz & Toldrá, 1997) and an arginyl aminopeptidase from *D. han*senii (Bolumar et al., 2003b) (see Table 9.1). Recently, an important generation of free amino acids has been reported in fermented sausages inoculated with *L. curvatus* and *S. xylosus* (Casaburi et al., 2007, 2008).

Transformation of Amino Acids

Amino acids have important contribution to taste (Talon, Leroy-Sétrin, & Fadda, 2004) and also to aroma by the generation of branched-chain aldehydes and its secondary products like acids, alcohols, and esters through microbial metabolism of leucine, valine, & isoleucine (Toldrá, 2006b). The *Micrococcaceae* family (*Staphylococcus* and *Kocuria*) and *D. hansenii* have good ability for such type of metabolism (Demeyer et al., 2000). The generation of compounds like 3-methylbutanoic acid and α-hydroxy isocaproic acid have been reported after incubation of leucine with *S. xylosus* and *S. carnosus* (Olesen et al., 2004b) although these reactions are somehow affected by curing salts (Olesen et al., 2004a). Other authors (Herranz, Fernández, de la Hoz, & Ordoñez, 2006) have studied the addition of intracellular cell-free extracts of *L. sakei* and *B. pumilis* to enhance amino acid catabolism, specially of valine, leucine, and isoleucine even though the sensory quality of the final dry fermented sausage was not really improved.

Other enzymes with activity against amino acids are of interest. For instance, a glutaminase from *D. hansenii* was purified and characterized (Durá et al., 2002). This glutaminase was able to act on L-glutamine and generate ammonia that neutralized the sour taste and L-glutamate that is a flavor enhancer.

Lipolysis in Fermented Meat

Lipids constitute the major fraction of dry fermented sausages that are subject to lipolytic and oxidative reactions. Lipolysis consists of the enzymatic breakdown of tri-, di- and mono-acylglycerols and phospholipids to release free fatty acids. Lipases and phospholipases, respectively, are the responsible enzymes. Once released, free fatty acids exert a direct effect on taste (slight contribution to sourness) and an indirect effect on aroma through the generation of aroma volatile compounds by oxidative reactions. Lipolysis in meat is the result of the action of muscle and adipose tissue lipases as well as muscle phospholipases (Toldrá & Navarro, 2002), while in fermented meat it is the combined result of muscle and microbial enzymes (Hierro, de la Hoz, & Ordoñez, 1997). The main muscle lipases in the muscle are the lysosomal acid lipase and acid phospholipase, while the hormone sensitive lipase and the monoacylglycerol lipase are the most important in adipose tissue (Toldrá, 1992, 1998). These enzymes show good stability in drying processes (Toldrá, 2006b), but their activity vary depending on the final pH, salt content, ripening temperature, and water activity (Toldrá, 2007b). There was some controversy in the literature about the percentage of contribution of both muscle and microbial lipolytic enzymes to total fat hydrolysis. It was estimated that muscle lipases contribution was around 60 to 80% and the rest being due to microbial lipases (Molly, Demeyer, Civera, & Verplaetse, 1996; Molly, Demeyer, Civera, & Verplaetse, 1997). In general, a noticeable generation of free fatty acids is usually observed along the ripening of the sausages (Casaburi et al., 2007; Molly et al., 1996).

Starter cultures have a definitive effect on the formation of flavor in dry sausages (Berdagué et al., 1993) since lipolytic activity depends on the type of microorganism. So, *Micrococcaceae* are considered responsible for lipid breakdown (Ordoñez et al., 1999), even though molds and yeasts can also contribute due to their high lipolytic activity (Hammes & Knauf, 1994). *Staphylococcus* have been reported to hydrolyze mono-, di- and tri-acylglycerols and short-chain fatty tri-acylglycerols (Casaburi et al., 2007) while lactobacilli strains can only hydrolyze them at a lower rate (Sanz, Selgas, Parejo, & Ordoñez, 1998). Staphylococci have also shown a high esterase activity in whole cell suspensions and cell-free extracts, being lower in the extracellular medium (Casaburi, Villani, Toldrá, & Sanz, 2006).

Oxidation in Fermented Meat

Fatty acids with double bonds (mono and polyunsaturated fatty acids) are susceptible to oxidative reactions with the production of peroxides and carbonyl compounds and final generation of volatile compounds. Oxidation can be initiated through external catalyzers like light, heat, presence of moisture and/or metallic cations or internal muscle oxidative enzymes, like peroxidases and cyclooxygenases (Toldrá et al., 2001). Primary oxidation products are formed (hydroxyperoxides) but are very reactive and generate secondary oxidation products that contribute to flavor. Off-flavor development is typical of excessive oxidation (Skibsted, Mikkelsen, & Bertelsen, 1998). *Staphylococcus* species contribute to oxygen consumption and catalase activity that reduces the rancidity and improves color stability (Berdagué et al., 1993). Production of catalase by Staphylococci has been reported to be maximal at the end of the exponential growth phase (Talon et al., 1999) but depends on the type of strain. So, *S. warneri* was found to be low in catalase activity and thus caused rancidity in the sausages while *S. carnosus* had a higher catalase activity which is better suited for ideal fermented sausage production (Talon et al., 1999).

The main products from lipid oxidation are aliphatic hydrocarbons, alcohols, aldehydes, and ketones. Esters may be produced in the absence of nitrite by the reaction of alcohols with free carboxylic fatty acids (Toldrá, 2006b).

Generation of Volatile Compounds

Usual starter cultures used for meat fermentation are known to contribute to the generation of a large amount of volatile compounds. Depending on the type, amount and balance of volatile compounds, pleasant or unpleasant aroma will be generated (Stahnke, 1994, 1995, 1999). A good example is the production of large amounts of hexanal that is associated to rancidity (Montel et al., 1998). In addition, sausages produced with different starter cultures show different volatile patterns associated to different sensory profiles (Montel et al., 1996). Staphylococci have been reported to contribute to the typical sausage flavor and enhance its perception as well as reduce the maturation time (Stahnke et al., 2002). In fact, *S. xylosus* and *S. carnosus* are able to produce large amounts of branched aldehydes, methyl ketones, and ethyl esters. The latter compounds contribute to a fruity note to the flavor even though it has a low sensory threshold (Demeyer et al., 2000). Both micro-organisms can degrade the branched-chain amino acids leucine, isoleucine and valine into methyl-branched aldehydes, alcohols and acids (Beck, Hansen, & Lauritsen, 2002a, 2002b). For instance, the degradation of leucine produces not only 3-methyl butanal (Demeyer et al., 2000) as the main product but also 3-methylbutanoic acid as well as some amounts of α-hydroxy isocaproic acid (Olesen et al., 2004b). In any case, the volatile profile depends on whether *S. xylsous* or *S. carnosus* are used as starter cultures (Olesen et al., 2004a). Since major volatile compounds proceed from amino acid degradation, it has been suggested that flavor quality is closely related to these compounds (Stahnke, 1999). The levels of most of these volatile compounds increase with the inoculation level and over the time of ripening (Tjener, Stahnke, Andersen, & Martinussen, 2004a). Staphylococci and *Kocuria* are known to improve the color and flavor of fermented meats and influence the composition of volatile compounds by inhibition of the unsaturated fatty acids oxidation (Talon, Leroy-Sétrin, & Fadda, 2002). In fact, various staphylococci and LAB have been studied for its effects on unsaturated fatty acids. Staphylococci limited the oxidation of linoleic acid but LAB was unable when manganesum was lacking in the media (Talon, Walter, & Montel, 2000).

Debaryomices spp. has been used in combination with starter cultures (LAB and staphylocci) and its effects on dry-fermented sausages have been studied. *Debaryomices* spp showed an important effect on the generation of volatile compounds by inhibition of the generation of lipid oxidation products and by generation of ethyl esters that contributed to an appropriate flavor (Flores, Durá, Marco, & Toldrá, 2004). In the same study, it was reported that an excess of *Debaryomices* spp produced large amounts of acids that masked the positive effect so that it is important to use the right amount of this yeast.

Conclusion

There are important interactions between micro-organisms, meat and fat enzymes and processing conditions, all of them determining the sensory quality of fermented sausages. The use of starter cultures contribute to an improvement in the development of color, texture, and flavor of the final dry-fermented sausage but all these sensory characteristics strongly depend on the type and amount of strains used in the starter culture and the processing conditions.

References

Beck, H. C., Hansen, A. M., & Lauritsen, F. R. (2002a). Metabolite production and kinetics of branched chain acid formation in *Staphylococcus xylosus*. *Enzyme and Microbial Technology, 31*, 94–101.
Beck, H. C., Hansen, A. M., & Lauritsen, F. R. (2002b). Catabolism of leucine to branched chain acids in *Staphylococcus xylosus*. *Journal of Applied Microbiology, 91*, 514–519.
Berdagué, J. L., Monteil, P., Montel, M. C., & Talon, R. (1993). Effects of starter cultures on the formation of flavour compounds in dry sausage. *Meat Science, 35*, 275–287.
Bolumar, T., Sanz, Y., Aristoy, M. C., Flores, M., & Toldrá, F. (2006). Sensory improvement of dry-fermented sausages by the addition of cell-free extracts from *Debaryomices hansenii* and *Lactobacillus sakei*. *Meat Science, 72*, 457–466.
Bolumar, T., Sanz, Y., Aristoy, M. C., & Toldrá, F. (2003a). Purification and characterization of a prolyl aminopeptidase from *Debaryomyces hansenii*. *Applied and Environmental Microbiology, 69*, 227–232.
Bolumar, T., Sanz, Y., Aristoy, M. C., & Toldrá, F. (2003b). Purification and properties of an arginyl aminopeptidase from *Debaryomyces hansenii*. *International Journal of Food Microbiology, 86*, 141–151.
Bolumar, T., Sanz, Y., Aristoy, M. C., & Toldrá, F. (2005). Protease B from *Debaryomyces hansenii*. Purification and biochemical properties. *International Journal of Food Microbiology, 98*, 167–177.
Bolumar, T., Sanz, Y., Aristoy, M. C., & Toldrá, F. (2008). Purification and characterization of proteases A and D from *Debaryomyces hansenii*. *International Journal of Food Microbiology*. Manuscript submitted for publication.
Bruna, J. M., Fernández, M., Hierro, E. M., Ordoñez, J. A., & de la Hoz, L. (2000). Improvement of the sensory properties of dry fermented sausages by the superficial inoculation and/or the addition of intracellular extracts of Mucor racemosus. *Journal of Food Science, 65*, 731–738.
Casaburi, A., Aristoy, M. C., Cavella, S., Di Monaco, R., Ercolini, D., Toldrá, F., et al. (2007). Biochemical and sensory characteristics of traditional fermented sausages of Vallo di Diano (Southern Italy) as affected by use of starter cultures. *Meat Science, 76*, 295–307.
Casaburi, A., Di Monaco, R., Cavella, S., Toldrá, F., Ercolini, D., & Villani, F. (2008). Proteolytic and lipolytic starter cultures and their effect on traditional fermented sausages ripening and sensory traits. *Food Microbiology, 25*, 335–347.

Casaburi, A., Villani, F., Toldrá, F., & Sanz, Y. (2006). Protease and esterase activity of Staphylococci. *International Journal of Food Microbiology, 112,* 223–229.

Chi, S-P., & Wu, Y-C. (2007). Spices and seasonings. In F. Toldrá, Y. H. Hui, I. Astiasarán, W. K. Nip, J. G. Sebranek, E. T. F. Silveira, L. H. Stahnke, & R. Talon (Eds.), *Handbook of fermented meat and poultry* (pp. 87–100). Ames, IA: Blackwell Publishing.

Cocolin, L., Manzano, M., Aggio, D., Cantoni, C., & Comi, G. (2001). A novel polymerase chain reaction (PCR)-denaturing gradient gel electrophoresis (DGGE) for the identification of Micrococacceae strains involved in meat fermentations. Its application to naturally fermented Italian sausages. *Meat Science, 58,* 59–64.

Cocolin, L., Manzano, M., Cantoni, C., & Comi, G. (2001). Denaturing gradient gel electrophoresis analysis of the 16S rRNA gene VI region to monitor dynamic changes in the bacterial population during fermentation of Italian sausages. *Applied and Environmental Microbiology, 67,* 5113–5121.

Comi, G., Urso, R., Iacumin, L., Rantsiou, K., Cattaneo, P., Cantoni, C., et al. (2005). Characterisation of naturally fermented sausages produced in the North East of Italy. *Meat Science, 69,* 381–392.

Demeyer, D. (1992). Meat fermentation as an integrated process. In F. J. M. Smulders, F. Toldrá, J. Flores, & M. Prieto (Eds.), *New Technologies for Meat and Meat Products* (pp. 21–36). Nijmegen (The Netherlands): Audet.

Demeyer, D. I., & Toldrá, F. (2004). Fermentation. In W. Jensen, C. Devine, & M. Dikemann (Eds.), *Encyclopedia of Meat Sciences* (pp. 467–474). London, UK: Elsevier Science Ltd.

Demeyer, D., Raemaekers, M., Rizzo, A., Holck, A., De Smedt, A., ten Brink, B., et al. (2000). Control of bioflavour and safety in fermented sausages: first results of a European project. *Food Research International, 33,* 171–180.

Demeyer, D., & Stahnke, L. (2002). Quality control of fermented meat products. In J. Kerry & D. Ledward (Eds.), *Meat processing: Improving quality* (pp. 359–393).Cambridge (UK): Woodhead Pub. Co.

Durá, M. A., Flores, M., & Toldrá, F. (2002). Purification and characterisation of a glutaminase from *Debaryomices* spp. *International Journal of Food Microbiology, 76,* 117–126.

Durá, M. A., Flores, M., & Toldrá, F. (2004a). Effect of growth phase and dry-cured sausage processing conditions on Debaryomyces spp. generation of volatile compounds from branched-chain amino acids. *Food Chemistry, 86,* 391–399.

Durá, M. A., Flores, M., & Toldrá, F. (2004b). Effect of *Debaryomyces* spp. on the proteolysis of dry-fermented sausages. *Meat Science, 68,* 319–328.

Durá, M. A., Flores, M., & Toldrá, F. (2004c). Effect of *Debaryomyces* spp on aroma formation and sensory quality of dry-fermented sausages. *Meat Science, 68,* 439–446.

Ellis, D. F. (2001). Meat smoking technology. In Y. H. Hui, H. Nip, R. W. Rogers & O. Young (Eds.), *Meat Science and Applications*. (pp. 509–519). New York: Marcel Dekker Inc.

Fadda, S., Sanz, Y., Vignolo, G., Aristoy, M. C., Oliver, G., & Toldrá, F. (1999a). Hydrolysis of pork muscle sarcoplasmic proteins by *Lactobacillus curvatus* and *Lactobacillus sake*. *Applied and Environmental Microbiology, 65,* 578–584.

Fadda, S., Sanz, Y., Vignolo, G., Aristoy, M. C., Oliver, G., & Toldrá, F. (1999b). Characterization of muscle sarcoplasmic and myofibrillar protein hydrolysis caused by *Lactobacillus plantarum*. *Applied and Environmental Microbiology, 65,* 3540–3546.

Flores, J., Marcus, J. R., Nieto, P., & Navarro, J. L. (1997). Effect of processing conditions on proteolysis and taste of dry-cured sausages. *Zetiscrift fürLebensmittel Untersuchung und Forschung A, 204,* 168–172.

Flores, M., Durá, M. A., Marco, A., & Toldrá, F. (2004). Effect of Debaryomyces spp. on aroma formation and sensory quality of dry-fermented sausages. *Meat Science, 68,* 439–446.

Hagen, B. F., Berdagué, J. L., Holck, A. L., Naes, H., & Blom, H. (1996). Bacterial proteinase reduces maturation time of dry fermented sausages. *Journal of Food Science, 61,* 1024–1029.

Hammes, W. P., & Knauf, H. J. (1994). Starters in the processing of meat products. *Meat Science, 36,* 155–168.

Herranz, B., de la Hoz, L., Hierro, E., Fernández, M., & Ordoñez, J. A. (2005). Improvement of the sensory properties of dry-fermented sausages by the addition of free amino acids. *Food Chemistry, 91,* 673–682.

Herranz, B., Fernández, M., de la Hoz, L., & Ordoñez, J. A. (2006). Use of bacterial extracts to enhance amino acid breakdown in dry fermented sausages. *Meat Science, 72,* 318–325.

Hierro, E., De la Hoz, L., Ordoñez, J. A. (1997). Contribution of microbial and meat endogenous enzymes to the lipolysis of dry fermented sausages. *Journal of Agricultural and Food Chemistry 45,* 2989–2995.

Hughes, M. C., Kerry, J. P., Arendt, E. K., Kenneally, P. M., McSweeney, P. L. H., & O'Neill, E. E. (2002). Characterization of proteolysis during the ripening of semidry fermented sausages. *Meat Science, 62,* 205–216.

Leistner, F. (1992). The essentials of producing stable and safe raw fermented sausages. In F. J. M. Smulders, F. Toldrá, J. Flores & M. Prieto (Eds.), *New Technologies for Meat and Meat Products* (pp. 1–19). Nijmegen (The Netherlands): Audet.

MacLeod, G., & Seyyedain-Ardebili, M. (1981). Natural and simulated meat flavors (with particular reference to beef). *CRC Critical Reviews in Food Science and Nutrition, 14,* 309–437.

Molly, K., Demeyer, D. I., Civera, T., & Verplaetse, A. (1996). Lipolysis in Belgian sausages: Relative importance of endogenous and bacterial enzymes. *Meat Science, 43,* 235–244.

Molly, K., Demeyer, D. I., Johansson, G., Raemaekers, M., Ghistelinck, M., & Geenen, I. (1997). The importance of meat enzymes in ripening and flavor generation in dry fermented sausages. First results of a European project. *Food Chemistry, 54,* 539–545.

Montel, M. C., Masson, F., & Talon, R. (1998). Bacterial role in flavour development, *Meat Science, 49,* S111–S123.

Montel, M. C., Reitz, J., Talon, R., Berdagué, J. L., & Rousset-Akrim, S. (1996). Biochemical activities of Micrococcaceae and their effects on the aromatic profiles and odours of a dry sausage model. *Food Microbiology, 13,* 489–499.

Montel, M. C., Seronine, M. P., Talon, R., & Hebraud, M. (1995). Purification and characterization of a dipeptidase from *Lactobacillus sake. Applied and Environmental Microbiology, 61,* 837–839.

Olesen, P. T., Meyer, A. S., & Stahnke, L. H. (2004b). Generation of flavour compounds in fermented sausages-the influence of curing ingredients, Staphylococcus starter culture and ripening time. *Meat Science, 66,* 675–687.

Olesen, P. T., Stahnke, L.,H., & Talon, R. (2004a). Effect of ascorbate, nitrate and nitrite on the amount of flavour compounds produced from leucine by *Staphylococcus xylosus* and *Staphylococcus carnosus. Meat Science, 68,* 193–200.

Ordoñez, J. A., Hierro, E. M., Bruna, J. M., & de la Hoz, L. (1999). Changes in the components of dry-fermented sausages during ripening. *Critical Reviews in Food Science and Nutrition, 39,* 329–367.

Santos, N. N., Santos, R. C., Sanz, Y., Bolumar, T., Aristoy, M-C., & Toldrá, F. (2001). Hydrolysis of muscle sarcoplasmic proteins by Debaryomyces hansenii. *International Journal of Food Microbiology, 68,* 199–206.

Sanz, B., Selgas, D., Parejo, I., & Ordoñez, J. A. (1998). Characteristics of lactobacilli isolated from dry fermented sausages. *International Journal of Food Microbiology, 6,* 199–205.

Sanz, Y., Fadda, S., Vignolo, G., Aristoy, M. C., Oliver, G., & Toldrá, F. (1999a). Hydrolytic action of *Lactobacillus casei* CRL 705 on pork muscle sarcoplasmic and myofibrillar proteins. *Journal of Agricultural and Food Chemistry, 47,* 3441–3448.

Sanz, Y., Fadda, S., Vignolo, G., Aristoy, M. C., Oliver, G., & Toldrá, F. (1999b). Hydrolysis of muscle myofibrillar proteins by *Lactobacillus curvatus* and *Lactobacillus sake. International Journal of Food Microbiology, 53,* 115–125.

Sanz, Y., Mulholland, F., & Toldrá, F. (1998). Purification and characterization of a tripeptidase from *Lactobacillus sake. Journal of Agricultural and Food Chemistry, 46,* 349–353.

Sanz, Y., Sentandreu, M. A., & Toldrá, F. (2002). Role of muscle and bacterial exopeptidases in meat fermentation. In F. Toldrá (Ed.), *Research advances in the quality of meat and meat products* (pp. 143–155). Trivandrum, India: Research Signpost.

Sanz, Y., & Toldrá, F. (1997). Purification and characterization of an aminopeptidase from *Lactobacillus sake*. *Journal of Agricultural and Food Chemistry, 45,* 1552–1558.

Sanz, Y., & Toldrá, F. (1999). The role of exopeptidases from Lactobacillus sake in dry fermented sausages. *Recent Research Developments in Agricultural & Food Chemistry, 3,* 11–21.

Sanz, Y., & Toldrá, F. (2001). Purification and characterization of an X-prolyl-dipeptidyl peptidase from *Lactobacillus sake*. *Applied and Environmental Microbiology, 67,* 1815–1820.

Sanz, Y., & Toldrá, F. (2002). Purification and characterization of an arginine aminopeptidase from *Lactobacillus sakei*. *Applied and Environmental Microbiology, 68,* 1980–1987.

Sentandreu, M. A., & Toldrá, F. (2001). Dipeptidylpeptidase activities along the processing of Serrano dry-cured ham. *European Food Research Technology, 213,* 83–87.

Skibsted, L. H., Mikkelsen, A., & Bertelsen, G. (1998). Lipid-derived off-flavours in meat. In F. Shahidi (Ed.), *Flavor of meat, meat products and seafoods* (pp. 216–256). London, UK: Blackie Academic & Professional.

Stahnke, L. H. (1994). Aroma components from dried sausages fermented with *Staphylococcus-xylosus*. *Meat Science, 38,* 39–53.

Stahnke, L. H. (1995). Dried Sausages Fermented with *Staphylococcus xylosus* at Different Temperatures and with Different Ingredient Levels .2. Volatile Components. *Meat Science, 41,* 193–209.

Stahnke, L. H. (1999). Volatiles produced by Staphylococcus xylosus and Staphylococcus carnosus during growth in sausage minces - Part I. Collection and identification. Food Science and Technology. *Lebensmittel-Wissenschaft & Technologie, 32,* 357–364.

Stahnke, L. H., Holck, A., Jensen, A., Nilsen, A., & Zanardi, E. (2002). Maturity acceleration of Italian dried sausage by Staphylococcus carnosus - Relationship between maturity and flavor compounds. *Journal of Food Science, 67,* 1914–1921.

Sunesen, L. O., Stahnke, L. H. (2003). Mould starter cultures for dry sausages-selection, application and effects. *Meat Science, 65,* 935–948.

Talon, R., Leroy-Sétrin, S., & Fadda, S. (2002). Bacterial starters involved in the quality of fermented meat products. In F. Toldrá (Ed.), *Research advances in the quality of meat and meat products* (pp. 175–191). Trivandrum (India): Research Signpost.

Talon, R., Leroy-Sétrin, S., & Fadda, S. (2004). Dry fermented sausages. In Y. H. Hui, L. M. Goddik, J. Josephsen, P. S. Stanfield, A. S. Hansen, W. K. Nip & F Toldrá (Eds.), *Handbook of food and beverage fermentation technology* (pp. 397–416). New York: Marcel-Dekker.

Talon, R., Walter, D., Chartier, S., Barriere, C., & Montel, M. C. (1999). Effect of nitrate and incubation conditions on the production of catalase and nitrate reductase by staphylococci. *International Journal of Food Microbiology, 52,* 47–56.

Talon, R., Walter, D., & Montel, M. C. (2000). Growth and effect of staphylococci and lactic acid bacteria on unsaturated free fatty acids. *Meat Science, 54,* 41–47.

Tjener, K., Stahnke, L. H., Andersen, L., & Martinussen, J. (2003). A fermented meat model system for studies of microbial aroma formation. *Meat Science, 66,* 211–218.

Tjener, K., Stahnke, L. H., Andersen, L., & Martinussen, J. (2004a). Growth and production of volatiles by *Staphylococcus carnosus* in dry sausages: Influence of inoculation level and ripening time. *Meat Science, 67,* 447–452.

Tjener, K., Stahnke, L. H., Andersen, L., & Martinussen, J. (2004b). Addition of alpha-ketoglutarate enhances formation of volatiles by Staphylococcus carnosus during sausage fermentation. *Meat Science, 67,* 711–719.

Toldrá F. (1992). The enzymology of dry-curing of meat products. In F. J. M. smulders, F. Toldrá, J. Flores & M. Prieto (Eds.) New Technologies for Meat and Meat Products (pp. 209–231). Nijmegen (The Netherlands): Audet.

Toldrá, F. (1998). Proteolysis and lipolysis in flavour development of dry-cured meat products. *Meat Science, 49,* s101–s110.

Toldrá, F. (2002). *Dry-cured meat products* (pp. 1–38). Trumbull, CT: Food & Nutrition Press.

Toldrá, F. (2004). Fermented meats. In Y. H. Hui, J. S. Smith (Eds.), *Food processing: Principles and applications* (pp. 399–415). Ames, IA: Blackwell Publishing.

Toldrá, F. (2006a). Meat fermentation. In Y. H. Hui, E. Castell-Perez, L. M. Cunha, I. Guerrero-Legarreta, H. H. Liang, Y. M. Lo, D. L. Marshall, W. K. Nip, F. Shahidi, F. Sherkat, R. J. Winger & K. L. Yam (Eds.), *Handbook of Food Science, Technology and Engineering* (Vol. 4, pp. 181-1 to 181-12). Boca Raton, FL: CRC Press.

Toldrá, F. (2006b). Biochemistry of fermented meat. In Y. H. Hui, W. K. Nip, M. L. Nollet, G. Paliyath & B. K. Simpson (Eds.), *Food biochemistry and food processing* (pp. 641–658). Ames, IA: Blackwell Publishing.

Toldrá, F. (2006c). Biochemical proteolysis basis for improved processing of dry-cured meats. In L. M. L. Nollet & F. Toldrá (Eds.), *Advanced technologies for meat processing* (pp. 329–351). Boca Raton, FL: CRC Press.

Toldrá, F. (2007a). Fermented meat production. In Y. H. Hui, R. Chandan, S. Clark, N. Cross, J. Dobbs, W. J. Hurst, L. M. L. Nollet, E. Shimoni, N. Sinha, E. B. Smith, S. Surapat, A. Titchenal & F. Toldrá (Eds.), *Handbook of Food Product Manufacturing* (Vol. 2, pp. 263–277). Hoboken, NJ: John Wiley Interscience of NY.

Toldrá, F. (2007b). Biochemistry of muscle and fat. In F. Toldrá, Y. H. Hui, I. Astiasarán, W. K. Nip, J. G. Sebranek, E. T. F. Silveira, L. H. Stahnke, & R. Talon (Eds.), *Handbook of fermented meat and poultry* (pp. 51–58). Ames, IA: Blackwell Publishing

Toldrá, F., & Flores, M. (2007). Processed pork meat flavors. In Y. H. Hui, R. Chandan, S. Clark, N. Cross, J. Dobbs, W. J. Hurst, L. M. L. Nollet, E. Shimoni, N. Sinha, E. B. Smith, S. Surapat, A. Titchenal & F. Toldrá (Eds), *Handbook of Food Product Manufacturing* (Vol. 2, pp. 279–299). Hoboken, NJ: John Wiley Interscience of NY.

Toldrá, F., Flores, M., & Sanz, Y. (2001). Meat fermentation technology. In Y. H. Hui, W. K. Nip, R. W. Rogers, & O. A. Young (Eds.), *Meat Science and applications* (pp. 537–561). New York: Marcel Dekker Inc.

Toldrá, F., & Navarro, J. L. (2002). Action of muscle lipases during the processing of dry-cured ham. In F. Toldrá (Ed.), *Research advances in the quality of meat and meat products* (pp. 249–254). Trivandrum, India: Research Signpost.

Toldrá, F., Nip, W. K., & Hui, Y. H. (2007). Dry fermented sausages: An overview. In F. Toldrá, Y. H. Hui, I. Astiasarán, W. K. Nip, J. G. Sebranek, E. T. F. Silveira, L. H. Stahnke, & R. Talon (Eds.), *Handbook of fermented meat and poultry* (pp. 321–332). Ames, IA: Blackwell Publishing.

Toldrá, F., & Reig, M. (2007). Sausages. In Y. H. Hui, R. Chandan, S. Clark, N. Cross, J. Dobbs, W. J. Hurst, L. M. L. Nollet, E. Shimoni, N. Sinha, E. B. Smith, S. Surapat, A. Titchenal, & F. Toldrá (Eds.), *Handbook of Food Product Manufacturing* (Vol. 2, pp. 249–262). Hoboken, NJ: John Wiley Interscience of NY.

Chapter 10
Latest Developments in Probiotics

Frédéric Leroy, Gwen Falony, and Luc de Vuyst

Introduction

Probiotic foods are a group of health-promoting, so-called functional foods, with large commercial interest and growing market shares (Arvanitoyannis & van Houwelingen-Koukaliaroglou, 2005). In general, their health benefits are based on the presence of selected strains of lactic acid bacteria (LAB), that, when taken up in adequate amounts, confer a health benefit on the host. They are administered mostly through the consumption of fermented milks or yoghurts (Mercenier, Pavan, & Pot, 2003). In addition to their common use in the dairy industry, probiotic LAB strains may be used in other food products too, including fermented meats (Hammes & Hertel, 1998; Incze, 1998; Kröckel, 2006; Työppönen, Petäjä, & Mattila-Sandholm, 2003). Although the concept is not new, only a few manufacturers offer fermented sausages with probiotic LAB. This is probably due to the more artisan orientation of sausage manufacturers as compared to the dairy industry, a larger variety of products, as well as a number of uncertainties concerning technological, microbiological, and regulatory aspects (Kröckel, 2006). The application of probiotic LAB must in all cases be based on a careful selection procedure, if any health claims are to be taken into account. The present chapter gives an overview of research activities that have previously explored the potential of probiotic LAB strains in fermented meats and aims at giving a critical interpretation of the results obtained.

Probiotics

History and Definitions

Although probiotics are usually linked with gut health, the first suggestion of a beneficial association between microorganisms and the human host can probably

---

Luc de Vuyst
Research Group of Industrial Microbiology and Food Biotechnology (IMDO), Vrije Universiteit Brussel (VUB), Pleinlaan 2, B-1050 Brussels, Belgium

be attributed to Albert Döderlein, who proposed in 1892 that lactic acid production by vaginal bacteria prevented or inhibited the growth of pathogenic bacteria (Döderlein, 1892). Most cited as the founding father of the probiotic concept, however, is Ilya Metchnikoff. In his work, "The Prolongation of Life – Optimistic Studies", published in 1908, he implicated a lactic acid bacterium found in Bulgarian yoghurts as the agent responsible for deterring intestinal putrefaction and ageing (Metchnikoff, 1908). Hence, he became the first to speculate on the potential health-promoting and even life-lengthening properties of LAB. Another milestone in the history of probiotics is undoubtedly the work of Minoru Shirota, who was the first to actually cultivate a beneficial intestinal bacterium, *Lactobacillus casei* Shirota, and distribute it in a dairy drink that was introduced to the market in 1935 (Yakult Central Institute for Microbiological Research, 1999).

The word "probiotic" stems from the Greek πρo βιoς (pro bios, "for life") and was originally proposed to describe growth-promoting substances produced by one protozoan for the benefit of another (Lilly & Stillwell, 1965). In 1974, it was first linked with the intestinal microbial balance by Parker (1974) in his work on animal feed supplements with beneficial effect on the host. In the following decades, the concept of probiotics was regularly defined and redefined. Nowadays, a widely accepted definition is a rather broad one that was proposed by a Joint Expert Consultation of the Food and Agricultural Organization of the United Nations (FAO) and the World Health Organization (WHO). It classifies probiotics as "life microogranisms that, when consumed in an adequate amount, confer a health benefit on the host" (FAO/WHO, 2001). This definition does not emphasize on the nature of the host (animal or human), the origin of the microorganisms (human or non-human), or the ability to adhere to body surfaces. Also, although stressed elsewhere in the FAO/WHO report, the viability at the target site is not considered a restriction, thus leaving the door open for the acceptance of the yoghurt bacteria *Streptococcus thermophilus* and *Lactobacillus delbrueckii* subsp. *bulgaricus* as probiotic microorganisms as well as of the health effects ascribed to certain cell components. As no site of action is precise, probiotic preparations can not only be targeted towards the benefit of the gut, but, for instance, also towards that of the oral cavity, the nasopharynx, the respiratory tract, the stomach, the vagina, the bladder, and the skin (Reid, 2005). Of course, no definition is final, and, amongst others, the viability of the probiotics, or better, their effectiveness after food processing and storage, their site of activity, the numbers necessary to exert a beneficial effect, the format of intake, and the nature of the carrier remain critical issues in a vivid discussion (Makras, Avonts, & De Vuyst, 2004; Mercenier et al., 2003; Senok, Ismaeel, & Botta, 2005).

As probiotics are mainly directed to alter the composition and/or metabolic activity of the gut microbiota towards what is generally believed to be a healthy or balanced one, that is, being predominantly saccharolytic and comprising significant numbers of bifidobacteria and lactobacilli (Picard et al., 2005), pre- and synbiotics need to be mentioned as alternative or complementary strategies to achieve this goal. Prebiotics are selectively fermented, non-digestible, food ingredients that allow specific changes, both in the composition and/or activity of the gastrointestinal microbiota, that confer benefits upon host well-being and health (Gibson, Probert, Van Loo, Rastall, & Roberfroid, 2004; Gibson & Roberfroid, 1995). To this day, sufficient scientific evidence exists to recognise three types of carbohydrates as prebiotics, namely inulin-type fructans, transgalacto-oligosaccharides, and lactulose (Gibson et al., 2004), although the latter should be considered as a laxative drug (Bass & Dennis, 1981). For some other candidate prebiotics, promising data have been published but further investigation is required (Gibson et al., 2004). Finally, synbiotics are mixtures of pro- and prebiotics, wherein, the latter is thought to improve survival and implantation of the former, either by stimulating growth or by metabolically activating the health-promoting bacteria, thereby taking advantage of the individual and possibly the synergistic health effects of both components (Gibson & Roberfroid, 1995; Rastall & Maitin, 2002).

Probiotic Microorganisms

Different types of food products or food supplements containing viable probiotic microorganisms with health-promoting properties are commercially available, either as fermented or non-fermented food commodities, or as specific food supplements and pharmaceutical preparations in the form of powders, tablets, or capsules. The first step in the development of such products is the selection of an appropriate microbial strain. Throughout the years, several selection criteria for probiotics have been formulated, including human origin, non-pathogenic behaviour, safety, resistance to gastric acidity and bile toxicity, adhesion to or interaction with the gut epithelial tissue, ability to persist within the gastrointestinal tract, production of antimicrobial substances and nutraceuticals, evidence of beneficial health effects, ability to influence metabolic activities, and resistance to technological processes (Dunne et al., 2001; Maldonado Galdeano, de Moreno de LeBlanc, Vinderola, Bibas Bonet, & Perdigón, 2007; Ross, Desmond, Fitzgerald, & Stanton, 2005). Although some of the present criteria appear obvious, others are directly related with the definition of probiotics handled by the authors. It should, however, be stressed that characteristics such as survival of the passage through the upper gastrointestinal tract, thus resisting the action of gastric juice, bile salts, and proteolytic enzymes, are not sufficient to call a certain microbial strain as probiotic. According to the WHO/FAO definition, the main criterion for a probiotic strain should be the fact that it confers a health benefit on the host (FAO/WHO, 2001). This benefit can only be demonstrated through well-designed, randomized, double blind, placebo-controlled, multi-centre human trials, the results of which are published in peer-reviewed international scientific journals (Guarner & Schaafsma, 1998; Salminen et al., 1998). Furthermore, it should be stressed that probiotic effects are strain-dependent, and extrapolation of the existing data from closely related microorganisms is not sufficient to identify a strain as probiotic (Mercenier et al., 2003). Exact identification and characterization of the potential probiotic strain used at the genus, species, and even strain level, using internationally accepted methodologies, should be the

Table 10.1 Microorganisms whose strains are used or considered for use as probiotics [adapted from (Collins & Gibson, 1999; Makras, 2004; Senok et al., 2005)]

Lactobacillus sp.	Bifidobacterium sp.	Other Lactic Acid Bacteria	Other microorganisms
L. acidophilus	B. adolescentis	Enterococcus faecalis [a]	Bacillus cereus [a,b]
L. amylovorus	B. animalis subsp. animalis	Enterococcus faecium [a]	Bacillus subtilis [b]
L. brevis	B. animalis subsp. lactis	Lactococcus lactis	Clostridium butyricum
L. casei	B. bifidum	Leuconostoc mesenteroides	Escherichia coli [b]
L. crispatus	B. breve	Sporolactobacillus inulinus [a]	Propionibacterium freudenreichii [a,b]
L. curvatus	B. longum	Streptococcus thermophilus	Saccharomyces cerevisiae [b]
L. delbrueckii subsp. bulgaricus			Saccharomyces boulardii [b]
L. fermentum			
L. gallinarum [a]			
L. gasseri			
L. johnsonii			
L. paracasei			
L. plantarum			
L. reuteri			
L. rhamnosus			
L. salivarius			

[a] mainly applied in animals
[b] mainly applied in pharmaceutical preparations

first step in every process of development of probiotic food product (Reid, 2005). It is generally believed that rational selection of probiotics will be facilitated by the recent availability of genome sequences of some probiotic and candidate probiotic strains, allowing the prediction of their physiological profiles (Klaenhammer, Barrangou, Buck, Azcarate-Peril, & Altermann, 2005; Leahy, Higgins, Fitzgerald, & van Sinderen, 2005).

Up to now, mainly bacteria belonging to the genera *Lactobacillus* and *Bifidobacterium* have been used or considered as probiotics, besides other bacteria (mostly belonging to the group of LAB) and some yeasts (Table 10.1).

Health Benefits: Prophylactic and Therapeutic Effects of Probiotics

A wide variety of potential beneficial health effects have been attributed to probiotics (Table 10.2). Claimed effects range from the alleviation of constipation to the prevention of major life-threatening diseases such as inflammatory bowel disease, cancer, and cardiovascular incidents. Some of these claims, such as the effects of probiotics on the shortening of intestinal transit time or the relief from

Table 10.2 Potential and established health benefits associated with the usage of probiotics [adapted from (FAO/WHO, 2001; Mercenier et al., 2003; Naidu, Bidlack, & Clemens, 1999; Parvez, Malik, Ah Kang, & Kim, 2006; Sanders, 1998; Sanders, & Huis in 't Veld, 1999)]

Health benefit	Proposed mechanism(s)
Cancer prevention	Inhibition of the transformation of pro-carcinogens into active carcinogens, binding/inactivation of mutagenic compounds, production of anti-mutagenic compounds, suppression of growth of pro-carcinogenic bacteria, reduction of the absorption of carcinogens, enhancment of immune function, influence on bile salt concentrations
Control of irritable bowel syndrome	Modulation of gut microbiota, reduction of intestinal gas production
Management and prevention of atopic diseases	Modulation of immune response
Management of inflammatory bowel diseases (Crohn's disease, ulcerative colitis, pouchitis)	Modulation of immune response, modulation of gut microbiota
Prevention of heart diseases/influence on blood cholesterol levels	Assimilation of cholesterol by bacterial cells, deconjugation of bile acids by bacterial acid hydrolases, cholesterol-binding to bacterial cell walls, reduction of hepatic cholesterol synthesis and/or redistribution of cholesterol from plasma to liver through influence of the bacterial production of short-chain fatty acids
Prevention of urogenital tract disorders	Production of antimicrobial substances, competition for adhesion sites, competitive exclusion of pathogens
Prevention/alleviation of diarrhoea caused by bacteria/viruses	Modulation of gut microbiota, production of antimicrobial substances, competition for adhesion sites, stimulation of mucus secretion, modulation of immune response
Prevention/treatment of *Helicobacter pylori* infections	Production of antimicrobial substances, stimulation of the mucus secretion, competition for adhesion sites, stimulation of specific and non-specific immune responses
Relief of lactose indigestion	Action of bacterial β-galactosidase(s) on lactose
Shortening of colonic transit time	Influence on peristalsis through bacterial metabolite production

lactose maldigestion, are considered well-established, while others, such as cancer prevention or the effect on blood cholesterol levels, need further scientific backup (Gill & Guarner, 2004). The mechanisms of action may vary from one probiotic strain to another and are, in most cases, probably a combination of activities, thus making the investigation of the responsible mechanisms a very difficult and complex task. In general, three levels of action can be distinguished: probiotics can influence human health by interacting with other microorganisms present on the site of action, by strengthening mucosal barriers, and by affecting the immune system of the host (Marteau & Shanahan, 2003). Again, the strain specificity of each probiotic effect must be stressed; concerning the prevention and treatment of diarrhoea, for example, only indicative evidence of an overall protective effect against travellers' and antibiotic-associated diarrhoea exists, while the efficacy of *L. rhamnosus* GG in treating rotaviral diarrhoea has extensively been demonstrated

(Gill & Guarner, 2004; Santosa, Farnworth, & Jones, 2006). Further research to support the health claims attributed to probiotics and to unravell the mechanisms behind them is needed.

Safety Considerations

As viable probiotic bacteria have to be consumed in large quantities, over an extended period of time, to exert beneficial effects, the issue of the safety of these microorganisms is of primary concern (Senok et al., 2005). Historical data indicate that lactobacilli and bifidobacteria are safe for human use (Reid, 2005). It has been suggested that the human origin of a strain confirms its normal commensal nature, and, therefore, its safety. However, it remains difficult to establish the origin of a bacterial species, and the fact that infants are born with sterile intestines raises the question whether "human origin" is an appropriate classification for bacteria (Reid, 2005).

Although minor side effects of the use of probiotics have been reported, infections with probiotic bacteria rarely occur and invariably only in immuno-compromised patients or those with intestinal bleeding (Gueimonde, Frias, & Ouwehand, 2006; Marteau, 2002; Reid, 2005).

An issue of concern regarding the use of probiotics is the presence of chromosomal, transposon-, or plasmid-located antibiotic resistance genes amongst the probiotic microorganisms. At this moment, insufficient information is available on situations in which these genetic elements could be mobilised, and it is not known if situations could arise where this would become a clinical problem. When dealing with the selection of probiotic strains, the FAO/WHO Consultancy recommends that probiotic microorganisms should not harbour transmissible drug resistance genes encoding resistance to clinically used drugs (FAO/WHO, 2001).

For the assessment of the safety of probiotic microorganisms and products, FAO/WHO has formulated guidelines, recommending that probiotic strains should be evaluated for a number of parameters, including antibiotic susceptibility patterns, toxin production, metabolic and haemolytic activities, and infectivity in immuno-compromised animals (FAO/WHO, 2002; Reid, 2005; Senok et al., 2005).

Application of Probiotics in Fermented Meat Products

Fermented Meat as a Carrier for Probiotic Bacteria

Dry fermented meat products are usually not or only mildly heated, which is adequate for the carriage of probiotic bacteria (Ammor & Mayo, 2007; Arihara, 2006). Although there are in principle no major reasons preventing application of probiotic LAB strains in meat, several points have to be carefully addressed.

Although meat is a food with high nutritional value, some consumers may perceive meat products as unhealthy (Arihara, 2006). This can be ascribed to the image of meat as such, in combination with the presence of nitrite, salt, and fat. Meat products are seldomly perceived as "healthy foods", which may compromise their marketing potential (Lücke, 2000). However, adding nutritional assets to meat products could be a strategy to promote them as valuable elements of a high quality diet and to meet the trend for healthier meat products (Arihara, 2006; Jiménez-Colmenero, Carballo, & Cofrades, 2001).

In addition, the impact of the meat environment, with its high content in curing salt and its low water activity and pH, and of meat fermentation technology, based on acidification and drying, on the viability of the cells must be taken into account.

The approaches followed up till now can be summarized as follows: 1) screening for probiotic properties among bacteria that are naturally present in the meat or that originate from meat starter cultures, 2) application of existing probiotic LAB in meat products, 3) evaluating the impact of probiotic sausages on humans during clinical studies, and 4) assessment of the technological suitability of probiotic LAB during sausage-making, in particular with respect to sensory deviations (Leroy, Verluyten, & De Vuyst, 2006).

Screening for Probiotic Properties Among Meat-Associated Bacteria

A promising strategy for the development of probiotic fermented sausages consists of using bacteria that are commonly associated with the meat environment and that possess probiotic properties. In this way, sausage isolates (Klingberg, Axelsson, Naterstad, Elsser, & Budde, 2005; Papamanoli, Tzanetakis, Litopoulou-Tzanetaki, & Kotzekidou, 2003; Pennacchia et al., 2004; Pennacchia, Vaughan, & Villani, 2006; Rebucci et al., 2007) or existing commercial meat starter cultures (Erkkilä & Petäjä, 2000) are screened for probiotic properties.

Frequently, the following characteristics are mentioned as indicators for probiotic activity: tolerance to the low pH of gastric juice, resistance to the detergent-like action of bile salts, adhesion to the intestinal mucosa for temporary ileum colonisation, growth capability in the presence of prebiotic carbohydrates, antimicrobial activity towards intestinal pathogens, and nutraceutical properties such as the production of vitamins and conjugated linoleic acid (Ammor & Mayo, 2007; Pennacchia et al., 2006). However, the relevance of at least some of these properties can be questioned and conclusions about true probiotic qualities require caution (see below).

Following this approach, the commercial meat starter strains *L. sakei* Lb3 and *Pediococcus acidilactici* PA-2 have been proposed as potential probiotic starter cultures because of their survival capacities under simulated gastrointestinal conditions (Erkkilä & Petäjä, 2000). Also, isolates of *L. casei/paracasei* from sausages fermented with *L. casei*, *L. paracasei*, *L. rhamnosus*, and *L. sakei* were screened for viability in artificial gastric juice, artificial intestinal fluid, in vitro adhesion to human intestinal cell lines, organic acid production, and pathogen inactivation (Rebucci et al., 2007). Several *L. plantarum* sausage isolates were found to have appreciable adhesion rates towards Caco-2 cell lines and were considered as better adhesive bacteria than *L. brevis* and *L. paracasei*–group sausage isolates (Pennacchia et al., 2006).

It is important to note that the obtained results of the latter studies are preliminary and that further research is needed to prove the true probiotic health nature of the candidate strains obtained.

Use/Application of Known Probiotic Strains

As an alternative to the approach mentioned above, it may be investigated if strains with (presumed) probiotic properties perform well in a fermented meat environment. Such strains are usually human intestinal isolates and hence not from meat origin. Therefore, they should be able to compete with the natural meat microbiota in an environment which is not their natural habitat, be able to survive the fermentation and drying process, and, preferably, be able to grow to numbers that display health-promoting effects. Alternatively, micro-encapsulation in alginate beads may be used to increase survival (Muthukumarasamy, & Holley, 2006, 2007). In this way, several lactobacilli of human intestinal origin have been shown to survive the sausage manufacturing process and can be detected in high numbers in the end-product (Arihara et al. 1998; Erkkilä, Petäjä, et al., 2001; Erkkilä, Suihko, Eerola, Petäjä, & Mattila-Sandholm, 2001; Pidcock, Heard, & Henriksson, 2002; Sameshima et al., 1998).

It is certainly an asset if these new meat starter cultures also contribute to food safety. Probiotic strains, with additional food safety assets, could contribute a high added value to healthy fermented meat products. For instance, *L. reuteri* ATCC 55730 and *Bifidobacterium longum* ATCC 15708 increased the inactivation of *Escherichia coli* O157:H7 during sausage manufacturing (Muthukumarasamy & Holley, 2007). *L. rhamnosus* FERM P-15120 and *L. paracasei* subsp. *paracasei* FERM P-15121 inhibited the growth and enterotoxin production of *Staphylococcus aureus* to the same extent as a commercial *L. sakei* starter culture (Sameshima et al., 1998). On the other hand, *L. acidophilus* FERM P-15119 could not satisfactorily decrease *Staph. aureus* numbers, indicating the importance of careful strain selection with respect to both probiotic and food safety properties.

Human Studies

Ultimately, human studies should confirm the functionality of probiotic fermented sausages. In contrast to the dairy industry, such studies are very scarce till date and results have been moderately successful. One study deals with the effect of probiotic sausages on immunity and blood serum lipids. The daily consumption of 50 g of probiotic sausage by healthy volunteers, containing *L. paracasei* LTH 2579, during several weeks has been shown to modulate various aspects of host immunity but there was no significant influence on the serum concentration of different cholesterol fractions and triacylglycerides (Jahreis et al., 2002). In faecal samples, there was a statistically significant increase in the numbers of *L. paracasei* LTH 2579, but not in the faeces of all volunteers. It is interesting to mention that the sausage matrix seems to protect the survival of probiotic lactobacilli through the gastrointestinal tract (Klingberg & Budde, 2006).

Technological Suitability

In all cases, it should be checked that the sensory properties of the end-products are not negatively affected, especially when strains from non-meat origin are used. The (potential) probiotic strains *L. rhamnosus* GG, *L. rhamnosus* LC-705, *L. rhamnosus* E-97800 and *L. plantarum* E-98098 have been tested as functional starter culture strains in Northern European sausage fermentation without negatively affecting the technological or sensory properties, with a (minor) exception for *L. rhamnosus* LC-705 (Erkkilä, Petäjä, et al., 2001; Erkkilä, Suihko, et al., 2001). Similarly, the intestinal isolates *L. paracasei* L26 and *B. lactis* B94 had no negative impact on the sensory properties of the product when applied in conjunction with a traditional meat starter culture (Pidcock et al., 2002). Also, the use of alginate-microencapsulation of *L. reuteri* was not resulting in differences concerning sensory quality (Muthukumarasamy, & Holley, 2006).

Conclusions and Critical Remarks

Although meat products containing probiotic LAB are already being marketed since 1998 by German and Japanese producers (Arihara, 2006), most scientific results obtained until now are rather preliminary and mostly based on incomplete approaches, not permitting full assessment of the probiotic effects of fermented sausages on human health.

In vitro studies are valuable tools to asses the survival of potential probiotic strains in the human gastrointestinal tract. Furthermore, they can provide insight into the abilities of a strain to adhere to surfaces and to inhibit growth or adhesion of pathogens. However, the mere ability of a strain to survive the passage through the human intestinal tract does not qualify a microorganism as a probiotic. Inhibition studies might help to elucidate the mechanisms behind a probiotic effect, but in vitro inhibition of pathogens by a potential probiotic by no means guarantees that the same will occur in the complexity of the colon ecosystem. When only based on results of in vitro studies, the use of the term "potential" probiotic is questionable. Research regarding the launch of new probiotic strains should rigorously follow the guidelines formulated by the FAO/WHO (2002), including detailed identification of the strain and an approved beneficial health effect.

Nevertheless, the addition of microorganisms with known probiotic characteristics to a meat fermentation process seems an elegant solution for the development of probiotic meat products. Most research concerning this strategy focuses on the survival of the added species in the meat matrix and its influence on the technological and sensory characteristics of the final product. However, the influence of the carrier (meat matrix) and its interactions with the microbial cells on the beneficial effects exerted by a probiotic strain must be assessed. It is recommended that the functionality of each probiotic strain is documented independently in each final formulation (Mercenier et al., 2003).

Finally, one must not overlook that meat, in particular, cured meat, might not be the most obvious carrier for probiotic microorganisms, as compared to dairy products, because of its negative connotations and potential health implications in the Western diet.

References

Ammor, M. S., & Mayo, B. (2007). Selection criteria for lactic acid bacteria to be used as functional starter cultures in dry sausage production: An update. *Meat Science, 76*, 138–146.

Arihara, K. (2006). Strategies for designing novel functional meat products. *Meat Science, 74*, 219–229.

Arihara, K., Ota, H., Itoh, M., Kondo, Y., Sameshima, T., Yamanaka, H., et al. (1998). *Lactobacillus acidophilus* group lactic acid bacteria applied to meat fermentation. *Journal of Food Science, 63*, 544–547.

Arvanitoyannis, I. S., & van Houwelingen-Koukaliaroglou, M. (2005). Functional foods: A survey of health claims, pros and cons, and current legislation. *Critical Reviews in Food Science and Nutrition, 45*, 385–404.

Bass, P., & Dennis, S. (1981). The laxative effects of lactulose in normal and constipated subjects. *Journal of Clinical Gastroenterology, 3*, 23–28.

Collins, M. D., & Gibson, G. R. (1999). Probiotics, prebiotics, and synbiotics: Approaches for modulating the microbial ecology of the gut. *American Journal of Clinical Nutrition, 69*, 1052S–1057S.

Döderlein, A. (1892). Das Scheidensecret und seine Bedeutung für das Puerperalfieber. *Centralblatt für Bacteriologie, 11*, 699–700.

Dunne, C., O'Mahony, L., Murphy, L., Thornton, G., Morrissey, D., O'Halloran, S., et al. (2001). *In vitro* selection criteria for probiotic bacteria of human origin: Correlation with *in vivo* findings. *American Journal of Clinical Nutrition, 73*, 386S–392S.

Erkkilä, S., & Petäjä, E. (2000). Screening of commercial meat starter cultures at low pH and in the presence of bile salts for potential probiotic use. *Meat Science, 55*, 297–300.

Erkkilä, S., Petäjä, E., Eerola, S., Lilleberg, L., Mattila-Sandholm, T., & Suihko, M. L. (2001). Flavour profiles of dry sausages fermented by selected novel meat starter cultures. *Meat Science, 58*, 111–116.

Erkkilä, S., Suihko, M. L., Eerola, S., Petäjä, E., & Mattila-Sandholm, T. (2001). Dry sausage fermented by *Lactobacillus rhamnosus* strains. *International Journal of Food Microbiology, 64*, 205–210.

FAO/WHO. (2001). *Health and nutritional properties of probiotics in food including powder milk with live lactic acid bacteria – Joint Food and Agricultural Organization of the United Nations and World Health Organization Expert Consultation Report*. Córdoba, Argentina: http://www.who.int/foodsafety/publications/fs_management/probiotics/ en/index.html.

FAO/WHO. (2002). *Guidelines for the evaluation of probiotics in food – Joint Food and Agricultural Organization of the United Nations and World Health Organization Working Group Meeting Report*. London Ontario, Canada: http://www.who.int/foodsafety/ publications/fs_management/probiotics2/en/index.html.

Gibson, G. R., Probert, H. M., Van Loo, J. A. E., Rastall, R. A., & Roberfroid, M. B. (2004). Dietary modulation of the human colonic microbiota: Updating the concept of prebiotics. *Nutrition Research Reviews, 17*, 259–275.

Gibson, G. R., & Roberfroid, M. B. (1995). Dietary modulation of the human colonic microbiota – Introducing the concept of prebiotics. *Journal of Nutrition, 125*, 1401–1412.

Gill, H. S., & Guarner, F. (2004). Probiotics and human health: A clinical perspective. *Postgraduate Medical Journal, 80*, 516–526.

Guarner, F., & Schaafsma, G. J. (1998). Probiotics. *International Journal of Food Microbiology, 39*, 237–238.

Gueimonde, M., Frias, R., & Ouwehand, A. C. (2006). Assuring the continued safety of lactic acid bacteria used as probiotics. *Biologia, 61*, 755–760.

Hammes, W. P., & Hertel, C. (1998). New developments in meat starter cultures. *Meat Science, 49*, S125–S138.

Incze, K. (1998). Dry fermented sausages. *Meat Science, 49*, S169–S177.

Jahreis, G., Vogelsang, H., Kiessling, G., Schubert, R., Bunte, C., & Hammes, W. P. (2002). Influence of probiotic sausage (*Lactobacillus paracasei*) on blood lipids and immunological parameters of healthy volunteers. *Food Research International, 35*, 133–138.

Jiménez-Colmenero, F., Carballo, J., & Cofrades, S. (2001). Healthier meat and meat products: Their role as functional foods. *Meat Science, 59*, 5–13.

Klaenhammer, T. R., Barrangou, R., Buck, B. L., Azcarate-Peril, M. A., & Altermann, E. (2005). Genomic features of lactic acid bacteria effecting bioprocessing and health. *FEMS Microbiology Reviews, 29*, 393–409.

Klingberg, T. D., Axelsson, L., Naterstad, K., Elsser, D., & Budde, B. B. (2005). Identification of potential probiotic starter cultures for Scandinavian-type fermented sausages. *International Journal of Food Microbiology, 105*, 419–431.

Klingberg, T. D., & Budde, B. B. (2006). The survival and persistence in the human gastrointestinal tract of five potential probiotic lactobacilli consumed as freeze-dried cultures or as probiotic sausage. *International Journal of Food Microbiology, 109*, 157–159.

Kröckel, L. (2006). Use of probiotic bacteria in meat products. *Fleischwirtschaft, 86*, 109–113.

Leahy, S. C., Higgins, D. G., Fitzgerald, G. F., & van Sinderen, D. (2005). Getting better with bifidobacteria. *Journal of Applied Microbiology, 98*, 1303–1315.

Leroy, F., Verluyten, J., & De Vuyst, L. (2006). Functional meat starter cultures for improved sausage fermentation. *International Journal of Food Microbiology, 106*, 270–285.

Lilly, D. M., & Stillwell, R. H. (1965). Probiotics: Growth-promoting factors produced by microorganisms. *Science, 147*, 747–748.

Lücke, F. K. (2000). Utilization of microbes to process and preserve meat. *Meat Science, 56*, 105–115.

Makras, L., Avonts, L., & De Vuyst, L. (2004). Probiotics, prebiotics, and gut health. In C. Remacle & B. Reusens (Eds.), *Functional foods: Ageing and degenerative disease* (pp. 416–482). Cambridge, United Kingdom: Woodhead Publishing Ltd.

Maldonado Galdeano, C., de Moreno de LeBlanc, A., Vinderola, G., Bibas Bonet, M. E., & Perdigón, G. (2007). Proposed model: Mechanisms of immunomodulation induced by probiotic bacteria. *Clinical and Vaccine Immunology, 14*, 485–492.

Marteau, P. (2002). Probiotics in clinical conditions. *Clinical Reviews in Allergy and Immunology, 22*, 255–273.

Marteau, P. & Shanahan, F. (2003). Basic aspects and pharmacology of probiotics: An overview of pharmacokinetics, mechanisms of action and side-effects. *Best Practice and Research in Clinical Gastroenterology, 17*, 725–740.

Mercenier, A., Pavan, S., & Pot, B. (2003). Probiotics as biotherapeutic agents: Present knowledge and future prospects. *Current Pharmaceutical Design, 9*, 175–191.

Metchnikoff, E. (1908). *The prolongation of life – Optimistic studies*. London, United Kingdom: Butterworth-Heinemann.

Muthukumarasamy, P., & Holley, R. A. (2006). Microbiological and sensory quality of dry fermented sausages containing alginate-microencapsulated *Lactobacillus reuteri*. *International Journal of Food Microbiology, 111*, 164–169.

Muthukumarasamy, P., & Holley, R. A. (2007). Survival of *Escherichia coli* O157:H7 in dry fermented sausages containing micro-encapsulated probiotic lactic acid bacteria. *Food Microbiology, 24*, 82–88.

Naidu, A. S., Bidlack, W. R., & Clemens, R. A. (1999). Probiotic spectra of lactic acid bacteria. *Critical Reviews in Food Science and Nutrition, 39*, 13–126.

Papamanoli, E., Tzanetakis, N., Litopoulou-Tzanetaki, E., & Kotzekidou, P. (2003). Characterization of lactic acid bacteria isolated from a Greek dry-fermented sausage in respect of their technological and probiotic properties. *Meat Science, 65*, 859–867.

Parker, R. B. (1974). Probiotics: The other half of the antibiotics story. *Animal Nutrition and Health, 29*, 4–8.

Parvez, S., Malik, K. A., Ah Kang, S., & Kim, H. Y. (2006). Probiotics and their fermented food products are beneficial for health. *Journal of Applied Microbiology, 100*, 1171–1185.

Pennacchia, C., Ercolini, D., Blaiotta, G., Pepe, O., Mauriello, G., & Villani, F. (2004). Selection of *Lactobacillus* strains from fermented sausages for their potential use as probiotics. *Meat Science, 67*, 309–317.

Pennacchia, C., Vaughan, E. E., & Villani, F. (2006). Potential probiotic *Lactobacillus* strains from fermented sausages: Further investigations on their probiotic properties. *Meat Science, 73*, 90–101.

Picard, C., Fioramonti, J., Francois, A., Robinson, T., Neant, F., & Matuchansky, C. (2005). Bifidobacteria as probiotic agents – Physiological effects and clinical benefits. *Alimentary Pharmacology and Therapeutics, 22*, 495–512.

Pidcock, K., Heard, G. M., & Henriksson, A. (2002). Application of nontraditional meat starter cultures in production of Hungarian salami. *International Journal of Food Microbiology, 76*, 75–81.

Rastall, R. A., & Maitin, V. (2002). Prebiotics and synbiotics: Towards the next generation. *Current Opinion in Biotechnology, 13*, 490–496.

Rebucci, R., Sangalli, L., Fava, M., Bersani, C., Cantoni, C., & Baldi, A. (2007). Evaluation of functional aspects in *Lactobacillus* strains isolated from dry fermented sausages. *Journal of Food Quality, 30*, 187–201.

Reid, G. (2005). The importance of guidelines in the development and application of probiotics. *Current Pharmaceutical Design, 11*, 11–16.

Ross, R. P., Desmond, C., Fitzgerald, G. F., & Stanton, C. (2005). Overcoming the technological hurdles in the development of probiotic foods. *Journal of Applied Microbiology, 98*, 1410–1417.

Salminen, S., Bouley, C., Boutron-Ruault, M. C., Cummings, J. H., Franck, A., Gibson, G. R., et al. (1998). Functional food science and gastrointestinal physiology and function. *British Journal of Nutrition, 80*, S147–S171.

Sameshima, T., Magome, C., Takeshita, K., Arihara, K., Itoh, M., & Kondo, Y. (1998). Effect of intestinal *Lactobacillus* starter cultures on the behaviour of *Staphylococcus aureus* in fermented sausage. *International Journal of Food Microbiology, 41*, 1–7.

Sanders, M. E. (1998). Overview of functional foods: Emphasis on probiotic bacteria. *International Dairy Journal, 8*, 341–347.

Sanders, M. E., & Huis in 't Veld, J. (1999). Bringing a probiotic-containing functional food to the market: Microbiological, product, regulatory and labeling issues. *Antonie Van Leeuwenhoek International Journal of General and Molecular Microbiology, 76*, 293–315.

Santosa, S., Farnworth, E., & Jones, P. J. H. (2006). Probiotics and their potential health claims. *Nutrition Reviews, 64*, 265–274.

Senok, A. C., Ismaeel, A. Y., & Botta, G. A. (2005). Probiotics: Facts and myths. *Clinical Microbiology and Infection, 11*, 958–966.

Työppönen, S., Petäjä, E., & Mattila-Sandholm, T. (2003). Bioprotectives and probiotics for dry sausages. *International Journal of Food Microbiology, 83*, 233–244.

Yakult Central Institute for Microbiological Research. (1999). *Lactobacillus casei Shirota – Intestinal flora and human health.* Tokyo, Japan: Yakult Honsha Co., Ltd.

Chapter 11
Bioactive Compounds in Meat

Keizo Arihara and Motoko Ohata

Introduction

Since health-conscious consumers have made functional foods the leading trends in the food industry, efforts have been taken in many countries to develop new functional foods and to establish regulations for functional foods (Arihara, 2004; Dentali, 2002; Eve, 2000; Hutt, 2000). For example, in 1991, the concept of foods for specified health use (FOSHU) was established by the Japanese Ministry of Health and Welfare (Arihara, 2004, 2006b). FOSHU are foods that, based on the knowledge of the relationship between foods or food components and health, are expected to have certain health benefits and have been licensed to bear the label claiming that a person using them may expect to obtain that health use through the consumption of these foods. As of June 2008, 786 FOSHU products have been approved in Japan. Also, in the United States and European countries, markets for functional foods have been expanding rapidly.

Chemicals found as natural components of foods that have been determined to be beneficial to the human body in preventing or treating one or more diseases or improving physiological performance are known as nutraceuticals (Wildman, 2000a, 2000b). Numerous food components with such physiological functions have been isolated and characterized (Hasler, 1998). Many vegetables, for example, have been shown to contain a variety of biologically active phytochemicals (Lindsay, 2000). There has been an accumulation of scientific findings regarding the roles of such components in the prevention of diseases. Rapid progress has been made in the development of functional foods based on the results of studies on food components that have positive health benefits other than the normal nutritional benefits (Arihara, 2004; Heasman & Mellentin, 2001).

In addition to various nutraceutical compounds found in vegetables (Lindsay, 2000) and milk (Chandan, 2007; Chandan & Shah, 2007), several attractive meat-based bioactive substances have been studied for their physiological properties (Arihara, 2004, 2006b; Williams, 2007). Such substances include conjugated

K. Arihara
Department of Animal Science, Kitasato University, Towada-shi, Aomori 034-8628, Japan

linoleic acid (CLA), carnosine, anserine, L-carnitine, glutathione, taurine, coenzyme Q10 and creatine. Utilizing or emphasizing these physiological activities originating from meat is one possible approach for designing healthier meat and meat products, including functional foods. The composition of animal products could be improved through manipulation of animal feed. Several studies have shown that the feeding conditions of animals affect the contents of bioactive components, such as CLA and L-carnitine, in animal products (Krajcovicova-Kudlackova, Simoncic, Bederova, Babinska, & Beder, 2000; Mir et al., 2004). Various aspects of product processing, such as modification of constituents and incorporation of ingredients, are also important for developing functional meat products. Such efforts could lead to the creation of differentiated meat and meat products.

This chapter provides a brief overview of the potential benefits of representative meat-based bioactive compounds (e.g., CLA, carnosine, L-carnitine) on human health. Along with these compounds, this chapter focuses on the properties of meat protein-derived bioactive peptides (e.g., antihypertensive peptides), which have a potential for the development of functional meat products.

Meat-Based Bioactive Compounds

In the food guide pyramid, meat is categorized as a protein food group along with poultry, fish, and eggs (Lachance & Fisher, 2005). Since meat contains an abundance of proteins with high biological value, regarded nutritional, meat is a fundamental source of essential amino acids. Meat is also an excellent source of some valuable minerals and vitamins (Biesalski, 2005; Mulvihill, 2004). Some of these nutrients are either not present or have inferior bioavailability in other foods. In addition to these basic nutrients, much attention has recently been paid to meat-based bioactive compounds, such as conjugated linoleic acid.

Minerals and Vitamins

Meat plays an important role in supplying iron, zinc, selenium, and B vitamins to the diet. The contributions of meat and meat products to total dietary intakes of selected micronutrients are: 14% iron, 30% zinc, 14% vitamin B2, 21% vitamin B6, 22% vitamin B12, 19% vitamin D, and 37% niacin (Mulvihill, 2004). Red meat is rich in iron (e.g., 2.1 mg iron per 100 g of fillet steak). A large proportion of iron in meat is haem, which is a high absorbable form of iron, and meat proteins enhance the absorption of iron. Also, zinc in meat is highly bioavailable. For a detailed information on minerals and vitamins in meat, refer to other articles (Biesalski, 2005; Higgs, 2000; Mulvihill, 2004).

Fat and Fatty Acids

Unfortunately, consumers often associate meat and meat products with a negative health image. The **regretable** image of meat **is** mainly **due to** the content of fat, saturated fatty acids, and cholesterol and their association with chronic diseases, such as cardiovascular diseases, some types of cancer, and obesity (Chan, 2004; Fernández-Ginés, Fernández-López, Sayas-Barberá, & Pérez-Alvarez, 2005; Ovesen, 2004a, 2004b; Valsta, Tapanainen, & Mannisto, 2005). Dietary fat should provide between 15 and 30% of the total diet energy. Less than 10% of calorie intake should be from saturated fatty acids (SFA), 6–10% and 10–15% of that should be from polyunsaturated fatty acids (PUFA) and from monounsaturated fatty acids (MUFA) respectively. Furthermore, less than 1% of that should be from trans fatty acids, and cholesterol intake should be limited to less than 300 mg per day.

Meat fat contains less than 50% SFA and up to 65–70% unsaturated fatty acids (Jiménez-Colmenero, 2007a). Various modifications, including reduced fatty acid levels, raised MUFA and PUFA levels, improved n-6:n-3 PUFA balances, and limited cholesterol contents, can be achieved by animal breeding and feeding, material formulation, and technological processing (Jiménez-Colmenero, Reig, & Toldrá, 2006). Also, numerous studies have demonstrated the possibility of changing the image of meat and meat products by the addition, elimination and reduction of fat and fatty acids (Jiménez-Colmenero, 2007b).

Conjugated Linoleic Acid

Conjugated linoleic acids (CLA) are a group of fatty acids found in meat and milk of ruminants (Gnadig, Xue, Berdeaux, Chardigny, & Sebedio, 2000; Nagao & Yanagita, 2005; Watkins & Yong, 2001). Since rumen bacteria convert linoleic acid to CLA by their isomerase, it is most abundant in the fat of ruminant animals. After its absorption in a ruminant animal, CLA is transported to the mammary tissue and muscles. For example, beef fat contains 3–8 mg of CLA per g of fat (Table 11.1). CLA, which was initially identified as an anti-carcinogenic compound in the extracts of grilled beef, is composed of a group of positional and geometric isomers of octadecadienoic acid.

The CLA content in meat is affected by several factors, such as breed, age and feed composition (Dhiman, Nam, & Ure, 2005). For example, products of grass-fed animals have 3–5 times more CLA than in products of animals fed on the typical diet of 50% hay and silage with 50% grain. Interestingly, the CLA content of foods is also increased by heating (cooking and processing). Also, lactic acid

Table 11.1 Contents of conjugated linoleic acid (CLA) in animal products

Product	CLA (mg per g of fat)
Beef	2.9–8.0
Pork	0.6
Chicken	0.9
Turkey	2.5
Bovine milk	5.4–7.0
Egg Yolk	0.6

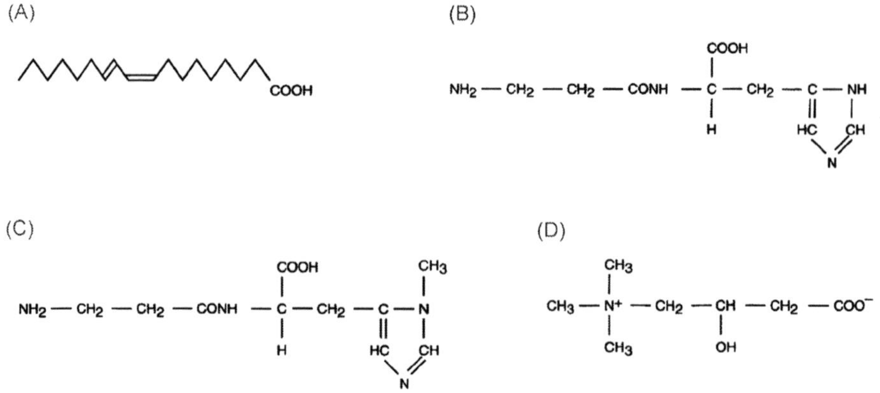

Fig. 11.1 Representative meat-based bioactive compounds (a) Conjugated Linoleic Acids (c9, t11-C18:2); (b) Carnosine; (c) Anserine; (d) L-Carnitine

bacteria promote the formation of CLA in fermented milk products (Alonso, Cuesta, & Gilliland, 2003; Coakley et al., 2003; Sieber, Collomb, Aeschlimann, Jelen, & Eyer, 2004; Xu, Boylston, & Glatz, 2005). Such conversion would be expected in fermented meat products. The most common CLA isomer found in beef is octadeca-c9, t11-dienoic acid (Fig. 11.1a). Since this fatty acid has an anti-carcinogenic activity, much interest has been shown in this compound. Recent epidemiological studies have suggested that high intakes of high-fat dairy foods and CLA may reduce the risk of colorectal cancer (Larsson, Bergkvist, & Wolk, 2005). In addition to its anti-carcinogenic property, CLA has anti-artheriosclerotic, antioxidative, and immunomodulative properties (Azain, 2003). CLA may also play a role in the control of obesity, reduction of the risk of diabetes and modulation of bone metabolism.

Histidyl Dipeptides

Consumption of antioxidant-rich foods, such as fruits and vegetables, has been shown to have a preventative effect on oxidative damage in the body (Lindsay, 2000). This beneficial action of food is attributed to neutralization and reduced release of free radicals by the antioxidant potency of various compounds (Langseth, 2000). Ascorbic acid, vitamin E, β-carotene and polyphenolic compounds are representative food-derived antioxidants. Such compounds may decrease the risk of many diseases, including cancer. Several endogenous antioxidants (e.g., tocopherols, ubiquinone, carotenoids, ascorbic acid, glutathione, lipoic acid, uric acid, spermine, carnosine, anserine) in meats have been studied (Decker, Livisay, & Zhou, 2000).

Both carnosine (β-alanyl-L-histidine; Fig. 11.1b) and anserine (N-β-alanyl-1-methyl-L-histidine; Fig. 11.1c) are antioxidative histidyl dipeptides and the most abundant antioxidatives in meats. The concentration of carnosine in meat ranges from 500 mg per kg of chicken thigh to 2700 mg per kg of pork shoulder. On the other hand, anserine is especially abundant in chicken muscle. Their antioxidant activities may result from their ability to chelate transition metals such as copper (Brown, 1981) These antioxidative peptides have been reported to play a role in wound healing, recovery from fatigue and prevention of diseases related to stress. A recent study demonstrated the bioavailability of carnosine by determining its concentration in human plasma after ingestion of beef (Park, Volpe, & Decker, 2005). Increasing attention to these meat-based bioactive compounds has resulted in the development of a new sensitive procedure for determining these compounds including carnosine (Mora, Sentandreu, & Toldrá, 2007).

L-Carnitine

L-Carnitine, β-hydroxy-gamma-trimethyl amino butyric acid (Fig. 11.1d), is detected in the skeletal muscle of various animals (Shimada et al., 2005). L-Carnitine is especially abundant in beef (e.g., 1300 mg per kg of the thigh). It assists the human body in producing energy and in lowering the levels of cholesterol. Also, it helps the body to absorb calcium to improve skeletal strength and chromium picolinate to help build lean muscle mass. A recent study demonstrated that L-carnitine blocked apotosis and prevented skeletal muscle myopathy in heart failure (Vescovo et al., 2002). A drink product containing L-carnitine, which is marketed in the United States, is advertised as having several beneficial effects, such as maintenance of stamina and fast recovery from fatigue. Also, a product containing **a good amout of** L-carnitine and carnosine, which is used as a functional food ingredient, has been marketed in Japan. This product is made from a by-product of corned beef.

Other Bioactive Components

Glutahione is an important antioxidative compound providing cellular defense against toxicological and pathological processes. Red meat is a good source of glutathione (12–26 mg per 100 g of beef; Jones et al., 1992). Taurine is a conditionally essential amino acid during lactation and at times of immune challenge (Bouckenooghe, Remacle, & Reusens, 2006). Also, taurine may protect our body from oxidative stress. Meat is the most excellent dietary source of taurine (77 mg per 100 g of beef; Purchas, Rutherfurd, Pearce, Vather, & Wilkinson, 2004). Coenzyme Q10 (ubiquinone) shows antioxidative activity and its content in beef is 2 mg per 100 g (Purchas & Busboom, 2005). Creatine and creatine phosphate have a critical role in muscle energy metabolism. Beef contains 350 mg of creatine per 100 g (Purchas & Busboom, 2005). Other components such as choline, balenine, creatinine, lipoic acid, putrescine, spermidine, and spermine should also be listed as meat-based bioactive compounds.

Meat Protein-Derived Bioactive Peptides

In addition to the meat-based bioactive compounds described above, meat protein-derived peptides are another group of promising bioactive components of meat (Arihara, 2004, 2006a, 2006b). Several bioactive peptides from hydrolyzates of muscle proteins have been found. Although the activities of these peptides in the sequences of the parent proteins are latent, they are released by proteolytic enzymes. In this aspect, meat proteins have possible bioactivities beyond a nutritional source of amino acids alone.

Bioactive Peptides Derived from Food Proteins

Information on bioactive peptides generated from meat proteins is still limited. However, various physiologically functional peptides **have** been found from enzymatic hydrolyzates of food proteins, such as milk and soy proteins (Arihara, 2006a; Korhonen & Pihlanto, 2007; Meisel, 1998; Mine & Shahidi, 2005; Pihlanto & Korhonen, 2003). Thus, bioactive peptides generated from food proteins are covered briefly at the beginning. Bioactive peptides from food proteins were first reported by Mellander (1950). He described the effect of casein-derived phospho-related peptides on vitamin D-independent bone calcification of rachitic infants. Since then, there have been numerous studies on bioactive peptides generated from food proteins.

Angiotensin I-Converting Enzyme Inhibitory Peptides

Angiotensin I-converting enzyme (ACE) inhibitory peptides generated from food proteins have been studied most extensively (Meisel, Walsh, Murry, & FitzGerald, 2005; Vermeirssen, Camp, & Verstraete, 2004). ACE inhibitory peptides have been shown to have antihypertensive effects and have been utilized for pharmaceuticals and physiologically functional foods. ACE plays an important physiological role in the regulation of blood pressure (Fig. 11.2). ACE is a dipeptidyl carboxypeptidase that converts an inactive form of the decapeptide, angiotensin I, to a potent vasoconstrictor, octapeptide angiotensin II, and inactivates bradykinin, which has a depressor action (Li, Le, Shi, & Shrestha, 2004). Therefore, by inhibiting the catalytic action of ACE, the elevation of blood pressure can be suppressed.

ACE inhibitory peptides from food protein were first identified in the hydrolyzate of gelatin by Oshima, Shimabukuro, and Nagasawa (1979). Since then, ACE inhibitory peptides have been found in the hydrolyzates of many proteins from milk, fish, meat, eggs, soybean, corn, wheat, seaweed, and others (Arihara, 2006a; Vermeirssen et al., 2004). Some of these peptides have been reported to show antihypertensive effects in spontaneously hypertensive rats (SHR) by oral administration. ACE inhibitory peptides isolated from meat proteins are described later.

11 Bioactive Compounds in Meat

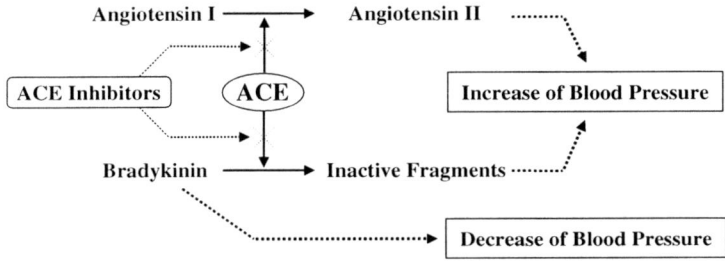

Fig. 11.2 Blood pressure regulation by Angiotensin I-Converting Enzyme (ACE)

Other Bioactive Peptides Derived from Food Proteins

In addition to ACE inhibitory (antihypertensive) peptides, it has been found that various bioactive peptides, such as opioid, immuno-modulating, antimicrobial, prebiotic, mineral-binding, antithrombotic, hypocholesterolaemic, and antioxidative peptides, are generated from food proteins (Arihara, 2006a; Korhonen & Pihlanto, 2007; Pihlanto & Korhonen, 2003). Since milk proteins are the main source of a range of these bioactive peptides (Gobbetti, Minervini, & Rizzello, 2007; Meisel, 1998; Silva & Malcata, 2005), representative examples of bioactive peptides generated from milk proteins are summarized in Table 11.2.

Opioid peptides have an affinity for an opioid receptor and have effect on the nerve system (Guesdon, Pichon, & Tomé, 2005). Immuno-modulating peptides stimulate the proliferation of lymphocytes and phagocytic activities of macrophages (Pihlanto & Korhonen, 2003). Antimicrobial peptides inhibit the growth of pathogenic bacteria (Chan & Li-Chan, 2005). Mineral-binding peptides, such as caseino-phosphopeptides (CPP) generated from milk proteins, function as carriers for minerals, including calcium (Bougle & Bouhallab, 2005). CPP also anti-carcinogenic activity (Cross, Huq, & Reynolds, 2005). Antithrombotic peptides inhibit fibrinogen binding to a specific

Table 11.2 Examples of milk protein-derived bioactive peptides

Bioactivity	Sequence[a]	Preparation	Reference
ACE-inhibitory Antihypertensive	IPP, VPP	Fermentation	Nakamura et al., 1995
Opioid agonistic	YIPIQYVLSR	Trypsin	Chiba, H., Tani, F., & Yoshikawa, 1989
Immunomodulating	VGPIPY	Trypsin-Chymotrypsin	Fiat et al., 1993
Antimicrobial	FVAPFPEVFG	Trypsin	Rizzello et al., 2004
Mineral-binding	Casein phosphopeptides	Trypsin	Gagnaire et al., 1996
Anticariogenic	Casein phosphopeptides	Trypsin	Cross et al., 2005
Antithrombotic	MAIPPKKNDQDK	Synthesis	Jollès et al., 1986
Hypocholesteromic	IIAEK	Trypsin	Nagaoka 2005
Antioxidative	YFYPEL	Pepsin	Suetsuna, Ukeda, & Ochi, 2000
Prebiotic	CAVGGCIAL	Synthesis	Lieple et al., 2002

[a] The one-letter amino acid codes were used.

receptor region on the platelet surface (Jollès et al., 1986). Hypocholesteromic peptides isolated from the hydrolyzate of milk β-lactoglobulin have a strong effect on serum cholesterol level (Nagaoka, 2005). Antioxidative and prebiotic peptides are described as meat protein-derived bioactive peptides later in this article.

Peptide Generation from Meat Proteins

Since many bioactive sequences of food proteins are inactive within the parent proteins, food proteins have to be digested for liberating **the** bioactive peptides. Gastrointestinal proteolysis, aging, fermentation, and enzymatic treatment are the principal means for digestion of meat proteins to generate bioactivities (Fig. 11.3).

Gastrointestinal Proteolysis

Bioactive peptides are thought to be generated from food (meat) proteins during gastrointestinal digestion. Ingested proteins are attacked by various digestive enzymes, such as pepsin, trypsin, chymotrypsin, elastase and carboxypeptidase (Pihlanto & Korhonen, 2003). Digestive enzymes (i.e., pepsin, trypsin, and α-chymotrypsin) in the gastrointestinal tracts generated ACE inhibitory activity from porcine skeletal muscle proteins (Arihara, Nakashima, Mukai, Ishikawa, & Itoh, 2001). Also, ACE inhibitory peptides were generated from meat proteins, such as myosin, actin, tropomyosin, and troponin, by pancreatic protease treatment (Katayama, Fuchu, et al., 2003). Since denatured proteins are particularly liable to attack by proteolytic enzymes, peptides would be easily generated from cooked meat and meat products.

Aging of Meats

After slaughter of a domestic animal, the skeletal muscle is converted to meat via aging. During aging or storage, meat proteins are hydrolyzed by muscle endogenous proteases, such as calpains and cathepsins (Etherington, 1984; Koohmaraie, 1994).

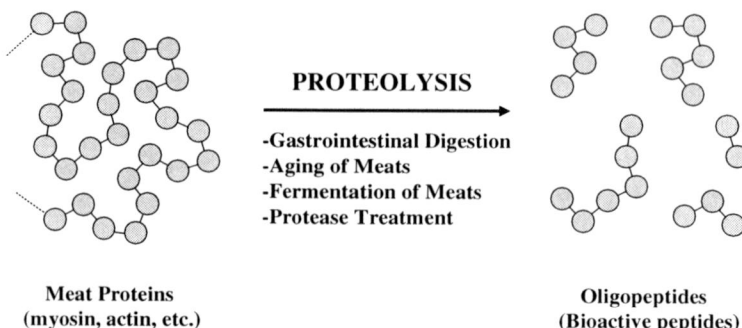

Fig. 11.3 Generation of bioactive peptides from meat proteins

Thus, the content of peptides in meat increases during post-mortem aging. Such changes in the levels of oligopeptides occur during storage of beef, pork, and chicken (Nishimura, Rhue, Okitani & Kato, 1988). For example, the content of peptide in pork increased from 2.40 mg at day 1 to 3.05 mg per g meat at day 6. Also, Mikami, Nagao, Sekikawa, and Miura (1995) reported that peptide contents of beef varied widely, ranging from 0.69 to 1.44 mg and from 2.64 to 4.65 mg per g meat 2 and 21 days after slaughter, respectively.

It is well recognized that enzymatic hydrolysis of meat proteins during aging or storage contributes to improvement of sensory properties of meat, such as texture, taste, and flavor. However, there has been no report about the generation of bioactive peptides in meat during post-mortem aging. Our preliminary study showed an increase in ACE inhibitory activity of beef during storage (unpublished data). Further studies are expected to reveal the novel meaning of meat aging.

Fermentation of Meats

Components generated from meat proteins by proteolytic reactions during fermentation of raw sausages and dry-cured hams are important for the development of sensory characteristics of fermented meat products (Hammes, Haller, & Gänzle, 2003; Toldrá, 2004; Toldrá & Flores, 1998). The content of peptides and amino acids of fermented sausages reaches about 1% dry matter of products during fermentation (Dainty & Blom, 1995). Meat proteins are mainly degraded into peptides by endogenous enzymes (cathepsin B, D, H, and L) during the fermentation process. Although microbial proteolytic enzymes are involved in meat fermentation, most bacteria such as lactic acid bacteria grown in fermented meat products have only weak proteolytic activity (Hierro, de la Hoz, & Ordonez, 1999). However, lactic acid bacteria contribute to degradation of meat proteins by causing a decrease in pH, which increases the activity of muscle proteolytic enzymes (Kato et al., 1994).

There has been no report on the generation of bioactive peptides from meat proteins in fermented meat products. However, we measured the ACE inhibitory activities of extracts of several fermented meat products (i.e., raw sausages and dry-cured ham) and found that activity levels of all extracts were higher than those of the extracts obtained from non-fermented pork products (unpublished data). Also, ACE inhibitory and antihypertensive activities were experimentally generated from porcine skeletal muscle proteins by lactic acid bacteria (Arihara, Nakashima, Ishikawa, & Itoh, 2004). Sentandreu et al. (2003) identified several small peptides in dry-cured ham. Such small peptides generated from meat proteins could have some bioactivities in fermented meat products. Sentandreu and Toldrá (2007a, 2007b) suggested that the proteolytic action of porcine muscle dipeptidyl peptidases during the ripening period of dry-cured ham could contribute to the generation of ACE inhibitory peptides.

Protease Treatment

In the food industry, many proteolytic enzymes are utilized for food processing, such as tenderization. Utilizing commercial proteases is an efficient method for releasing bioactive peptides from food proteins. Many bioactive peptides have been experimentally generated by commercial proteases (Korhonen & Pihlanto, 2003; Pihlanto & Korhonen, 2003). Proteases from animal, plant, and microbial origins have been used for the digestion of food proteins. In the meat industry, proteolytic enzymes have been used for meat tenderization (Dransfield & Etherington, 1981). The most commonly used enzymes for meat tenderization are the plant enzymes papain, bromelain, and ficin. In meat treated with enzymatic tenderization, peptides having bioactivities could be generated. On the other hand, effects of commercial proteases on protein breakdown and sensory characteristics of dry fermented sausages have been investigated (Bruna, Fernandez, Hierro, Ordonez, & de la Hoz, 2000). Such treatment would also generate bioactive peptides in meat products.

ACE Inhibitory Peptides from Meat Proteins

As stated above, the most extensively studied bioactive peptides derived from food proteins are ACE inhibitory peptides. Also, among the bioactive peptides derived from meat proteins, ACE inhibitory peptides have been studied extensively (Arihara, 2006a, 2006b; Vercruysse, Van Camp, & Smagghe, 2005). ACE inhibitory peptides generated from meat proteins are summarized in Table 11.3.

Table 11.3 ACE inhibitory peptides derived from meat proteins

Sequence[a]	Source	IC50[b] (μM)	SHR[c]	References
IKW	Chicken muscle	0.2	+	Fujita et al., 2000
LKA	Chicken muscle (creatine kinase)	8.5	nt	Fujita et al., 2000
LKP	Chicken muscle (aldolase)	0.3	+	Fujita et al., 2000
LAP	Chicken muscle	3.5	+	Fujita et al., 2000
VWI	Porcine muscle (actin)	1.1	+	Arihara et al., 2005
ITTNP	Porcine myosin	549.0	+	Arihara et al., 2001 Nakashima et al., 2002
MNPPK	Porcine myosin	945.5	+	Arihara et al., 2001 Nakashima et al., 2002 Fujita et al., 2000
FQKPKR	Chicken muscle (myosin)	14.0	nt	Jang & Lee, 2005
VLAQYK	Bovine muscle	23.2	+	Fujita et al., 2000
FKGRYYP	Chicken muscle (creatine kinase)	0.6	-	Arihara et al., 2004
VFPMNPPK	Fermented pork (myosin)	66.0	nt	Fujita et al., 2000
IVGRPRHQG	Chicken muscle (actin)	2.4	-	Katayama, Tomatsu et al., 2003
RMLGQTPTK	Porcine troponin C	34.0	nt	Katayama et al., 2004 Saiga, Okumura et al., 2003
GFXGTXGLXGF	Chicken muscle (collagen)	42.4	nt	

[a] The one-letter amino acid codes were used.
[b] The concentration of peptide needed to inhibit 50% of the ACE activity.
[c] Antihypertensive activities in spontaneously hypertensive rats (+, positive activity; -, no activity; nt, not tested).

Fujita, Yokoyama, & Yoshikawa (2000) isolated the ACE inhibitory peptides (Leu-Lys-Ala, Leu-Lys-Pro, Leu-Ala-Pro, Phe-Gln-Lys-Pro-Lys-Arg, Ile-Val-Gly-Arg-Arg-Arg-His-Gln-Gly, Phe-Lys-Gly-Arg-Tyr-Tyr-Pro, Ile-Lys-Trp) generated from chicken muscle proteins by thermolysin treatment. Arihara et al. (2001) reported ACE inhibitory peptides in hydrolyzates of porcine skeletal muscle proteins. Two ACE inhibitory peptides (Met-Asn-Pro-Pro-Lys and Ile-Thr-Thr-Asn-Pro), which are found in the sequence of the myosin heavy chain, have been identified in the thermolysin digest of porcine muscle proteins. These peptides showed antihypertensive activity when administered orally to SHR (Nakashima, Arihara, Sasaki, Ishikawa, & Itoh, 2002). Katayama, Fuchu, et al. (2003), Katayama, Tomatsu et al. (2003), Katayama et al. (2004) utilized porcine skeletal muscle and respective muscle proteins for proteolytic digestion. They isolated a corresponding peptide (Arg-Met-Leu-Gly-Gln-Thr-Pro-Thr-Lys) from hydrolyzed porcine troponin C with pepsin. Saiga, Okumura, et al. (2003) reported antihypertensive activity of *Aspergillus* protease-treated chicken muscle extract in SHR. They also isolated four ACE inhibitory peptides from the hydrolyzate. Three of those four peptides possessed a common sequence, Gly-X-X-Gly-X-X-Gly-X-X, which is homologous with that of collagen. Jang & Lee (2005) assayed ACE inhibitory activities of several enzymatic hydrolyzates of sarcoplasmic protein extracts from beef rump. An ACE inhibitory peptide (Val-Leu-Ala-Gln-Tyr-Lys) was isolated from the hydrolyzate with the highest level of ACE inhibitory activity obtained by using the combination of thermolysin and proteinase A.

ACE inhibitory peptides have shown antihypertensive effects by oral administration in animal experiments using SHR. However, the inhibitory potencies of peptides do not always correlate with their antihypertensive effects in vivo. Some peptides with potent ACE inhibitory activities in vitro are inactive by oral administration. Such a phenomenon between the inhibitory activity and antihypertensive effect and the structure-activity relationships of peptides has been reviewed (Arihara, 2006a; Li et al., 2004; Meisel et al., 2005).

Antioxidative Peptides from Meat Proteins

Information on meat protein-derived bioactive peptides other than **the** ACE inhibitory peptides is still limited. Several antioxidative peptides have been reported to be generated from meat proteins by enzymatic digestion. Saiga, Tanabe, and Nishimura (2003) reported that hydrolyzates obtained from porcine myofibrillar proteins by protease treatment (papain or actinase E) exhibited high levels of antioxidant activity in a linolenic acid peroxidation system. Among **the** five antioxidative peptides identified from papain hydrolyzate, Asp-Ala-Gln-Glu-Lys-Leu-Glu, corresponding to a part of the sequence of porcine actin, showed the highest level of activity. Arihara et al. (2005) investigated antioxidative activities of enzymatic hydrolyzates of porcine skeletal muscle actomyosin using a hypoxanthine-xanthine oxidase system as the source of **the** superoxide anion. Three antioxidative peptides were isolated from a papain-treated hydrolyzate of pork actomyosin and they were sequenced as Asp-Leu-Tyr-Ala, Ser-Leu-Tyr-Ala, and Val-Trp. In addition to the antioxidative activity in vitro, these peptides showed physiological activity in vivo. Each of these peptides had an anti-fatigue effect when orally administered to mice in an experiment using a treadmill.

Prebiotic Peptides from Meat Proteins

Prebiotics are defined as "non-digestible food ingredients that beneficially affect the host by selectively stimulating the growth and/or activity of one or a limited number of bacteria in the colon and thus improve the health of the host" (Gibson, & Roberfroid, 1995). Oligosaccharides are representative prebiotic substances, which are known to enhance the activity of probiotic bacteria (i.e., intestinal lactobacilli and bifidobacteria). The presence of prebiotic peptides has been suggested. Lieple et al. (2002) first reported nonglycosylated peptides that selectively stimulate the growth of bifidobacteria. Recently, we found that the hydrolyzate of porcine skeletal muscle actomyosin, digested by papain, enhanced the growth of *Bifidobacterium* strains (Arihara, Ishikawa, & Itoh, 2006). One corresponding prebiotic peptide was identified as Glu-Leu-Met.

Other Promising Peptides from Meat Proteins

Opioid peptides are defined as peptides that have an affinity for an opiate receptor as well as opiate-like effects (Pihlanto & Korhonen, 2003). Opioid peptides have effects on the nerve system and gastrointestinal functions. Typical opioid peptides (e.g., endorphins, enkephalin, and prodynorphin) have the same N-terminal sequence, Tyr-Gly-Gly-Phe. Several opioid peptides derived from food proteins have been reported. The N-terminal sequence of most of these peptides is Tyr-X-Phe or Tyr-X1-X2-Phe. The N-terminal tyrosine residue and the presence of an aromatic amino acid at the third or fourth position form an important structure that fits with the binding site of opioid receptors (Pihlanto-Leppälä, 2001). There has been no report on the generation of opioid peptides from meat proteins. However, since opioid sequences are thought to be present in the sequences of muscle proteins, it should be possible to find opioid peptides in meat proteins by proteolytic treatment. Although bovine blood hemoglobin is regarded as a minor component of meat and meat products, in some meat products, such as blood sausage, hemoglobin is a major component. Investigation of hemoglobin peptic hydrolyzate has revealed the presence of biologically active peptides with affinity for opioid receptors (Nyberg, Sanderson, & Glämsta, 1997; Zhao, Garreau, Sannier, & Piot, 1997).

Morimatsu et al. (1996) demonstrated hypocholesterolemic effects of papain-hydrolyzed pork meat. However, the corresponding peptides were not identified. Immuno-modulating peptides have been discovered in enzymatic hydrolyzates of proteins from various foods, such as milk, eggs, soybeans, and rice. However, to date, meat protein-derived immuno-modulating peptides have not been reported. In addition to bioactive peptides described here, several bioactive peptides (e.g., antimicrobial, mineral-binding, antithrombotic, and hypocholesterolemic peptides) have been found from enzymatic hydrolyzates of various food proteins (Mine & Shahidi, 2005). It is expected that interest will be directed to research aimed at finding such meat protein-derived peptides.

Apart from bioactivities, meat protein-derived peptides also contribute to organo-leptic properties of meat (Nishimura & Kato, 1988; Nishimura et al., 1988; Arihara, 2006a). Meat products can be modified by adding functional ingredients beneficial for health or by eliminating or reducing components that are harmful. However, these modifications could result in products with inferior sensory properties. Therefore, generation of peptides from meat proteins has a great potential to produce novel functional foods with good organo-leptic properties. Also, from the aspect of food allergies, enzymatic treatment of food proteins, including meat proteins, warrants attention. Protein hydrolyzate based products have been produced and considered suitable for the diet of allergic patients. Although meat is less allergic than common allergy-inducing foods, such as milk, eggs, and soya, meat proteins (e.g., serum albumin, gamma globulin, actin, myoglobin and tropomyosin) sometimes cause allergic reactions (Tanabe & Nishimura, 2005).

Utilization of Peptides for Functional Meat Products

Although bioactive peptides have not yet been utilized in the meat industry, such peptides are promising candidates for ingredients of functional foods. Several food products containing ACE inhibitory peptides have been successfully marketed for hypertensives (Arihara, 2006b). There are two commercial dairy products containing Ile-Pro-Pro and Val-Pro-Pro, which are generated from milk protein by fermentation (Nakamura, Yamamoto, Sakai, & Takano, 1995; Seppo, Jauhiaine, Poussa, & Korpela, 2003). Calpis Amiel-S drink has been approved as a FOSHU in Japan. The Finnish fermented milk drink Evolus, developed by Valio Ltd., contains the same tripeptides as those in Amiel-S. The enzymatic digest of dried bonito containing antihypertensive peptides has also been used in a soup product in Japan. A FOSHU sour milk product utilizing casein-phosphopeptide (CPP) has been developed in Japan. CPP milk protein-derived peptides act as mineral trappers resulting in enhancement of the absorption efficiency of calcium (Gagnaire, Pierre, Molle, & Leonil, 1996). Hydrolyzates of meat proteins and their corresponding bioactive peptides would be utilized for functional foods.

Accumulation of bioactive peptides in meat products by fermentation is a good strategy for developing functional meat products. Bioactive peptides would be generated in fermented meat products, since meat proteins are hydrolyzed by proteolytic enzymes during fermentation and storage. Rediscovery of traditional fermented meats as functional foods is also an interesting direction. In the dairy industry, many traditional fermented foods, such as fermented dairy products, have been rediscovered as functional foods (Farnworth, 2003). Numerous physiologically active components, including bioactive peptides, have been discovered in these traditional fermented foods. Thus, traditionally fermented meats are attractive targets for finding new functional meat products.

Concluding Remarks

Although there has been extensive research and development of functional foods in the dairy industry (Chandan, 2007; Chandan & Shah, 2007; Mattila-Sandholm & Saarela, 2000; Playne, Bennett, & Smithers, 2003), little attention has been paid to functional meat products until recently. However, efforts have been directed in recent years to research and development of functional meat products (Arihara, 2004, 2006a, 2006b; Fernández-Ginés et al., 2005; Jiménez-Colmenero, 2007a, 2007b; Jiménez-Colmenero, Carballo, & Cofrades, 2001; Jiménez-Colmenero et al., 2006). Since meat and meat products are important in the diet in most developed countries, healthier meat and meat products would contribute to human health. Utilizing or emphasizing meat-based bioactive compounds, including bioactive peptides generated from meat proteins, is a promising means for developing attractive functional meat products. Such efforts would also contribute to the demonstration for consumers of the scientific benefits of meat and meat components for human health.

References

Alonso, L., Cuesta, E. P., & Gilliland, S. E. (2003). Production of free linoleic acid by *Lactobacillus acidophilus* and *Lactobacillus casei* of human intestinal origin. *Journal of Dairy Science*, 86, 1941–1946.
Arihara, K. (2004). Functional foods. In W. K. Jensen, C. Devine, & M. Dikeman (Eds.), *Encyclopedia of meat sciences* (pp. 492–499). Oxford: Elsevier.
Arihara, K. (2006a). Functional properties of bioactive peptides derived from meat proteins. In N. M. L. Nollet, & F. Toldrá (Eds.), *Advanced technologies for meat processing* (pp. 245–274). Boca Raton, FL: CRC Press.
Arihara, K. (2006b). Strategies for designing novel functional meat products. *Meat Science*, 74, 219–229.
Arihara, K., Ishikawa, S., & Itoh, M. (2006). *Bifidobacterium* growth promoting peptides derived from meat proteins. *Japan patent* (submitted to government).
Arihara, K., Nakashima, Y., Mukai, T., Ishikawa, S., & Itoh, M. (2001). Peptide inhibitors for angiotensin I-converting enzyme from enzymatic hydrolysates of porcine skeletal muscle proteins. *Meat Science*, 57, 319–324.
Arihara, K., Nakashima, Y., Ishikawa, S., & Itoh, M. (2004). Antihypertensive activities generated from porcine skeletal muscle proteins by lactic acid bacteria. *Abstracts of 50th International Congress of Meat Science and Technology* (p. 236), 8–13 August 2004, Helsinki, Finland.
Arihara, K., Tomita, K., Ishikawa, S., Itoh, M., Akimoto, M., & Sameshima, T. (2005). Anti-fatigue peptides derived from meat proteins. *Japan patent* (submitted to government).
Azain, M. J. (2003). Conjugated linoleic acid and its effects on animal products and health in single-stomached animals. *Proceedings of the Nutrition Society*, 62, 319–328.

Biesalski, H.-K. (2005). Meat as a component of a healthy diet – are there any risks or benefits if meat is avoided in the diet? *Meat Science*, 70, 509–524.

Bouckenooghe, T., Remacle, C., & Reusens, B. (2006). Is taurine a functional nutrient? *Current Opinion in Clinical Nutrition and Metabolic Care*, 9, 728–733.

Bougle, D., & Bouhallab, S. (2005). Mineral-binding proteins and peptides and bioavailability of trace elements. In Y. Mine, & F. Shahidi (Eds.), *Nutraceutical proteins and peptides in health and disease* (pp. 29–40). Boca Raton, FL: CRC Press.

Brown, C. E. (1981). Interactions among carnosine, anserine, ophidine and copper in biochemical adaptation. *Journal of Theoretical Biology*, 88, 245–256.

Bruna, J. M., Fernandez, M., Hierro, E. M., Ordonez, J. A., & de la Hoz, L. (2000). Combined use of pronase E and a fungal extract (*Penicillium aurantiogriseum*) to potentiate the sensory characteristics of dry fermented sausages. *Meat Science*, 54, 135–145.

Chan, J. C. K., & Li-Chan, E. C. Y. (2005). Antimicrobial peptides. In Y. Mine, & F. Shahidi (Eds.), *Nutraceutical proteins and peptides in health and disease* (pp. 99–136). Boca Raton, FL: CRC Press.

Chan, W. (2004). Macronutrients in meat. In W. K. Jensen, C. Devine, & M. Dikeman (Eds.), *Encyclopedia of meat sciences* (pp. 614–618). Oxford: Elsevier.

Chandan, R. C. (2007). Functional properties of milk constituents. In Y. H. Hui (Ed.), *Handbook of food products manufacturing – Principles, bakery, beverages, cereals, cheese, confectionary, fats, fruits, and functional foods* (pp. 971–987). Hoboken, NJ: John Wiley & Sons.

Chandan, R. C., & Shah, N. P. (2007). Functional foods based on dairy ingredients. In Y. H. Hui (Ed.), *Handbook of food products manufacturing – Principles, bakery, beverages, cereals, cheese, confectionary, fats, fruits, and functional foods* (pp. 957–970). Hoboken, NJ: John Wiley & Sons.

Chiba, H., Tani, F., & Yoshikawa, M. (1989). Opioid antagonist peptides derived from κ-casein. *Journal of Dairy Research*, 56, 363–366.

Coakley, M., Ross, R. P., Nordgren, M., Fitzerald, G., Devery, R., & Stanton, C. (2003). Conjugated linoleic acid biosynthesis by human-derived *Bifidobacterium* species. *Journal of Applied Microbiology*, 94, 138–145.

Cross, K. J., Huq, N. L., & Reynolds, E. C. (2005). In Y. Mine, & F. Shahidi (Eds.), *Nutraceutical proteins and peptides in health and disease* (pp. 335–351). Boca Raton, FL: CRC Press.

Dainty, R., & Blom, H. (1995). Flavor chemistry of fermented sausages. In G. Campbell-Plattand, & P. E. Cook (Eds.), *Fermented meats* (pp. 176–193). Glasgow, Scotland: Blackie Academic & Professional.

Decker, E. A., Livisay, S. A., & Zhou, S. (2000). Mechanisms of endogenous skeletal muscle antioxidants: Chemical and physical aspects. In E. A. Decker, C. Faustman, & C. J. Lopez-Bote (Eds.), *Antioxidants in muscle foods* (pp. 25–60). New York: Wiley-Interscience.

Dentali, S. (2002). Regulation of functional foods and dietary supplements. *Food Technology*, 56(6), 89–94.

Dhiman, T. R., Nam, S. H., & Ure, A. L. (2005). Factors affecting conjugated linoleic acid content in milk and meat. *Critical Reviews of Food Science and Nutrition*, 45, 463–482.

Dransfield, E., & Etherington, D. (1981). Enzymes in the tenderization of meat. In G. G. Birch, N. Blakebrough, & K. J. Parker (Eds.), *Enzymes and food processing* (pp. 177–194). London: Applied Science Publishers.

Etherington, D. J. (1984). The contribution of proteolytic enzymes to postmortem changes in muscle. *Journal of Animal Science*, 59, 1644–1650.

Eve, L. (2000). Regulatory issues: Europe and Japan. In M. K. Schmidl, & T. P. Labuza (Eds.), *Essentials of functional foods* (pp. 363–384). Gaithersburg, MD: Aspen Publication.

Farnworth, E. R. (2003). *Handbook of fermented functional foods*. Boca Raton, FL: CRC Press.

Fernández-Ginés, J. M., Fernández-López, J., Sayas-Barberá, E., & Pérez-Alvarez, J. A. (2005). Meat products as functional foods: A review. *Journal of Food Science*, 70, R37–R43.

Fiat, A. M., Migliore-Samour, D., Jollès, P., Drouet, L., Collier, C., & Caen, J. (1993). Biologically active peptides from milk proteins with emphasis on two example concerning antithrombotic and immuno-modulating activities. *Journal of Dairy Science*, 76, 301–310.

Fujita, H., Yokoyama, K., & Yoshikawa, M. (2000). Classification and antihypertensive activity of angiotensin I-converting enzyme inhibitory peptides derived from food proteins. *Journal of Food Science*, 65, 564–569.

Gagnaire, V., Pierre, A., Molle, D., & Leonil, J. (1996). Phosphopeptides interacting with colloidal calcium phosphate isolated by tryptic hydrolysis of bovine casein micelles. *Journal of Dairy Research*, 63, 405–422.

Gibson, G. R., & Roberfroid, M. B. (1995). Dietary modulation of the human colonic microbiota: Introducing the concept of prebiotics. *Journal of Nutrition*, 125, 1401–1412.

Gnadig, S., Xue, Y., Berdeaux, O., Chardigny, J. M., & Sebedio, J.-L. (2000). Conjugated linoleic acid (CLA) as a functional ingredient. In T. Mattila-Sandholm, & M. Saarela (Eds.), *Functional dairy products* (pp. 263–298). Boca Raton, FL: CRC Press.

Gobbetti, M., Minervini, F., & Rizzello, C. G. (2007). Bioactive peptides in dairy products. In Y. H. Hui (Ed.), *Handbook of food products manufacturing – Health, meat, milk, poultry, seafood, and vegetables* (pp. 489–517). Hoboken, NJ: John Wiley & Sons.

Guesdon, B., Pichon, L., & Tomé, D. (2005). Opioid peptides. In Y. Mine, & F. Shahidi (Eds.), *Nutraceutical proteins and peptides in health and disease* (pp. 367–376). Boca Raton, FL: CRC Press.

Hammes, W. P., Haller, D., & Gänzle, M. G. (2003). Fermented meat. In E. R. Farnworth (Ed.), *Handbook of fermented functional foods* (pp. 251–275). Boca Raton, FL: CRC Press.

Hasler, C. M. (1988). Functional foods: Their role in disease prevention and health promotion. *Food Technology*, 52(10), 63–70.

Heasman, M., & Mellentin, J. (2001). *The functional foods revolution*. London: Earthscan Publications.

Hierro, E., de la Hoz, L., & Ordonez, J. A. (1999). Contribution of the microbial and meat endogenous enzymes to the free amino acid and amine contents of dry fermented sausages. *Journal of Agricultural and Food Chemistry*, 47, 1156–1161.

Higgs, J. D. (2000). The changing nature of red meat: 20 years of improving nutritional quality. *Trends in Food Science and Technology*, 11, 85–95.

Hutt, P. B. (2000). U.S. Government regulation of food with claims for special physiological value. In M. K. Schmidl, & T. P. Labuza (Eds.), *Essentials of functional foods* (pp. 339–352). Gaithersburg, MD: Aspen Publication.

Jang, A., & Lee, M. (2005). Purification and identification of angiotensin converting enzyme inhibitory peptides from beef hydrolysates. *Meat Science*, 69, 653–661.

Jiménez-Colmenero, F. (2007a). Functional foods based on meat products. In Y. H. Hui (Ed.), *Handbook of food products manufacturing – Principles, bakery, beverages, cereals, cheese, confectionary, fats, fruits, and functional foods* (pp. 989–1015). Hoboken, NJ: John Wiley & Sons.

Jiménez-Colmenero, F. (2007b). Healthier lipid formulation approaches in meat-based functional foods. Technological options for replacement of meat fats by non-meat fats. *Trends in Food Science and Technology*, 18, 567–578.

Jiménez-Colmenero, F., Carballo, J., & Cofrades, S. (2001). Healthier meat and meat products: Their role as functional foods. *Meat Science*, 59, 5–13.

Jiménez-Colmenero, F., Reig, M., & Toldrá, F. (2006). New approaches for the development of functional meat products. In L. M. L. Nollet, & F. Toldrá (Eds.), *Advanced technologies for meat processing* (pp. 275–308). Boca Raton, FL: CRC Press.

Jollès, P. S., Levy-Toledano, S., Fiat, A. M., Soria, C., Gillessen, D., Thomaidis, A., et al. (1986). Analogy between fibrinogen and casein. *European Journal of Biochemistry*, 158, 379–384.

Jones, D. P., Coates, R. J., Flagg, E. W., Eley, J. W., Block, G., Greenberg, R. S., et al. (1992). Glutathione in foods listed in the National Cancer Institute's Health Habits and History Food Frequency Questionnaire. *Nutrition and Cancer*, 17, 57–75.

Katayama, K., Fuchu, H., Sakata, A., Kawahara, S., Yamauchi, K., Kawahara, Y., et al. (2003). Angiotensin I-converting enzyme inhibitory activities of porcine skeletal muscle proteins following enzyme digestion. *Asian-Australian Journal of Animal Science*, 16, 417–424.

Katayama, K., Tomatsu, M., Fuchu, H., Sugiyama, M., Kawahara, S., Yamauchi, K., et al. (2003). Purification and characterization of an angiotensin I-converting enzyme inhibitory peptide derived from porcine troponin C. *Animal Science Journal*, 74, 53–58.

Katayama, K., Tomatsu, M., Kawahara, S., Yamauchi, K., Fuchu, H., Kodama, Y., et al. (2004). Inhibitory profile of nonapeptide derived from porcine troponin C against angiotensin I-converting enzyme. *Journal of Agricultural and Food Chemistry*, 52, 771–775.

Kato, T., Matsuda, T., Tahara, T., Sugimoto, M., Sato, Y., & Nakamura, R. (1994). Effects of meat conditioning and lactic fermentation on pork muscle protein degradation. *Bioscience, Biotechnology and Biochemistry*, 58, 408–410.

Koohmaraie, M. (1994). Muscle proteinases and meat aging. *Meat Science*, 36, 93–104.

Korhonen, H., & Pihlanto, A. (2003). Food-derived bioactive peptides: Opportunities for designing future foods. *Current Pharmaceutical Design*, 9, 1297–1308.

Korhonen, H., & Pihlanto, A. (2007). Bioactive peptides from food proteins. In Y. H. Hui (Ed.), *Handbook of food products manufacturing – Health, meat, milk, poultry, seafood, and vegetables* (pp. 5–37). Hoboken, NJ: John Wiley & Sons.

Krajcovicova-Kudlackova, M., Simoncic, R., Bederova, A., Babinska, K., & Beder, I. (2000). Correlation of carnitine levels to methionine and lysine intake. *Physiological Research*, 49, 399–402.

Lachance, P. A., & Fisher, M. C. (2005). Reinvention of the food guide pyramid to promote health. *Advances in Food and Nutrition Research*, 49, 1–39.

Langseth, L. (2000). Antioxidants and their effect on health. In M. K. Schmidl, & T. P. Labuza (Eds.), *Essentials of functional foods* (pp. 303–317). Gaithersburg, MD: Aspen Publication.

Larsson, S. C., Bergkvist, L., & Wolk, A. (2005). High-fat dairy food and conjugated linoleic acid intakes in relation to colorectal cancer incidence in the Swedish Mammography Cohort. *American Journal of Clinical Nutrition*, 82, 894–900.

Li, G. H., Le, G. W., Shi, Y. H., & Shrestha, S. (2004). Angiotensin I-converting enzyme inhibitory peptides derived from food proteins and their physiological and pharmacological effects. *Nutrition Research*, 24, 469–486.

Lieple, C., Adermann, K., Raida, M., Magert, H.-J., Forssman, W.-G., & Zucht, H.-D. (2002). Human milk provides peptides highly stimulating the growth of bifidobacteria. *European Journal of Biochemistry*, 269, 712–718.

Lindsay, D. G. (2000). Maximizing the functional benefits of plants foods. In G. R. Gibson, & C. M. Williams (Eds.), *Functional foods* (pp. 183–208). Boca Raton, FL: CRC Press.

Mattila-Sandholm, T., & Saarela, M. (2000). *Functional dairy products*. Boca Raton, FL: CRC Press.

Meisel, H. (1998). Overview on milk protein-derived peptides. *International Dairy Journal*, 8, 363–373.

Meisel, H., Walsh, D. J., Murry, B., & FitzGerald, R. J. (2005). ACE inhibitory peptides. In Y. Mine, & F. Shahidi (Eds.), *Nutraceutical proteins and peptides in health and disease* (pp. 269–315). Boca Raton, FL: CRC Press.

Mellander, O. (1950). The physiological importance of the casein phosphopeptide calcium salts II: Peroral calcium dosage of infants. *Acta Society Medicine Uppsala*, 55, 247–255.

Mikami, M., Nagao, M., Sekikawa, M., & Miura, H. (1995). Changes in peptide and free amino acid contents of different bovine muscle homogenate during storage. *Animal Science and Technology (Japan)*, 66, 630–638.

Mine, Y., & Shahidi, F. (2005). *Nutraceutical proteins and peptides in health and disease*. Boca Raton, FL: CRC Press

Mir, P. S., McAllister, T. A., Scott, S., Aalhus, J., Baron, V., McCartney, D., et al. (2004). Conjugated linoleic acid-enriched beef production. *American Journal of Clinical Nutrition*, 79, 1207S–1211S.

Mora, L., Sentandreu, M. A., & Toldrá, F. (2007). Hydrophilic chromatographic determination of carnosine, anserine, balenine, creatine, and creatinine. *Journal of Agricultural Food Chemistry*, 55, 4664–4669.

Morimatsu, F., Ito, M., Budijanto, S., Watanabe, I., Furukawa, Y., & Kimura, S. (1996). Plasma cholesterol-suppressing effect of papain-hydrolyzed pork meat in rats fed hypercholesterolemic diet. *Journal of Nutritional Science and Vitaminology*, 42, 145–153.

Mulvihill, B. (2004). Micronutrients in meat. In W. K. Jensen, C. Devine, & M. Dikeman (Eds.), *Encyclopedia of meat sciences* (pp. 618–623). Oxford: Elsevier.

Nagao, K., & Yanagita, T. (2005). Conjugated fatty acids in food and their health benefits. *Journal of Bioscience and Bioengineering*, 100, 152–157.

Nagaoka, S. (2005). Cholesterol-lowering proteins and peptides. In Y. Mine, & F. Shahidi (Eds.), *Nutraceutical proteins and peptides in health and disease* (pp. 29–40). Boca Raton, FL: CRC Press.

Nakamura, Y., Yamamoto, N., Sakai, K., & Takano, T. (1995). Antihypertensive effect of sour milk and peptides isolated from it that are inhibitors to angiotensin I-converting enzyme. *Journal of Dairy Science*, 78, 1253–1257.

Nakashima, Y., Arihara, K., Sasaki, A., Ishikawa, S., & Itoh, M. (2002). Antihypertensive activities of peptides derived from porcine skeletal muscle myosin in spontaneously hypertensive rats. *Journal of Food Science*, 67, 434–437.

Nishimura, T., & Kato, H. (1988). Mechanisms involved in the improvement of meat taste during postmortem aging. *Food Science and Technology International Tokyo*, 4, 241–249.

Nishimura, T., Rhue, M. R., Okitani, A., & Kato, H. (1988). Components contributing to the improvement of meat taste during storage. *Agricultural and Biological Chemistry*, 52, 2323–2330.

Nyberg, F., Sanderson, K., & Glämsta, E.-L. (1997). The hemorphins: A new class of opioid peptides derived from the blood protein hemoglobin. *Biopolymers*, 43, 147–156.

Oshima, G., Shimabukuro, H., & Nagasawa, K. (1979). Peptide inhibitors of angiotensin I-converting enzyme in digests of gelatin by bacterial collagenase. *Biochimica et Biophysica Acta*, 566, 128–137.

Ovesen, L. (2004a). Cardiovascular and obesity health concerns. In W. K. Jensen, C. Devine, & M. Dikeman (Eds.), *Encyclopedia of meat sciences* (pp. 623–628). Oxford: Elsevier.

Ovesen, L. (2004b). Cancer health concerns. In W. K. Jensen, C. Devine, & M. Dikeman (Eds.), *Encyclopedia of meat sciences* (pp. 628–633). Oxford: Elsevier.

Park, Y. J., Volpe, S. L., & Decker, E. A. (2005). Quantitation of carnosine in humans plasma after dietary consumption of beef. *Journal of Agricultural and Food Chemistry*, 53, 4736–4739.

Pihlanto-Leppälä, A. (2001). Bioactive peptides derived from bovine whey proteins: Opioid and ACE inhibitory peptides. *Trends in Food Science and Technology*, 11, 347–356.

Pihlanto, A., & Korhonen, H. (2003). Bioactive peptides and proteins. *Advances in Food and Nutrition Research*, 47, 175–276,

Playne, M. J., Bennett, L. E., & Smithers, G. W. (2003). Functional dairy foods and ingredients. *Australian Journal of Dairy Technology*, 58, 242–263.

Purchas, R. W., Rutherfurd, S. M., Pearce, P. D., Vather, R., & Wilkinson, B. H. P. (2004). Concentrations in beef and lamb of taurine, carnosine, coenzyme Q10, and creatine. *Meat Science*, 66, 629–637.

Purchas, R. W., & Busboom, J. R. (2005). The effect of production system and age on levels of iron, taurine, carnosine, coenzyme Q10, and creatine in beef muscles and liver. *Meat Science*, 70, 589–596.

Rizzello, C. G., Losito, I., Gobbetti, M., Carbonara, T., De Bari, M. D., & Zambonin, P. G. (2005). Antibacterial activities of peptides from the water-soluble extracts of Italian cheese varieties. *Journal of Dairy Science*, 88, 2348–2360.

Saiga, A., Okumura, T., Makihara, T., Katsuta, S., Shimizu, T., Yamada, R., et al. (2003). Angiotensin I-converting enzyme inhibitory peptides in a hydrolyzed chicken breast muscle extract. *Journal of Agricultural and Food Chemistry*, 51, 1741–1745.

Saiga, A., Tanabe, S., & Nishimura, T. (2003). Antioxidant activity of peptides obtained from porcine myofibrillar proteins by protease treatment. *Journal of Agricultural and Food Chemistry*, 51, 3661–3667.

Sentandreu, M. A., & Toldrá, F. (2007a). Oligopeptides hydrolysed by muscle dipeptidyl peptidases can generate angiotensin-I converting enzyme inhibitory dipeptides. *European Food Research and Technology*, 224, 785–790.

Sentandreu, M. A., & Toldrá, F. (2007b). Evaluation of ACE inhibitory activity of dipeptides generated by the action of porcine muscle dipeptidyl peptidases. *Food Chemistry*, 102, 511–515.
Sentandreu, M. A., Stoeva, S., Aristoy, M. C., Laib, K., Voelter, W., & Toldrá, F. (2003). Identification of small peptides generated in Spanish dry-cured ham. *Journal of Food Science*, 68, 64–69.
Seppo, L., Jauhiaine, T., Poussa, T., & Korpela, R. (2003). A fermented milk high in bioactive peptides has a blood-pressure lowering effect in hypertensive subjects. *American Journal of Clinical Nutrition*, 77, 326–330.
Shimada, K., Sakura, Y., Fukushima, M., Sekikawa, M., Kuchida, K., Mikami, M., et al. (2005). Species and muscle differences in L-carnitine levels in skeletal muscles based on a new simple assay. *Meat Science*, 68, 357–362.
Sieber, R., Collomb, M., Aeschlimann, A., Jelen, P., & Eyer, H. (2004). Impact of microbial cultures on conjugated linoleic acid in dairy products – a review. *International Dairy Journal*, 14, 1–15.
Silva, S. V., & Malcata, F. X. (2005). Caseins as source of bioactive peptides. *International Dairy Journal*, 15, 1–15.
Suetsuna, K., Ukeda, H., & Ochi, H. (2000). Isolation and characterization of free radical scavenging activities peptides derived from casein. *Journal of Nutritional Biochemistry*, 11, 128–131.
Tanabe, S., & Nishimura, T. (2005). Meat allergy. In Y. Mine, & F. Shahidi (Eds.), *Nutraceutical proteins and peptides in health and disease* (pp. 481–491). Boca Raton, FL: CRC Press.
Toldrá, F. (2004). Dry. In W. K. Jensen, C. Devine, & M. Dikeman (Eds.), *Encyclopedia of meat sciences* (pp. 360–365). Oxford: Elsevier.
Toldrá, F., & Flores, M. (1998). The role of muscle proteases and lipases in flavor development during the processing of dry-cured ham. *Critical Reviews of Food Science and Nutrition*, 38, 331–352.
Valsta, L. M., Tapanainen, H., & Mannisto, S. (2005). Meat fats in nutrition. *Meat Science*, 70, 525–530.
Vercruysse, L., Van Camp, J., & Smagghe, G. (2005). ACE inhibitory peptides derived from enzymatic hydrolysates of animal protein: A review. *Journal of Agricultural and Food Chemistry*, 53, 8106-8115.
Vermeirssen, V., Camp, J. V., & Verstraete, W. (2004). Bioavailability of angiotensin I converting enzyme inhibitory peptide (review article). *British Journal of Nutrition*, 92, 357–366.
Vescovo, G., Ravara, B., Gobbo, V., Sandri, M., Angelini, A., Dalla Libera, L., et al. (2002). L-Carnitine: A potential treatment for blocking apotosis and preventing skeletal muscle myopathy in heart failure. *American Journal of Physiology*, 283, C802–C810.
Watkins, B. A., & Yong, L. (2001). Conjugated linoleic acid: The present state of knowledge. In Wildman, R. E. C. (Ed.), *Handbook of nutraceuticals and functional foods* (pp. 445–476). Boca Raton, FL: CRC Press.
Wildman, R. E. C. (2000a). Nutraceuticals: A brief review of historical and teleological aspects. In R. E. C. Wildman (Ed.), *Handbook of nutraceuticals and functional foods* (pp. 1–12). Boca Raton, FL: CRC Press.
Wildman, R. E. C. (2000b). Classifying nutraceuticals. In R. E. C. Wildman (Ed.), *Handbook of nutraceuticals and functional foods* (pp. 13–30). Boca Raton, FL: CRC Press.
Williams, P. G. (2007). Nutritional composition of red meat. *Nutrition and Dietetics*, 64(Suppl. 4), S113–S119.
Xu, S., Boylston, T. D., & Glatz, B. A. (2005). Conjugated linoleic acid content and organoleptic attributes of fermented milk products produced with probiotic bacteria. *Journal of Agricultural and Food Chemistry*, 53, 9064–9072.
Zhao, Q., Garreau, I., Sannier, F., & Piot, J. M. (1997). Opioid peptides derived from hemoglobin: Hemorphins. *Biopolymers*, 43, 75–98.

Part IV
Biotechnology for Safer Meat and Meat Products

Chapter 12
Biocontrol of Pathogens in the Meat Chain

Catherine M. Burgess, Lucia Rivas, Mary J. McDonnell, and Geraldine Duffy

Introduction

Bacterial foodborne zoonotic diseases are of major concern, impacting public health and causing economic losses for the agricultural-food sector and the wider society. In the United States (US) alone foodborne illness from pathogens is responsible for 76 million cases of illnesses each year (Mead et al., 1999). *Salmonella*, *Campylobacter jejuni* and Enterohaemorraghic *Escherichia coli* (EHEC; predominately serotype O157:H7) and *Listeria monocytogenes* are the most predominant foodborne bacterial pathogens reported in the developed world (United States Department of Agriculture, 2001). The importance of meat and meat products as a vehicle of foodborne zoonotic pathogens cannot be underestimated (Center for Disease Control, 2006; Gillespie, O'Brien, Adak, Cheasty, & Willshaw, 2005; Mazick, Ethelberg, Nielsen, Molbak, & Lisby, 2006; Mead et al., 2006). Pathogen carriage in food animals, such as livestock and poultry can lead to both direct and indirect contamination of raw and processed meats. Hide contamination and fecal pathogen shedding contribute to the contamination of the beef carcass (Elder et al., 2000; Koohmaraie et al., 2005), while skin and feathers contaminated with feces serve as major sources of poultry contamination (Doyle & Erickson, 2006). Processing of meat can further spread microbial contamination, while inadequate temperature control can allow pathogens to increase in numbers. Eradication of pathogens from farm livestock and the environment is not yet an achievable goal. However, risk reduction measures can be implemented on the farm to minimize the risk of infection. During meat slaughter and processing, methods to ensure food safety and preservation may include a range of chemical preservative agents and/or physical processing intervention strategies. However, increased consumer demand for healthier and minimally processed food with lower amounts of additives as well as concerns regarding antibiotic resistance in foodborne bacteria has led to a greater interest and demand for natural, biological methods of food preservation and safety.

G. Duffy
Ashtown Food Research Centre, Teagasc, Ashtown, Dublin 15, Ireland

Extensive research on alternative biological methods (biocontrol) for biopreservation and reduction of foodborne pathogens is an active area of research. Numerous methods have been reported for the reduction of food pathogens at the pre-harvest stage (in the animal) and post-harvest and processing stages of the meat chain. These approaches include the use of organic compounds, vaccines and bacteriophages as well as the use of antagonistic bacteria (Callaway et al., 2004; Joerger, 2003; Koohmaraie et al., 2005; LeJeune & Wetzel, 2007). Antagonism refers to the inhibition of other (e.g., undesired or pathogenic) microorganisms through either competitive exclusion or by the production of one or more antimicrobial active metabolites such as organic (lactic and acetic) acids, hydrogen peroxide, or bacteriocins (Holzapfel, Geisen, & Schillinger, 1995; Kostrzynska & Bachand, 2006). The criteria for biocontrol agents is that they must be (1) efficacious, (2) practical, and (3) safe and not interfere with animal growth or development (Doyle & Erickson, 2006). This chapter reviews various biocontrol agents and their applications to reduce the carriage of key food pathogens *Salmonella*, EHEC, *Listeria* and *Campylobacter* in animals or levels of these pathogens on carcasses or processed meats.

Organic Compounds

Organic Acids

Organic acid solutions are the most frequently used decontaminants in meat processing. Organic acids including acetic and lactic acid are widely used in the USA to decontaminate carcasses. The use of these organic acid solutions at concentrations of up to 2.5% has been approved by the US Department of Agriculture's Food Safety and Inspection service (USDA-FSIS) (1996). Such solutions are usually applied to a carcass following hide removal when the carcass is still warm. Numerous studies have reported the effects of various organic acid solutions on general microflora and pathogenic organisms on meat products (Table 12.1). The effectiveness of organic acid solutions in inactivating or removing bacterial pathogens on a carcass and meat product varies considerably (Table 12.1) (Dorsa, 1997; Huffman, 2002; Smulders & Greer, 1998). Hardin et al. (1995) reported that a beef carcass wash followed by a 2% acid spray was more effective than either trimming or washing with water alone in the reduction of *E. coli* O157:H7 and *S.* Typhimurium. In many studies, however, sanitizing rinses were either ineffective in reducing the level of *E. coli* O157:H7 on beef tissues or only reduced the bacterial counts by 1 to 2 log CFU cm^{-2} (Brackett, Hao, & Doyle, 1994; Fratamico, Schultz, Benedict, & Buchanan, 1996).

Recently, concerns have been expressed about whether acidic decontamination may induce acid resistance in pathogens (Samelis, Sofos, Kendall, & Smith, 2002). The efficiency of such compounds can also be dependent on a number of environmental factors which can influence application in various food products. For example, the antimicrobial effect of organic acids can be influenced by pH, tissue type and

Table 12.1 Examples of organic acids used in the decontamination of meat products

Organic acid	Concentration tested (%)	Meat type	Target pathogen(s)	Reduction ($\log_{10}$ CFU)	Reference
Acetic acid	2	Chicken breast	L. monocytogenes, S. enteritidis, E. coli O157:H7	1.8–2.0, 0.9–1.7, 1.8–2.6	Anang, Rusul, Bakar, & Ling (2007)
Acetic acid, lactic acid, citric acid	1.5	Lean beef	E. coli O157:H7	0.3–0.5	Brackett et al. (1994)
Acetic acid	2	Beef tenderloin and adipose tissue	E. coli O157:H7 and E. coli K12	No effect compared to control	Fratamico et al. (1996)
Acetic or lactic acid	3	Lean beef	L. monocytogenes, S. Typhimurium, E. coli, C. jejuni	~1.0–1.2, ~0.5–1.0, ~0.5, ~1.0–1.2	Greer and Dilts (1992)
Lactic acid	4	Chilled beef carcasses	Natual microflora, Total coliforms	3.0–3.3 on aerobic plate counts. Undetectable levels	Castillo, Lucia, Mercado, and Acuff (2001)
Lactic acid	2.5–5	Lean beef	E. coli O157:H7	0.9–1.1	Castillo, Lucia, Mercado, and Acuff (2001)
Fumaric acid	1	Lean beef	L. monocytogenes	0.1–0.9	Heller et al. (2007)
Lactic acid	1			0.1–0.5	Podolak, Zayas, and Kastner (1996)
Acetic acid	1			0.1–0.5	
Fumaric acid	1		E. coli O157:H7	0.4–1.3	
Lactic acid	1			0.1–0.8	
Acetic acid	1			0.1–0.8	

bacterial microorganisms (Smulders & Greer, 1998). Sensory effects such as color, flavor or odor can also be a major concern. Dilute organic acid solutions of up to 3%, generally, have no effect on the desirable sensory properties of meat when used as a carcass decontaminant. However, some treatment conditions using lactic and acetic acid can produce adverse sensory and appearance changes when applied directly to meat cuts or products (Smulders & Greer, 1998). Further evaluation of organic acids and their effects in meat processing and their wider impact on pathogens continues to be researched and improved.

Essential Oils

Essential oils (EOs) are aromatic oily liquids obtained from plant materials. The greatest use of EOs is as flavorings in food; however, their antimicrobial potential has attracted increased attention and research. EOs can comprise more than 60 individual components, where the major components make up 85% of the EO and other components are present as trace (Burt, 2004). The mode of action of EOs is relatively unknown (Burt, 2004; Lambert, Skandamis, Coote, & Nychas, 2001). However, the extensive range of EOs and components suggest that there is most likely to be many different modes of action (Burt, 2004). The phenolic components of EOs are chiefly responsible for the antibacterial properties of the EOs and usually EOs containing a high percentage of phenolic compounds have the strongest activity (Lambert et al., 2001). Many studies have reported the antimicrobial effect of EOs or their components against foodborne pathogens (Burt, 2004; Burt & Reinders, 2003; Hammer, Carson, & Riley, 1999; Lambert et al., 2001; Si et al., 2006). EOs have a broad spectrum of activity, but are slightly more active against Gram-positive than Gram-negative bacteria (Lambert et al., 2001). Application of EOs and their components have been reported in many foods, including meat products (Table 12.2). The physical structure and composition of the food matrix as well as the environmental conditions of the food (e.g. temperature, vacuum/gas/air packaging) can affect the efficiency of the EO against pathogens in foods (Tassou, Drosinos, & Nychas, 1995; Tsigarida, Skandamis, & Nychas, 2000). For example, high levels of fat and/or protein in foods such as meat products can protect the bacteria from the antimicrobial activity of the EO, whereby the EO can dissolve in the lipid phase of the food and will therefore be less available against bacteria (Gill, Delaquis, Russo, & Holley, 2002; Tassou et al., 1995). Many studies have attempted to improve microbial quality of meats by combining the use of EOs with other preservation techniques, such as different packaging environments, radiosensitization, bacteriocins, and incorporation of EOs into packaging films (Brashears, Reilly, & Gilliland, 1998; Chiasson, Borsa, Ouattara, & Lacroix, 2004; Ghalfi, Benkerroum, Doguiet, Bensaid, & Thonart, 2007; Gill et al., 2002; Tsigarida et al., 2000). An important aspect to consider in the application of EOs to foods is the effect on the organo-leptic properties of the food product. Low concentrations of EOs in meat products have been reported to be acceptable after storage and

Table 12.2 Examples of essential oils used to control foodborne pathogens in meat products

Essential oil or component	Meat type	Concentration tested	Target pathogen(s)	Storage conditions	Reduction ($\log_{10}$ CFU g^{-1})	Reference
Carvacrol	Ground beef	1.0%	E. coli S. Typhimurium	Treated with radiosensitization for 24 h at 4°C	Relative sensitivity increased by 2.2 times.	Chiasson et al. (2004)
Oregano oil Savory oil	Pork meat	50 ul/100 g^{-1}	L. monocytogenes	42 d, 4°C	0.12–4.8 0.2–2.7	Ghalfi et al. (2007)
Cilantro oil	Ham	0.1–0.6%	L. monocytogenes	Vacuum-packed. Oil incorporated into gelatin film 28 d, 10°C.	1.3 in first week but no effect after this	Gill et al. (2002)
Eugenol	Cooked chicken breast	0.1 ml spread onto 25 g chicken pieces	L. monocytogenes	14 d, 5°C 15°C	0.79–1.73 0.7–2.2	Hao, Brackett, and Doyle (1998a)
	Cooked beef	0.1 ml spread onto 25 g beef pieces	L. monocytogenes	14 d, 5°C 15°C	0.61–0.96 0.60–1.77	Hao, Brackett, and Doyle (1998a)
Mustard oil	Cooked acidified chicken meat	0.10%	E. coli	14 d, 22°C	~2.0	Lemay et al. (2002)
Oregano oil	Beef fillets	0.80%	S. Typhimurium	Aerobic, vacuum package or MAP, 20–25 d, 5°C.	1–2 all conditions.	Skandamis, Tsigarida, and Nychas (2002)
		0.80%	L. monocytogenes	Aerobic, vacuum package or MAP, 12–15d, 5°C.	2–3 all conditions.	Tsigarida et al. (2000)
Clove oil	Minced mutton	0.5–1.0%	L. monocytogenes	15 d, 7°C 30°C	0.2–2.5 0.5–2.7	Vrinda Menon and Garg (2001)

cooking (Tsigarida et al., 2000). Oregano oil (1% v/w) was also found to improve the flavor and quality of minced meat following storage in modified atmospheres (Skandamis & Nychas, 2001). Further studies with respect to the interactions of EOs and their components with food constituents is required to improve the efficacy of EOs against food spoilage and pathogenic organisms whilst minimizing impact on the organo-leptic properties of the product.

Antagonistic Bacteria

Bacterial Metabolites

The production of one or more antimicrobial active metabolites is part of the complex mechanisms by which a culture becomes established in the presence of other competing organisms (Holzapfel et al., 1995). Along with bacteriocins, bacteria can produce many types of substances or metabolites that are inhibitory to other bacteria. These can include clinical or therapeutic low-molecular weight antibiotics, lytic agents, toxins, bacteriolytic enzymes, bacteriophages and other metabolic products such as hydrogen peroxide and diacetyl (Holzapfel et al., 1995; Kostrzynska & Bachand, 2006). These substances can act as bio-preservatives by inhibiting spoilage or pathogenic microorganisms (Deegan, Cotter, Hill, & Ross, 2006). The main mechanism by which lactic acid bacteria (LAB) inhibit microorganisms is through the production of organic acids such as lactic acid and acetic acid. Organic acids produced by LAB, including lactic, acetic and propionic acid, exert antimicrobial effects due to their action on the bacterial cytoplasmic membrane, which interferes with the maintenance of membrane potential and inhibits active transport (Kostrzynska & Bachand, 2006). The use of LAB is common in sausage fermentations, in which accumulating lactic acid levels inhibit meat-borne pathogenic bacteria (Lucke, 2000). In addition, the inhibitory property of hydrogen peroxide has also been reported. *Lactobacillus lactis* can produce hydrogen peroxide which in effect can reduce the numbers of *E. coli* O157 (0.37–1.09 $\log_{10}$ CFU ml^{-1} lower counts compared to controls) on refrigerated raw chicken (Brashears et al., 1998). Senne & Gilliland (2003) reported that the application of *Lb. delbrueckii* subsp. *lactis* (also found to produce hydrogen peroxide) could reduce the numbers of *E. coli* O157:H7 (0.8–1.3 $\log_{10}$ CFU cm^{-2}) and *S.* Typhimurium (0.8–1.5 $\log_{10}$ CFU cm^{-2}) on pork and beef carcasses kept in refrigerated storage. Reuterin, a broad-spectrum low-molecular weight antimicrobial substance produced by *Lb. reuteri* during glycerol conversion has also been reported as a potential biopreservative for food. One study found that *Lb. reuteri* in the presence of glycerol was highly effective against inocula of *E. coli* O157:H7 of 3 $\log_{10}$ CFU g^{-1} and 6 $\log_{10}$ CFU g^{-1} levels in ground beef during refrigerated storage in modified atmosphere packages (Muthukumarasamy, Han, & Holley, 2003). Reuterin was also reported to inhibit **the** growth of *L. monocytogenes* but not *Salmonella* spp. on the surface of sausages (Kuleasan & Cakmakci, 2002).

Limitations of Using Metabolites

Although some metabolites of LAB have been shown to inhibit Gram-negative bacteria their use in maintaining the safety and stability of meat products is not ideal as some metabolites can interfere with the sensory properties (e.g. hydrogen peroxide) of the food or may not be produced in sufficient amounts (e.g. reuterin). Research may provide methods to engineer metabolic pathways to give better control of the rate and extent of formation of lactic and acetic acids, and to eliminate unwanted properties such as formation of biogenic amines. Depending on the product and processing situation, one or more of these metabolites may constitute a basis for the selection of a protective culture (Holzapfel et al., (1995). It must be emphasized that LAB used to reduce pathogens in foods should not affect the sensory characteristic of the foods, ensuring that foods are palatable to consumers (Kostrzynska & Bachand, 2006).

Competitive Exclusion Technology

Competitive exclusion (CE) as a technology, involves the addition of a non-pathogenic bacterial culture to the intestinal tract of food animals in order to reduce colonization or decrease populations of pathogenic bacteria in the gastrointestinal tract (Callaway et al., 2004). A CE culture may be composed of one or many strains or species of bacteria, but should ideally be composed of species normally resident in the animal intestinal microflora. The use of CE is similar to using probiotics which are defined as 'a preparation of or a product containing viable, defined microorganisms in sufficient numbers which alter the microflora of the host and exert beneficial health effects in this host' (de Vrese, & Schrezenmeir, 2002). In contrast to CE, probiotic preparations generally consist of individual species or mixtures of LAB or yeasts that are not necessarily of animal origin (Callaway et al., 2004). The use of probiotics in food animals have been extensively reviewed (Callaway et al., 2004; Nava, Bielke, Callaway, & Castaneda, 2005; Wagner, 2006). This section will only focus on the use of CE cultures as a food safety strategy in food animals.

The precise mechanism by which CE microorganisms reduce pathogens in the animal intestine is unclear, however, the main role of CE cultures is to attach to the surface of the intestinal epithelium and establish itself within the gut. This direct binding of the CE culture to the intestinal wall prevents potentially pathogenic strains from attaching to it. Some of the bacteria may produce antimicrobial compounds such as acids or bacteriocins to eliminate species competing within the same niche (Callaway et al., 2004). Some of these antimicrobial compounds have been specifically investigated for use as biopreservatives and for food safety applications in meat and their products and will be discussed later in this chapter. CE cultures may be composed of **defined** microbial strains (known and characterized) or undefined (incompletely characterized or unknown) microbial strains. The use of CE cultures has been used extensively in poultry to reduce *Salmonella* and *Campylobacter* carriage (Chen & Stern, 2001; La Ragione & Woodward, 2003;

Wagner, 2006; Zhang, Ma, & Doyle, 2007a, 2007b). The use of CE cultures, including commercial products in poultry has been reported to reduce the colonization of poultry with *Salmonella* spp. by up to 70% or by 7–9 $\log_{10}$ cycles (Davies & Breslin, 2003; Hoszowski & Truszczynski, 1997; Schneitz & Hakkinen, 1998). Reductions of between 3–100% of *Campylobacter* spp. colonization on poultry has also been reported (Schoeni & Wong, 1994). The use of CE cultures in cattle and pigs to eliminate *E. coli* O157:H7 and/or *Salmonella* from rumen and gastrointestinal tract have also shown potential for further commercial development (Brashears, Jaroni, & Trimble, 2003; Genovese et al., 2003; Zhao et al., 2003).

Limitations of Using CE Cultures

The most important property of a CE culture is the establishment of a complex intestinal microbiota that resists colonization by human and animal pathogens. While CE has been shown to work in several animal species, the benefits have not been consistent. The differing results between studies involving CE in animals may be due to the difference between host animals, cultures, or experimental designs. Although the benefit of using CE to reduce human pathogen carriage in animals has enormous potential, many of the effective and commercially available CE products have their disadvantages. Firstly, the use of CE culture in which the microbial isolates are unknown requires regulation as an animal drug due to the cultures' effects on animal health and the risk of transfer of undesirable bacteria to humans. There is also the potential of these component bacteria of CE to transfer virulence genes onto the animal or human microbiota resulting in enhanced antimicrobial resistance (Wagner, 2006). Ongoing research in this area aims to characterize the bacteria present in CE products which will lead to safe assurance of use of the product and will allow optimization of their efficacy.

Bacteriocin-Producing Bacteria and Bacteriocins

Bacteriocins are ribosomally synthesized antimicrobial peptides or proteins produced by bacteria that kill or inhibit the growth of other bacteria, either in the same species (narrow spectrum), or across genera (broad spectrum) (Cotter, Hill, & Ross, 2005). The use of bacteriocins in biopreservation is a growing area of research. Bacteriocins are a heterogeneous group, characteristically selected for evaluation and used as specific antagonists against problematic bacteria. However, their effectiveness in foods can become limited for various reasons, and cost remains an issue impeding the broader use of bacteriocins as food additives. Evaluation of methods to improve the use of bacteriocins in various food products is continually being investigated and new developments are frequently reported. Although, Gram-negative and Gram-positive bacteria can produce bacteriocins, LAB continues to be the preferred source of food-use bacteriocins as they are generally regarded as safe (GRAS) bacteria (Chen & Hoover, 2003). The structure, biosynthesis, and application of LAB bacteriocins in many different food products have been reviewed by others (Cleveland, Montville, Nes, & Chikindas, 2001; Cotter et al., 2005; Deegan et al., 2006; Galvez, Abriouel, Lopez, & Omar, 2007; Kostrzynska & Bachand, 2006; O'Sullivan, Ross, & Hill, 2002). The use of bacteriocin-producing bacteria (BPB) or the derived bacteriocins in food safety strategies in meat applications will be further discussed below.

Bacteriocins generally act by creating pores in the membrane of their target cells and may be cytotoxic to the target cell as a result of disturbance of the bacterial inner or outer membranes. Alternatively, bacteriocins may pass through the membrane to reach a target inside the cell causing major disruptions in cell functions. Different mechanisms of action by bacteriocins have been described but interaction with the bacterial membrane is an important requirement for most, if not all, bacteriocins (Cleveland et al., 2001).

Application of BPB and Bacteriocins in Animals

Rather than applying the bacteriocin directly as a biocontrol, the bacteriocin-producing bacteria (BPB) may be applied. BPB have been isolated from rumen environments and some have been applied to manipulate the rumen environment in poultry and cattle (Cole, Farnell, Donoghue, Stern, & Evetoch, 2006; Diez-Gonzalez, 2007; Etcheverria, Arroyo, Perdigon, & Parma, 2005; Nava et al., 2005; Russell & Mantovan, 2002; Svetoch et al., 2005; Zhao et al., 1998). BPB can be administered to animals by mixing dried or wet cultures with feed or drinking water, and depending on the ability of the bacterial strain to colonize the gastrointestinal tract may be fed sporadically or continuously. The feeding of BPB can have a direct effect on reducing the existing populations of foodborne pathogens such as *Salmonella* and *E. coli* O157:H7 and long-term colonization with BPB would prevent further re-introduction of the pathogenic bacteria. Few studies have addressed the fate of bacteriocins in the intestinal tract, but some data suggests that some of the low molecular weight bacteriocins can survive at least some of the intestinal environments and possibly could be administered through feed (Ganzle, Hertel, van der Vossen, & Hammes, 1999).

The use of colicins (bacteriocin produced by *E. coli*), as a pre-harvest control, has been actively evaluated to reduce pathogenic *E. coli* in cattle populations (Diez-Gonzalez, 2007; Schamberger, Phillips, Jacobs, Diez-Gonzalez, 2004). Calves fed with a colicin-producing *E. coli* yielded an overall reduction of 1.1 $\log_{10}$ CFUg^{-1} of *E. coli* O157:H7 and a maximum decrease of 1.8 $\log_{10}$ CFUg^{-1} of *E. coli* O157:H7 over 24 days (Schamberger et al., 2004). Colicins specific for *E. coli* are particularly advantageous in the rumen intestine in that it will only inhibit one type of bacterial strain or species while not disrupting the other microbial populations in the intestine (Diez-Gonzalez, 2007). Studies continue to investigate whether expression of colicins could be incorporated into bacteria which are normally present in the animal rumen, which can decrease the chances of transfer of colicin genes to other potentially pathogenic *E. coli* strains, such as *E. coli* O157:H7 (McCormick, Klaenhammer, & Stiles, 1999). In poultry, a few studies have investigated the use of BPB to control foodborne pathogens. Bacteriocin-like compounds were shown to have direct antimicrobial activity, in vitro against *Campylobacter* (Morency, Mota-Meira, LaPointe, Lacroix, & Lavoie, 2001; Schoenis & Doyle, 1992; Svetoch et al., 2005). A purified bacteriocin produced by *Paenibacillus polymyxa* microencapsulated and administered via feed was reported to reduce cecal *Campylobacter* colonization in young broiler chickens and turkeys (Cole et al., 2006; Stern et al., 2005). Treatment with bacteriocin eliminated detectable *Campylobacter* concentrations in turkey cecal contents (detection limit, 1×10^2 CFU g^{-1}) compared to controls (1.0×10^6 CFU g^{-1} of cacal contents) (Cole et al., 2006). In chicken cacel samples, use **of bacteriocin** resulted in significant reductions in colonization by *C. jejuni* for *C. jejuni*. These studies reported significant reductions of both intestinal levels and frequency of chicken and turkey colonization by *C. jejuni*, suggesting that this on-farm application would be an alternative to chemical disinfection of contaminated carcasses (Stern et al., 2005). In order for BPB and their bacteriocins to be used successfully in animal production, more research is required to improve their efficacy in different animal systems.

Application of Bacteriocin-Producing Bacteria and Bacteriocins in Meat Products

Bacteriocins are potentially valuable biological tools to improve food preservation and food safety by reducing the prevalence of undesirable spoilage microorganisms or foodborne pathogens in a food product (Deegan et al., 2006). Although the use of LAB is common in many food processes, the use of these microorganisms in different meat products and storage conditions continues to be evaluated (Kostrzynska & Bachand, 2006). The application of bacteriocins is not recommended as a primary processing step or barrier to prevent the growth or survival of pathogens, but should form part of a system with multiple hurdles. Bacteriocins can be incorporated into food as an ingredient in the form of a purified/semi-purified bacteriocin preparation (Table 12.3). Alternatively, a BPB can be introduced as a 'protective culture' in the form of a live culture which produces the bacteriocins *in situ* in the food (Table 12.4). In this case, the BPB is either substituted for all or part of the starter or is subsequently applied to the food to improve the safety of the culture (O'Sullivan et al., 2002). A major criterion for a protective culture in meat products is to inhibit pathogens and/or prolong shelf life, while not changing the sensory properties of the product (Lucke, 2000). The use of purified bacteriocins is not always attractive as some can be inactivated in meats and may also required regulatory approval or be labeled as an additive on the meat product (Aasen et al., 2003). The use of BPB as a protective culture is a more practical approach as it does not require regulatory approval or preservative label declarations and can be substituted into the product as a starter culture (Deegan et al., 2006; Jacobsen, Budde, & Koch, 2003; Katla et al., 2002).

Numerous studies have attempted to isolate BPB, particularly LAB from meat products (Albano et al., 2007; Arlindo et al., 2006; Budde, Hornbaek, Jacobsen,

Table 12.3 Examples of bacteriocins used to control pathogens in meat products

Bacteriocin	Producer strain	Meat product	Target pathogen(s)	Reduction $\log_{10}$ CFU g^{-1}	Reference
Sakacin P	Lb. sakei	Chicken cuts, fillet	L. monocytogenes	3.0–5.1 (from 35 µg g^{-1})	Aasen et al. (2003)
Enterocin AS-48	Enterococcus faecalis	Pork sausage mixture	S. aureus	5.31 (from 40 µg g^{-1})	Ananou, Maqueda, Martinez-Bueno, Galvez, and Valdivia (2005)
Piscicocin CS526	Carnobacterium piscicola CS526	Ground meat (beef and pork mixture)	L. monocytogenes	to <3.0×10^3 at 4 and 12°C.	Azuma, Bagenda, Yamamoto, Kawai, and Yamazaki (2006)
Lactocin 705 and Lactocin AL705	Lb. curvatus CRL705	Beef meat	L. innocua B. thermosphata	no reduction over 36 d (2°C). 2.8	Castellano and Vignolo (2006)
Plantaricin UG1	Lb. plantarum UG1	Cooked and uncooked chicken, turkey and beef and dry mortadella.	L. monocytogenes C. perfringens	<1(3 d, 15°C). zero within 7–14 d.	Enan (2006a, 2006b)
Sakacin K	Lb. sakei CTC494	Raw minced pork	L. innocua	from 50 MPN g^{-1} to <3MPN g^{-1} (8 d, 7C).	Hugas, Pages, Garriga, and Monfort (1998)
Sakacin P	Lb. sakei	Chicken cold cuts	L. monocytogenes	Inhibition of growth (28 d, 10°C).	Katla, Moretro, Sreen, Aasen, Axelsson, Rolvik and Waterstad (2002)
Pediocin AcH	Lb. plantarum	Sliced cooked sausage	L. monocytogenes	~0.7(6 d, 6°C).	Mattila, Saris, and Tyopponen (2003)
Bacteriocin not identified	Pediococcus acidilactici	Raw pork meat	L. monocytogenes C. perfringens	3 (over 72 h, 15°C). 0.5	Nieto-Lozano, Reguera-Useros, Pelaez-Martinez, and Hardisson de la Torre (2006)
Lactocin 705, Enterocin CRL35 and Nisin.	Lb. casei CRL35, Enterococcus faecium CRL35	Mince meat	L. monocytogenes L. innocua	~4.8–5.2 (from individual bacteriocins) ~7.0–9.9 (from bacteriocin combinations)	Vignolo et al. (2000)

Table 12.4 Examples of use of bacteriocin-producing bacteria as protective cultures to control pathogens in meat products

Culture	Meat product	Target pathogen(s)	Reduction ($Log_{10}CFU\ g^{-1}$)	Reference
Lb. sakei CECT 4808 and/or Lb. curvatus CECT904	Pork sausage mixture	L. monocytogenes	to $<1\times10^2$ (over 9 d, 22°C).	Ananou et al. (2005)
Enterococcus faecalis A-48-32	Dry-fermented sausage	L. monocytogenes	to <1 (over 15 d ~15°C).	Benkerroum et al. (2005)
Lactococcus lactis subsp. lactis LMG21206 and Lb. curvatus LBPE	Raw sausage (mergeuz)	L. monocytogenes	2.4 during fermentation	Benkerroum et al. (2003)
Five uncharacterised Lactic acid bacteria	Ham and servelat sausage	L. monocytogenes	Inhibited growth (at 8 and 4°C over 28 d).	Bredholt, Nesbakken, and Holck (2001)
Leuconostoc carnosum 4010	Pork meat sausage slices	L. monocytogenes	to <10 CFU (21 d, 5°C).	Budde et al. (2003)
Lb. sakei TH1	Beef meat	L. innocua B. thermosphacta	No reduction (36 d, 2°C). 2.8 (14 d).	Castellano and Vignolo (2006)
Lb. casei CRL705	Beef meat slurry	L. innocua Lb sakei	No reduction (21 d, 4°C)	Castellano, Holzapfel, and Vignolo (2004)
Lb. curvatus CRL705	Pork adipose tissue	B. thermosphacta L. monocytogenes	~2–3 less than controls (7 or 13 d, 4°C) ~2–3 less than controls (7 or 13 d, 4°C)	Greer and Dilts (2006)
Lb. sakei CTC494	Poultry breast	L. monocytogenes	0.8 (7 d, 7°C)	
Leuconostoc carnosum 4010	Sliced pork saveloys	L. monocytogenes	No reduction spray (28 d, 10°C).	Jacobsen et al. (2003)
	Cooked pork		No reduction.	

Barkholt, & Koch, 2003; Noonpakdee, Santivarangkna, Jumriangrit, Sonomoto, & Panyim, 2003; Prema, Bharathy, Palavesam, Sivasubramanian, & Immanuel, 2006; Schneider et al., 2006; Yin, Wu, & Jiang, 2003), with the aim of identifying a potential protective culture that can not only be used as a starter culture but also possess bacteriocin activity to eliminate or reduce foodborne pathogens (Benkerroum et al., 2005; Benkerroum, Daoudi, & Kamal, 2003; Leroy & De Vuyst, 2005). Many LABs are used to combat other spoilage organisms in aerobic or vacuum packed meat products (Barakat, Griffiths, & Harris, 2000; Lucke, 2000). With respect to meat safety, majority of **the** studies investigating the use of bacteriocins in meat products have targeted the pathogen *L. monocytogenes* due to its ability to grow at refrigeration temperature and survive in fermented foods (Tables 12.3 and 12.4). The disadvantage of many LAB bacteriocins is that although they are active against Gram-positive organisms they are not as effective against Gram-negative foodborne pathogens such as *E. coli* (particularly *E. coli* O157:H7) or *Salmonella* spp. which are also a concern in meat products (Lucke, 2000). This is due to the fact that Gram-negative bacteria are protected by their outer membrane, which prevents bacteriocins from reaching the cytoplasmic membrane (Abee, Krockel, & Hill, 1995). Studies have therefore attempted to find alternative bacteriocins with broader spectrums or have used other combinations of strategies to improve bacteriocin activity.

The most commonly used bacteriocin in foods, including meat products is Nisin (produced by *Lactococcus lactis* subsp. *lactis*). Nisin is one of the commercially available bacteriocins with US Food and Drug Administration (FDA) approval. It inhibits the growth of a wide range of Gram-positive bacteria including *L. monocytogenes*. Nisin has been shown to be effective in a number of food systems but is predominantly used in canned foods and dairy products, particularly in cheese production, where it protects against heat-resistant, spore forming organisms such as *Bacillus* and *Clostridium* spp (Cotter et al., 2005; Deegan et al., 2006). Nisin has been used as a food safety agent in beef (Ariyapitipun, Mustapha, & Clarke, 2000; Barboza De Martinez, Ferrer, & Marquez Salas, 2002; Zhang & Mustapha, 1999), sausages (Patel, Sanglay, Sharma, & Solomon, 2007), ground mince (Castillo, Meszaros, & Kiss, 2004) and poultry (Yuste, Pla, Capellas and Mor-Mur, 2002; Zuckerman & Abraham, 2002), but is not as effective in the preservation of meat as it is in dairy products. The inhibitory activity of Nisin is reduced by interference from meat components such as phospholipids, especially where there may be a high fat content (Leroy & De Vuyst, 2005). In addition, Nisin has a low solubility at normal meat pH and its interaction with phospholipids results in the uneven distribution of the bacteriocin in the meat (De Martinis, Publio, Santarosa, & Freitas, 2001; Stergiou, Thomas, & Adams, 2006).

Substantial research has been done to evaluate and improve the effectiveness of Nisin and other bacteriocins activity against Gram-negative pathogens and its applicability in different food products. Several bacteriocins show additive or synergistic effects when used in combination with other antimicrobial agents including, organic acids, other antimicrobials or sublethal treatments such as mild heat or high pressure (Ananou, Galvez, Martinez-Bueno, Maqueda, & Valdivia, 2005;

Ariyapitipun et al., 2000; Arques et al., 2004; Ganzle, Weber, & Hammes, 1999; Rodriguez, Nunez, Gaya, & Medina, 2005).

An alternative use of bacteriocins is as the agent incorporated into packaging materials (Cooksey, 2005; Quintavalla & Vicini, 2002). Combining the bacteriocin directly into a plastic material provides a number of advantages for delivery of the bacteriocin to the food product. Firstly, only the necessary amount of bacteriocin would be required. Secondly, the agent would not be a direct additive to the food product and would therefore avoid labeling and regulatory approval. Thirdly, if the plastic material were made from an edible and/or biodegradable plastic benefits for the environment would be apparent (Siragusa, Cutter, & Willett, 1999). Bacteriocins such as Nisin and Pediocin among others have been incorporated into different films and have been successful in inhibiting spoilage microorganisms, such as LAB and *Brochothrix thermosphacta* on beef carcass tissue, as well as *Listeria* spp. on meat and poultry samples (Marcos, Aymerich, Monfort, & Garriga, 2007; Mauriello, Ercolini, La Storia, Casaburi, & Villani, 2004; Ming, Weber, Ayres, & Sandine, 1997; Scannell et al., 2000; Siragusa et al., 1999). Further development of packaging technologies may prove to be an effective way of delivering bacteriocins to the surface of food to improve food safety and preservation.

Limitations of Using Bacteriocins and Future Prospects

The extensive use of bacteriocins in foods is hindered by a number of factors. One of the main limitations is the narrow spectrum of activity of most bacteriocins, though the specificity of a bacteriocin can be advantageous for applications in which a single bacterial strain of species is targeted without disrupting other microbial populations. Research continues to identify alternative bacteriocins which may extend **the** bacteriocin applications either through their use alone or in combination with other antimicrobials or hurdle technologies. Considerable studies are also required to investigate the factors influencing the applicability of certain bacteriocins in various food systems. Bacteriocin activity is difficult to maintain in a range of foods particularly in meat products. Another potential problem associated with using bacteriocins in foods is the development of resistant populations of problematic bacteria. Resistance can occur naturally and it has been reported especially with regard to Class IIa bacteriocins (Naghmouchi, Kheadr, Lacroix, & Fliss, 2007). Consequently, studies have examined the possibility of generating multiple bacteriocin producers to limit the potential of bacteriocin-resistant populations (O'Sullivan, Ryan, Ross, & Hill, 2003). Some BPB strains may also spontaneously loss their ability to produce bacteriocin(s) due to genetic instability (Holzapfel et al., 1995; Riley & Wertz, 2002) and can also become ineffective if the cell membrane of a target organism changes in response to a particular environmental condition. Although there are some limitations for the use of bacteriocins in meat applications, the ongoing study of existing bacteriocins as well as the identification of new bacteriocins and improvements of application will only optimize the potential of these agents in many different food applications which will lead to further improvements of food safety and quality of meat products.

Antimicrobial Peptides

Antimicrobial peptides (AMPs) are generally short cationic peptides produced by both animals and plants that have potent killing activity against a range of bacteria, fungi, viruses, and protozoa (Higgs et al., 2007). They are ubiquitous in nature and they play an important role in host defence and microbial control. Their exact mode of action is not completely understood, but it is thought that cationic AMPs are attracted to negatively charged phospholipids in the cell membrane and interact with the membrane, displacing lipids and altering the membrane structure. A number of hypotheses have been put forward as to how AMPs kill microbes and these are discussed elsewhere (Zasloff, 2002). Over 850 AMPs have been identified from a host of species and a catalogue of these can be found online (http://www.bbcm.univ.trieste.it/~tossi/amsdb.html). AMPs can also be designed and chemically synthesized and have been shown to possess antimicrobial activity (Anzai et al., 1991; Appendini & Hotchkiss, 1999; Haynie, Crum, & Doele, 1995).

Many studies have demonstrated the activity of AMPs and derivatives from a variety of species, including host species, against pathogens including *Salmonella*, *E. coli* O157, and *Listeria*. A selection of these studies is shown in Table 12.5. AMPs are now emerging as a solution to the development of antibiotic resistance by bacterial pathogens, and therefore, the focus at present is on understanding their mode of action and potential application from a clinical point of view (Hancock & Sahl, 2006). Investigations of AMPs in food systems to date are limited. Some studies have been performed in food products or suggest the potential of the AMP in a food product (Table 12.5). Yaron, Rydlo, Shachar, and Mor (2003) studied the antimicrobial activity of dermaseptin S4 and its derivative K_4-S4 which come from tree frogs. It was found that K_4-S4 reduced the population of *E. coli* O157:H7 in apple juice by more than 7 log units in less than 2 hours, indicating its potential usefulness as a biocontrol agent. The authors put forward potential advantages for the use of animal derived AMPs in food safety and preservation measures, such as their activity over a wide range of conditions. Another study in apple juice showed that a synthetic AMP could reduce the *E. coli* O157:H7 numbers by 3.5 log units in 8 hours. However, the same peptide had no effect on *E. coli* O157:H7 when grown in skim milk indicating that components from the milk could be interacting with the peptide causing it to lose its antibacterial activity (Appendini & Hotchkiss, 1999). A different study showed that peptides produced by *Lb. acidophilus* fermentation of sodium caseinate had bacteriocidal activity against *Enterobacter sakazakii*, *L. innocua* and *E. coli*, and it was suggested that such peptides could be used as a protection mechanism against *E. sakazakii* in infant formula by producing a casein-cased milk ingredient by fermentation (Hayes, Ross, Fitzgerald, Hill, & Stanton, 2006).

Table 12.5 Examples of antimicrobial peptides (AMPs) which have antimicrobial activity against pathogens

AMP	Source	Antimicrobial activity	Reference
Synthetic cathelicidins	ovine cathelicidins	E. coli O157:H7	Anderson (2005)
Synthetic PR-26	porcine neutrophils AMP	E. coli O157:H7 and L. monocytogenes	Annamalai, Venkitanarayanan, Hoagland, and Khan (2001)
6K8L	Synthetic	B. subtilis, E. coli O157:H7, L. monocytogenes, Pseudomonas fluorescens, S. Typhimurium, Serratia liquefasciens, Staphylococcus aureus, Kluyveromyces marxianus	Appendini & Hotchkiss (1999)
L. acidophilus fermentation of sodium caseinate	Bovine α_{s1}-casein	Enterobacter sakazakaii, L. innocua, E. coli	Hayes et al. (2006)
Synthetic modified AvBD8	chicken avian β-defensin-8	E. coli, S. Typhimurium and L. monocytogenes	Higgs et al. (2007)
Pepsin digest of casein	Bovine kappa-casein	E. coli, L. innocua	Lopez-Exposito, Minervini, Amigo, and Recio (2006)
Recombinant gallinacins	Chicken	S. Typhimurium, S. enteriditis	Milona, Townes, Bevan, and Hall (2007)
Purified aurelin	Aurelia aurita jellyfish	E. coli, L. monocytogenes	(2006)
Synthetic cathelicidin Bac7 and derivatives	Granulocytes of mammalian species	S. Typhimurium, E. coli	Podda et al. (2006)
Synthetic cathelicidin K9CATH	Canine	L. monocytogenes, S. aureus, E. coli, Klebsiella pneumoniae, S. Typhimurium, Ps. aeroginosa, Proteus mirabilis, S. enteriditis, Neisseria gonorrhoeae	Sang et al. (2007)
Lactoferrin, pepsin hydrolysed lactoferrin, Lactoferricin®	Bovine	E. coli O157:H7	Shin et al. (1998)
Synthetic dermaseptin S4 and derivative K$_4$-S4	Tree frogs of the Phyllomedusa species	E. coli, E. coli O157:H7, S. Typhimurium, S. Stanley, L. monocytogenes, L. innocua	Yaron et al. (2003)

Very little research has focused on the use of AMPs in the meat industry. One study looked at the effect of a synthetic peptide on the microflora in meat exudates, but aerobic and anaerobic counts were reduced by less than one log indicating a small degree of inhibition (Appendini & Hotchkiss, 1999). Although studies in this area are limited at the moment, the potential advantages provided by AMPs, such as their wide spectrum of activity and lack of bacterial resistance to AMPs, suggests that the interest in their use as biocontrol agents in foods can only grow in the years to come.

Bacteriophage

Bacteriophages are viruses that infect and kill bacterial cells by reproducing within the bacteria and disrupting the host metabolic pathways, causing the bacterium to lyse. Bacteriophage are ubiquitous in the environment; they specifically target bacterial cells and do not infect mammalian cells, hence the reason why they are proposed as biocontrol agents in human, animal, clinical and industrial applications.

Classification and Mode of Action

Most bacteriophages range in size from 24–200 nm in length and are classified into 13 families. The bacteriophage structure consists of a head or capsid; composed of protein that acts as a protective barrier in which either DNA or RNA is stored. Some bacteriophages possess a tail through which the nucleic acid is delivered to the bacterial cell during infection.

Bacteriophages are classified as either lytic or lysogenic. Lytic bacteriophages bring about rapid lysis and death of the bacterial cell, whereas a lysogenic bacteriophage does not result in immediate lysis, but instead enters a quiescent state and is known as a prophage during this period (Hanlon, 2007). A lysogenic bacteriophage integrates into the genome of the host bacterial cell, it undergoes replication with the host chromosome and the viral DNA is passed onto the daughter cells. Lysogenic bacteriophages have the ability to transfer genes for toxin production or pathogenicity factors between bacterial communities (Wagner & Waldor, 2002). Therefore, lytic bacteriophages are preferred for the purpose of bacteriophage therapy.

Bacteriophage first come in contact with bacterial host cells during Brownian motion. Infection of a bacterial cell by a bacteriophage involves attachment of the bacteriophage to the bacterial membrane. This is accomplished by the tail fibers or an equivalent structure, attaching to specific receptors on the surface of the bacterial cell. These receptors may be protein, peptidoglycan, teichoic acid, lipopolysaccharide and oligosaccharide (Lenski, 1988).

Application of Bacteriophage

There has been a vast amount of research into the application of bacteriophage to control foodborne pathogens such as *Campylobacter*, *E. coli* O157:H7, *Listeria* and *Salmonella* in animal food products, and food processing environments. Bacteriophage has also been applied to fruit and vegetables to control against bacterial pathogens pre-harvest, (Balogh et al., 2003; Pao, Randolph, Westbrook, & Shen, 2004). Another aspect of bacteriophage biocontrol is the use of bacteriophages as indicators for detection of pathogens in foods and fecal contamination of animal feeds (Goodridge, Chen, & Griffiths, 1999; Hsu, Shieh, & Sobsey, 2002; Maciorowski, Pillai, & Ricke, 2001). Detection of bacterial contamination can be achieved quickly by using bacteriophage rather than using bacteria (Hsu et al., 2002). An area of interest is the use of bacteriophage enzymes in foods and food grade bacteria (LAB). Endolysins are bacteriophage enzymes that are synthesized late during virus replication. They target bacterial peptidoglycan which results in the release of progeny virions (Gaeng, Scherer, Neve, & Loessner, 2000; Loessner, 2005). Gaeng et al. (2000) applied endolysin encoding genes from *Listeria* bacteriophage to lactococcal starter organisms to obtain organisms with biopreservation properties against *L. monocytogenes*.

Application of Bacteriophage in Animals: Preharvest Control

Wagenaar, Van Bergen, Mueller, Wassenaar, and Carlton (2005) applied a bacteriophage (bacteriophage strain 71 and 69) with a wide host range against *C. jejuni* strains to control *C. jejuni* colonization in broiler chickens. They observed a 3 $\log_{10}$ CFUg^{-1} reduction in *C. jejuni* counts initially in the therapeutic group; however, counts stabilized after 5 days to 1 $\log_{10}$ CFUg^{-1} lower than the control group. They concluded that this bacteriophage treatment could be an alternative method for reducing *C. jejuni* colonisation in broiler chickens. Loc Carrillo et al. (2005) reported a reduction in *C. jejuni* counts of between 0.5 and 5 $\log_{10}$ CFUg^{-1} when broiler chickens were orally administered bacteriophage CP8 and CP34 in an antacid suspension to reduce *C. jejuni* colonization in broiler chickens.

The literature, till date, suggests that application of a cocktail of bacteriophage yields more successful results in biocontrol of *E. coli* than bacteriophage administered singly (Bach, McAllister, Viera, Gannon, & Holley, 2002). Waddell et al. (2000) administered a cocktail of bacteriophage to control shedding of *E. coli* O157:H7 in calves and showed that shedding of the pathogen was observed for 6 to 8 days in bacteriophage treated calves, compared to 6 to 14 days in the control calves. Kudva, Jelacic, Tarr, Youderian, and Hovde (1999) isolated three coliphages (KH1, KH4, and KH5) and applied them to a number of O157 and non-O157 strains to determine their ability to lyse laboratory cultures. The three coliphages were capable of lysing the O157 serotype, and did not have any effect on the non-O157 strains. A high multiplicity of infection (MOI) of 10^3 plaque forming units (PFU) and aeration were required for successful control of *E. coli* O157:H7. Multiplicity of infection is the ratio of infectious agents to infection targets. A difference between the three coliphages in their ability to kill host cells was observed. KH1 was the most effective coliphage in reducing host cell numbers; however, the three coliphages were unable to eliminate *E. coli* O157:H7.

Sheng, Knecht, Kudva, & Hovde (2006) applied bacteriophage to determine their ability to control intestinal *E. coli* O157:H7 in ruminants. Bacteriophage KH1 and SH1 were applied orally and rectally to sheep and cattle respectively. No reduction in intestinal carriage of *E. coli* O157:H7 in sheep was observed when KH1 was administered orally. An equal mixture of KH1 and SH1 were administered rectally to cattle. Combination of the two bacteriophages reduced the numbers of *E. coli* O157:H7; however, they were unable to clear the *E. coli* O157:H7 infection from the cattle.

Raya et al. (2006) demonstrated that a single oral dose of *E. coli* O157:H7 specific bacteriophage (CEV1) applied to sheep reduced shedding of the pathogen by 2 $\log_{10}$ CFUg^{-1}. Recently, the US Department of Agriculture approved a bacteriophage for hide washing. The product produced by OmniLytics is administered on the hides of live animals prior to slaughter to minimize contamination of *E. coli* O157:H7 onto beef carcasses.

Application of Bacteriophage in Meat and Meat Products

The application of bacteriophage has been reported for a range of pathogens on poultry and meat. Atterbury, Connerton, Dodd, Rees, and Connerton (2003) showed that when bacteriophage Φ 02 was applied to chicken skin inoculated with *C. jejuni* and stored at 4 and −20°C a reduction of approximately 1 $\log_{10}$ CFU ml^{-1} was obtained. They also observed a reduction in Campylobacteraceae on the frozen samples by 2 $\log_{10}$ CFU ml^{-1} after day 1, and remained at similar levels thereafter. Goode, Allen, and Barrow (2003), also showed that three bacteriophages, specific for *C. jejuni* and *Salmonella* gave a 1 $\log_{10}$ CFU ml^{-1} reduction of both pathogens, at a MOI of 1. When other bacteriophages were applied at a MOI of 100 to 1000, a reduction of 2 $\log_{10}$ CFU ml^{-1} in *S.* Enteritidis was observed over 48 hours.

The effectiveness of a three bacteriophage cocktail in reducing *E. coli* O157:H7 on inoculated meat samples has been demonstrated by O'Flynn et al. (2004). The three bacteriophages (e11/2, pp01 e41c) reduced *E. coli* O157:H7 from initial numbers of approximately 3 $\log_{10}$ CFU ml^{-1} to undetectable levels during a 2 hour enrichment **process**. The effectiveness may be due to lysis from outside the cell as the MOI used was 10^6 − fold, the bacterial cells may have been overwhelmed by the number of bacteriophage attaching to the cell surface causing the bacterial cell to lyse.

Dykes and Moorhead (2002), applied listeriophage LH7 to beef inoculated with *L. monocytogenes*, which was vacuum packed and stored at 4°C. They also applied LH7 to mixed population of *L. monocytogenes* stored in PBS. A combination of listeriophage and nisin had no effect when applied to vacuum packed beef. Listeriophage alone had no effect when applied to broth, however, when nisin was combined with LH7 and applied to broth a decrease in *L. monocytogenes* counts was observed. In 2006, the US Food and Drug Administration approved a bacteriophage preparation to be applied on Ready-To-Eat (RTE) meat and poultry products as an antimicrobial agent against *L. monocytogenes* (Federal Register, 2006). The bacteriophage preparation consists of six *Listeria* specific bacteriophages, which were combined to reduce the possibility of *L. monocytogenes* developing resistance to the agent. Once *L. monocytogenes* is no longer present in the product the bacteriophage remains dormant. The regulation specifies that the cocktail of bacteriophage must be negative for *L. monocytogenes* and listeriolysin O, a toxin produced by *L. monocytogenes*. Another commercially available product is Listex^{TM}P100, produced by EBI Food Safety. It is a bacteriophage **used for the** control of *L.* monocytogenes in meat and cheese products. It is recognized as GRAS by the FDA and has a wide host range against *Listeria* strains (Carlton, Noordman, Biswas, de Meester, & Loessner, 2005).

Whichard, Sriranganathan, & Pierson (2003) investigated the ability of bacteriophage Felix 01 (wild type) and a variant of Felix 01, to control *Salmonella* growth on chicken frankfurters. A 1.8 and 1.2 $\log_{10}$ CFU ml^{-1} reduction of *S.* Typhimurium in artificially contaminated chicken frankfurters was reported for the wild type Felix 01 and the variant respectively. Another study artificially inoculated broiler carcasses with *S.* Enteritidis and stored them at 4°C for 2 hours. Carcasses were then sprayed with 5.5 ml of saline containing bacteriophage PHL 4 at various concentrations. This resulted in a 93% reduction in recovery of *S.* Enteritidis when inoculated with 10^8 or 10^{10} PFU. They also applied PHL 4 to carcass rinse water at concentrations of 10^6 and 10^{10} PFU. When bacteriophage was applied at 10^{10} PFU recovery of *S.* Enteritidis was reduced to between 50 and 100% (Higgins et al., 2005).

Limitations of Bacteriophage Use

Adsorption of a bacteriophage to receptors on a bacterial cell occurs during Brownian motion. This process may be obstructed by the existence of considerable numbers of non host bacterial cells. This initial interaction may also be hindered by a viscous environment, for example, the rumen environment (Joerger, 2003). This type of environment may protect the bacterial cell from bacteriophage infection. Host bacterial cells may also be protected by biofilms present in a food environment as bacteriophage would be incapable of penetrating and accessing the bacterial cell within the biofilm. Many meat products are distributed and stored at refrigeration temperatures, conditions under which many pathogens may not grow. This poses a problem with bacteriophage use as replication can not occur under conditions where the host is not dividing. In addition, other food conditions can also affect bacteriophage use, these include visible and UV light, osmotic shock and pressure and thermotolerence. Bacteriophages may be destroyed in food processing environments as they are effectively cleaned and sanitized. **There are r**eports of dairy bacteriophages be**ing** destroyed by sodium hypochlorite (100 ppm free chlorine) and peracetic acid (Quiberoni, Suarez, & Reinheimer, 1999). Some pathogens can survive **the** low pH environments while bacteriophages would be instable at such low acidity. Smith, Huggins, and Shaw (1987) reported the stability of bacteriophage over a pH range of 3.5 to 6.8, however a decrease in bacteriophage titer was observed at pH 3, and a large decrease was observed below pH 3.

The need for high bacterial populations for phage replication to occur and lyse the bacterial cells has been reported in the literature. Berchieri, Lovell, and Barrow (1991), stated that approximately 10^4 CFU ml^{-1} of the host cell was required for the bacteriophage to have an effect on the host cell.

Bacteriophages also have the capability to transfer unfavorable genes from one bacterium to another; these could be virulence genes or antibiotic resistance genes (Alisky, Iczkowski, Rapoport, & Troitsky, 1998; Figueroa-Bossi & Bossi, 1999; Mirold, Rabsch, Tschape, & Hardt, 2001; Schmieger & Schicklmaier, 1999). This is a major concern and much effort must be applied to characterize and determine the potential of a bacteriophage to transfer virulence factors and antibiotic resistance genes to commensal or pathogenic bacteria before it is commercialised.

The emergence of host resistance to bacteriophages due to DNA mutations has been reported in pathogens in pre-harvest (Sklar & Joerger, 2001) and post-harvest environments (Greer & Dilts, 2002). A cocktail of bacteriophage may settle the issues of host resistance (Barrow & Soothill, 1997; Leverentz et al., 2003; Tanji et al., 2004). Desirably, bacteriophages with a broad host spectrum are favourable for biocontrol as there are many subtypes in each pathogen species that may possess different cell surface receptors. Finally, if bacteriophages are to be used in food products, consumer acceptance will be a major factor. The marketing strategy involved for bacteriophage products will be a critical area to ensure consumer acceptance.

Vaccines

An option which has been investigated as a potential biocontrol agent at the pre-harvest stage is vaccination of food animals. In this way the animal's own immune system is used to reduce pathogen loads by producing antigens against particular pathogens. A successful vaccine would prevent colonization of the host by the pathogen; i.e. when the animal ingests the pathogen it can not colonize and multiply and therefore less of the pathogen would be present at slaughter or in the feces thus reducing the likelihood of the pathogen entering the food chain. There are obstacles to be overcome; however, a major challenge being the ability to prime the mucosal immune response of animals to mount a protective response against an otherwise commensal organism (LeJeune & Wetzel, 2007). Vaccinations have been developed with varying degrees of success for a number of zoonotic pathogens. The following is by no means an exhaustive list but provides an overview of the different types of vaccine, currently either in development or in use.

Intimin from *E. coli* O157:H7 has been identified as a potential vaccine candidate. Intimin is an outer membrane protein encoded by the *eae* gene that is required for intestinal colonization and attaching and effacing activity of *E. coli* O157:H7 in piglets and calves (Dean-Nystrom, Bosworth, Moon, & O'Brien, 1998). A vaccine containing intimin$_{O157}$ was tested using neonatal piglets as a challenge model and it was demonstrated that piglets that ingested maternal antibodies against intimin$_{O157}$ were protected from colonization with an intimin producing *E. coli* O157:H7 strain (Dean-Nystrom, Gansheroff, Mills, Moon, & O'Brien, 2002). This provides evidence that this may be a viable candidate for an anti-*E. coli* O157:H7 vaccine in cattle, the main reservoir of enterohaemorrhagic *E. coli*. Another study looked at the use of the cell-binding domain of intimin or a truncated EHEC factor for adherence (Efa-1) as potential vaccines. Both were found to induce humoral immunity in calves but did not protect against intestinal colonization by *E. coli* O157:H7 and O26:H- upon subsequent challenge (van Diemen et al., 2007). Similarly, the same study showed that an inactivated vaccine comprising of formalin-killed *E. coli* O157:H7 was ineffective, despite IgG responses. Nonetheless, the authors concluded that it may be possible to use these antigens, provided appropriate exposure to the intestinal immune system can be achieved.

Potter et al. (2004) looked at the use of proteins involved in intestinal colonization as possible vaccine targets. In this case, the authors used the proteins Tir, EspA and EspB, which are part of a type III secretion system involved in bovine intestinal colonization. Cattle were immunised with supernatant proteins containing Esps and Tir and subsequently challenged with *E. coli* O157:H7 and the fecal shedding was monitored. It was found that fecal shedding of *E. coli* O157:H7 was significantly reduced following vaccination, and the number of animals in the group shedding the pathogen and the duration of shedding were also reduced. It was also suggested that this vaccine could be used as a vaccine for non-O157 serotypes as the type III secreted antigens are relatively conserved among non-O157 serotypes. However, a subsequent field trial of the vaccine in nine feedlots showed no significant association between vaccination and pen prevalence of fecal *E. coli* O157:H7 (Van Donkersgoed, Hancock, Rogan, & Potter, 2005). A number of possible reasons were suggested for this, including different preparation of the vaccine, different vaccination strategies, different pen sizes and different timeframes. Another study also investigated the use of highly purified recombinant EspA as a vaccine for calves, and while this induced antigen specific IgG and salivary IgA responses, these responses did not protect against intestinal colonization upon subsequent challenge with *E. coli* O157:H7 (Dziva et al., 2007).

Salmonella is another zoonotic pathogen where substantial efforts have been invested to develop an effective vaccine. The success of this is dependent on understanding how *Salmonella* infect their hosts and the host response. However, a major obstacle is that *Salmonella* pathogenicity is both serotype-dependent and host-dependent and the factors influencing serotype-host specificity are not well known (Barrow, 2007).

Salenvac® is a commercially available, killed iron-restricted *S. enteriditis* PT4 vaccine which has been used as part of control programmes to reduce the burden of *S. enteriditis* infection of poultry flocks. A laboratory trial showed that the vaccine was successful in decreasing egg contamination (5.4–7.4% had culture positive shells in comparison to 16.7% for unvaccinated birds) and tissue colonization subsequent to intravenous *S. enteriditis* challenge (Woodward, Gettinby, Breslin, Corkish, & Houghton, 2002). However, it has been suggested that oral or respiratory challenge would have been more relevant (Barrow, 2007). Another study evaluated the efficacy of Salenvac® T which is made up of inactivated *S.* Typhimurium and *S. enteriditis* which have been grown under iron restriction. In this case the chickens were orally challenged with *S.* Typhimurium. Vaccination resulted in a significant reduction in the shedding of *S.* Typhimurium (Clifton-Hadley et al., 2002). Liu, Yang, Chung, & Kwang (2001) used formalin-inactivated *S. enteriditis* encapsulated in biodegradable microspheres as a vaccine and dosed chickens either orally or via an intramuscular route and also challenged them by these routes. It was found that shedding and colonization by *S. enteriditis* was significantly decreased in the vaccinated birds in comparison to the control birds. When challenged intramuscularly vaccinated birds were 27.9% feces positive and 18.7% organ positive for *S. enteriditis* in comparison to 59.3% feces positive and 44% organ positive for nonvaccinated chickens.

A number of live *Salmonella* vaccines have also been developed, some of which are commercially available. TAD Salmonella vac® E and TAD Salmonella vac® T are metabolic drift mutants of *S. enteriditis* and *S.* Typhimurium respectively, which were produced by chemical mutagenesis (Linde, Beer, & Bondarenko, 1990). It was shown in laboratory studies that these vaccines used either singly or in combination reduced organ and reproductive tract colonization and internal egg contamination, in comparison to control birds. This indicates cross protection provided by the TAD Salmonella vac® T vaccine (Gantois et al., 2006). Another example of a live *Salmonella* vaccine for use in chickens was described by Cerquetti and Gherardi (2000). Trials showed that in the case of *S. enteriditis* and *S.* Gallinarum there was a significant decrease in colonization of the cecum and also reduced colonization by *S.* Typhimurium but not to a significant degree. A recent study has looked at using strains harbouring mutants in the *Salmonella* pathogenicity islands 1 and 2 (either *hilA, sipA,* or *ssrA*) as protective vaccines against *S. enteriditis* challenge in newly hatched chicks. While the *sipA* and *ssrA* mutants were found to protect against challenge strain colonization of the cecum and internal organs they were not deemed to be useful due to the vaccine strains' persistent colonization throughout the study. However, the *hlyA* mutant strain was found to confer protection against colonization by the challenge strain and the vaccine strain could not be detected in the cecum four weeks post inoculation. The authors proposed that the longer the vaccine strain colonized the intestine, the longer the protection against virulent *Salmonella*. Therefore, the challenge is to provide a balance between colonization of the vaccine and clearance of the vaccine before slaughter to exploit colonization inhibition as a protection mechanism (Bohez et al., 2007).

The investigation of *Salmonella* vaccines has not just been limited to poultry with numerous studies being carried out in other animal models. One such example is the use of an attenuated *S.* Typhimurium strain which has a mutation in DNA adenine methylase in calves. It was found that it provided protection against subsequent virulent *S.* Typhimurium challenge via adaptive immunity and competitive exclusion (Dueger, House, Heithoff, & Mahan, 2003a). This strain was also found to provide protection in avian models (Dueger, House, Heithoff, & Mahan, 2001, 2003b). Pigs usually do not develop clinical salmonellosis, but as they can be carriers and shedders they can be a reservoir for the disease in humans. A metabolic drift mutant of *S*. Typhimurium was used as an oral vaccine for piglets that were subsequently challenged with a highly virulent *S*. Typhimurium strain. It was found that vaccinated animals shed substantially smaller amounts of the challenge strain and for a shorter period of time (Roesler et al., 2004). A similar trend was seen when the same strain was tested in poultry (Linde, Hahn, & Vielitz, 1996).

Subunit vaccines, i.e., vaccines which only contain individual proteins which will act as antigens, have received much less attention as a control agent for *Salmonella* and appear to have focused on using outer membrane proteins. Two studies of such vaccines were shown to either reduce colonization of *S. enteriditis* to the chicken intestinal mucosa following challenge (Khan, Fadl, & Venkitanarayanan, 2003) or had significantly reduced shedding of the challenge organism (Meenakshi et al., 1999).

There are presently no commercially available vaccines against *Campylobacter* in poultry. The development of such vaccines is hampered by a number of factors including the antigenic variety of strains and the lack of knowledge of antigens which induce a protective immune response. Also, there is a need to provide protection in the very early days of life as *Campylobacter* infection can occur at a very early stage (Wagenaar, Mevius, & Havelaar, 2006). However, a number of strategies have been or are being investigated and are reviewed by de Zoete, van Putten, & Wagenaar (2007). Killed whole cell vaccines have been examined and have shown mixed results. In one study, formalin-inactivated *C. jejuni* was administered with and without *E. coli* heat-labile toxin and after challenge by the homologous strain it was found that lower numbers of *C. jejuni* were isolated from the cecum (Rice, Rollins, Mallinson, Carr, & Joseph, 1997). However, another study which also used formalin-inactivated *C. jejuni* did not result in reduced cecal colonization upon subsequent homologous strain challenge (Cawthraw, Ayling, Nuijten, Wassenaar, & Newell, 1998). Subunit vaccines have also been looked at, and again have had variable levels of success. A number of studies have looked at using flagellin, which is involved in colonization of *Campylobacter* in the chicken gut. One such study by Widders et al. (1998) found that birds immunized twice intraperitoneally with killed *C. jejuni* and purified flagellin showed a significant reduction in cecal colonization, whereas, birds that received a second immunization orally, or birds immunized twice with flagellin alone did not show significant reductions in cecal colonization of *C. jejuni*. Another study created a fusion protein where flagellin was fused to the B-subunit of the labile toxin of *E. coli* and was administered as a vaccine to chickens that were subsequently challenged with *C. jejuni*. It was found that there was significantly less colonization in comparison to the control birds (Khoury & Meinersmann, 1995). A cocktail of attenuated live *C. jejuni* strains has also been used, but subsequent challenge with the parent strain did not result in reduced colonization compared to control birds (Ziprin, Hume, Young, & Harvey, 2002). These studies clearly indicate that there is a long way to go in designing an efficient *Campylobacter* vaccine.

A major advantage of vaccines as a biocontrol agent is that they can be used as a pre-harvest intervention. Their use has the potential to reduce pathogen carriage, thereby lessening the likelihood of contamination and occurrence of horizontal transfer which allows the pathogen into the food chain. However, the overall results from vaccine trials to reduce the pathogen load in food animals have been mixed and results appear to be very specific to the host used. Arguments also exist about what type of vaccine is best to use-live, killed or subunit vaccines. In the case of *Salmonella*, live vaccines are seen to be more efficacious but concerns such as public acceptability and safety do exist. Killed vaccines are considered safer but appear be less efficient at prompting a protective immune response. Subunit vaccines require an indepth understanding of the pathogen's interaction with its host, in order to choose an effective antigen. Overall, the development of effective vaccines against foodborne pathogens for animals still faces many hurdles.

Concluding Remarks

The key role which meat and meat products play as a vehicle of foodborne zoonotic pathogens is in no doubt. Microbial contamination can occur at all areas of the farm to fork chain. Numerous chemical methods can be employed to remove microbial contamination but more and more there is an increased consumer demand for foods with less additives and an interest in the use of alternative biological agents. It has been shown in this review that numerous potential options are available which can be applied at various points in the food chain and there is substantial research being invested in the area of biocontrol agents. Many of these are in the early stages of development, but some have been employed in food products or in animal trials, with varying degrees of success. What is very clear is that the success of a particular biocontrol agent depends very much on the food matrix, the target pathogen and the conditions used, and a biocontrol which works successfully in one food or animal environment cannot necessarily be extrapolated to another food product type. From a meat perspective, the potential for the use of biocontrol agents is huge as has been exemplified in this review. However, much work remains to be done to tailor these agents for particular products and pathogens and will no doubt be the focus of much research in the years to come.

References

Aasen, I. M., Markussen, S., Moretro, T., Katla, T., Axelsson, L., & Naterstad, K. (2003). Interactions of the bacteriocins sakacin P and nisin with food constituents. *International Journal of Food Microbiology, 87*, 35–43.

Abee, T., Krockel, L., & Hill, C. (1995). Bacteriocins: Modes of action and potentials in food preservation and control of food poisoning. *International Journal of Food Microbiology, 28*, 169–185.

Albano, H., Todorov, S. D., van Reenen, C. A., Hogg, T., Dicks, T. M. T., & Teixeira, P. (2007). Characterization of two bacteriocins produced by *Pediococcus acidilactici* isolated

from "Alheira", a fermented sausage traditionally produced in Portugal. *International Journal of Food Microbiology, 116*, 239–247.
Alisky, J., Iczkowski, K., Rapoport, A., & Troitsky, N. (1998). Bacteriophages show promise as antimicrobial agents. *The Journal of Infection, 36*, 5–15.
Alves, V. F., Martinez, R. C. R., Lavrador, M. A. S., & De Martinis, E. C. P. (2006). Antilisterial activity of lactic acid bacteria inoculated on cooked ham. *Meat Science, 74*, 623–627.
Anang, D. M., Rusul, G., Bakar, J., & Ling, F. H. (2007). Effects of lactic acid and lauricidin on the survival of *Listeria monocytogenes, Salmonella enteritidis* and *Escherichia coli* O157:H7 in chicken breast stored at 4C. *Food Control, 18*, 961–969.
Ananou, S., Galvez, A., Martinez-Bueno, M., Maqueda, M., & Valdivia, E. (2005). Synergistic effect of enterocin AS-48 in combination with outer membrane permeabilizing against *Escherichia coli* O157:H7. *Journal Applied Microbiology, 99*, 1364–1372.
Ananou, S., Garriga, M., Hugas, M., Maqueda, M., Martinez-Bueno, M., Galvez, A., et al. (2005). Control of *Listeria monocytogenes* in model sausages by enterocin AS-48. *International Journal of Food Microbiology, 103*, 179–190.
Ananou, S., Maqueda, M., Martinez-Bueno, M., Galvez, A., & Valdivia, E. (2005). Control of *Staphylococcus aureus* in sausages by enterocin AS-48. *Meat Science, 71*, 549–556.
Anderson, R. C., & Yu, P.-L. (2005). Factors affecting the antimicrobial activity of ovine-derived cathelicidins against *E. coli* O157:H7. *International Journal of Antimicrobial Agents, 25*, 205–210.
Annamalai, T., Venkitanarayanan, K. S., Hoagland, T. A., & Khan, M. I. (2001). Inactivation of *Escherichia coli* O157: H7 and *Listeria monocytogenes* by PR-26, a synthetic antibacterial peptide. *Journal of Food Protection, 64*, 1929–1934.
Anzai, K., Hamasuna, M., Kadono, H., Lee, S., Aoyagi, H., & Kirino, Y. (1991). Formation of ion channels in planar lipid bilayer membranes by synthetic basic peptides. *Biochimica et Biophysica Acta, 1064*, 256–266.
Appendini, P., & Hotchkiss, J. H. (1999). Antimicrobial activity of a 14-residue peptide against *Escherichia coli* O157:H7. *Journal of Applied Microbiology, 87*, 750–756.
Ariyapitipun, T., Mustapha, A., & Clarke, A. D. (2000). Survival of *Listeria monocytogenes* Scott A on vacuum-packaged raw beef treated with polylactic acid, lactic acid and nisin. *Journal of Food Protection, 63*, 131–136.
Arlindo, S., Calo, P., Franco, C., Prado, M., Cepeda, A., & Barros-Velazquez, J. (2006). Single nucleotide polymorphism analysis of the enterocin P structural gene of *Enterococcus faecium* strains isolated from nonfermented animal foods. *Molecular Nutrition and Food Research, 50*, 1229–1238.
Arques, J. L., Fernandez, J., Gaya, P., Nunez, M., Rodrigues, E., & Medina, M. (2004). Antimicrobial activity of reuterin in combination with nisin against food-borne pathogens. *International Journal of Food Microbiology, 95*, 225–229.
Atterbury, R. J., Connerton, P. L., Dodd, C. E., Rees, C. E., & Connerton, I. F. (2003). Application of host-specific bacteriophages to the surface of chicken skin leads to a reduction in recovery of *Campylobacter jejuni*. *Applied and Environmental Microbiology, 69*, 6302–6306.
Azuma, T., Bagenda, D. K., Yamamoto, T., Kawai, Y., & Yamazaki, K. (2006). Inhibition of *Listeria monocytogenes* by freeze-dried Piscicocin CS526 fermentate in food. *Letters in Applied Microbiology, 44*, 138–144.
Bach, S. J., McAllister, T. A., Viera, D. M., Gannon, V. P. J., & Holley, R. A. (2002). Transmission and control of *Escherichia coli* O157:H7. A review. *Canadian Journal of Animal Science, 82*, 475–490.
Balogh, B., Jones, J. B., Momol, M. T., Olson, S. M., Obradovic, A., King, P., et al. (2003). Improved efficacy of newly formulated bacteriophages for management of bacterial spot on tomato. *Plant Disease, 87*, 949–954.
Barakat, R. K., Griffiths, M. W., & Harris, L. J. (2000). Isolation and characterization of *Carnobacterium, Lactococcus* and *Enterococcus* spp. from cooked, modified atmosphere packaged, refrigerated, poultry meat. *International Journal of Food Microbiology, 62*, 83–94.

Barboza De Martinez, Y., Ferrer, K., & Marquez Salas, E. (2002). Combined effects of lactic acid and nisin solution in reducing levels of microbiological contamination in red meat carcasses. *Journal of Food Protection, 65*, 1780–1783.

Barrow, P. A. (2007). *Salmonella* infections: Immune and non-immune protection with vaccines. *Avian Pathology, 36*, 1–13.

Barrow, P. A., & Soothill, J. S. (1997). Bacteriophage therapy and prophylaxis: Rediscovery and renewed assessment of potential. *Trends in Microbiology, 5*, 268–271.

Benkerroum, N., Daoudi, A., & Kamal, M. (2003). Behaviour of *Listeria monocytogenes* in raw sausages (merguez) in presence of a bacteriocin-producing lactococcal strain as a protective culture. *Meat Science, 63*, 479–484.

Benkerroum, N., Daoudi, A., Thamraoui, T., Ghalfi, H., Thiry, G., Duroy, M., et al. (2005). Lyophilized preparation of bacteriocinogenic *Lactobacillus curvatus* and *Lactococcus lactis* subsp. *lactis* as potential protective adjuncts to control *Listeria monocytogenes* in dry-fermented sausages. *Journal of Applied Microbiology, 98*, 56–63.

Berchieri, A., Jr., Lovell, M. A., & Barrow, P. A. (1991). The activity in the chicken alimentary tract of bacteriophages lytic for *Salmonella typhimurium*. *Research in Microbiology, 142*, 541–549.

Bohez, L., Ducatelle, R., Pasmans, F., Haesebrouck, F., & Van Immerseel, F. (2007). Long-term colonisation-inhibition studies to protect broilers against colonisation with *Salmonella* Enteritidis, using *Salmonella* Pathogenicity Island 1 and 2 mutants. *Vaccine, 25*, 4235–4243.

Brackett, R. E., Hao, Y. Y., & Doyle, M. P. (1994). Ineffectiveness of hot acid sprays to decontaminate *Escherichia coli* O157:H7 on beef. *Journal of Food Protection, 57*, 198–203.

Brashears, M. M., Jaroni, D., & Trimble, J. (2003). Isolation, selection, and characterization of Lactic acid bacteria for a competitive exclusion product to reduce shedding of *Escherichia coli* O157:H7 in cattle. *Journal of Food Protection, 66*, 355–363.

Brashears, M. M., Reilly, S. S., & Gilliland, S. E. (1998). Antagonistic action of cells of *Lactobacillus lactis* toward *Escherichia coli* O157:H7 on refrigerated raw chicken meat. *Journal of Food Protection, 61*, 166–170.

Bredholt, S., Nesbakken, T., & Holck, A. (1999). Protective cultures inhibit growth of *Listeria monocytogenes* and *Escherichia coli* O157:H7 in cooked, sliced, vacuum- and gas- packaged meat. *International Journal of Food Microbiology, 53*, 43–52.

Bredholt, S., Nesbakken, T., & Holck, A. (2001). Industrial application of an antilisterial strain of *Lactobacillus sakei* as a protective culture and its effect on the sensory acceptibilty of cooked, sliced, vacuum-packaged meats. *International Journal of Food Microbiology, 66*, 191–196.

Budde, B. B., Hornbaek, T., Jacobsen, T., Barkholt, V., & Koch, A. G. (2003). *Leuconostoc carnosum* 4010 has the potential for use as a protective culture for vacuum-packed meats: Culture isolation, bacteriocin identification and meat application experiments. *International Journal of Food Microbiology, 83*, 171–184.

Burt, S. (2004). Essential oils: Their antibacterial properties and potential applications in foods – A review. *International Journal of Food Microbiology, 94*, 223–253.

Burt, S. A., & Reinders, R. D. (2003). Antibacterial activity of selected plant essential oils against *Escherichia coli* O157:H7. *Letters in Applied Microbiology, 36*, 162–167.

Callaway, T. R., Anderson, R. C., Edrington, T. S., Genovese, K. J., Harvey, R. B., & Poole, T. L., et al. (2004). Recent pre-harvest supplementation strategies to reduce carriage and shedding of zoonotic enteric bacterial pathogens in food animals. *Animal Health Research Reviews, 5*, 35–47.

Carlton, R. M., Noordman, W. H., Biswas, B., de Meester, E. D., & Loessner, M. J. (2005). Bacteriophage P100 for control of *Listeria monocytogenes* in foods: Genome sequence, bioinformatic analyses, oral toxicity study, and application. *Regulatory Toxicology and Pharmacology, 43*, 301–312.

Castellano, P. H., Holzapfel, W. H., & Vignolo, G. (2004). The control of *Listeria innocua* and *Lactobacillus sakei* in broth and meat slurry with the bacteriocinogenic strain *Lactobacillus casei* CRL705. *Food Microbiology, 21*, 291–298.

Castellano, P., & Vignolo, G. (2006). Inhibition of *Listeria innocua* and *Brochothrix thermosphacta* in vacuum-pakced meat by addition of bacteriocinogenic *Lactobacillus curvatus* CRL705 and its bacteriocins. *Letters in Applied Microbiology, 43*, 194–199.

Castillo, A., Lucia, L. M., Mercado, I., & Acuff, G. R. (2001). In-plant evaluation of a lactic acid treatment for reduction of bacteria on chilled beef carcasses. *Journal of Food Protection, 64*, 738–740.

Castillo, L. A., Meszaros, L., & Kiss, I. F. (2004). Effect of high hydrostatic pressure and nisin on microorganisms in minced meats. *Acta Alimentaria, 33*, 183–190.

Cawthraw, S., Ayling, R., Nuijten, P., Wassenaar, T., & Newell, D. G. (1998). Prior infection, but not a killed vaccine, reduces colonization of chickens by *Campylobacter jejuni*. In A. J. Lastovica, D. G. Newell, and E. E. Lastovica (Eds.), *Campylobacter, Helicobacter and related organisms* (pp. 364–372). Cape Town: Institute of Child Health, University of Cape Town.

Center for Disease Control and Prevention (2006). Multistate outbreak of *Salmonella typhimurium* infections associated with eating ground beef–United States, 2004. *MMWR. Morbidity and Mortality Weekly Report, 55*, 180–182.

Cerquetti, M. C., & Gherardi, M. M. (2000). Orally administered attenuated *Salmonella* enteritidis reduces chicken cecal carriage of virulent *Salmonella* challenge organisms. *Veterinary Microbiology, 76*, 185–192.

Chen, H.-C., & Stern, N. J. (2001). Competitive exclusion of heterologous *Campylobacter* spp. in chicks. *Applied And Environmental Microbiology, 67*, 848–851.

Chen, H., & Hoover, D. G. (2003). Bacteriocins and their food applications. *Comprehensive Reviews in Food Science and Food Safety, 2*, 82–100.

Chiasson, F., Borsa, J., Ouattara, B., & Lacroix, M. (2004). Radiosensitization of *Escherichia coli* and *Salmonella* Typhi in ground beef. *Journal of Food Protection, 67*, 1157–1162.

Cleveland, J., Montville, T. J., Nes, I. F., & Chikindas, M. L. (2001). Bacteriocins: Safe, natural antimicrobials for food preservation. *International Journal of Food Microbiology, 71*, 1–20.

Clifton-Hadley, F. A., Breslin, M., Venables, L. M., Sprigings, K. A., Cooles, S. W., Houghton, S., et al. (2002). A laboratory study of an inactivated bivalent iron restricted *Salmonella enterica* serovars Enteritidis and Typhimurium dual vaccine against Typhimurium challenge in chickens. *Veterinary Microbiology, 89*, 167–179.

Cole, K., Farnell, M. B., Donoghue, A. M., Stern, N. J., Svetoch, E. A., Eruslanov, B. N., Volodina, L. I., Kovalev, N., Perelygin V. V., Mitsevich, E. V., Mitserich, P., Levchuk, V. P., Pokhilenkov, D., Borzenkov, V. N., Svetoch, O. E., Kudrgavtseva, T. Y., Reyes-Herreia, I., Blore, P. J., Solis De Los Santos, F., & Donoghue, D. J. (2006). Bacteriocins reduce *Campylobacter* colonization and alter gut morphology in turkey poults. *Poultry Science, 85*, 1570–1575.

Cooksey, K. (2005). Effectiveness of antimicrobial food packaging materials. *Food Additives and Contaminants, 22*, 980–987.

Cotter, P. D., Hill, C., & Ross, R. P. (2005). Bacteriocins: Developing innate immunity for food. *Nature Reviews. Microbiology, 3*, 777–788.

Davies, R. H., & Breslin, M. F. (2003). Observations on the distribution and persistence of *Salmonella enterica* serovar Enteritidis phage type 29 on a cage layer farm before and after the use of competitive exclusion treatment. *British Poultry Science, 44*, 551–557.

De Martinis, E. C. P., Publio, M. R. P., Santarosa, P. R., & Freitas, F. Z. (2001). Antilisterial activity of lactic acid bacteria isolated from vacuum-packaged Brazilian meat and meat products. *Brazilian Journal of Microbiology, 32*, 32–37.

de Vrese, M., & Schrezenmeir, J. (2002). Probiotics and non-intestinal infectious conditions. *The British Journal of Nutrition, 88*, S59–S66.

de Zoete, M. R., van Putten, J. P. M., & Wagenaar, J. A. (2007). Vaccination of chickens against *Campylobacter*. *Vaccine, 25*, 5548–5557.

Dean-Nystrom, E. A., Bosworth, B. T., Moon, H. W., & O'Brien, A. D. (1998). *Escherichia coli* O157:H7 requires intimin for enteropathogenicity in calves. *Infection and Immunity, 66*, 4560–4563.

Dean-Nystrom, E. A., Gansheroff, L. J., Mills, M., Moon, H. W., & O'Brien, A. D. (2002). Vaccination of pregnant dams with intimin (O157) protects suckling piglets from *Escherichia coli* O157:H7 infection. *Infection and Immunity, 70*, 2414–2418.

Deegan, L. H., Cotter, P. D., Hill, C., & Ross, R. P. (2006). Bacteriocins: Biological tools for bio-preservation and shelf-life extension. *International Dairy Journal, 16*, 1058–1071.

Diez-Gonzalez, F. (2007). Applications of bacteriocins in livestock. *Current Issues in Intestinal Microbiology, 8*, 15–24.

Dorsa, W. J. (1997). New and established carcass decontamination procedures commonly used in the beef-processing industry. *Journal of Food Protection, 60*, 1146–1151.

Doyle, M. P., & Erickson, M. C. (2006). Reducing the carriage of foodborne pathogens in livestock and poultry. *Poultry Science, 85*, 960–973.

Dueger, E. L., House, J. K., Heithoff, D. M., & Mahan, M. J. (2001). *Salmonella* DNA adenine methylase mutants elicit protective immune responses to homologous and heterologous serovars in chickens. *Infection and Immunity, 69*, 7950–7954.

Dueger, E. L., House, J. K., Heithoff, D. M., & Mahan, M. J. (2003a). *Salmonella* DNA adenine methylase mutants elicit early and late onset protective immune responses in calves. *Vaccine, 21*, 3249–3258.

Dueger, E. L., House, J. K., Heithoff, D. M., & Mahan, M. J. (2003b). *Salmonella* DNA adenine methylase mutants prevent colonization of newly hatched chickens by homologous and heterologous serovars. *International Journal of Food Microbiology, 80*, 153–159.

Dykes, G. A., & Moorhead, S. M. (2002). Combined antimicrobial effect of nisin and a listeriophage against *Listeria monocytogenes* in broth but not in buffer or on raw beef. *International Journal of Food Microbiology, 73*, 71–81.

Dziva, F., Vlisidou, I., Crepin, V. F., Wallis, T. S., Frankel, G., & Stevens, M. P. (2007). Vaccination of calves with EspA, a key colonisation factor of *Escherichia coli* O157:H7, induces antigen-specific humoral responses but does not confer protection against intestinal colonisation. *Veterinary Microbiology, 123*, 254–261.

Elder, R. O., Keen, J. E., Siragusa, G. R., Brarkocy-Gallagher, G. A., Koohmaraie, M., & Lagreid, W. W. (2000). Correlation of enterohemorrhagic *Escherichia coli* O157 prevalance in feces, hides and carcasses of beef cattle during processing. *Proceedings of the National Academy of Sciences of the United States of America, 97*, 2999–3003.

Enan, G. (2006a). Inhibition of *Clostridium perfringens* LMG 11264 in meat samples of chicken, turkey and beef by the bacteriocin Plantarcin UG1. *International Journal OF Poultry Science, 5*, 195–200.

Enan, G. (2006b). Behaviour of *Listeria monocytogenes* LMG 10470 in poultry meat and its control by the bacteriocin Plantaricin UG 1. *International Journal of Poultry Science, 5*, 355–359.

Etcheverria, A. I., Arroyo, G., Perdigon, G., & Parma, A. E. (2005). *Escherichia coli* with anti-O157:H7 activity isolated from bovine colon. *Journal of Applied Microbiology, 100*, 384–389.

Federal Register (2006). Food additives Permitted for Direct Addition to Food for Human Consumption; Bacteriophage Preparation. *Federal Register, 71*, 47729–47732.

Figueroa-Bossi, N., & Bossi, L. (1999). Inducible prophages contribute to *Salmonella* virulence in mice. *Molecular Microbiology, 33*, 167–176.

Fratamico, P. M., Schultz, F. J., Benedict, R. C., & Buchanan, R. L. (1996). Factors influencing attachment of *Escherichia coli* O157:H7 to beef tissues and removal using selected sanitizing rinses. *Journal of Food Protection, 59*, 453–459.

Gaeng, S., Scherer, S., Neve, H., & Loessner, M. J. (2000). Gene cloning and expression and secretion of *Listeria monocytogenes* bacteriophage-lytic enzymes in *Lactococcus lactis*. *Applied Environmental Microbiology, 66*, 2951–2958.

Galvez, A., Abriouel, H., Lopez, R. L., & Omar, N. B. (2007). Bacteriocin-based strategies for food biopreservation. *International Journal of Food Microbiology*, In press.

Gantois, I., Ducatelle, R., Timbermont, L., Boyen, F., Bohez, L., Haesebrouck, F., et al. (2006). Oral immunisation of laying hens with the live vaccine strains of TAD Salmonella vac E and

TAD Salmonella vac T reduces internal egg contamination with *Salmonella* Enteritidis. *Vaccine, 24,* 6250–6255.

Ganzle, M. G., Hertel, C., van der Vossen, J. M. B. M., & Hammes, W. P. (1999). Effect of bacteriocin-producing lactobacilli on the survival of *Escherichia coli* and *Listeria* in a dynamic model of the stomach and the small intestine. . *International Journal of Food Microbiology, 48,* 21–35.

Ganzle, M. G., Weber, S., & Hammes, W. P. (1999). Effect of ecological factors on the inhibtory spectrum and activity of bacteriocins. *International Journal of Food Microbiology, 46,* 207–217.

Genovese, K. J., Anderson, R. C., Harvey, R. B., Callaway, T. R., Poole, T. L., Edrington, T. S., et al. (2003). Competitive exclusion of *Salmonella* from the cut of neonatal and weaned pigs. *Journal of Food Protection, 66,* 1353–1359.

Ghalfi, H., Benkerroum, N., Doguiet, D. D. K., Bensaid, M., & Thonart, P. (2007). Effectiveness of cell-adsorbed bacteriocin produced by *Lactobacillus curvatus* CWBI-B28 and selected essential oils to control *Listeria monocytogenes* in pork meat during cold storage. *Letters in Applied Microbiology, 44,* 268–273.

Gill, A. O., Delaquis, P. J., Russo, P., & Holley, R. A. (2002). Evaluation of antilisterial action of cilantro oil on vacuum packed ham. *International Journal of Food Microbiology, 73,* 83–92.

Gillespie, I. A., O'Brien, S. J., Adak, G. K., Cheasty, T., & Willshaw, G. (2005). Foodborne general outbreaks of Shiga toxin-producing *Escherichia coli* O157 in England and Wales 1992–2002: where are the risks? *Epidemiology and Infection, 133,* 803–808.

Goode, D., Allen, V. M., & Barrow, P. A. (2003). Reduction of experimental *Salmonella* and *Campylobacter* contamination of chicken skin by application of lytic bacteriophages. *Applied Environmental Microbiology, 69,* 5032–5036.

Goodridge, L., Chen, J., & Griffiths, M. (1999). The use of a fluorescent bacteriophage assay for detection of *Escherichia coli* O157:H7 in inoculated ground beef and raw milk. *International Journal of Food Microbiology, 47,* 43–50.

Greer, G. G., & Dilts, B. D. (1992). Factors affecting the susceptibilitiy of meatborne pathogens and spoilage bacteria to organic acids. *Food Research International, 25,* 355–364.

Greer, G. G., & Dilts, B. D. (2002). Control of *Brochothrix thermosphacta* spoilage of pork adipose tissue using bacteriophages. *Journal of Food Protection, 65,* 861–863.

Greer, G. G., & Dilts, B. D. (2006). Control of meatborne *Listeria monocytogenes* and *Brochothrix thermosphacta* by a bacteriocinogenic *Brochothrix campestris* ATCC 43754. *Food Microbiol, 23,* 785–790.

Hammer, K. A., Carson, C. F., & Riley, T. V. (1999). Antimicrobial activity of essential oils and other plant extracts. *Journal of Applied Microbiology, 86,* 985–990.

Hancock, R. E., & Sahl, H. G. (2006). Antimicrobial and host-defense peptides as new anti-infective therapeutic strategies. *Nature Biotechnology, 24,* 1551–1557.

Hanlon, G. W. (2007). Bacteriophages: An appraisal of their role in the treatment of bacterial infections. *International Journal of Antimicrobial Agents, 30,* 118–128.

Hao, Y. Y., Brackett, R. E., & Doyle, M. P. (1998a). Efficacy of plants extracts in inhibiting *Aeromonas hydrophilia* and *Listeria monocytogenes* in refrigerated cooked poultry. *Food Microbiol, 15,* 367–378.

Hao, Y. Y., Brackett, R. E., & Doyle, M. P. (1998b). Inhibition of *Listeria monocytogenes* and *Aeromonas hydrophila* by plant extracts in refrigerated cooked beef. *Journal of Food Protection, 61,* 307–312.

Hardin, M. D., Acuff, G. R., Lucia, L. M., Oman, J. S., & Savell, J. W. (1995). Comparison of methods for decontamination from beef carcass surfaces. *Journal of Food Protection, 58,* 368–374.

Hayes, M., Ross, R. P., Fitzgerald, G. F., Hill, C., & Stanton, C. (2006). Casein-derived antimicrobial peptides generated by *Lactobacillus acidophilus* DPC6026. *Applied and Environmental Microbiology, 72,* 2260–2264.

Haynie, S. L., Crum, G. A., & Doele, B. A. (1995). Antimicrobial activities of amphiphilic peptides covalently bonded to a water-insoluble resin. *Antimicrobial Agents and Chemotherapy, 39,* 301–307.

Heller, C. E., Scanga, J. A., Sofos, J. N., Belk, K. E., Warren-Serna, W., Bellinger, G. R., et al. (2007). Decontamination of beef subprimal cuts intended for blade tenderization or moisture enhancement. . *Journal of Food Protection, 70*, 1174–1180.

Higgins, J. P., Higgins, S. E., Guenther, K. L., Huff, W., Donoghue, A. M., Donoghue, D. J., et al. (2005). Use of specific bacteriophage treatment to reduce *Salmonella* in poultry products. *Poultry Science, 84*, 1141–1145.

Higgs, R., Lynn, D., Cahalane, S., Alaña, I., Hewage, C., James, T., et al. (2007). Modification of chicken avian β-defensin-8 at positively selected amino acid sites enhances specific antimicrobial activity. *Immunogenetics, 59*, 573–580.

Holzapfel, W. H., Geisen, R., & Schillinger, U. (1995). Biological preservation of foods with reference to protective cultures, bacteriocins and food-grade enzymes. *International Journal of Food Microbiology, 24*, 343–362.

Hoszowski, A., & Truszczynski, M. (1997). Prevention of *Salmonella typhimurium* caecal colonisation by different preparations for competitive exclusion. *Comparative Immunology, Microbiology and Infectious Diseases, 20*, 111–117.

Hsu, F. C., Shieh, Y. S., & Sobsey, M. D. (2002). Enteric bacteriophages as potential fecal indicators in ground beef and poultry meat. *Journal of Food Protection, 65*, 93–99.

Huffman, R. D. (2002). Current and future technologies for the decontamination of carcasses and fresh meat. *Meat Science, 62*, 285–294.

Hugas, M., Pages, F., Garriga, M., & Monfort, J. M. (1998). Application of the bacteriocinogenic *Lactobacillus sakei* CTC494 to prevent growth of *Listeria* in fresh and cooked meat products packed with different atmospheres. *Food Microbiology, 15*, 639–650.

Jacobsen, T., Budde, B. B., & Koch, A. G. (2003). Application of *Leuconostoc carnosum* for biopreservation of cooked meat products. *Journal of Applied Microbiology, 95*, 242–249.

Joerger, R. D. (2003). Alternatives to antibiotics: Bacteriocins, antimicrobial peptides and bacteriophages. *Poultry Science, 82*, 640–647.

Katla, T., Moretro, T., Sveen, I., Assen, I. M., Axelsson, L., Rorvik, L. M., & Naterstad, K. (2002). Inhibition of *Listeria monocytogenes* by addition of Sakacin P and Sakacin P-producing *Lactobacillus sakei*. *Journal of Applied Microbiology, 83*, 191–196.

Khan, M. I., Fadl, A. A., & Venkitanarayanan, K. S. (2003). Reducing colonization of *Salmonella* Enteritidis in chicken by targeting outer membrane proteins. *Journal of Applied Microbiology, 95*, 142–145.

Khoury, C. A., & Meinersmann, R. J. (1995). A genetic hybrid of the *Campylobacter jejuni flaA* gene with LT-B of *Escherichia coli* and assessment of the efficacy of the hybrid protein as an oral chicken vaccine. *Avian Diseases, 39*, 812–820.

Koohmaraie, M., Arthur, T. M., Bosilevac, J. M., Guerini, M., Shackelford, S. D., & Wheeler, T. L. (2005). Post-harvest interventins to reduce/eliminate pathogens in beef. *Meat Science, 71*, 79–91.

Kostrzynska, M., & Bachand, A. (2006). Use of microbial antagonism to reduce pathogen levels on produce and meat products: A review. *Canadian Journal of Microbiology, 52*, 1017–1026.

Kudva, I. T., Jelacic, S., Tarr, P. I., Youderian, P., & Hovde, C. J. (1999). Biocontrol of *Escherichia coli* O157 with O157-specific bacteriophages. *Applied Environmental Microbiology, 65*, 3767–3773.

Kuleasan, H., & Cakmakci, M. L. (2002). Effect of reuterin produced by *Lactobacillus reuteri* on the surface of sausages to inhibit the growth of *Listeria monocytogenes* and *Salmonella* spp. *Nahrung/Food, 46*, 408–410.

La Ragione, R. M., & Woodward, M. J. (2003). Competitive exclusion by *Bacillus subtilis* spores of *Salmonella enterica* serotype Enteritidis and *Clostridium perfringens* in young chickens. *Veterinary Microbiology, 94*, 245–256.

Lambert, R. J. W., Skandamis, P., Coote, P. J., & Nychas, G.-J. E. (2001). A study of the minimum inhibitory concentration and mode of action of oregano essential oil, thymol and carvacrol. *Journal of Applied Microbiology, 91*, 453–462.

LeJeune, J. T., & Wetzel, A. N. (2007). Preharvest control of *Escherichia coli* O157 in cattle. *Journal of Animal Science, 85*, E73–80.

Lemay, M. J., Choquette, J., Delaquis, P. J., Claude, G., Rodrigue, N., & Saucier, L. (2002). Antimicrobial effect of natural preservatives in a cooked and acidified chicken meat model. *International Journal of Food Microbiology, 78,* 217–226.

Lenski, R. E. (1988). Dynamics of interactions between bacteria and virulent bacteriophage *Advances in Microbial Ecology, 10,* 1–44.

Leroy, F., & De Vuyst, L. (2005). Simulation of the effect of sausage ingredients and technology on the functionality of the bacteriocin-producing *Lactobacillus sakei* CTC 494 strain. *International Journal of Food Microbiology, 100,* 141–152.

Leverentz, B., Conway, W. S., Camp, M. J., Janisiewicz, W. J., Abuladze, T., Yang, M., et al. (2003). Biocontrol of *Listeria monocytogenes* on fresh-cut produce by treatment with lytic bacteriophages and a bacteriocin. *Applied Environmental Microbiolgy, 69,* 4519–4526.

Linde, K., Beer, J., & Bondarenko, V. (1990). Stable *Salmonella* live vaccine strains with two or more attenuating mutations and any desired level of attenuation. *Vaccine, 8,* 278–282.

Linde, K., Hahn, I., & Vielitz, E. (1996). Entwicklung von optimal fur das Huhn attenuierten *Salmonella*-Lebendimpfstoffen. *Tierärztliche Umschau, 51,* 23–31.

Liu, W., Yang, Y., Chung, N., & Kwang, J. (2001). Induction of humoral immune response and protective immunity in chickens against *Salmonella* Enteritidis after a single dose of killed bacterium-loaded microspheres. *Avian Diseases, 45,* 797–806.

Loc Carrillo, C., Atterbury, R. J., El-Shibiny, A., Connerton, P. L., Dillon, E., Scott, A., et al. (2005). Bacteriophage Therapy To Reduce *Campylobacter jejuni* Colonization of Broiler Chickens. *Applied Environmental Microbiology, 66:* 220–225.

Loessner, M. J. (2005). Bacteriophage endolysins – current state of research and applications. *Current Opinion in Microbiology, 8,* 480–487.

Lopez-Exposito, I., Minervini, F., Amigo, L., & Recio, I. (2006). Identification of antibacterial peptides from bovine kappa-casein. *Journal of Food Protection, 69,* 2992–2997.

Lucke, F. K. (2000). Utilization of microbes to process and preserve meat. *Meat Science, 56,* 105–115.

Maciorowski, K. G., Pillai, S. D., & Ricke, S. C. (2001). Presence of bacteriophages in animal feed as indicators of fecal contamination. *Journal of Environmental Science and Health, 36,* 699–708.

Marcos, B., Aymerich, T., Monfort, J. M., & Garriga, M. (2007). Use of antimicrobial biodegradable packaging to control *Listeria monocytogenes* during storage of cooked ham. *International Journal of Food Microbiology,* In press.

Mattila, K., Saris, P., & Tyopponen, S. (2003). Survival of *Listeria monocytogenes* on sliced cooked sausage after treatment with pediocin AcH. *International Journal of Food Microbiology, 89,* 281–286.

Mauriello, G., Ercolini, D., La Storia, A., Casaburi, A., & Villani, F. (2004). Development of polythene films for food packaging activated with an antilisterial bacteriocin from *Lactobacillus curvatus* 32Y. *Journal of Applied Microbiology, 97,* 314–322.

Mazick, A., Ethelberg, S., Nielsen, E. M., Molbak, K., & Lisby, M. (2006). An outbreak of *Campylobacter jejuni* associated with consumption of chicken, Copenhagen, 2005. *Euro Surveillance, 11,* 137–139.

McCormick, J. K., Klaenhammer, T. R., & Stiles, M. E. (1999). Colicin V can be produced by lactic acid bacteria. *Letters in Applied Microbiology, 29,* 37–41.

Mead, P. S., Dunne, E. F., Graves, L., Wiedmann, M., Patrick, M., Hunter, S., et al. (2006). Nationwide outbreak of listeriosis due to contaminated meat. *Epidemiology and Infection, 134,* 744–751.

Mead, P. S., Slutsker, L., Dietz, V., McCaig, L. F., Bresee, J. S., Shapiro, C., et al. (1999). Food-related illness and death in the United States. *Emerging Infectious Diseases, 5,* 607–625.

Meenakshi, M., Bakshi, C. S., Butchaiah, G., Bansal, M. P., Siddiqui, M. Z., & Singh, V. P. (1999). Adjuvanted outer membrane protein vaccine protects poultry against infection with *Salmonella* Enteritidis. *Veterinary Research Communications, 23,* 81–90.

Milona, P., Townes, C. L., Bevan, R. M., & Hall, J. (2007). The chicken host peptides, gallinacins 4, 7, and 9 have antimicrobial activity against *Salmonella* serovars. *Biochemical and Biophysical Research Communications, 356,* 169–174.

Ming, X., Weber, G. H., Ayres, J. W., & Sandine, W. E. (1997). Bacteriocins applied to food packaging material to inhibit *Listeria monocytogenes* on meats. *Journal of Food Science, 62*, 413–415.
Mirold, S., Rabsch, W., Tschape, H., & Hardt, W. D. (2001). Transfer of the *Salmonella* type III effector *sopE* between unrelated phage families. *Journal of Molecular Biology, 312*, 7–16.
Morency, H., Mota-Meira, M., LaPointe, G., Lacroix, C., & Lavoie, M. C. (2001). Comparison of the activity spectra against pathogens of bacterial strains producing a mutacin or a lantibiotic. *Canadian Journal of Microbiology, 47*, 322–331.
Muthukumarasamy, P., Han, J. H., & Holley, R. A. (2003). Bactericidal effects of *Lactobacillus reuteri* and allyl isothiocyanate on *Escherichia coli* O157:H7 in refrigerated ground beef. *Journal of Food Protection, 66*, 2038–2044.
Naghmouchi, K., Kheadr, E., Lacroix, C., & Fliss, I. (2007). Class I/IIa bacteriocin cross-resistance phenomenon in *Listeria monocytogenes*. *Food Microbiology, 24*, 718–727.
Nava, G. M., Bielke, L. R., Callaway, T. R., & Castaneda, M. P. (2005). Probiotic alternatives to reduce gastrointestinal infections: The poultry experience. *Animal Health Research Reviews, 6*, 105–118.
Nieto-Lozano, J. C., Reguera-Useros, J. I., Pelaez-Martinez, M. C., & Hardisson de la Torre, A. (2006). Effect of a bacteriocin produced by *Pediococcus acidilactici* against *Listeria monocytogenes* and *Clostridium perfringenes* on Spanish raw meat. *Meat Science, 72*, 57–61.
Noonpakdee, W., Santivarangkna, C., Jumriangrit, P., Sonomoto, K., & Panyim, S. (2003). Isolation of nisin-producing *Lactococcus lactis* WNC20 strain from *nham*, a traditional Thai fermented sausage. *International Journal of Food Microbiology, 81*, 137–145.
O'Flynn, G., R. P.Ross, Fitzgerald, G. F., & Coffey, A. (2004). Evaluation of a Cocktail of Three Bacteriophages for Biocontrol of *Escherichia coli* O157:H7. *Applied Environmental Microbiology, 70*, 3417–3424.
O'Sullivan, L., Ross, R. P., & Hill, C. (2002). Potential of bacteriocin-producing lactic acid bacteria form improvements in food safety and quality. *Biochimie, 84*, 593–604.
O'Sullivan, L., Ryan, M. P., Ross, R. P., & Hill, C. (2003). Generation of food-grade Lactococcal starters which produce the Lantibiotics Lacticin 3147 and Lacticin 481. *Applied Environmental Microbiology, 68*, 3681–3685.
Ovchinnikova, T. V., Balandin, S. V., Aleshina, G. M., Tagaev, A. A., Leonova, Y. F., Krasnodembsky, E. D., et al. (2006). Aurelin, a novel antimicrobial peptide from jellyfish *Aurelia aurita* with structural features of defensins and channel-blocking toxins. *Biochemical and Biophysical Research Communications, 348*, 514–523.
Pao, S., Randolph, S. P., Westbrook, W., & Shen, H. (2004). Use of bacteriophages to control *Salmonella* in experimentally contaminated sprout seeds. *Journal of Food Science, 69*, 127–130.
Patel, J. R., Sanglay, G. C., Sharma, M., & Solomon, M. B. (2007). Combining antimicrobials and hydrodynamic pressure processing for control of *Listeria monocytogenes* in frankfurters. *Journal of Muscle Foods, 18*, 1—18.
Podda, E., Benincasa, M., Pacor, S., Micali, F., Mattiuzzo, M., Gennaro, R., et al. (2006). Dual mode of action of Bac7, a proline-rich antibacterial peptide. *Biochimica Et Biophysica Acta, 1760*, 1732–1740.
Podolak, R. K., Zayas, J. F., Kastner, C. L., & Fung, D. Y. C. (1996). Inhibition of *Listeria monocytogenes* and *Escherichia coli* O157:H7 on beef by application of organic acids. *Journal of Food Protection, 59*, 370–373.
Potter, A. A., Klashinsky, S., Li, Y., Frey, E., Townsend, H., Rogan, D., et al. (2004). Decreased shedding of *Escherichia coli* O157:H7 by cattle following vaccination with type III secreted proteins. *Vaccine, 22*, 362–369.
Prema, P., Bharathy, S., Palavesam, A., Sivasubramanian, M., & Immanuel, G. (2006). Detection, purification and efficacy of warnerin produced by *Staphylococcus warneri*. *World Journal of Microbiology & Biotechnology, 22*, 865–872.
Quiberoni, A., Suarez, V. B., & Reinheimer, J. A. (1999). Inactivation of *Lactobacillus helveticus* bacteriophages by thermal and chemical treatments. *Journal of Food Protection, 62*, 894–898.
Quintavalla, S., & Vicini, L. (2002). Antimicrobial food packaging in meat industry. *Meat Science, 62*, 373–380.

Raya, R. R., Varey, P., Oot, R. A., Dyen, M. R., Callaway, T. R., Edrington, T. S., et al. (2006). Isolation and characterization of a new T-even bacteriophage, CEV1, and determination of its potential to reduce *Escherichia coli* O157:H7 levels in sheep. *Applied Environmental Microbiology, 72,* 6405–6410.

Rice, B. E., Rollins, D. M., Mallinson, E. T., Carr, L., & Joseph, S. W. (1997). *Campylobacter jejuni* in broiler chickens: Colonization and humoral immunity following oral vaccination and experimental infection. *Vaccine, 15,* 1922–1932.

Riley, M. A., & Wertz, J. E. (2002). Bacteriocins: Evolution, ecology and application. *Annual Review of Microbiology, 56,* 117–137.

Rodriguez, E. J. L. A., Nunez, M., Gaya, P., & Medina, M. (2005). Combined effect of high-pressure treatments and bacteriocin-producing lactic acid bacteria on inactivation of *Escherichia coli* O157:H7 in raw-milk cheese. *Applied And Environmental Microbiology, 71,* 3399–3404.

Roesler, U., Marg, H., Schroder, I., Mauer, S., Arnold, T., Lehmann, J., et al. (2004). Oral vaccination of pigs with an invasive *gyrA-cpxA-rpoB Salmonella* Typhimurium mutant. *Vaccine, 23,* 595–603.

Russell, J. B., and Mantovan, H. C. (2002). The bacteriocins of ruminal bacteria and their potential as an alternative to antibiotics. *Journal of Molecular Microbiology and Biotechnology, 4,* 347–355.

Samelis, J., Sofos, J. N., Kendall, P. A., 7 Smith, G. C. (2002). Effect of acid adaptation on survival of *Escherichia coli* O157:H7 in meat decontamination washing fluids and potential effects of organic acid interventions on the microbial ecology of the meat plant environment. *Journal of food protection, 65,* 33–40.

Sang, Y., Teresa Ortega, M., Rune, K., Xiau, W., Zhang, G., Soulages, J. L., et al. (2007). Canine cathelicidin (K9CATH): Gene cloning, expression, and biochemical activity of a novel promyeloid antimicrobial peptide. *Developmental And Comparative Immunology,* In Press, Corrected Proof.

Scannell, A. G. M., Hill, C., Ross, R. P., Marx, S., Hartmeier, W., & Arendt, E. K. (2000). Development of bioactive food packaging materials using immobilised bacteriocins Lacticin 3147 and Nisaplin. *International journal of food microbiology, 60,* 241–249.

Schamberger, G. P., Phillips, R. L., Jacobs, J. L., & Diez-Gonzalez, F. (2004). Reduction of *Escherichia coli* O157:H7 populations in cattle by feeding colicin E7-producing *E. coli*. *Applied and Environmental Microbiology, i,* 6053–6060.

Schmieger, H., & Schicklmaier, P. (1999). Transduction of multiple drug resistance of *Salmonella enterica* serovar Typhimurium DT104. *FEMS Microbiology Letters, 170,* 251–256.

Schneider, R., Fernandez, F. J., Aguilar, M. B., Guerrero-Legerreta, I., Alpuche-Solis, A., & Ponce-Alquicira, E. (2006). Partial characterization of a class IIa pediocin produced by *Pediococcus parvulus* 133 strain isolated from meat (Mexican 'chorizo'). *Food Control, 17,* 909–915.

Schneitz, C., & Hakkinen, M. (1998). Comparison of two different types of competitive exclusion products. *Letters in Applied Microbiology, 26,* 338–341.

Schoeni, J. L., & Wong, A. C. L. (1994). Inhibition of *Campylobacter jejuni* colonization in chicks by defined competitive exclusion bacteria. *Applied And Environmental Microbiology, 60,* 1191–1197.

Schoenis, J. L., & Doyle, M. P. (1992). Reduction of *Campylobacter jejuni* colonization of chicks by cecum-colonizing bacteria producing anti-*C. jejuni* metabolites. *Applied And Environmental Microbiology, 58,* 664–670.

Senne, M. M., & Gilliland, S. E. (2003). Antagonistic action of cells of *Lactobacillus delbrueckii* subsp. *lactis* against pathogenic and spoilage microorganisms in fresh meat systems. *Journal of food protection,*: 418–425.

Sheng, H., Knecht, H. J., Kudva, I. T., & Hovde, C. J. (2006). Application of bacteriophages to control intestinal *Escherichia coli* O157:H7 levels in ruminants. *Applied And Environmental Microbiology,*: 5359–5366.

Shin, K., Yamauchi, K., Teraguchi, S., Hayasawa, H., Tomita, M., Otsuka, Y., et al. (1998). Antibacterial activity of bovine lactoferrin and its peptides against enterohaemorrhagic *Escherichia coli* O157:H7. *Letters in Applied Microbiology,* 407–411.

Si, W., Gong, J., Tsao, R., Zhou, E., Yu, H., Poppe, C., et al. (2006). Antimicrobial activity of essential oils and structurally related synthetic food additives towards selected pathogenic and beneficial gut bacteria. *Journal of Applied Microbiology*, 296–305.

Siragusa, G. R., Cutter, C. N., & Willett, J. L. (1999). Incorporation of bacteriocin in plastic retains activity and inhibits surface growth of bacteria on meat. *Food Microbiology, 16*, 229–235.

Skandamis, P. N., & Nychas, G.-J. E. (2001). Effect of oregano essential oil on the microbiological and physico-chemical attributes of minced meat stored in air and modified atmospheres. . *Journal of Applied Microbiology, 91*, 1011.

Skandamis, P., Tsigarida, E., & Nychas, G.-J. E. (2002). The effect of oregano essential oil on survival/death of *Salmonella typhimurium* in meat stored at 5C under aerobic, VP/MAP conditions. *Food Microbiology, 19*, 97–103.

Sklar, I. B., & Joerger, R. D. (2001). Attempts to utilize bacteriophage to combat *Salmonella enterica* serovar Enteritidis infection in chickens. *J Food Safety, 21*, 15–29.

Smith, H. W., Huggins, M. B., & Shaw, K. M. (1987). Factors influencing the survival and multiplication of bacteriophages in calves and in their environment. *Journal of General Microbiology, 133*, 1127–1135.

Smulders, F. J., & Greer, G. G. (1998). Integrating microbial decontamination with organic acids in HACCP programmes for muscle foods: Prospects and controversies. *International Journal of Food Microbiology, 44*, 149–169.

Stergiou, V. A., Thomas, L. V., & Adams, M. R. (2006). Interactions of nisin with glutathione in a model protein system and meat. *Journal of Food Protection, 69*, 951–956.

Stern, N. J., Svetoch, E. A., Eruslanov, B. V., Kovalev, Y. N., Volodina, L. I., Perelygin, V. V., et al. (2005). *Paenibacillus polymyxa* purified bacteriocin to control *Campylobacter jejuni* in chickens. *Journal of Food Protection, 68*, 1450–1453.

Svetoch, E. A., Stern, N. J., Eruslanov, B. V., Kovalev, Y. N., Volodina, L. I., Perelygin, V. V., et al. (2005). Isolation of *Bacillus circulans* and *Paenibacillus polymyxa* strains inhibitory to *Campylobacter jejuni* and characterization of associated bacteriocins. *Journal of Food Protection, 68*, 11–17.

Tanji, Y., Shimada, T., Yoichi, M., Miyanaga, K., Hori, K., & Unno, H. (2004). Toward rational control of *Escherichia coli* O157:H7 by a phage cocktail. *Applied Microbiology and Biotechnology, 64*, 270–274.

Tassou, C. C., Drosinos, E. H., & Nychas, G. J. (1995). Effects of essential oil from mint (*Mentha piperita*) on *Salmonella enteritidis* and *Listeria monocytogenes* in model food systems at 4 degrees and 10 degrees C. *The Journal of Applied Bacteriology, 78*, 593–600.

Tsigarida, E., Skandamis, P., & Nychas, G.-J. E. (2000). Behaviour of *Listeria monocytogenes* and autochthonous flora on meat stored under aerobic, vacuum and modified atmosphere packaging conditions with or without the presence of oregano essential oil at 5C. *Journal of Applied Microbiology, 89*, 901–909.

United States Department of Agriculture, Economic Resource Service (2001). Economics of foodborne disease: Estimating the benefits of reducing foodborne disease. United States Department of Agriculture, E.R.S. (ed.) Washington DC.

United States Department of Agriculture-Food Safety and Inspection Service (1996). Notice of policy change; achieving the zero tolerance performances stard for beef carcasses by knife trimming and vacuuming with hot water or steam; use of acceptable carcass interventions for reducing carcass contamination without prior agency approval. In *Fed. Reg. 61:15024–15027*. United States Department of Agriculture, F.S.a.I.S. (ed.).

van Diemen, P. M., Dziva, F., Abu-Median, A., Wallis, T. S., van den Bosch, H., Dougan, G., et al. (2007). Subunit vaccines based on intimin and Efa-1 polypeptides induce humoral immunity in cattle but do not protect against intestinal colonisation by enterohaemorrhagic *Escherichia coli* O157:H7 or O26:H. *Veterinary Immunology and Immunopathology, 116*, 47–58.

Van Donkersgoed, J., Hancock, D., Rogan, D., & Potter, A. A. (2005). *Escherichia coli* O157:H7 vaccine field trial in 9 feedlots in Alberta and Saskatchewan. *The Canadian Veterinary Journal, 46*, 724–728.

Vignolo, G., Palacios, J., Farias, M. E., Sesma, F., Schillinger, U., Holzapfel, W. H., et al. (2000). Combined effect of bacteriocins on the survival of various *Listeria* species in broth and meat system. *Current Microbiology, 41*, 410–416.

Vrinda Menon, K., & Garg, S. R. (2001). Inhibitory effect of clove oil on *Listeria monocytogenes* in meat and cheese. *Food Microbiology, 18*, 647–650.

Waddell, T., Mazzocco, A., Johnson, R. P., Pacan, J., Campbell, S., Perets, A., et al. (2000). Control of *Escherichia coli* O157:H7 infection in calves by bacteriophages. *Presentation, 4th International Symposium and Workshop on Shiga toxin (Verocytotoxin)- Producing Escherichia coli Infections*, Kyoto, Japan.

Wagenaar, J. A., Mevius, D. J., & Havelaar, A. H. (2006). *Campylobacter* in primary animal production and control strategies to reduce the burden of human campylobacteriosis. *Revue Science et Technique, 25*, 581–594.

Wagenaar, J. A., Van Bergen, M. A., Mueller, M. A., Wassenaar, T. M., & Carlton, R. M. (2005). Phage therapy reduces *Campylobacter jejuni* colonization in broilers. *Veterinary Microbiology, 109*, 275–283.

Wagner, P. L., and Waldor, M. K. (2002). Bacteriophage control of bacterial virulence. *Infection and Immunity, 70*, 3985–3993.

Wagner, R. D. (2006). Efficacy and food safety consideration of poultry competitive exclusion products. *Molecular Nutrition & Food Research, 50*, 1061–1071.

Whichard, J. M., Sriranganathan, N., & Pierson, F. W. (2003). Suppression of *Salmonella* growth by wild-type and large-plaque variants of bacteriophage Felix O1 in liquid culture and on chicken frankfurters. *Journal of Food Protection*, 220–225.

Widders, P. R., Thomas, L. M., Long, K. A., Tokhi, M. A., Panaccio, M., & Apos, E. (1998). The specificity of antibody in chickens immunised to reduce intestinal colonisation with *Campylobacter jejuni*. *Veterinary Microbiology, 64*, 39–50.

Woodward, M. J., Gettinby, G., Breslin, M. F., Corkish, J. D., & Houghton, S. (2002). The efficacy of Salenvac, a *Salmonella enterica* subsp. Enterica serotype Enteritidis iron-restricted bacterin vaccine, in laying chickens. *Avian Pathology, 31*, 383–392.

Yaron, S., Rydlo, T., Shachar, D., & Mor, A. (2003). Activity of dermaseptin K4-S4 against foodborne pathogens. *Peptides, 24*, 1815–1821.

Yin, L. J., Wu, C. W., & Jiang, S. T. (2003). Bacteriocins from *Pediococcus pentosaceus* L and S from pork meat. *Journal of Agricultural and Food Chemistry, 51*, 1071–7076.

Yuste, J., Pla, R., Capellas, M., & Mor-Mur, M. (2002). Application of high-pressure processing and nisin to mechanically recovered poultry meat for microbial decontamination. *Food Control, 13*, 451–455.

Zasloff, M. (2002). Antimicrobial peptides of multicellular organisms. *Nature, 415*, 389–395.

Zhang, G., & Mustapha, A. (1999). Reduction of *Listeria monocytogenes* and *Escherichia coli* O157:H7 numbers on vacuum-packaged fresh beef treated with nisin or nisin combined with EDTA. *Journal of Food Protection, 62*, 1123–1127.

Zhang, G., Ma, L., & Doyle, M. P. (2007a). Salmonellae reduction in poultry by competitive exclusion bacteria *Lactobacillus salivarius* and *Streptococcus cristatus*. *Journal of Food Protection, 70*, 874–878.

Zhang, G., Ma, L., & Doyle, M. P. (2007b). Potential competitive exclusion bacteria from poultry inhibitory to *Campylobacter jejuni* and *Salmonella*. *Journal of Food Protection, 70*, 867–873.

Zhao, T., Doyle, M. P., Harmon, B. G., Brown, C. A., Mueller, P. O. E., & Parks, A. H. (1998). Reduction of carriage of Enterohemorrhagic *Escherichia coli* O157:H7 in cattle in inoculation with probiotic bacteria. *Journal of Clinical Microbiology, 36*: 641–647.

Zhao, T., Tkalcic, S., Doyle, M. P., Harmon, B. G., Brown, C. A., & Zhao, P. (2003). Pathogenicity of Enterohaemorrhagic *Escherichia coli* in neonatal calves and evaluation of fecal shedding by treatment with probiotic *Escherichia coli*. *Journal of Food Protection, 66*, 924–30.

Ziprin, R. L., Hume, M. E., Young, C. R., & Harvey, R. B. (2002). Inoculation of chicks with viable non-colonizing strains of *Campylobacter jejuni*: Evaluation of protection against a colonizing strain. *Current microbiology. 44*, 221–223.

Zuckerman, H., & Abraham, R. B. (2002). Quality improvement of kosher chilled poultry. *Poultry science, 81*, 1751–1757.

Chapter 13
At-Line Methods for Controlling Microbial Growth and Spoilage in Meat Processing Abattoirs

Daniel Y.C. Fung, Jessica R. Edwards, and Beth Ann Crozier-Dodson

Introduction

Many decontamination strategies are available to the meat industry for the control of spoilage and disease causing microorganisms. Most of these strategies are spray-wash methods, and a variety of other methods are becoming increasingly popular in the industry as new research and developments are made. Hide-on decontamination has been shown to be extremely effective for controlling pathogens and may become more commonly incorporated into the hazard analysis and critical control point (HACCP) plans. A comprehensive review of decontamination of *Escherichia coli* O157:H7 in meat processing was made by Edwards and Fung (2006). The current chapter is for at-line consideration of all types of microorganisms related to meat processing with *E. coli* O157:H7 as the main model. Another detailed analysis of the entire topic of meat safety was made by Fung et al. (2001). It covered the topics of history of meat industry safety, microbiological, chemical, physical hazards associated with meat and meat processing, regulatory policies and inspection, and the future in meat safety.

Beef Cattle Slaughter Process

The manner in which beef cattle are slaughtered has significant effects on the resultant quality of the meat produced from the carcasses. The Humane Slaughter Act of 1958 regulates the method in which animals may be treated at harvest. Animals must be kept calm, equipment must be properly functioning, the personnel must be fully trained in the process and animals must be rendered insensitive to pain. There are three approved methods for the stunning of an animal: mechanical, electrical, or chemical (Code of Federal Regulations [CFR], 2005d). Figure 13.1a and b (International Meat and Poultry HACCP Alliance [IMPHA], 1996) shows a general flow diagram typical of commercial beef slaughter facilities. After stunning the animal,

---

D.Y.C. Fung
207 Call Hall, Kansas State University, Manhattan, Kansas 66506, USA

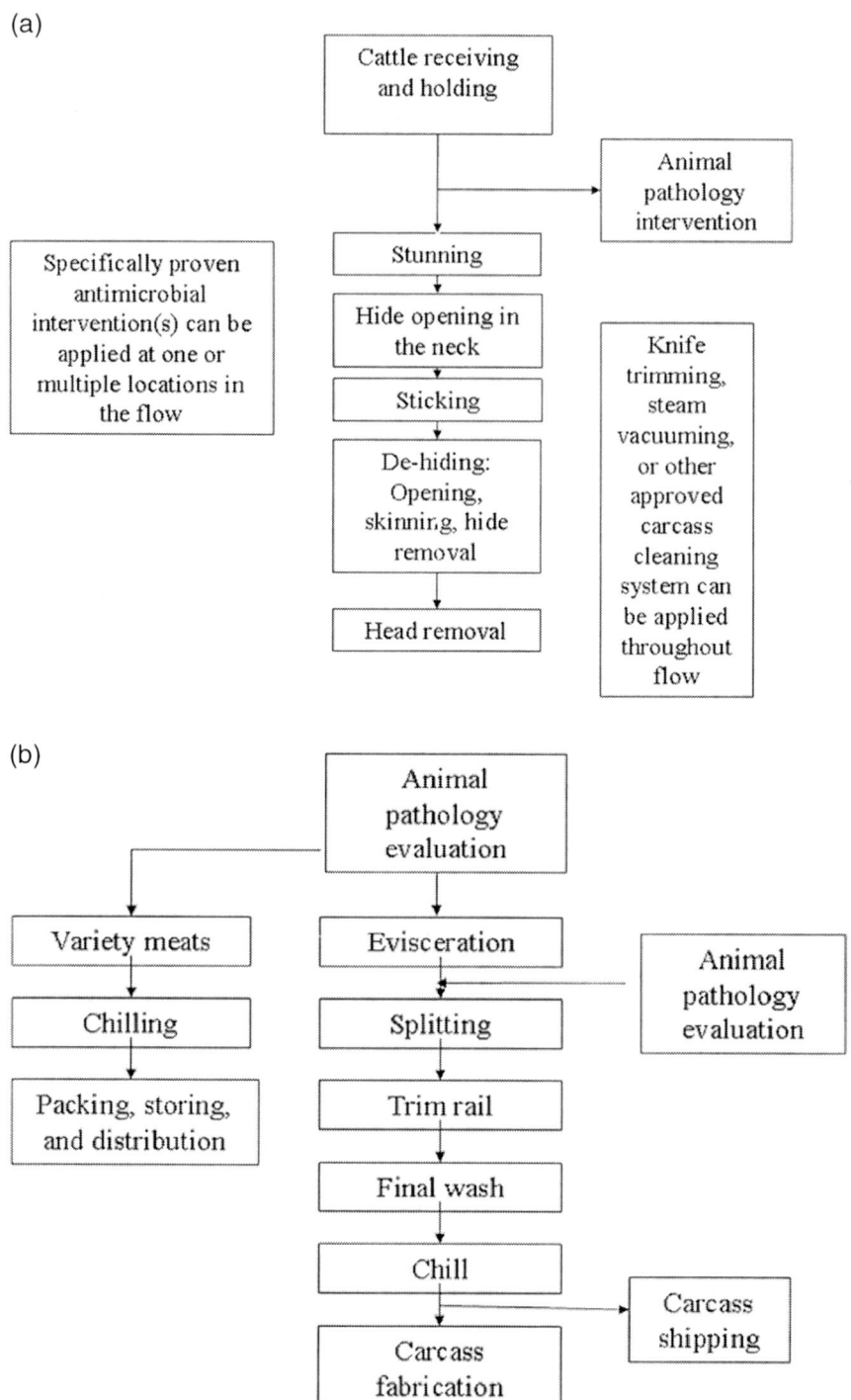

Fig. 13.1 (a) Beef slaughter process (Adapted from IMPHA, 1996). (b) Beef slaughter process (Adapted from IMPHA, 1996)

13 At-Line Methods for Controlling Microbial Growth and Spoilage

Fig. 13.2 Hide-on carcass wash and sanitizing assembly (Chad, 2004a)

the hide can be washed and sanitized mechanically (Fig. 13.2). The hide is then opened at the neck and the animal is exsanguinated by cutting the carotid arteries and the jugular vein. During this process, 85–95 % of the blood can be removed, which accounts for 7–8% of the total body weight. The hide is then removed, generally by using an automated hide puller (Fig. 13.3, Milmech, 2004). The next step is to eviscerate the carcass. The bung and esophagus must be tied off to prevent leakage and contamination, and the organs in the abdominal and thoracic cavities must be removed. Proper technique is critical to avoid contamination to the edible portion of the carcass.

The final steps in carcass preparation involve preparing the carcass for inspection, splitting the carcass and decontaminating any defects in microbiological quality. This may include trimming, steam vacuuming and/or application of decontamination methods such as chemical sanitizers, steam pasteurization or irradiation. The carcass is then weighed, given an identification number and sent to the chill room (Aberle et al., 2001).

Sources of Contamination

Although internal carcass muscles are essentially sterile (Gill & Penney, 1979), the contamination of carcasses during slaughter is unavoidable (Belk, 2001), and has since become one of the most significant challenges to food safety. Contamination events transfer approximately 3 log CFU/cm^2 to the carcass surfaces. The US Food

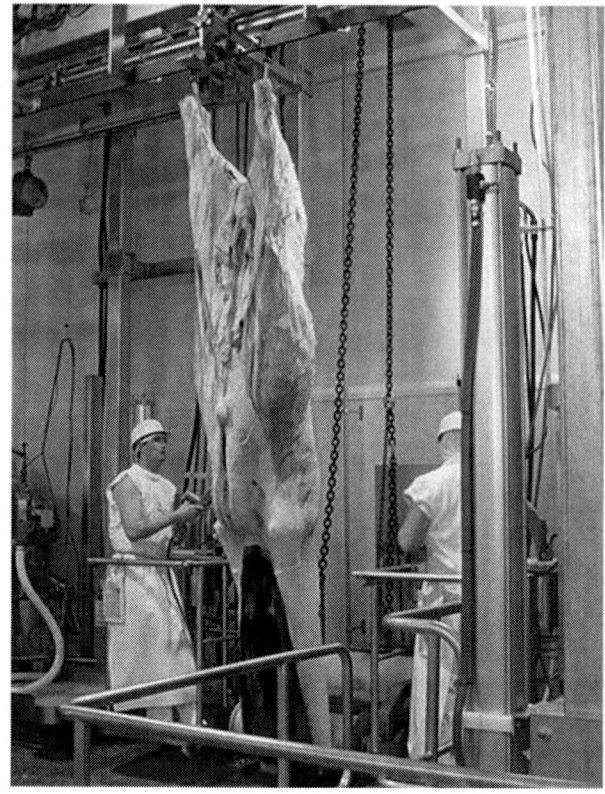

Fig. 13.3 Downward hide puller for the automated removal of hides (Milmech, 2004)

Safety and Inspection Service (FSIS) require the removal of visible fecal materials from the surfaces of beef carcasses during processing (Food safety and Inspection Service [FSIS], 1996b). In 1994, the FSIS declared that contamination of raw ground beef with *E. coli* O157:H7 would be classified as an adulterated product. Raw ground beef was specifically targeted because of a strong epidemiological link that exists between *E. coli* O157:H7 infection and ground beef, which can cause a serious life-threatening illness in humans (Food safety and Inspection Service [FSIS], 2002).

In 1999, the FSIS extended their scope to include intact raw beef products as well. It had been found that 28% of animals presented for slaughter harbored *E. coli* O157:H7 in their intestines, and after slaughter, 43% of carcasses were contaminated with the organisms (FSIS, 2002). Contamination originates from the slaughter of stock animals, and the process during slaughter and dressing may lead to transmission of pathogens and other organisms to the carcass.

In one study conducted at the Colorado State University, *E. coli* O157:H7 was found in 72% of fecal samples (fecal material removed from transected colon before evisceration) and 38% of hide samples (ventral brisket sterilely swabbed for a total of 450 cm^2) from cattle prepared for slaughter from a variety of facilities. Upon processing of these carcasses, 57% of post evisceration carcasses and 17% of post processing carcasses were positive for *E. coli* O157:H7 contamination (Elder et al., 2000).

Particular areas of the beef carcass are more prone to contamination with all kinds of microorganisms, with the brisket, fore-rib, flanks and round sites being the most common (Johanson et al., 1983). The inner round is characterized by substantial amount of lean tissue surrounded by a collar of fat. This ridge allows fecal material to become embedded in the junction and between muscle bundles, which results in difficult removal. This area is also the most difficult area to trim, because of the large collar of fat tissue, which also contributes to cross contamination (Hardin et al., 1995). Fecal contamination and pathogens are greater on areas of the beef carcass covered by subcutaneous fat (Hardin et al., 1995). The butt, anal area, and rump are areas of the beef carcass that remain moderately contaminated with organisms of fecal origin even after trimming and washing (Gill et al., 1995).

The abattoir contains many environments that can lead to cross contamination with pathogens (Belk, 2001; United States Department of Agriculture [USDA], 1993). Holding pens, slaughter and dressing processes, carcass skinning and evisceration have all been identified as points of entry for bacterial contamination (USDA, 1993). Contamination is also possible from walls, floors, air personnel, knives and protective garments (Fung et al., 2001). Carcasses may even contaminate each other if they make direct contact (Elder et al., 2000). The extent to which carcasses are contaminated is directly influenced by plant design, the speed of slaughter and the overall skill of employees (Belk, 2001). The season, type of animal being slaughtered and the specific site on the carcass can dramatically affect the level of contamination present on the carcass after slaughter (Sofos, Belk, & Smith, 1999). Despite the fact that processing methods and sanitation procedures may reduce the number of *E coli* O157:H7 positive carcasses, the amount of contamination after processing still poses a high food safety risk. Contamination of beef carcasses with *E. coli* O157:H7 designates a need for control on farms and in feedlots to reduce the risk of food-borne infection by this and other pathogens and even non-pathogens (Elder et al., 2000).

Microbial Considerations in Meat and Foods, and Food Processing Environments

Before a discussion on the occurrence of microbes on meat and decontamination strategies, it is appropriate to define the meaning of microbial counts in or on meat and food, air, food processing environments and food processing surfaces and utensils.

Table 13.1a and 13.1b is the Fung Applied Microbiology Scales listing the ranges of microorganisms for applied microbiology developed by the author after more than 30 years of research experience in food and meat microbiology.

Table 13.1a Fung scale for total viable cell counts in foods (Fung et al., 1980)

Total counts for spoilage considerations	Ranges
Low count	10^{0-2} CFU/ml, g, cm^2
Intermediate count	10^{3-4} CFU/ml, g, cm^2
High count	10^{5-6} CFU/ml, g, cm^2
Index of spoilage	10^{7} CFU/ml, g, cm^2
Odor development	10^{8} CFU/ml, g, cm^2
Slime development	10^{9} CFU/ml, g, cm^2
Unacceptable, too high	10^{10} CFU/ml, g, cm^2
No pathogens allowed in cooked ready to eat foods	

Table 13.1b Fung scale for air environment and food surfaces (Fung et al., 1995) (Al-Dagal & Fung 1993)

Total counts for air samples	Ranges
Acceptable	0–100 CFU/m^3
Intermediate	100–300 CFU/m^3
Too high, needs corrective action	>300 CFU/m^3
Total counts for food surfaces (knives, dishes, chopping blocks, etc.)	Ranges
Acceptable	0–10 CFU/cm^2
Intermediate	10–100 CFU/cm^2
Not Acceptable	>100 CFU/cm^2

The first scale is for total count for Spoilage Consideration of Foods in General (Fung et al., 1980)

0–2 log CFU/ml, g, and cm^2 Low count	Acceptable
3–4 log CFU/ml, g, and cm^2 Intermediate Count	Caution
5–6 log CFU/ml, g, and cm^2 High Count	Corrective actions
7 log CFU/ml, g, and cm^2	Index of Spoilage
Food should be discarded above 7 log CFU/ml, g, and cm^2	

This scale of viable cell count is useful for ascertaining spoilage potential of most foods. This scale is not for pathogenic bacteria in food. For cooked foods, no pathogens are allowed. For raw ground beef no *E. coli* O157:H7 is allowed.

Of course, for fermented solid and liquid foods high level of desirable organisms are needed to make excellent products and this scale does not apply.

The second scale is for total count for Air Samples (Al-Dagal & Fung, 1993)

0 to 100	CFU/m^3	Acceptable
100 to 300	CFU/m^3	Caution
>300	CFU/m^3	Corrective actions

Note: Singapore has a standard of >500 CFU/m^3 as unacceptable for food environments.

The third scale is for food contact surfaces (knives, spoons, forks, chop boards, table tops, etc. Fung et al., 1995)

0–10	CFU/cm^2	Acceptable
10–100	CFU/cm^2	Caution
>100	CFU/cm^2	Corrective actions

Hide Removal

Meat becomes contaminated with microorganisms when feces or contaminated hides contact the carcass during slaughter (Derfler, 2004; Elder et al., 2000; Gill et al., 1995). Elder et al. (2000) indicated that in cattle prepared for slaughter, 28% of carcasses had feces positive for *E. coli* O157:H7 and 11% of hides were positive. During processing, contamination spread to the carcass can range from 2 to 4 log CFU/cm^2 (Anderson, Sebaugh, Marshall, & Stringer, 1980). The bulk of microbial contamination occurs during hide removal (Bell, 1997; Buchanan & Doyle, 1997; Gill & Penney, 1979). When animals arrive at the abattoir, they carry a wide variety of microorganisms on the hides and hooves, and in their intestinal tracts (Belk, 2001; Derfler, 2004; McEvoy et al., 2003).

Bung Tying

Bung tying is a possible source of contamination in the slaughter process, and great care must be taken to prevent transfer of bacteria from the anus of the animal onto the edible adipose tissue (Gill et al., 1995). The bung tying process involves cutting to loosen the anus, and then bagging the bung and securing with either a tie of a clip (Food Safety and Inspection Service [FSIS], 1994). The bung is then pushed through to the abdominal cavity, where it can be removed during evisceration (Romans, Costello, Carlson, Greaser, & Jones, 2001). Tools or personnel that contact the bung may contribute to cross contamination (McEoy et al., 2003). Cross contamination, that is a direct result of manual bung tying, may be eliminated by using an automated system. Such systems have reported lower total of *E. coli* and coliform counts in the anal area than manual methods (Sheridan, 1998). Proper taping of the esophagus and enclosing the rectum to prevent leakage of the ingesta and feces, and ensuring that the gastrointestinal tract is removed without incident are essential for controlling contamination (Bell, 1997). Figure 13.4 shows pre-evisceration carcass washes.

Evisceration

The gastrointestinal tracts of cattle can carry a multitude of enteric pathogens. The evisceration process carries the potential of ingesta contamination to the

Fig. 13.4 Pre-evisceration carcass wash assembly and organic acid washer (Chad, 2004d)

carcass, environment or equipment. To prevent contamination, great care must be taken to minimize the potential for evisceration defects, such as puncturing or rupturing of the intestines (Hulebak & Schlosser, 2001). The gut must be removed entirely soon after slaughter, within 30 min for most operations (Romans et al., 2001). Swelling after death makes removal difficult and increases the chances of evisceration defects. Swelling of the gastrointestinal tract also puts pressure on the gallbladder forcing bile back into the liver and the surrounding muscles, causing green discoloration of the tissue (Gill, 1991). If evisceration defects occur, corrective actions must be in place to remove any contamination from the carcass. Such measures include trimming of visible contamination, reducing the line speed so that employees can exercise better caution and sanitizing tools in 82°C water. Through advances in modern processing, the respiratory and intestinal tracts are no longer major sources of contamination to the edible carcass tissue (Bell, 1997). Evisceration can be completed with minimal, if any, contamination if the intestinal tract is not ruptured or punctured, which can be facilitated by the skillfulness of the workers performing the evisceration (Nottingham, Penney, & Harrison, 1974).

Carcass-to-Carcass Transfer

McEoy et al. (2003) indicated that in a typical Irish beef abattoir, researchers examined the persistence of *E. coli* O157:H7 on beef carcasses, as well as surveyed the prevalence of the bacterium in post slaughter and post dressing carcasses. In the facilities chosen for the study, carcasses were slaughtered at a speed of 40–80 animals per hour. Carcasses were processed using an automatic hide puller and were washed with potable water at 35°C before chilling (McEvoy et al. 2003). Carcass sampling revealed that cross contamination does occur during chilling. Sampling of two carcasses before chilling resulted in one positive and one negative sample for *E. coli* O157:H7. However, after chilling both carcasses tested positive for the organism. These carcasses were not together on the slaughter line, but were side by side during chilling. Fecal rumen and carcass sampling were performed on 250 carcasses. Of those 2.4% of fecal samples, 0.8% of rumen samples and 3.2% of carcass samples were positive for *E. coli* O157:H7 (McEvoy et al., 2003).

Decontamination of Beef Carcasses

Governmental Regulations and HACCP

It is impossible to guarantee bacteria and pathogen-free meat in large abattoirs, therefore, decontamination step in the form of washing and sanitizing during the slaughter process is important. Decontamination methods can improve the microbiological safety and increase shelf life, and should be an integral part of the slaughter process (Dickson & Anderson, 1992). Prompt chilling of carcasses after slaughter to below optimal bacterial growth temperatures is a critical measure, and chilling (Abdul-Raouf, Beuchat, & Ammur, 1993). Romans et al. (2001) recommend that carcasses should be chilled 16–20 h at –1 to –2C. Chilling can reduce the prevalence of many cells and will stress cells. By closely adhering to the clean cattle policy and employing hygienic dressing and evisceration, contamination of carcasses can be minimized. In addition, ensuring that only clean, healthy animals are presented for slaughter and are processed correctly will reduce the incidence of contamination.

Governmental Regulations and HACCP

There has been great interests within the meat industry to develop slaughtering and dressing operations that reduce the incidence of contamination (Sofos, 1993). The first federal legislation concerning meat inspection dates back to 1890. In 1891, US Congress assigned responsibility for meeting European meat requirements to the USDA. The USDA then began conducting livestock inspections both before and after slaughter of meat animals intended for US consumers (Hulebak &

Schlosser, 2001). The well known story of Upton Sinclair's *The Jungle* which described the sickening and unsanitary conditions of the meat packing industry outraged the American public (Centers for Disease Control and Prevention [CDC], 1999) and resulted in the passing of the Federal Meat Inspection Act by US Congress. This act established sanitation standards for slaughter and processing and authorized ante-mortem inspection of animals and post-mortem inspection of every carcass for interstate commerce (Hulebak & Schlosser, 2001). Carcass inspection was intended to keep meat from diseased animals out of the food supply, with Veterinarians and food inspectors checking live animals and carcasses for signs of disease. The 1906 Meat Inspection Act allowed for continuous inspection of processing operations by the USDA. It was not until the 1950s that concern grew about the unseen threats to food safety, those that visual inspection could not identify. The wholesome Meat act of 1967 required that adequate inspection of all meat products was performed. In 1977, the Food Safety and Quality Service (FSQS) was established and given the responsibility of inspecting meat and poultry products. In 1981 the name was changed to Food Safety and Inspection Service (FSIS). The Hazard Analysis Critical Control Point concept, started in 1960, began as a joint effort between the National Aeronautics and Space Administration, the United States Army Natick Laboratories and the Pillsbury Company. Now it is being applied to many industries including the meat industry. Because of HACCP the food supply nationally and internationally are much safer than before.

Hide-on Decontamination

One method to reduce microbe and pathogen transfer during hide removal is to reduce the levels of bacteria present on animal hides before these animals enter the slaughterhouse. A hide-on decontamination cabinet has been designed and used in pilot plants for the removal of fecal contamination and debris before the dehiding process. The double-shackled cattle proceed into the cabinet after stunning and exsanguination (Fig. 13.4). Cattle are then vacuumed to remove excess liquid. Hide-on decontamination cabinets require an enormous amount of water, and therefore require recirculating systems to make such systems cost-effective (Bosilevac, Nou, Osborn, Allen, & Koohmaraie, 2005). Hide-on decontamination was studied using water and tri-sodium phosphate with and without vacuuming steps. Whole pulled hides were draped over barrels to simulate the hide-on carcass. Wash treatments were applied using an electric pressurized sprayer at a pressure of 1200 psi and at a distance of 65 cm from the hide. The treatments were applied for 20 s at 60°C. Half of the treated hides were then vacuumed using a steam vacuum without steam to remove residual liquid and visible contamination. Hide-on decontamination practices can reduce the bacteria on hides by about 5 log CFU/100 cm^2, and decrease the bacterial load of carcasses by 1 log CFU/100 cm^2. This in turn dramatically reduces the prevalence of *E. coli* O157:H7 on carcasses (Bosilevac et al., 2005).

Chemical Dehairing

Chemical dehairing is a process utilizing sodium sulfide to remove hair and debris from cattle hides. The solution is then neutralized by a combination of hydrogen peroxide spray and water washing. The cleaning of cattle hides results in a drastic reduction in subsequent carcass contamination (Nou et al., 2003). The object of chemical dehairing is to remove hair, dirt, and fecal materials from the hide before slaughtering in an attempt to reduce subsequent carcass contamination. In a study conducted by Nou et al. (2003), chemical dehairing was evaluated as a means to reduce carcass contamination. Carcass contamination can be nearly eliminated if the quality of the hide is improved before dehiding. Chemical dehairing is extremely effective at virtually eliminating carcass decontamination from the dehiding process. There are some additional considerations regarding the efficacy of chemical dehairing. Sodium sulfide wash solutions and dehairing waste create extensive environmental problems for the industry, as the chemical is damaging to the environment (Agricultural Marketing Service [AMS], 2000).

Cetylpyridinium Chloride (CPC)

CPC is a quaternary ammonium compound used as a common oral antimicrobial (Bosilevac, Wheeler, et al., 2004), which is currently approved for use on poultry carcasses as an antimicrobial agent for surface contamination. It is applied as a mist at room temperature prior to poultry chilling. Application cannot exceed 0.3 g CPC/lb of raw poultry (Code of Federal Regulations [CFR], 2005b). Many types of pathogens are sensitive to CPC and the chemical does not cause undesirable effects to flavor, texture, appearance or odor of foods. The antibacterial activity of CPC is facilitated by the cations it forms in solution which decrease surface tension and allows the molecules to penetrate the bacterial membrane to act on cellular components. CPC can reduce *E coli* O157:H7 by 2.78 log after 14 days of storage (Lim & Mustapha 2004). Since CPC is rapidly inactivated by organic materials, the degree to which a hide is soiled will directly impact the efficacy of CPC as a decontamination method. The most logical point at which to apply CPC treatment would be on stunned, shackled animals before dehiding (Bosilevac, Wheeler, et al., 2004). Currently, CPC is not an approved method of decontamination in the beef industry (Ransom et al., 2003). However, studies have shown that CPC is effective for reducing pathogens on beef hides (Cutter et al., 2000). One percent CPC is the most effective concentration for reducing bacterial population on hides. The greatest efficacy for CPC is as a high-pressure pre-wash at a concentration of 1% followed by a second high-pressure wash after stunning (Bosilevac, Wheeler, et al., 2004). Bosilevac Arthur, et al., (2004) studied CPC on live animals in holding pens and found that CPC significantly reduced *E. coli* O157:H7 on hides, resulting in a near-zero incidence on pre-eviscerated carcasses.

Knife Trimming

The knife trimming of carcasses is now required for slaughter facilities in an effort to remove all visible contaminants. Knife trimming has been written into many sanitation standard operating procedures and HACCP plans (Belk, 2001). The standard commercial practice among slaughter houses in the USA is to knife trim all visible contamination on the carcass, followed by a water wash (Acuff, Castillo, & Savell, 1997; Prasai et al., 1995). Trimming is far superior to washing, in order to remove visible contaminants, as washing may not remove attached organisms and may spread contamination over the surface of the carcass (Food Safety and Inspection Service [FSIS], 1996a). Trimming is especially useful at removing visible organic matters in areas that are difficult to decontaminate with water wash or chemical sprays (Hardin et al., 1995). Some problems exist in trimming of edible meat. Trimming is a highly variable process whose efficacy is related to the skill of the individual doing the trimming. The equipment used during trimming can potentially spread contamination if not properly sanitized. In addition, holding carcasses in the warm slaughter house environment can allow for attachment of bacteria before washing and chilling (Reagan et al., 1996). Meat inspection and trimming rely on removing visual contamination from the carcass, and may overlook low-level contamination. The knife may be used on multiple sites of a single carcass or on multiple carcasses. Because the use of sterile instruments is not practical, a washing procedure will always be part of the standard protocol (Prasai et al., 1995). There is always a chance that the knife may contaminate other parts of the meat when trimming is being done. Trimming must be employed to comply with the zero-tolerance guidelines for fecal contamination; however, FSIS recommends that a secondary decontamination step be employed at some point in the process.

Vacuuming

In 1993, following the outbreak of *E. coli* O157:H7 that left four children dead, the FSIS stressed that manual trimming was the only way to remove visible contamination. At that time, there were no firm data indicating that any other method was more efficacious (FSIS, 1996a). In 1995, the FSIS held a public meeting to consider other methods for removing visible contamination. Testing of steam vacuuming was approved as a possible alternative to achieve the zero-tolerance regulatory policy for fecal contamination. Fifty plants tested the new vacuum system (FSIS, 1996a). Spot vacuuming of carcasses involves the use of hot water or steam to inactivate bacteria and remove visible fecal contamination (Sofos et al., 1999). Each plant tested 120 samples, 60 were trimmed and 60 were vacuumed. All contamination over 1 inch in diameter was trimmed regardless of the treatment group. A second study was done taking 60 samples over 60 days to evaluate the consistency of the new method (FSIS, 1996a). These studies revealed, that carcasses that were vacuumed had, on the average, 0.69 log CFU/cm^2 less contamination than the trimmed samples. In addition, over the 60-day trial period, a 0.54 log CFU/cm^2 lower level on the vacuumed samples was maintained. The FSIS announced a policy change for the slaughter of beef cattle in 1996 to include vacuuming as an accepted method of removing contamination. Spot vacuuming is now used extensively through the industry (FSIS, 1996a). Steam vacuuming can improve the visual cleanliness of carcasses, reduce microbial contamination, reduce the need for manual knife trimming, and aid in meeting the zero-tolerance regulatory policy for visible contamination of beef carcasses. When visible contamination is <1 inch in diameter, steam or hot water vacuuming can be used as an alternative to the traditional knife trimming method (Fig. 13.5).

Spray Chilling

Chilling of carcasses is recommended to be completed in 18–36 h, and beef sides must reach a temperature below 7C. Chilling causes drying that may result in a 2–3% weight loss. To prevent this many processing plants applied chilled water to carcass sides in the form of a spray while in chill (Sofos et al., 1999). Spray chilling is the accepted practice among slaughterhouses in the USA. This evaporative cooling contributes to microbiological control without significant water loss and weight loss in the carcass (Dickson & Anderson, 1992). Chemicals approved by the USDA for use on carcasses during spray chilling include acidified sodium chlorite (ASC), tri-sodium phosphate, lactoferricin B, chlorine solutions and organic acids (Stopforth et al., 2004). During the first 12 h of carcass chilling, carcasses are sprayed with a mist of either water or a 50-ppm chlorine solution to minimize water

Fig. 13.5 Steam vacuuming of a section beef carcass (Weller, 2003)

loss and to further decontaminate the carcass. The carcass is generally sprayed for up to 2 min every 30 min (Stopforth et al., 2004).

Spray Washing

Research showed that spray washing is as effective as trimming in removing physical and microbiological contaminants from beef carcasses (Crouse, Anderson, & Naumann, 1988; Reagan et al., 1996) Spray washing focuses on sanitation of the carcass as a whole. Current methods include sanitation with organic acids, chlorinated water or hot water (Dickson & Anderson, 1992; Hardin et al., 1995). In most facilities, potable water spray systems are in place to mediate the removal of gross contamination (Bell, 1997). Examples of carcass spray wash cabinets are illustrated in Figs. 13.5 and 13.6.

Water

Potable water sprays have some effectiveness against microbial populations on beef carcasses. However, the efficacy of the water spray depends heavily on the parameters at which it is applied. Contact time, water pressure, water temperature and the type of equipment used for application can all affect the success of water

Fig. 13.6 Organic acid spray wash assembly (Chad, 2004c)

sprays (Fung et al., 2001). Hot water has been shown to be much more effective than cold water at decontamination. Hot water washing is affected by the same parameters as cold water sprays (i.e. contact time, temperature, pressure, etc.). Cold water sprays still have some applications in meat processing plants for the removal of debris from the hide of the animal, although there is limited effect on the microbial safety of the product (Fung et al., 2001). The use of hot water for decontamination is highly regulated. The USDA-FSIS approves of hot water as a decontamination method and recommends it to be used at 74–85°C (Sofos et al., 1999). Using hot water or steam creates the potential for sanitation problems by increased humidity and available moisture in the plant environment, thus creating an environmental niche for microorganisms. The equipment for decontamination must meet the requirements set forth in 9 CFR 308.5. This method is only allowed at the processing steps where water spray is allowed. Automated and handheld equipments are available (FSIS, 1996b). In one study, spray washing of edible beef tissues showed that spraying carcasses with hot pressurized water (74°C and 20.68 bar) can reduce the initial microbial populations by 1-log unit and extend the shelf life by 5 days (Gorman, Morgan, Sofos, & Smith, 1995; Gorman, Sofos, Morgan, Schmidt, & Smith, 1995). In many instances temperature, pressure, chemicals combinations in water spraying on carcasses have been studied throughout the world. Basically, a 2 log CFU/cm^2 reduction is considered a good method.

Acid Spray Washes

Organic acids are used in the beef industry as decontamination solutions during slaughter and are the classic method for controlling the growth of microorganisms on meat (Fang & Tsai, 2003). These spray washes reduce the microbial load and are intended to improve the safety of the meat, and some combinations resulted in reducing pathogenic contamination by up to 2 log CFU/cm^2 (Berry & Cutter, 2000). The FSIS approves the use of a water rinse followed by a rinse with food grade organic acids on pre-eviscerated carcasses. The water rinse is intended to remove any hair, dirt or dust that has accumulated on the surface (FSIS, 1996a). Acetic, lactic and citric acids have been generally recognized as safe (GRAS) by the FDA for multipurpose usage (Acuff et al., 1987; FSIS, 1996a). Two cabinet systems are approved for use, and the acids are applied as a fine mist or fog. Organic acid treatment can be used in aqueous solutions of 1.5–2.5% and can be sprayed on carcasses at any stage in which water spray is allowed. The use of approved automated spray cabinets or hand-operated equipment is allowed. Acid washes have been shown to effectively remove 2 log CFU/cm^2 of the general microflora from beef carcasses (Fung et al., 2001).

Acetic Acid

Acetic acid has been an effective spray chilling sanitizer on beef carcasses in reducing the number of pathogenic populations (Dickson, 1991). Anderson, Marshall,

Stringer, & Naumann, (1977) showed that acetic acid sprays (AASs) on beef carcasses could reduce aerobic bacteria by 99.9%. However, some research data shows that spray washing with acetic acid has similar results (no significant difference) to hot water spray washing. For example, hot water (74°C) reduced bacteria from 6 to 4.36 log CFU/cm^2 and acetic acid reduced *E. coli* to 4.31 log CFU/cm^2. Short-chain fatty acids have been shown to discolor meat to a higher degree than other GRAS acids. AASs have resulted in brown to yellow discoloration and offensive acid odors (Hamby, Savell, Acuff, Vanderzant, & Cross, 1987). In a study conducted by Brackett, Hao, & Doyle (1994), maximum reduction of *E. coli* O157:H7 on beef carcasses was achieved when tissues were sprayed with 3% acetic acid at 20°C. However, this treatment only resulted in a 0.61 log CFU/cm^2 decrease of the pathogen load. These researchers suggest that organic acid sprays are of little value against *E. coli* O157:H7.

Lactic Acid

Aqueous solutions of lactic acid can be used to successfully decontaminate beef carcasses (Kozempel et al., 2003). However, the acid concentration used on beef is limited because of the potential for adverse effects to the quality of the product. Lactic acid, specifically, can alter fresh meat color at blood spots, which can be prevented by removing blood spots with a water wash before treatment with lactic acid (Crozier-Dodson, 2000). Samelis, Sofos, Kendall, & Smith, (2002) recommended that lactic acid sprays be used, rather than AASs, for the decontamination of beef carcasses. It has been shown that acetic acid has a lower effectiveness against *E. coli* O157:H7 and a higher effectiveness against the natural flora of meat carcasses. Lactic acid inactivates *E. coli* O157:H7 at a faster rate than acetic acid at the same concentration in decontamination run-off fluids (Samelis, Sofos, Kendall, & Smith, 2001). Lactic acid is very effective for pathogen reduction as shown in a study conducted by Ransom et al. (2003). Lactic acid sprays were able to achieve reductions of pathogen numbers by 2.5 log CFU/cm^2. From this study, it was shown that lactic acid exceeds acetic acid reductions.

Fumaric Acid

Fumaric acid has been largely ignored by the processing industry as an organic acid capable of increasing the microbiological safety of meats. It has several benefits that make it superior to the conventional acids used, such as acetic and lactic acids. Fumaric acid is readily available in nature and is present in fresh meat at low concentrations. It is very cost-effective and more acidic than the other GRAS organic acids. It also has antioxidant properties that may make it beneficial in some industries. Research at the Kansas State University (Podolak, Zayas, Kastner, & Fung, 1996) compared the effectiveness of fumaric acid on the decontamination of lean beef

Table 13.2 Reductions (log CFU/cm^2) of *E. coli* O157:H7 on beef muscle samples treated with 1.0% acid and stored under vacuum at 4°C (Adapted from Podolak et al., 1996)

	Days of storage						
Treatment	0	7	14	21	30	60	90
Fumaric acid	0.57	0.58	1.30	1.19	0.77	0.54	0.42
Lactic acid	0.75	0.52	0.76	0.75	0.21	0.29	0.10
Acetic acid	0.77	0.64	0.63	0.68	0.21	0.29	0.11

tissue to that of acetic and lactic acids. Samples were treated with 1% solutions of fumaric, acetic, and lactic acids and the log CFU/cm^2 reduction of *E. coli* O157:H7 of the three acids are listed in Table 13.2.

Chlorine

Chlorine is a very effective and an inexpensive means of controlling contamination in the meat industry. However, its applications are mainly for the reduction of bacterial populations on food contact surfaces and equipment, rather than on meat carcass surfaces. Chlorine is quickly inactivated by organic materials, which makes disinfection of beef carcasses difficult. Effectiveness is also decrease by low temperatures and affected by the pH of the solution (Fung et al., 2001). Chlorine (20–50 ppm) in water is another approved decontamination method, which can be applied at any point during the dressing when the carcass is allowed to be sprayed with water (FSIS, 1996a). When used as a food contact sanitizer, chlorine can effect up to a 2-log CFU/cm^2 reduction of microorganisms on beef carcass (Fung et al., 2001); although other researchers have cited chlorine as one of the least effective methods available to decontaminate beef carcasses, when compared to other chemical and thermal approaches (Belk, 2001). In a study by the US Environmental protection Agency, seven strains of *E. coli* O157:H7 from cattle were analyzed for their sensitivity to chlorine. Chorine was able to, on average; reduce the pathogen from an initial level of 5.79 log CFU/mL to < 1 log after 2 min of exposure in the culture medium (Rice, Clark, & Johnson, 2003). However, these results are unable to be relayed to the beef carcass because of the organic materials associated with fresh beef. On beef carcass surfaces, chlorine has been unable to reduce *E. coli* O157:H7 by more than 1.3 log CFU/cm^2.

Chlorine Dioxide

Chlorine dioxide was first produced in 1811 through the reaction of potassium chlorate and hydrochloric acid. Today, it is made through a similar reaction using sodium chlorite. Chlorine dioxide has been used in pulp and paper industries, water treatment facilities and for application in food processing and medical sterilization (Haas, 2001). Chlorine dioxide may be better suited for decontamination of beef carcasses than aqueous chlorine. Chlorine dioxide does not react with nitrogenous compounds as chlorine does (Berg, Roberts, & Matin, 1986), is less reactive with organic compounds than chlorine (Shin, Chang, & Kang, 2004), has a greater oxidizing potential than chlorine, active at high pH (Berg et al., 1986), and has a higher antimicrobial activity than chlorine in food products (Shin et al., 2004). Chlorine dioxide is approved for use at 3 ppm as an antimicrobial agent for poultry chill water (Code of Federal Regulations [CFR], 2005c) and approval is awaited for direct food contact, although approval for use on beef carcasses is not yet permitted (Shin et al., 2004). Chlorine dioxide inactivates bacteria by causing a loss of permeability of the outer cellular membrane, thus disrupting the ionic gradient across the membrane (Berg et al., 1986). Cutter & Dorsa (1995) found that chlorine dioxide reduced aerobic plate counts of <1 log CFU/cm^2 at low pressure, and at high pressures with longer contact times, reductions were increased from 0.93 to 1.52 CFU/cm^2 and it was concluded that it is ineffective as a decontaminant of beef carcasses against bacteria of fecal origin.

Tri-sodium Phosphate

Tri-sodium phosphate has been approved for use on beef carcasses and has shown to inhibit bacterial attachment to carcass surfaces (Sofos et al., 1999). Food grade tri-sodium phosphate may be applied to meat carcasses under very specific conditions: the temperature must be maintained at a range of 32–43°C and be used at concentrations between 8 and 12%. Treatment cannot be applied for more than 30 s. Tri-sodium phosphate can be applied at any point where carcasses can be sprayed with water. Automated cabinets and hand-operated equipment are approved for use (FSIS, 1996a). The success of tri-sodium phosphate and other alkaline sanitizers in the poultry industry has lead to numerous studies for the application for beef. Dickson, Nettles Cutter, & Siragusa, (1994) investigated the effectiveness of tri-sodium phosphate as a sanitizer of beef tissue. Reduction of *E. coli* O157:H7 on beef was similar to those found in studies for reducing *Salmonella* sp. contamination on poultry carcasses. An average of 1.0–1.5 log of *E. coli* O157:H7 was successfully achieved with the application of tri-sodium phosphate at 0.25 and 1.0 min; longer application times generally resulted in larger log reduction of *E. coli* O157:H7. Temperature had no significant effect on the results. Tri-sodium phosphate is unable to remove 100% of pathogenic contamination. However, the reductions show promise for this sanitizer as part of a multi-hurdle system for increasing the safety of beef products (Dickson et al., 1994).

Acidified Sodium Chlorite (ASC)

The FDA approves the use of ASC for the decontamination of beef carcasses when it is combined with any organic acid with GRAS status. It can also be used in further processing of red meat as a dip. Sodium chlorite concentrations must be between

500 and 1200 ppm and acidified to a final pH of 2.5–2.9 (Code of Federal Regulations [CFR], 2005a). The antimicrobial activity of ASC is attributed to chlorine ions in the solution that are converted to chlorine dioxide, which inhibits protein synthesis in the cell. Some studies have shown a >4 log reduction of *E. coli* O156:H7 after 14 days of storage (Lim & Mustapha 2004). It has been shown that treatment with ASC does not adversely affect the color or odor of the treated meat samples. Researchers at the Texas A&M University set out to evaluate the effectiveness of sodium chlorite when mixed with either phosphoric or citric acid. Beef carcass surfaces were inoculated with 5.5 log CFU/cm^2 of *E. coli* O157:H7. Carcasses were then sprayed with water and sodium chlorite acidified with phosphoric acid or citric acid. Water alone significantly reduced the pathogen load by an average of 2.3 log CFU/cm^2. ASC had significantly greater reductions than that of water alone, with phosphoric acid–ASC reducing *E coli* O157:H7 by 3.8 log CFU/cm^2 and citric acid-ASC by 4.5 log CFU/cm^2 (Castillo, Lucia, Kemp, & Acuff, 1999).

Lactoferricin B

Lactoferrin is an 80-kDa iron-binding glycoprotein that is present in many body secretions such as milk, saliva and tears (Masson, Heremans, & Dive, 1966). It was once thought that the anti-microbial properties of Lactoferrin were solely because of its iron-binding abilities. However, a 25 amino acid peptide near the N-terminus was discovered and found to have higher antimicrobial activity than intact Lactoferrin (Bellamy et al., 1993). This peptide, called lactoferricin, is able to inactivate a variety of microorganisms, and is released when pepsin cleaves the peptide from intact Lactoferrin (Hwang, Zhou, Shan, Arrowsmith, & Vogel, 1998). Table 13.3 summarises the reduction of *E. coli* O157:H7 by various acid and liquid treatments.

Table 13.3 Reduction of E. coli O157:H7 on beef carcass tissue samples when challanged by various antimicrobials (Adapted from Ransom et al., 2003)

High level Inoculum (5–6 log CFU/cm^2)		Low level Inoculum (3–4 log CFU/cm^2)	
Treatment	Reduction (log CFU/cm^2)	Treatment	Reduction (log CFU/cm^2)
Water (25°C)	1.2	Water (25°C)	0.6
Acidified chlorine (0.001%)	0.4	Acidified chlorine (0.001%)	0.8
Acetic acid (2%)	1.6	Acetic acid (2%)	2.1
Lactic acid (2% at 55°C)	3.3	Lactic acid (2% at 55°C)	3.1
Lactoferricin B (1%)	0.7	Lactoferricin B (1%)	0.6
Peroxyacetic acid (0.02%)	1.4	Peroxyacetic acid (0.02%)	1.4
Acidified sodium chlorite (0.02%)	1.9	Acidified sodium chlorite (0.02%)	2.0
CPC (0.05%)	4.8	CPC (0.05%)	3.6

Steam Pasteurization

Pressurized steam, more commonly known as steam pasteurization, is approved for use by USDA-FSIS on red meat carcasses. It rapidly kills bacteria (Kozempel et al., 2003) and has been successful at removing contamination from beef carcasses (Dorsa, Cutter, Siragusa, & Koohmaraie, 1996). Commercial steam pasteurization units are now available (Steam Pasteurization System; Frigoscandia Food Process Systems, Bellevue, WA) for medium and large processing plants (Figs. 13.7 and 13.8). They are able to pasteurize up to 3400 carcasses per hour. Steam pasteurization is intended as the last step of the slaughter process, and can successfully reducebacterial contamination (Fung et al., 2001). The process is completely automated, consisting of three steps. The first step is to remove excess water using vertical air blowers (Retzlaff et al., 2004). Water on carcass surfaces can act as a protective barrier to the bacteria, preventing steam from adequate contact (Warriner et al., 2001). The surface is then pasteurized in a steam-flushed chamber with an application time of 6.5–10 s. The steam flush chamber saturates the carcass with steam, raising the surface temperature to 87.8°C (Retzlaff et al., 2004) to inactivate bacteria without affecting texture or color (Warriner et al., 2001). A second type of steam pasteurizer uses a static steam chamber, rather than a steam-flushed chamber. In the static steam system, a constant flow of steam flows through the unit, rather than refilling for each carcass (Retzlaff et al., 2004). The last step is to

Fig. 13.7 Side view of carcass pasteurization system (Chad, 2004b)

Fig. 13.8 Front view of carcass pasteurization system (Stanfos, 2004)

spray the carcass with cool water. This quickly reduces the surface temperature and prevents discoloration caused by prolonged heating (Retzlaff et al., 2004; Warriner et al., 2001). The process is applied to surfaces for 6 s and is able to reduce bacterial loads by 1–2 log. Discoloration is a problem if applied for more than 6 s (Sofos et al., 1999). In a study by Retzlaff et al. (2004), pasteurization treatments for 6–15 s at 93.3°C followed by a second treatment at 98.9°C for 3–15 s caused greater than a 1 log reduction of experimentally inoculated pathogens. The researchers recommended that steam pasteurization of beef carcasses be conducted at 98.9°C for at least 9 s.

Irradiation

Regulatory acceptance of irradiation in the meat industry began in the 1950's (Curry et al., 2000). The FDA approved its use on red meat in 1997 and USDA finalized the approval in 2000 (Centers for Disease Control and Prevention [CDC], 2005).

Over 23 countries are currently employing meat irradiation in processing plants. The USDA-FSIS has approved of a maximum level of 4.5 kGy for fresh read meat and 7.0 kGy for frozen meats (Code of Federal Regulations [CFR], 2005e). Despite the controversy over irradiated food products, surveys from the late 1990s indicated that 55–80% consumers were willing to purchase irradiated meat and poultry products (Fung et al., 2001). Although meat irradiation primarily focuses on ground products, there have been some efforts to provide irradiation technology for the carcass as a whole. Ionizing radiation of ground beef requires high-penetration, high-energy radiation to ensure that the product is irradiated, not only on the surface, but to the interior as well, These high doses often lead to the formation of off flavors and odors (Arthur et al., 2005). Low-dose, low penetration radiation can be used to irradiate a large surface to a depth of 15 mm. This is ideal for treating the entire carcass sides. The interior of the carcass should be relatively clean, with contamination issue on the surface (Arthur et al., 2005). At the Meat Industry Research Conference; scientists from MITEC Advanced Technologies presented their ideas on irradiation of entire carcasses as part of an online decontamination method. According to industry research, irradiation could be beneficial at controlling pathogens by penetrating into the crevice on the surface that heat and chemical treatments cannot reach (Olsen, 2004). Many nations have approved the use of gamma irradiation in the preservation of foods. Irradiation is known to improve the microbiological quality of foods, improve food safety, and increase shelf life (Clavero, Monk, Beuchat, Doyle, & Brackett, 1994).

Electron Beam Irradiation (High Dose-rate X-rays)

Electron beam irradiation can be used to reduce *E. coli* O157:H7 in ground beef. In a joint effort by the Russian Academy of Sciences and the University of Missouri-Columbia (UMC), a novel portable pasteurization system was completed that allows meat packing facilities to use X-rays to increase the safety of their meat at a reduced cost (Curry et al., 2000). Electron beams can be incorporated into processing plants at relatively low cost. Researchers at the UMC compared the efficacy of high dose-rate electron beams to cobalt-60 gamma irradiation. The thickness of a meat sample directly relates to the efficacy of electron beam decontamination. X-ray (5MeV) must be used to penetrate a sample to a depth of 400 cm (Curry et al., 2000). Low-dose, low penetration electron beams provided a broad decontamination method with limited effects to quality attributes (Arthur et al., 2005). Other studies have indicated similar results with carcass surface samples and have recorded reduction at or about 4.5 log (Fu et al., 1995). Low dose, low penetration electron beams could quickly, safely and completely eliminate <3 log CFU/100 cm^2 of *E. coli* O157:H7 from carcasses (Arthur et al., 2005). There is no question that radiation can provide safe meat products from carcasses, meat cuts and ground meats. The problem now it to have greater consumer education and acceptance of this relatively new technology.

Microwave and Combination Technology

At Kansas State University a novel approach of combining lactic acid dip, vacuum packaging and microwave treatment of sub primal meat cuts was tested. Microwaves heat food faster than conventional ovens due to rapid oscillation of microwaves through the food which generate friction of rapidly vibrating dielectric molecules such as water. In a household 2450 MHz microwave oven, water molecules will vibrate 2450 million times per second when the microwave is activated. Heat is generated by friction of the water molecules as they align with the oscillating microwaves. However, heating tends to be uneven, which may influence the effectiveness in reducing microbial populations (Sawyer, Biglari, & Thompson, 1984). Microwaves inactivate bacteria by a number of possible mechanisms. These may include electrical potential, differentials, and/or oscillating magnetic fields (Kozempel et al., 2003). Microbial inactivation by microwave cannot be totally attributed to temperature alone. Culkin & Fung (1975) showed that reductions of *Salmonella typhimurium* and *Escherichia coli* in microwave cooked soup were greatest at the surface, rather than at the bottom or middle portion of the soup, despite the fact that the temperature at the top was coolest compared to the middle and the bottom. The field intensity of microwaves hitting the soup from the top may explain these results. A comprehensive review of the effect of microwaves on microorganisms in foods as made by Fung and Cunningham (1980).

In this experiment Crozier-Dodson (2000) with auther Fung used 2.3 Kg portions of semitendinosus meat (eye of round), with and without experimentally inoculated *E. coli* O157:H7, first dipped in 80C, 2% lactic acid for 0, 2, or 4 s, vacuum packaged and then placed the packages in the microwave oven and irradiated them for 0, 50 s, 60 s, and 70 s. and then stored the meat (treated and non-treated) at 4C for 24 h before microbial analysis (Fig. 13.9). The data indicated that the best combination was dipping the meat in 2% lactic acid for 2 s at 80°C and treated with microwave for 70 seconds will reduce *E. coli* O157:H7 by ca. 1.5 log CFU/cm^2. This combination treatment will render the packaged meat with reduced microbial load but will not be re-contaminated until the packaged meat was opened either at the retail store or at home. In other at-line decontamination methods such as steam pasteurization, acid and water washes and irradiation, etc. the meat was treated and microorganisms were reduced but there was always a chance or re-contamination down the line after treatment.

Costs/Benefit Analysis for Decontamination Methods

Methods of decontamination of meat vary greatly in their effectiveness against *E coli* O157:H7. Recalling that contamination events generally result in 3 log CFU/cm^2 of *E. coli* O157:H7 being transferred to the carcass (Arthur et al., 2005). This can be set as the baseline of effectiveness for decontamination methods and the cost related to recalls. Detailed analysis of costs/benefit can be found in the paper by Edwards

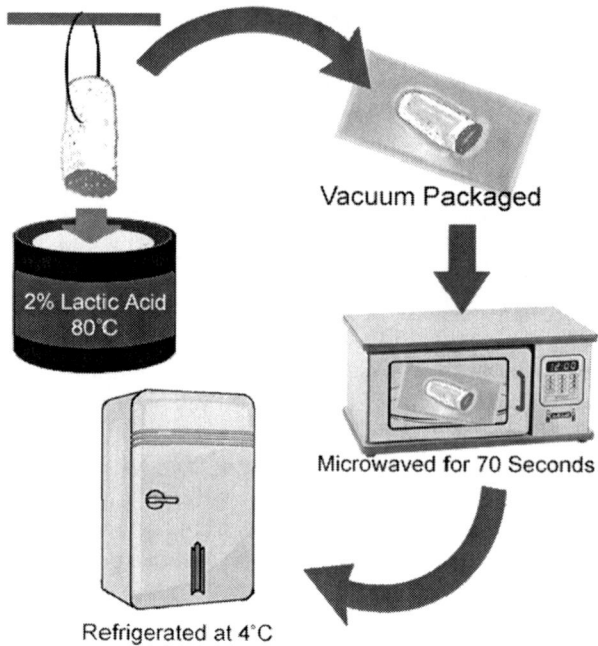

Fig. 13.9 Microwave and combination technology

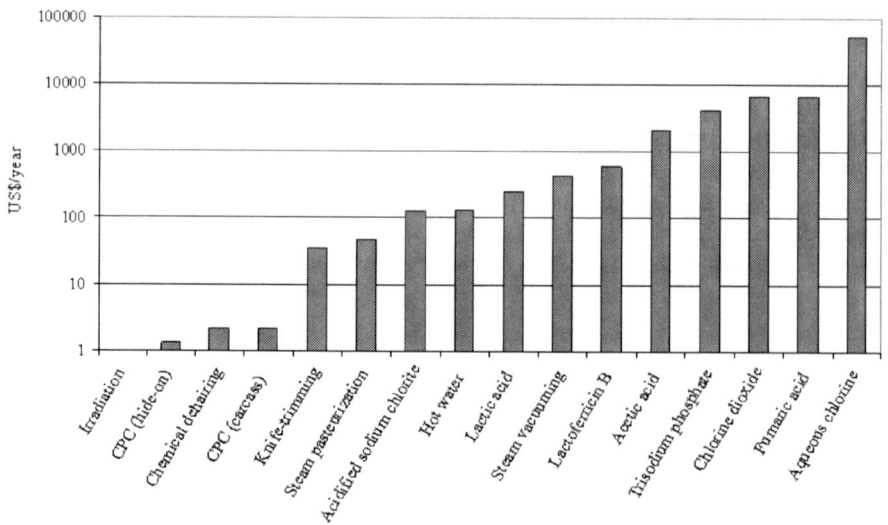

Fig. 13.10 Cost/benefit analysis for decontamination methods available for raw beef carcasses based on a yearly recall cost of $134,440 USD values based on research from peer-reviewed journals

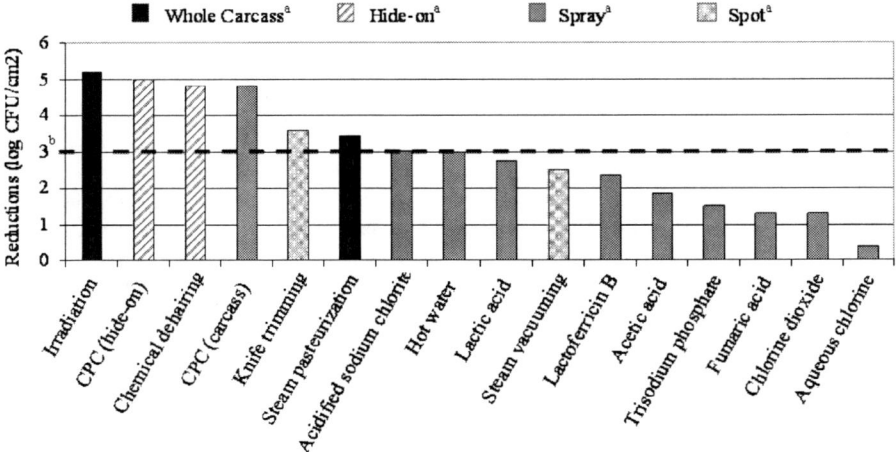

Fig. 13.11 Overall comparisons of decontamination methods for raw beef carcasses values based on research from peer-reviewed journals b Dashed line at 3.0 log CFU/cm^2 is the general level transferred to carcass in a contamination event, and is shown as the baseline of effectiveness for decontamination methods

and Fung (2006). Figure 13.10 provides the Cost/Benefit analysis for decontaminating methods available for raw beef carcasses based on a yearly total recall cost of $134,440 US$. The least cost to the most cost of yearly recall (Fig. 13.10) and also the most effective to least effective method (Fig. 13.11) are ranked in the following order: Irradiation < CPC (hide-on) < Chemical dehairing < CPC (carcass) < Knife-trimming < Steam pasteurization < Acidified sodium chlorate < Hot water < Lactic acid < Steam vacuuming < Lactoferricin B < Acetic acid, < Tri-sodium phosphate < Chlorine dioxide < Fumaric acid < Aqueous chlorine.

In conclusion, an at-line method for controlling microbial growth and spoilage is a very important topic for meat processors. This chapter provided an over view of the currently used methods for modern abattoir. Of course, many other methods are also available. Readers are advised to consult current journal articles, text books and magazines in meat science and food science for further study of this very exciting topic.

References

Abdul-Raouf, U. M., Beuchat, L. R., & Ammur, M. S. (1993). Survival and growth of *Escherichia coli* O157:H7 in ground, roasted beef as affected by pH, acidulant, and temperature. *Applied and Environmental Microbiology, 59*, 2364–2368.

Aberle, E. D., Forrest, J. C., Gerrard, D. E., Mills, E. W., Hedrick, H. B., Judge, M. D., et al. (2001). Conversion of muscle to meat and development of meat quality. *Principles of meat science* (4th ed., pp. 83–108). Dubuque, IA: Kendall Hunt.

Acuff, G. R., Castillo, A., & Savell, J. W. (1997). Hot water rinses. *Proceedings Reciprocal. Meat Conference, 49*, 125–128.

Acuff, G. R., Vanderzant, C., Savell, J. W., Jones, D. K., Griffin, D. B., Ehlers, J. G., et al. (1987). Effect of acid decontamination of beef subprimal cuts on the microbiological and sensory characteristics of steaks. *Meat Science, 19*, 217–226.

Agricultural Marketing Service (AMS). (2000). Leather meal. AMS–USDA. http://www.ams.usda.gov/nop/NationalList/TAPReviews/LeatherMeal.pdf. Accessed 27 Oct 2005.

Al-Dagal, M., & Fung, D. Y. C. (1993). Aeromicrobiology: An assessment of a new meat research complex. *J. Environmental Health, 56*, 7–14.

Anderson, M. E., Marshall, R. T., Stringer, W. C., & Naumann, H. D. (1977). Combined and individual effects of washing and sanitizing on bacterial counts of meat – A model system. *Journal of Food Protection, 40*, 373–376.

Anderson, M. E., Sebaugh, J. L., Marshall, R. T., & Stringer, W. C. (1980). A method for decreasing sampling variance in bacteriological analyses of meat surfaces. *Journal of Food Protection, 43*, 21–22.

Arthur, T. M., Wheeler, T. L., Shackelford, S. D., Bosilevac, J. M., Nou, X., & Koohmaraie, M. (2005). Effects of low-dose, low-penetration electron beam irradiation of chilled beef carcass surface cuts on *Escherichia coli* O157:H7 and meat quality. *Journal of Food Protection, 68*, 666–672.

Belk, K. E. (2001). Beef decontamination technologies. National Cattlemen's Beef Association. http://www.beef.org/documents/ACFFC.pdf. Accessed 12 Feb 2005.

Bell, R. G. (1997). Distribution and sources of microbial contamination on beef carcasses. *Journal of Applied Microbiology, 82*, 292–300.

Bellamy, W., Wakabayashi, H., Takase, M., Kawase, K., Shimamura, S., & Tomita, M. (1993). Killing of *Candida albicans* by lactoferricin B, a potent antimicrobial peptide derived from the N-terminal region of bovine lactoferrin. *Medical Microbiology Immunology, 182*, 97–105.

Berg, J. D., Roberts, P. V., & Matin, A. (1986). Effect of chlorine dioxide on selected membrane functions of *Escherichia coli*. *The Journal of Applied Bacteriology, 60*, 213–220.

Berry, E. D., & Cutter, C. N. (2000). Effects of acid adaptation of Escherichia coli O157:H7 on efficacy of acetic acid spray washes to decontaminate beef carcass tissue. *Applied Environmental Microbiology, 66*, 1493–1498.

Bosilevac, J. M., Arthur, T. M., Wheeler, T. L., Shackelford, S. D., Rossman, M., Reagan, J. O., et al. (2004). Prevalence of Escherichia coli O157 and levels of aerobic bacteria and Enterobacteriaceae are reduced when hides are washed and treated with cetylpyridinium chloride at a commercial beef processing plant. *Journal of Food Protection, 67*, 646–650.

Bosilevac, J. M., Nou, X., Osborn, M. S., Allen, D. M., & Koohmaraie, M. (2005). Development and evaluation of an on-line hide decontamination procedure for use in a commercial beef processing plant. *Journal of Food Protection, 68*, 265–272.

Bosilevac, J. M., Wheeler, T. L., Rivera-betancourt, M., Nou, X., Arthur, T. M., Shackelford, S. D., et al. (2004). Protocol for evaluating the efficacy of cetylpyridinium chloride as a beef hide intervention. *Journal of Food Protection, 67*, 303–309.

Brackett, R. E., Hao, Y. Y., & Doyle, M. P. (1994). Ineffectiveness of hot acid sprays to decontaminate Escherichia coli O157:H7 on beef. *Journal of Food Protection, 57*, 198–203.

Buchanan, R. L., & Doyle, M. P. (1997). Foodborne disease significance of Escherichia coli O157:H7 and other enterohemorrhagic E. coli. *Food Technology, 51*, 69–76.

Castillo, A., Lucia, L. M., Kemp, G. K., & Acuff, G. R. (1999). Reduction of Escherichia coli O157:H7 and Salmonella Typhimurium on beef carcass surfaces using acidified sodium chlorite. *Journal of Food Protection, 62*, 580–584.

Centers for Disease Control and Prevention (CDC). (1999). Safer and healthier foods – 1900–1999. *Morbidity and Mortality Weekly Report, 48*, 905–913.

Centers for Disease Control and Prevention (CDC). (2005). Food irradiation. CDC. http://www.cdc.gov/ncidod/dbmd/diseaseinfo/foodirradiation.htm. Accessed 27 Oct 2005.

Chad. (2004a). Hide-on carcass wash and sanitizing assembly. http://www.chadcompany.com/newproducts.html. Accessed 11 Oct 2005.

Chad. (2004b). Hot water pasteurization system. http://www.chadcompany.com/newproducts.html. Accessed 11 Oct 2005.

Chad. (2004c). Lactic acid spray assembly. http://www.chadcompany.com/ newproducts.html. Accessed 11 Oct 2005.
Chad. (2004d). Preevisceration carcass wash assembly and organic acid washer. http://www.chadcompany.com/newproducts.html. Accessed 11 Oct 2005.
Clavero, M. R. S., Monk, J. D., Beuchat, L. R., Doyle, M. P., & Brackett, R. E. (1994). Inactivation of Escherichia coli O157:H7, salmonellae, and Campylobacter jejuni in raw ground beef by gamma irradiation. *Applied Environmental of Microbiology, 60*, 2069–2075.
Code of Federal Regulations (CFR). (2005a). Acidified sodium chlorite solutions. *Code of Federal Regulations. 21 CFR 173, 325*, 140–142.
Code of Federal Regulations (CFR). (2005b). Cetylpyridinium chloride. *Code of Federal Regulations. 21 CFR 173, 375*, 148–149.
Code of Federal Regulations (CFR). (2005c). Chlorine dioxide. *Code of Federal Regulations. 21 CFR 173, 300*, 135–136.
Code of Federal Regulations (CFR). (2005d). Humane slaughter of livestock. *Code of Federal Regulations. 9 CFR, 313*, 145–151.
Code of Federal Regulations (CFR). (2005e). Ionizing radiation for the treatment of food. *Code of Federal Regulations. 21 CFR 179, 26*, 451–453.
Crouse, J. D., Anderson, M. D., & Naumann, H. D. (1988). Microbial decontamination and weight of carcass beef as affected by automated washing pressure and length of time of spray. *Journal of Food Protection, 51*, 471–474.
Crozier-Dodson, B. (2000). Combined treatments of 2% lactic acid (80°C), and microwaves for the reduction of natural microflora and Escherichia coli O157:H7 on vacuum packaged beef subprimals. MS Thesis, Hale Library, Kansas State University, Manhattan, KS.
Culkin, K. A., & Fung, D. Y. C. (1975). Destruction of Escherichia coli and Salmonella Typhimurium in microwave-cooked soups. *Journal of Milk and Food Technology, 38*, 8–15.
Curry, R. D., Unklesbay, K., Unklesbay, N., Clevenger, T. E., Brazos, B. J., Mesyats, G., et al. (2000). The effect of high-dose-rate X-rays on E. coli O157:H7 in ground beef. *IEEE Transactions on Plasma Science, 28*, 122–127.
Cutter, C. N., & Dorsa, W. J. (1995). Chlorine dioxide spray washes for reducing fecal contamination on beef. *Journal of Food Protection, 58*, 1294–1296.
Cutter, C. N., Dorsa, W. J., Handie, A., Rodriguez-morales, S., Zhou, X., Breen, P. J., et al. (2000). Antimicrobial activity of cetylpyridinium chloride washes against pathogenic bacteria on beef surfaces. *Journal of Food Protection, 63*, 593–600.
Derfler, P. S. (2004). FSIS directive: Verification of procedures for controlling fecal material, ingesta, and milk in slaughter operations. USDA FSIS. http://www.fsis.usda.gov/Frame/FrameRedirect.asp?main=/oppde/rdad/fsisdirectives/6420.2.pdf. Accessed 7 May 2005.
Dickson, J. S. (1991). Control of Salmonella Typhimurium, Listeria monocytogenes, and Escherichia coli O157:H7 on beef in a model spray chilling system. *Journal of Food Science, 56*, 191–193.
Dickson, J. S., & Anderson, M. E. (1992). Microbiological decontamination of food animal carcasses by washing and sanitizing systems: A review. *Journal of Food Protection, 55*, 133–140.
Dickson, J. S., Nettles Cutter, C. G., & Siragusa, G. R. (1994). Antimicrobial effects of tri-sodium phosphate against bacteria attached to beef tissue. *Journal of Food Protection, 57*, 952–955.
Dorsa, W. J., Cutter, C. N., Siragusa, G. R., & Koohmaraie, M. (1996). Microbial decontamination of beef and sheep carcasses by steam, hot water spray washes and a steam vacuum sanitizer. *Journal of Food Protection, 59*, 127–135.
Edwards, J. R., & Fung, D. Y. C. (2006). Prevention and decontamination of Escherichia coli O157:H7 on raw beef carcasses in commercial beef abattoirs. *Journal of Rapid Methods and Automation in Microbiology, 14*, 1–95.
Elder, R. O., Keen, J. E., Siragusa, G. R., Barkocy-Gallagher, G. A., Koohmaraie, M., & Laegreid, W. W. (2000). Correlation of enterohemorrhagic Escherichia coli O157:H7 prevalence in feces, hides, and carcasses of beef cattle during processing. *PNSA* 97, 2999–3003.

Fang, T. J., & Tsai, H. C. (2003). Growth patterns of Escherichia coli O157:H7 in ground beef treated with nisin, chelators, organic acids and their combinations immobilized in calcium alginate gels. *Food Microbiology, 20*, 243–253.

Food Safety and Inspection Service (FSIS). (1994). Post-mortem inspection. http://www.fsis.usda.gov/PDF/PHVg-Post-mortem_Inspection.pdf. Accessed 27 Oct 2005.

Food Safety and Inspection Service (FSIS). (1996a). Achieving the zero tolerance performance standard for beef carcasses by knife trimming and vacuuming with hot water and steam; use of acceptable carcass interventions for reducing carcass contamination without prior agency approval: Notice of policy change. *Federal Register, 61*, 15024–15027.

Food Safety and Inspection Service (FSIS). (1996b). Pathogen reduction: Hazard analysis and critical control point (HACCP) systems; final rule. *Federal Register, 61*, 38805–38989.

Food Safety and Inspection Service (FSIS). (2002). Backgrounder: New measures to address E. coli O157:H7 contamination. USDA FSIS. http://www.fsis.usda.gov/OA/background/ec0902.pdf. Accessed 27 March 2005.

Fu, A., Sarbanes, J. G., & Moreno, E. A. (1995). Survival of Listeria monocytogenes, Yersinia enterocolitica, and Escherichia coli O157:H7 and quality changes after irradiation of beefsteaks and ground beef. *Journal of Food Science, 60*, 972–977.

Fung, D. Y. C and Cunningham, F. E. (1980). Effects of microwave cooking on microorganisms in foods. *Journal of Food Protection, 43*, 641–650.

Fung, D. Y. C., Hajmeer, M. N., Kastner, C. L., Kastner, J. J., Marsden, J. L., Penner, K. P., et al. (2001). Meat safety. In Y. H. Hui, W-K. Nip, R. W. Rogers, & O. W. Young (Eds.), *Meat science applications* (pp. 171–205). New York, NY: Marcel Dekker.

Fung, D. Y. C., Kastner, C. L., Hunt, M. C., Dikeman, M. E., & Kropf, D. H. (1980). Mesophilic and psychrotrophic bacterial populations on hot-boned and conventionally processed beef. *Journal of. Food Protection, 43*, 547–550.

Fung, D. Y. C., Phebus, R. K., Kang, D. H., and Kastner, C. L. (1995). Effect of alcohol flaming on meat cutting knives. *Journal of Rapid Methods and Automation in Microbiology, 3*, 237–243.

Gill, C. O. (1991). Microbial principles in meat processing. In J. B. Woolcock (Ed.), *Microbiology of animals and animal products* (pp. 249–270). Amsterdam, The Netherlands: Elsevier.

Gill, C. O., McGinnis, J. C., & Badoni, M. (1995). Assessment of the hygienic characteristics of a beef carcass dressing process. *Journal of Food Protection, 59*, 136–140.

Gill, C. O., & Penney, N. (1979). Survival of bacteria in carcasses. *Applied of Environmental Microbiology, 37*, 667–669.

Gorman, B. M., Morgan, J. B., Sofos, J. N., & Smith, G. C. (1995). Microbiological and visual effects of trimming and/or spray washing for removal of fecal material from beef. *Journal of Food Protection, 58*, 984–989.

Gorman, B. M., Sofos, J. N., Morgan, J. B., Schmidt, G. R., & Smith, G. C. (1995). Evaluation of hand-trimming, various sanitizing agents, and hot water spray washing as decontamination interventions for beef brisket adipose tissue. *Journal of Food Protection, 58*, 899–907.

Haas, N. C. (2001). Decontamination using chlorine dioxide: Hearings on the decontamination of anthrax and other biological agents. U.S. House of Representatives Committee on Science. http://www.house.gov/science/full/nov08/haas.htm. Accessed 1 Sep 2005.

Hamby, P. L., Savell, J. W., Acuff, G. R., Vanderzant, C., & Cross, H. R. (1987). Spray chilling and carcass decontamination systems using lactic and acetic acids. *Meat Science, 21*, 1–14.

Hardin, M. D., Acuff, G. R., Lucia, L. M., Oman, J. S., & Savell, J. W. (1995). Comparison of methods for decontamination from beef carcass surfaces. *Journal Food Protection, 58*, 368–374.

Hulebak, K. L., & Schlosser, W. S. (2001). HACCP history and conceptual overview. USDA FSIS. http://www.ce.ncsu.edu/risk/pdf/hulebak.pdf (accessed April 6, 2005).

Hwang, P. M., Zhou, N., Shan, X., Arrowsmith, C. H., & Vogel, H. J. (1998). Three-dimensional solution structure of lactoferricin B, and antimicrobial peptide derived from bovine lactoferrin. *Biochemistry, 37*, 4288–4298.

International Meat and Poultry HACCP Alliance (IMPHA). (1996). Generic HACCP model for beef slaughter. USDA–FSIS. http://haccpalliance.org/alliance/haccpmodels/beefslaughter.pdf. Accessed 10 Oct 2005.

Johanson, L., Underdal, B., Grosland, K., Whelehan, O. P., & Roberts, T. A. (1983). A survey of the hygienic quality of beef and pork carcasses in Norway. *Acta veterinaria Scandinavica, 25*, 1–13.

Kozempel, M., Goldberg, N., & Craig, J. J. C. (2003). The vacuum/steam/vacuum process. *Food Technology, 57*, 30–33.

Lim, K., & Mustapha, A. (2004). Effects of cetylpyridinium chloride, acidified sodium chlorite, and potassium sorbate on populations of *Escherichia coli* O157:H7, *Listeria monocytogenes*, and *Staphylococcus aureus* on fresh beef. *Journal of Food Protection, 67*, 310–315.

Masson, P. L., Heremans, J. F., & Dive, C. (1966). An iron-binding protein common to many external secretions. *Clinica Chimica Acta, 14*, 735–739.

McEvoy, J. M., Doherty, A. M., Sheridan, J. J., Thomsoncarter, F. M., Garvey, P., McGuire, L., et al. (2003). The prevalence and spread of Escherichia coli O157:H7 at a commercial beef abattoir. *Journal of Applied Microbiology, 95*, 256–266.

Milmech. (2004). Milmech Beef Downward Hide Puller. http://www.milmech.com/custom-slaughter.html. Accessed Oct 2005.

Nottingham, P. M., Penney, N., & Harrison, J. C. L. (1974). Microbiology of beef processing. *New Zealand Journal of Agricultural Research, 17*, 79–83.

Nou, X., Rivera-Betancourt, M., Bosilevac, J. M., Wheeler, T. L., Shackelford, S. D., Gwartney, B. L., et al. (2003). Effect of chemical dehairing on the prevalence of Escherichia coli O157:H7 and the levels of aerobic bacteria and Enterobacteriaceae on carcasses in a commercial beef processing plant. *Journal of Food Protection, 66*, 2005–2009.

Olsen, D. (2004). Food Irradiation update. *Presented at the Meat Industry Research Conference* (pp. 1–28).

Phebus, R. K., Nutsch, A. L., Schafer, D. E., Wilson, R. C., Riemann, M. J., Leising, J. D., et al. (1997). Comparisons of steam pasteurization and other methods for reduction of pathogens on surfaces of freshly slaughtered beef. *Journal of Food Protection, 60*, 476–484.

Podolak, R. K., Zayas, J. F., Kastner, C. L., & Fung, D. Y. C. (1996). Inhibition of Listeria monocytogenes and Escherichia coli O157:H7 on beef by application of organic acids. *Journal of Food Protection, 59*, 370–373.

Prasai, R. K., Phebus, R. K., García Zepeda, C. M., Kastner, C. L., Boyle, A. E., & Fung, D. Y. C. (1995). Effectiveness of trimming and/or washing on microbiological quality of beef carcasses. *Journal of Food Protection, 58*, 1114–1117.

Ransom, J. R., Belk, K. E., Sofos, J. N., Stopforth, J. D., Scanga, J. A., & Smith, G. C. (2003). Comparison of intervention technologies for reducing Escherichia coli O157:H7 on beef cuts and trimmings. *Food Protection Trends, 23*, 24–34.

Reagan, J. O., Acuff, G. R., Buege, D. R., Buych, M. J., Dickson, J. S., Kastner, C. L., et al. (1996). Trimming and washing of beef carcasses as a method of improving the microbiological quality of meat. *Journal of Food Protection, 59*, 751–756.

Retzlaff, D., Phebus, R., Nutsch, A., Riemann, J., Kastner, C., & Marsden, J. (2004). Effectiveness of a laboratory-scale vertical tower static chamber steam pasteurization unit against Escherichia coli O157:H7, Salmonella Typhimurium, and Listeria innocua on prerigor beef tissue. *Journal of Food Protection, 67*, 1630–1633.

Rice, E. W., Clark, R. M., & Johnson, C. H. (2003). Chlorine inactivation of Escherichia coli O157:H7. *CDC Emerging Infectious Diseases*. http://www.cdc.gov/ncidod/EID/vol5no3/rice.htm. Accessed 12 Sep 2005.

Romans, J. R., Costello, W. J., Carlson, C. W., Greaser, M. L., & Jones, K. W. (2001). Cattle harvest. *The Meat We Eat* (14th ed., pp. 173–196). IL: Interstate Publishers.

Samelis, J., Sofos, J. N., Kendall, P. A., & Smith, G. C. (2001). Fate of Escherichia coli O157:H7, Salmonella Typhimurium DT 104, and Listeria monocytogenes in fresh meat decontamination fluids at 4 and 10°C. *Journal of. Food Protection, 64*, 950–957.

Samelis, J., Sofos, J. N., Kendall, P. A., & Smith, G. C. (2002). Effect of acid adaptation on survival of Escherichia coli O157:H7 in meat decontamination washing fluids and potential effects of organic acid interventions on the microbial ecology of the meat plant environment. *Journal of Food Protection, 65*, 33–40.

Sawyer, C. A., Biglari, S. D., & Thompson, S. S. (1984). Internal end temperature and survival of bacteria on meats with and without a polyvinyl chloride wrap during microwave cooking. *Journal of Food Science, 49*, 972–974.

Sheridan, J. J. (1998). Sources of contamination during slaughter and measures for control. *Journal of Food Safety, 18*, 321–339.

Shin, J., Chang, S., & Kang, D. (2004). Application of antimicrobial ice for reduction of foodborne pathogens (Escherichia coli O157:H7, Salmonella Typhimurium, Listeria monocytogenes) on the surface of fish. *Journal of Applied Microbiology, 97*, 916–922.

Sofos, J. N. (1993). The HACCP system in meat processing and inspection in the United States. *Meat Focus International, 2*, 217–225.

Sofos, J. N., Belk, K. E., & Smith, G. C. (1999). Processes to reduce contamination with pathogenic microorganisms in meat. *Congress of Meat Science and Technology, 45*, 596–605.

Stanfos. (2004). Carcass pasteurizer. Stanfos, Inc. http://www.stanfos.com. Accessed 8 Oct 2005.

Stopforth, J. D., Yoon, Y., Belk, K. E., Scanga, J. A., Kendall, P. A., Smith, G. C., et al. (2004). Effect of simulated spray chilling with chemical solutions on acid habituated and non-acid habituated Escherichia coli O157:H7 cells attached to beef carcass tissue. *Journal of Food Protection, 67*, 2099–2106.

United States Department of Agriculture (USDA). (1993). National advisory committee of microbiological criteria for foods, USDA – generic HACCP for raw beef. *Food Microbiology, 10*, 449–488.

Warriner, K., Eveleigh, K., Goodman, J., Betts, G., Gonzales, M., & Waites, W. M. (2001). Attachment of bacteria to beef from steam pasteurized carcasses. *Journal of Food Protection, 64*, 493–497.

Weller, K. (2003). ARS photo library. USDA Agriculture Research Service. http://www.ars.usda.gov/is/graphics/photos/oct00/k9082-14.htm. Accessed 8 Oct 2005.

Chapter 14
The Detection of Genetically Modified Organisms: An Overview

Jaroslava Ovesná, Kateřina Demnerová, and Vladimíra Pouchová

Introduction

Genetically modified organisms (GMOs) are those whose genetic material has been altered by the insertion of a new gene or by the deletion of an existing one(s). Modern biotechnology, in particular, the rise of genetic engineering, has supported the development of GMOs suitable for research purposes and practical applications (Gepts, 2002; Novoselova, Meuwissen, & Huirne, 2007; Sakakibara & Saito, 2006). For over 20 years GM bacteria and other GM organisms have been used in laboratories for the study of gene functions (Maliga & Small, 2007; Ratledge & Kristiansen, 2006). Agricultural plants were the first GMOs to be released into the environment and placed on the market. Farmers around the world use GM soybeans, GM corn and GM cotton that are herbicide tolerant, or insect resistant, or combine several traits that reduce the costs associated with crop production (Corinne, Fernandez-Cornejo, & Goodhue, 2004). Biotech crop coverage increased globally by 13% (12 million hectares) in 2005–06 (James, 2007), and, for example, in 2007 over 70% of all soybean-producing areas were covered by GM varieties.

Although transgenesis of livestock began around 20 years ago, GM farm animals, including fish, are still not as common as GM plants, the development of which began somewhat earlier. Transgenic plants are most often developed by the insertion of an alien (recombinant) gene using the soil bacteria, *Agrobacterium tumefaciens*, which is able to transfer a piece of its own genetic information into a plant cell. While GM plant development is at least partially based on naturally occurring mechanisms, the engineering of most transgenic livestock relies on highly technical approaches, such as pronuclear microinjection. However, newly developed techniques [sperm mediated gene transfer (SGMT), somatic cell nuclear transfer (SCNT)] have been recently introduced that enable transgenic animals to be produced more efficiently and more cheaply. These have been successfully applied to the development of several types of GM animals including cattle, sheep, pigs, chicken and fish. The potential benefits of GM animals include accelerated animal

---

J. Ovesná
Crop Research Institute, Drnovská 507, 161 06 Prague 6 – Ruzyně, Czech Republic

F. Toldrá (ed.), *Meat Biotechnology*,
© Springer Science+Business Media, LLC 2008

growth, enhanced resistance to disease, and better meat quality (Niemann, Kues, & Carnwath, 2005).

The biomedical and agricultural communities have called for more research into the use of viral vectors in the development of farmed GM animals in the hope that modified animals may prove more resistant to diseases (e.g. trypanosomiasis), as well as to the cold. GM animals are able to produce human proteins in their milk or eggs, for e.g. GM chickens may produce as much as 0.1 g of human protein in each egg. Apart from the potential pharmaceutical applications of such protein (e.g. US patent 20060185029), it is also expected that GM animal products will follow GM plants and GM foodstuffs onto the food market in the near future (Wheeler, 2007). With GM animal products likely to appear on the consumer's fork sometime soon, regulatory bodies have been set up to develop guidelines and discuss key issues (e.g. *Codex Alimentarius*) such as food safety, risk to the environment and ethical concerns. These discussions have resulted in regulations being developed for the protection of consumer rights.

However, the situation is proving complicated to manage because GM technologies and their derived products are perceived in different ways, both on a regional basis and by the various competing groups within a region. International organizations, such as WTO, OECD, FAO and *Codex Alimentarius*, are attempting to produce a harmonized approach, but there are differences between the legislative approaches followed in North America, Asia and Europe.

To protect consumer rights in the EU, the placing of GM foods on the market is subject to "the precautionary principle" in EC food law (Just, Alston & Zilberman, 2006). Based on this principle, all GM products have to be proven safe in accordance with the complex procedure prescribed by EC Regulation 1829/2003 on genetically modified food and feed. The safety of each GM product is assessed by the EFSA (European Food Safety Authority) GMO panel of experts, with information submitted by each applicant being subject to detailed risk analysis. The availability of validated detection and quantification methods is one of the prerequisites for the acceptance of GM products. Applicants are required to submit a specific method for the identification and quantification of the particular GMO they wish to market. The method supplied must be verified by the Community Reference Laboratory (CRL) at the Institute of Consumer Health and Protection in Ispra, Italy (http://ihcp.jrc.ec.europa.eu), and validated in collaborative trials conducted by laboratories affiliated to the European Network of GMO Laboratories (ENGL). Once the method has been validated it is published at http://gmocrl.jrc.it/statusofdoss.htm. Apart from the method, an applicant must also provide a sample of the control material to the CRL, together with, on a confidential basis, information about the DNA sequence inserted into the genome of the modified organism. Additional legislation covers GM handling procedures (Table 14.1); the whole system being based on the traceability and appropriate labeling of approved GMO products. (In the EU, all products containing more than 0.9% of GM material as 'unavoidable contamination of a constituent' have to be labeled.)

The need to be able to trace GMOs and their derived products on the market has generated a demand for analytical methods capable of reliably detecting, identifying

Table 14.1 Basic EU legislation driving GM handling

Directive 2001/18/EC	On the deliberate release into the environment of genetically modified organisms and repealing Council Directive 90/220/EEC
Regulation (EC) 1829/2003	On genetically modified food and feed
Regulation (EC) 1946/2003	On the transboundary movements of genetically modified organisms (GMOs)
Regulation (EC) 1830/2003	On traceability and labelling of genetically modified organisms and the traceability of food and feed products produced from genetically modified organisms and amending Directive 2001/18/EC
Regulation (EC) 65/2004	On establishing a systém for the development and assignment of unique identifiers for genetically modified organisms

and quantifying them; such methods provided the basis for the implementation of appropriate labeling rules.

Basic Approaches in GMO Analysis

When information about a modified sequence is available, unambiguous identification of the GMOs and transgenic elements used in a food product is, in most instances, easy to realize. At present, hundreds of GMOs around the world have been assessed with respect to food safety, some of which have also been approved by the EU.

The detection method selected is a reflection of the way in which a particular GMO was developed. The development of a GMO generally consists of the following steps: preparation of a suitable recombinant construct; delivery of the construct into the target cell(s); selection and further development of an entire transgenic organism. The following elements constitute the basis of a recombinant gene construct: a promoter that drives gene expression; a DNA sequence coding for the protein; and a terminator that halts gene expression. These elements (e.g. a virus promoter, followed by an animal/plant gene, and a bacterial terminator) may all originate from different species. Usually, the construct delivered into the cell also contains a selectable marker that enables identification of the transgenic cells. Other types of constructs, e.g. Cre/lox system (Sauer, 2002), allow the deletion of the gene of interest from the genetic equipment of a cell.

Selectable markers are used with antibiotic resistance [e.g. neomycin phosphotransferase II (NPTII)] or herbicide resistance genes, the latter currently only being used in the case of transgenic plants. In the case of animals either antibiotic resistance genes [e.g. chloramphenicol acetyl transferase (CAT), NPTII] or other appropriate markers, such as green fluorescent protein (GFP), β-galactosidase (beta-gal) or secreted alkaline phosphatase (AP) are used. Marker genes, used for the selection of transformed cell lines, are also often used for the detection of GMOs.

GMOs and their derivatives are identified by detecting the DNA molecules, RNA molecules, proteins or other metabolites associated with, or derived from, a specific genetic modification. As with any analytical procedure, a precise sequence must be followed, which involves a representative sample being screened for the presence of GM materials, and, in the case of a positive result, the appropriate identification and quantification method being applied (Fig. 14.1).

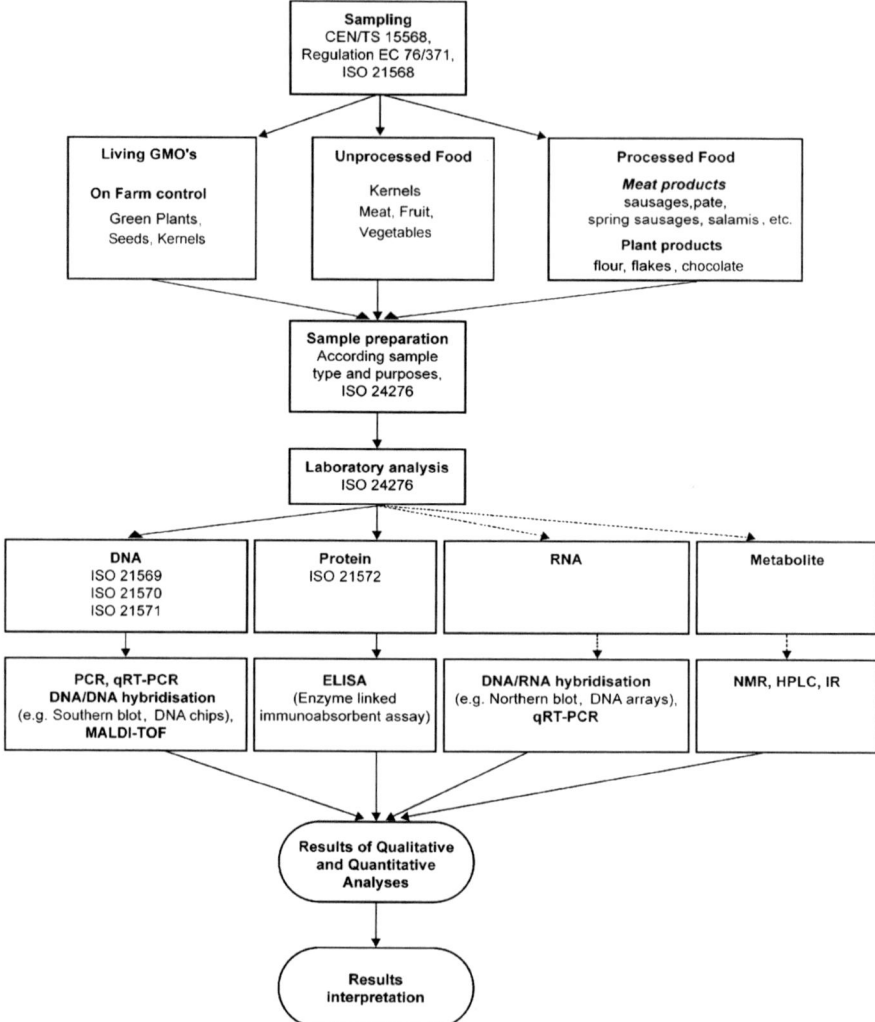

Fig. 14.1 Flowchart of DNA detection process. Detection of GMOs is based on EC regulation 1829/2004 and relevant ISO standards. Beginning with sampling, the process is then followed by the preparation of a test portion, by analyte extraction and by analysis. Results should be issued in accordance with ISO standards

Sampling

Sampling and sample preparation are crucial steps in the process of GMO detection (Terry, Harris & Parkes, 2002). While comprehensive analytical procedures exist to deliver precise results, the ability of a sampling procedure in general, and a sample size in particular, to erroneously influence final values should not be underestimated. The sampling procedure determines the representativeness of a result. In the case of GM material, because the distribution of GM particles in a lot is usually random and heterogeneous, a distribution-free statistical model has been suggested (Paoletti et al., 2006), and models have been developed that combine information about all stages of GMO event detection (Macarthur et al., 2007).

Being closely connected with statistical requirements, sample size has often been the subject of considerable discussion. It must be large enough to enable reliable GM detection at the desired sensitivity. A sample containing at least 3,500 particles is necessary to ensure that a 1% concentration of GMO in a batch will be reliably detected with a probability of $p \leq 0.05$. To obtain this probability, in the case of low level contamination, e.g. 0.1%, a test sample should contain 10 times more particles. Processed foods and complex matrices contain GM materials as one of several ingredients, so strongly stratified variance distribution can be expected. This is why it not always possible to use existing sampling strategies for the detection of other components in food commodities (Berger & von Holst, 2001).

General principles and rules for the sampling of GM foodstuffs are described in more detail in *ISO 21568, Foodstuffs - Methods of analysis for the detection of genetically modified organisms and derived products -Sampling*. In accordance with this procedure, samples delivered to laboratories must be stored under appropriate conditions, and further test portions prepared. A test portion should reflect the concentration of GM material in the sample. A laboratory will analyze at least two test portions prepared from the identical sample. Analyte processing then follows.

Analysis of GM Material

The detection of GM materials is based on the identification of the DNA sequence, RNA sequence, protein sequence, or even metabolites, all of which result from genetic modification (Table 14.2). Methods differ with respect to their analytical parameters, feasibility and cost-effectiveness. Only DNA- and protein-based methods are currently used for control purposes.

Protein-Based Analysis

Typically, the production of specific proteins distinguishes GMOs from non-GMOs. Intact proteins are the required target for protein-based analysis; in the case of GMOs, by using immunoassays. Immunoassays are analytical measurement

Table 14.2 Basic overwiev of detection methods

Target	Method	Matrix (example)
DNA	PCR, qPCRm DNA/DNA hybridisation (e.g. Southern blot, DNA arrays), MALDI-TOF	Processed/unprocessed food__ Meat products as sausages, paté, spring sausages, salami
RNA	DNA/RNA hybridisation e.g. Northern blot), qRT-PCR	Living GMOs
Protein	ELISA (Enzyme Linked Immonoabsorbent Assay)	Unprocessed food
Metabolite	NMR, HPLC, NIRS	Containing lipids and fatty acid

systems that use antibodies as test reagents. Antibodies, produced in the serum of animals in response to foreign substances (antigens), are proteins that specifically bind to the substance that elicited their production. ELISA (Enzyme Linked Immunosorbent Assay) is the technique most commonly used for the specific binding of an antibody to a target protein. The principal advantage of ELISA lies in the high specificity of the ensuing immunological reaction, which allows antigenic substances to be accurately identified, even in the presence of interfering compounds. It also offers a high degree of automation and sample throughput. In addition, easy-to-use variations of this technique, such as lateral flow strips or dipstick kits, may offer semi-quantitative tests of considerable practical value, for example, in the detection of transgenic plants, such as 'Roundup Ready' soybeans. While such tests (described in detail in international standard ISO 21572, *Foodstuffs - Methods for the detection of genetically modified organisms and derived products - Protein based methods*) cannot be used to reliably analyze processed materials, they are well suited to the inspection of raw materials, and, therefore, might be of use in the identification of live transgenic animals. Because they offer speedy turnaround times, and require relatively small investment in both equipment and personnel, ELISA and protein strip tests are the methods of choice for differentiating between GM and non-GM entities, and for identifying the modification event.

DNA-Based Methods

Unlike protein, DNA is a stable molecule capable of being identified even when broken, degraded or denatured to some extent. In even highly-processed food matrices it is possible to detect, isolate and further analyze DNA. Polymerase Chain Reaction (PCR), which can be used for processed material in which DNA has been degraded, is currently the method of choice for such analysis.

Polymerase Chain Reaction

PCR (Mullis & Falloona, 1987) allows the multiplication of certain gene sequences. The whole procedure is based on the activity of *Taq* polymerase (DNA-dependent DNA polymerase), which was originally isolated from the bacterial species

Thermus aquaticus. This enzyme, which basically requires a DNA template (as short as 200 pb is sufficient) and pieces of single-stranded DNA (primers) surrounding the target sequence, is able to produce (amplify) millions of copies of the part of the genetic information that is to be analyzed. PCR is a cyclical process, each cycle consisting of three successive steps running at different temperatures. The first step in a cycle involves the separation of the two strands of the original DNA molecule. The second step involves the binding of the two primers to their complementary strands. Using the complementary strands as a template, the third step involves making two copies of the original double-stranded DNA molecule by adding the right nucleotides. Once a cycle is completed, it can be repeated, typically between 30–50 times. With the number of copies doubling in each cycle, the number of target sequences grows exponentially according to the number of cycles undergone (Holst Jensen, 2004).

If an analyte contains a target sequence, for e.g. part of the transgene- or species-specific internal gene, amplification products can be subsequently visualized and/or quantified. The whole procedure consists of: (1) DNA extraction and purification, (2) amplification of the target sequence, (3) visualization by electrophoresis or fluorogenes detection, (4) quantification and (5) interpretation of results. The application of PCR to GMO analysis has been described in the literature (Lipp et al., 2005; Miraglia et al., 2004).

DNA Isolation

DNA quality, in particular its purity, and DNA quantity are critical parameters for GMO analysis. DNA can be isolated either as high molecular weight DNA obtained from fresh material (e.g. blood, grains) or as fragmented DNA obtained from elderly or processed matrices (e.g. bone parts, flour, sausages, pâté).

The basic steps necessary to accomplish DNA isolation have been described by several authors (Somma 2004; Spoth & Strauss, 1999). DNA isolation begins with homogenization of the test portion, grinding and cell disruption. Cell disruption, which results from homogenization followed by the addition of a buffer consisting, amongst other reagents, of a detergent (e.g. SDS or CTAB) and DNase inhibitors (β-mercaptoethanol and EDTA), enables the transfer of DNA from a cell into the buffer. The mixture is further purified as cell debris, proteins and RNA from the extract are all removed. DNA is then precipitated. DNA can be further purified by appropriate procedures, such as the use of commercial kits based, for example, on gel filtration.

Several variations of this basic procedure exist. For example, PVP-based methods are particularly suitable for DNA extraction from raw or boiled meat, sausages, chopped meat and processed meat products containing soybean protein. CTAB-based methods have been shown to be efficient for the isolation of DNA from gravy, hamburgers, fatty salami and ham. All of these methods have been validated in collaborative trials, and are described in detail in *ISO 21 571 "Foodstuffs - Methods of analysis for the detection of genetically modified organisms and derived products - Nucleic acid extraction"*.

The resulting DNA should meet the quality criteria. Typically, its absorbance is measured in UV light, which enables the presence of any contaminating proteins and/or phenolic compounds to be identified. Gel electrophoresis is used to detect the presence of contaminating RNAs that may interfere with upstream reactions.

PCR Analysis

In order to use PCR, the precise nucleotide sequences flanking both ends of the target DNA region must be known. Any PCR-based detection strategy is dependent on a detailed knowledge of the transgenic DNA sequence, which, in the case of authorized GMOs, can be found in the published databases (e.g. http://www.agbios.com/dbase.php). It also depends on the selection of the appropriate oligonucleotide primers. In addition to the well-documented issues concerning primer selection (Burpo, 2001), the choice of primers depends on the objectives of PCR analysis.

Once the DNA is isolated, its amplificability must be ascertained. Usually, a species-specific sequence is used. Species-specific genes (e.g. lectin from soybeans, zein from corn) should be represented by one copy per haploid genome. Protocols exist for the amplification of many species-specific sequences, and may be used for control purposes (Von Holst, Baeten, Berben, & Brambilla, 2004)

If the isolated DNA is amplifiable, screening methods are usually applied to exclude negative samples from subsequent analysis. Screening tests for the common transgenic are executed elements (e.g. promoters, terminators, selectable markers) present in some GMOs. In the case of GM plants, the 35S CaMV promoter (35S subunit of the cauliflower mosaic virus) and the T-NOS terminator (from the nopaline strain of *Agrobacterium tumefaciens*) are used for screening. However, this method is not GMO-specific, and contamination of the commodities by either of the two organisms (CaMV and *A. tumefaciens*) may lead to false positives. Therefore, in case of a positive result, further specific tests must be performed to verify the findings.

With the increasing number of GMOs in the market, not all kinds of GMOs can be reliably detected using existing screening methods. Some transgenic coding sequences might also occur across multiple GM varieties or species. If the primers are specific to a transgene occurring in multiple GMs it becomes difficult to locate the original source of the GM material in complex matrices. Construct-specific methods, involving, for example, one primer specific to the promoter and another specific to the coding sequence, may help to identify GMOs more accurately. However, event-specific methods, spanning transgene- and species-specific sequences, remain desirable for GM plant identification Fig. 14.2 (Yang et al., 2007).

Separation (either on agarose gel or by using PAGE) followed by visualization by UV light enable the amplification products to be easily identified and evaluated (Fig. 14.3). Because each amplification product is characterized by its specific size, it can be identified by checking it against a molecular ladder and by using positive controls. The most respected control materials are the certified reference materials produced by the Institute for Reference Material and Measurement (IRMM), an EU Joint Research Institution (JRC) located in Geel, Belgium. Also,

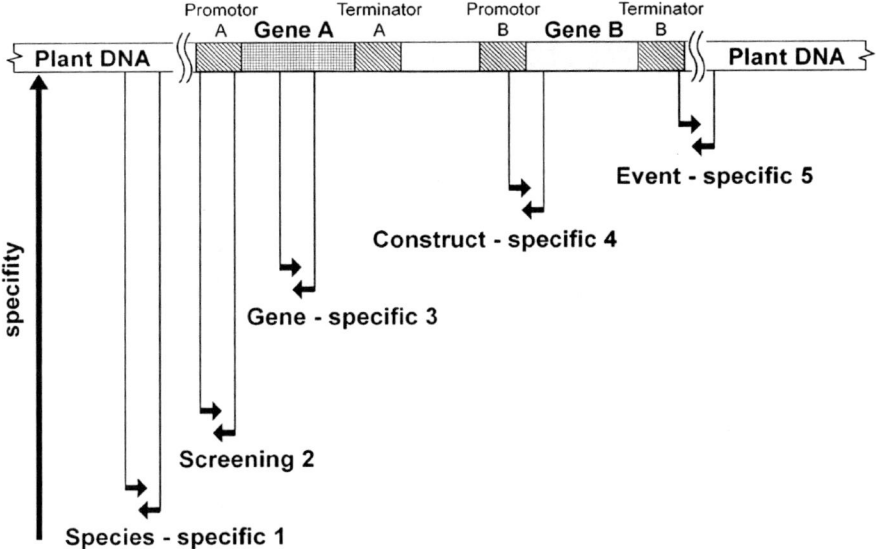

Fig. 14.2 Plots show possible accommodation of PCR primers within recombinant sequence. If common elements (e.g. 25s CaMV, T.NOS) are amplified, it can be described as screening. (a). Screening eliminates samples that probably do not contain GMOs. When transgenic elements are common across several GMOs (e.g. Bt gene) the assay is referred to as transgene-specific (b). When other primers, spanning two different elements (e.g. promoter, coding sequence), are common for several GMOs, the system/assay is referred to as construct-specific (c). Primers spanning specifies-specific and transgene-specific are unique for an event which is referred as event-specific (d).

CRL produces plasmid control materials containing GM-specific amplicons for the use of EU controlled laboratories. While the most reliable method for confirming the authenticity of a PCR product is still by its sequencing, only a few laboratories are equipped to carry out this approach routinely. Hence, few authors have reported on its use (Byrne, 2002; DMIF-GEN Final Report, 1999; Ehlers et al., 1997). Other methods used to confirm the authenticity of the amplicon include restriction analysis and Southern hybridization (Einspanier, 2006).

Several modifications/enhancements have been made to the PCR method, one of which, multiplex PCR, involves using several primers in the reaction mixture. However, multiplex PCR is not widely used as it has a higher risk of producing false results.

PCR is a highly sensitive method, theoretically able to efficiently amplify a single DNA molecule. In practice, reliable results are obtained when the reaction mixture contains 25–100 molecules of template DNA (Berdal & Holst-Jensen, 2001; Kay & van den Eede, 2001). Consequently, PCR is susceptible to external contamination. Therefore, several controls must be employed during each test to minimize false-positive results caused by contamination. As a minimum, these tests must include: positive control (tube containing reagents and DNA with target sequence, e.g. from positive standard); negative master mix control (control containing reagents and water, instead of analyzed DNA); and extraction control (control containing reagents

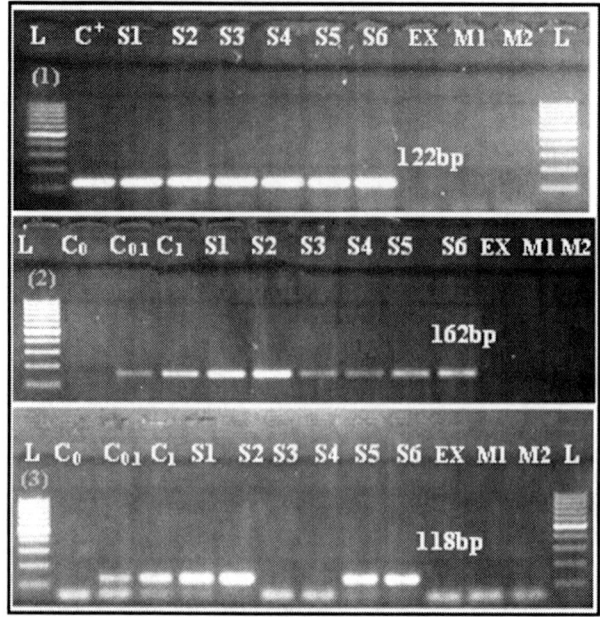

Fig. 14.3 DNA-based GMO analysis consists of several steps. (1) The quality of the isolated DNA is checked by amplification (PCR) of the internal gene, which is then followed by, (2) screening for common DNA promoters (e.g. 36S CaMV in the case of transgenic plants), and (3) by identification of the specific transgene in cases of positive screening. All reactions involve negative PCR controls (Neg), Negative extraction controls (Next), positive controls (P)

and solution resulting from extraction blank). Another often-used improvement is the running of PCR in real-time format, rather than using the more cost-effective, but less reliable, conventional PCR.

Quantification by PCR

Real-time PCR analysis, the most precise method of gene quantification currently available, requires special equipment able to trap and process fluorescent signals. This method is used to measure the amount of fluorescence in a PCR reaction, and from that, to estimate the amount of product synthesized during PCR. Internal probes, coupled to labels that are able to emit fluorescent light, are added to each reaction. As the probes hybridize to the amplification products (one probe per molecule), the intensity of the fluorescent light produced corresponds to the amount of newly synthesized DNA molecules. Thus, real-time PCR resolves the problem of establishing a relationship between the initial concentration of target DNA and the total amount of PCR product synthesized during the reaction. Several types of hybridization probes are available, with TaqMan probes currently being the most widely used in GMO diagnostics. Other types of probes exists (e.g. LUX, FRET, etc.) but are not currently validated for GMO detection (Bowyer, 2007).

14 The Detection of Genetically Modified Organisms: An Overview

The aim of quantification is to determine the ratio (expressed as a percentage) of two DNA targets. The first target is a species-specific single copy sequence (e.g. lectin in the case of soybeans) and the second is the transgene of the analyzed GMO. The comparative and standard curve methods are the two main approaches used to estimate the target molecule concentration from the fluorescence intensities recorded for the tested samples (Applied Biosystems, 2001). The standard curve method is more robust because it takes into account the amplification efficiency of different amplicons. Being dependent on reaction setup and DNA purity, the amplification efficiency of both targets may differ (Cankar, Štebih, Dreo, Žel, & Gruden, 2006). Technically, either one multiplex (duplex) reaction or two simplex reactions may be applied. Each reaction type has its advantages and disadvantages. For instance, multiplex reactions minimize pipetting errors, whereas simplex reactions reduce the problem of overlapping fluorescent signals coming from different internal probes.

For each target, a calibration curve is measured. After calibration curve and test sample data have been collected, the concentration of GM material is calculated (Fig. 14.4). Results should be expressed in compliance with current legislation.

Apart from increasing the likelihood of accurate quantification, the other main advantages of real-time PCR are in its speeding up of sample throughput (Kubista et al., 2006) and minimization of false results (Vaïtilingom et al., 1999).

Currently, real-time PCR is considered to be the most powerful tool for the detection and quantification of GMOs across a wide range of agricultural and food products (Bonfini, Heinze, Kay, & Van den Eede, 2001).

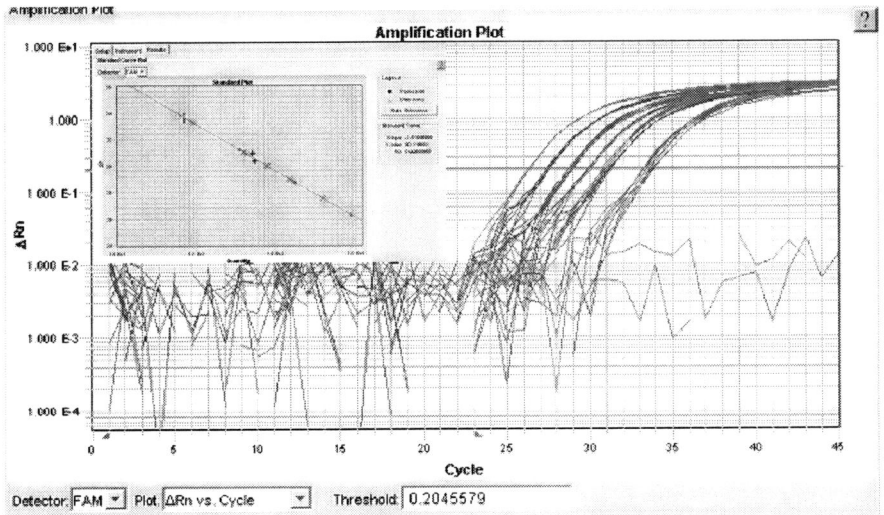

Fig. 14.4 An example of the result of quantitative analysis. Real-Time PCR results are shown in the form of curves that identify a Ct value (a PCR cycle, in which the amount of amplified DNA is higher than the background). Ct values enable the content of GM to be calculated from a calibration curve (b).

Other DNA-Based Methods

About 10 years ago, GM-based food/feed production was rare, but today the number of GM products approved or awaiting approval is growing rapidly. While a few years ago it was enough to use the PCR method for the detection of single GM events, such as 'Roundup Ready' soybean, nowadays we must employ tens of them, and in the near future it is likely that hundreds of such methods will be needed (Table 14.3). Already there is an increasing demand for more sophisticated methods capable of capturing different GMOs in a single reaction. One such option may be the use of DNA chips (arrays) that are based on the hybridization of analyzed DNA with specific probes on a glass support (Deisingh & Badrie, 2005; Nesvold, H., Kristoffersen, Holst-Jensen, & Berdal, 2005). Thousands of DNA sequences can be accommodated on one glass support giving rise to the possibility of being able to simultaneously identify multiple GMOs. The DNA chips (arrays) currently available have primarily been used for research purposes: to evaluate gene expression (e.g. Affymetix DNA chips); to identify the methylation status of a genome (e.g. TILLING arrays); and to differentiate among genotype [DArT (Diversity Array Technology) arrays recognizing SNPs (single nucleotide polymorphism), e.g. APEX arrays] (Khan et al., 1998; Roy & Sen, 2006; Schrijver, Külm, Gardner, Pergament, & Fiddler, 2007; Wenzl et al., 2004). Such analytical procedures require costly equipment, and even with it, analyte processing is not fully resolved in the case of GMO. However, Chen et al. (2006) described a method for the realization of full analyte processing using a cDNA array as a prototype for the detection of multiple GMOs. Their procedure requires further testing and validation. Although a cheaper version of the DNA array technique recently appeared, utilizing scaled-down equipment and multiplex PCR to process the analyte (Kok, Aarts, Van Hoef, & Kuiper, 2002; Wenijn, Siyang, Minnan, & Guangming, 2003), the very use of multiplex PCR constitutes the limitation of this approach.

Table 14.3 Basic websites addresses of international bodies dealing with GMOs

Institution	URL
European Food Safety Authority (EFSA)	http://www.efsa.europa.eu
Institute for Consumer Health Protection in Ispra, Italy	http://ihcp.jrc.ec.europa.eu
Europen Network of GMO Laboratories. (ENGL)	http://gmo-crl.jrc.it/statusofdoss.htm
World Trade Organization (WTO)	http://www.wto.org
Organization for Eonomic Co-operation and Development (OECD)	http://www.oecd.org
Food and Agriculture Organization of the United Nations (FAO)	http://www.fao.org
Codex Alimentarius (FAO/WHO Food Standards)	http://www.codexalimentarius.net
Agbios database	http://www.agbios.com/dbase.php
International Organization for Standardization	http://www.iso.org
United States Department of Agriculture	http://www.usda.gov
U.S.Food and Drug Administration	http://www.fda.gov/
U.S.Environmental Protection Agency	http://www.epa.gov/
Institute for Reference Materials and Measurements	http://www.irmm.jrc.be

Another approach, based on Ligation Probe Amplification (LPA), has been recently introduced. LPA allows simultaneous event-specific detection and the relative quantification of DNA from several GMOs in a single reaction. However, this approach requires careful primer setup and the optimization of reaction conditions just for the system to detect a mere 40–50 targets per reaction (Carrino, 1996; Mezzelani et al., 2002).

Novel approaches to DNA analysis based on micro and nanoparticles, as well as on mass spectrometry, may open new possibilities in DNA detection. The possible advantages of some of these approaches have been described by Sato, Hosokawa, and Maeda (2007); Einspanier (2006), amongst others.

Other Methods of GMO Detection

Where the composition of GMO, e.g. fatty acids or triglycerides, has been modified, conventional chemical methods based on chromatography can be used to detect differences in the chemical profile. Metabolites, where they are the product of transgenic protein activity, can be detected by physical methods, such as NMR (Nuclear Magnetic Resonance), HPLC (High Performance Liquid Chromatography) or Near Infrared Spectroscopy (NIRS) (Heller, 2006; Roussel, Hardy, Hurburgh, & Rippke, 2001; Von Holst et al., 2004).

Data Quality and Interpretation

Any procedure used in GMO analysis for control purposes should be based on validated methods, some of which have been published by ISO (International Organization for Standardization http://www.iso.org). All methods should be reproducible, repeatable, precise and accurate and should reflect the true value. These analytical parameters are critical when applied to any method used for the detection of GMOs in foodstuffs. The limits of detection (LOD) and quantification (LOQ) should be determined in accordance with the method being used and the matrix of the sample being analyzed (Berdal & Holst-Jensen, 2001). Full details of how to report and issue results are described in *ISO 24276: "Foodstuffs - Methods of analysis for the detection of genetically modified organisms and derived products - General requirements and definitions"*. To ensure data quality, all control laboratories should be accredited according to the ISO 17025:2005 (Žel, Cankar, Ravnikar, Camloh, & Gruden, 2006).

Despite the prescribed procedures, the Food Analysis Performance Scheme (FAPAS), which operates a proficiency testing scheme for GM food analyses in the UK, has reported the following findings: a tendency for laboratories to overestimate GM levels; significant differences between the results obtained using PCR and ELISA; data being skewed and not normally distributed until undergoing mathematical treatment (Powell & Owen, 2002).

These findings indicate the need to better understand the processes underlying GM detection methods, as well as the need for considerable investment in the research of more advanced techniques. Further development of methods used for the detection and quantification of GMOs in agricultural commodities is essential to monitor the implementation of regulatory compliance concerning the use of GM food/feed. Such research should focus on improving the analytical parameters of high-throughput methods so that they are fit for the purpose.

It is likely that new methods will not only be used to control GMOs and accommodate consumer requirements, in some countries, but also to protect breeders' rights and intellectual properties worldwide.

Acknowledgements This work was supported by the project CZ0002700602 of the Czech Ministry of Agriculture and by project the 1B44068 of the National Agency for Agricultural Research.

References

Applied Biosystems. (2001). *Relative Quantification of Gene Expression*. Applied Biosystems. *User Bulletin 2*.
Berdal, K. G., & Holst-Jensen, A. (2001). Effect of food components and processing parameters on DNA degradation in food. European Food Research and Technology, 213, 432–438.
Berger, B., & von Holst C. (2001). Pesticides residues in products of plant origin in the European Union: Sampling strategy and results from the co-ordinated EU monitoring programme in 1996 and 1997. *Environmental Science and Pollution Research International, 8*, 109–112.
Bonfini, L., Heinze, P., Kay, S., & Van den Eede, G. L. M. (2002). *Review of GMO Detection and Quantification Techniques*. EUR 20384/EN.
Bowyer, V. L. (2007). Real-time PCR. *Forensic Science, Medicine, and Pathology, 3*, 61–63.
Burpo F. J. (2001). A critical review of PCR primer design algorithms and crosshybridizationcase study, Biochemistry 218 at Stanford University. Retrieved August 11, 2001, from http://cmgm.stanford.edu/biochem218/Projects%202001/Burpo.pdf.
Byrne, D. (2002). Commission Recommendation of 25 January 2002 concerning a coordinated programme for the official control of foodstuffs for 2002. *Official Journal of the European Communities 30.1.2002 L 26/8*. Retrieved January 30, 2002, from http://www.sapidlife.org/documenti_repository/scarica.php?file=en200612071039152002_66_CEEN.pdf.
Cankar, K., Štebih, D., Dreo, T., Žel, J., & Gruden, K. (2006). Critical points of DNA quantification by real-time PCR – effects of DNA extraction method and sample matrix on quantification of genetically modified organisms. *BMC Biotechnology, 6*: 37.
Carrino, J. J. (1996). Multiplex ligations-dependent amplication. *Patent aplication WO 96/15271*.
Chen, T. L., Sanjaya, V., Prasad, C. H., Lee, K. H., Lin, H. C., Chiueh,V., et al. (2006). Validation of cDNA microarray as a prototype for throughput detection of GMOs. *Botanical Studies, 47*, 1–11.
Corinne, A., Fernandez-Cornejo, J., & Goodhue, R. E. (2004). *Farmers' Adoption of Genetically Modified Varieties with Input Traits*. Giannini Foundation Research Report 347 (October 2003). University of California: Division of Natural Resources. Retrieved October, 2003, from http://giannini.ucop.edu/researchreports/347-goodhue.pdf.
Deisingh, A. K., & Badrie, N. (2005). Detection approaches for genetically modified organisms in foods. *Food Research International, 38*, 639–649.
DMIF-GEN Final Report. (1999). Retrieved February 1, 2000, from http://www.dmif-gen.bats.ch/dmif-gen/.

Ehlers, B., Strauch, E., Goltz, M., Kubsch, D., Wagner, H., Maidhof, H., et al. (1997). *Nachweis gentechnischer Veränderungen in Mais mittels PCR. Bundesgesundhbl.*, 4, 118–121.
Einspanier, R. (2006). DNA based methods for detection of genetic modification. In K. J. Heller (Ed.), *Genetically engineered food: Methods and detection* (pp. 163–185, 2nd ed.). Wiley-VCH Verlag, GmbH & Co. KGaA., Weinheim, Germany
Gepts, P. (2002). A comparison between crop domestication, classical plant breeding, and genetic engineering. *Crop Science, 42*, 1780–1790.
Heller, K. J. (2006). Detection of Genetic Modifications-Some Basic Considerations *Genetically engineered food methods and detection* (pp. 155–162, 2nd ed.) Wiley-VCH Verlag GmbH., Weinheim, Germany
Holst-Jensen, A. (2004). Detecting genetically modified food. New Food, 7, 16, 18–20, 22.
James, C. (2007). *Global Status of Commercialized Biotech.* ISAAA Briefs No. 35-2006: /GM Crops: 2006. Retrieved October 16, 2007, from http://www.isaaa.org/Resources/ Publications/briefs/35/executivesummary/default.html.
Just, R. E., Alston J. M., & Zilberman. (2006). Regulating agricultural biotechnology: Economics and policy. *Natural Resource Management and Policy, 30*, 392–402.
Kay, S., & Van Den Eede, G. (2001). The limits of GMO detection. *National Biotechnology, 19*, 405.
Khan, et al. (1998). Gene expression profiling of alveolar rhabdomyosarcoma with cDNA microarrays. Cancer Research, 58, 5009–5013.
Kok, E. J., Aarts, H. J., Van Hoef, A. M., & Kuiper, H. A. (2002). DNA methods: Critical review of innovative approaches. *Journal of AOC International, 85*, 797–800.
Kubista, M., Andrade, J. M., Bengtsson, M., Forootan, A., Jonák, J., Lind, K., et al. (2006). The real-time polymerase chain reaction. *Molecular Aspects of Medicine, 27*, 95–125.
Lipp, M., Shillito, R., Giroux, R., Spiegelhalter, F., Charlton, S., Pinero, D., et al. (2005). Polymerase chain reaction technology as analytical tool in agricultural biotechnology. *Journal of AOC International, 88*, 136–155.
Macarthur, R., Murray, A. W. A., Allnutt, T. R., Deppe, C., Hird, H. J., Kerins, G.M., et al. (2007). Model for tuning GMO detection in seed and grain. *National Biotechnology, 25*, 2, 169–170.
Maliga, P., & Small, I. (2007). Plant biotechnology: All three genomes make contributions to progress. *Current Opinion in Biotechnology, 18*, 97–99.
Mezzelani, A., Bordoni, R., Consolandi, C., Rossi Bernardi, L., Rosini, A., Castiglioni, B., et al. (2002). Ligation detection reaction and universal array for detection and identification of genetically modified organisms (GMOs). *Minerva Biotechnology, 14*, 269–271.
Miraglia, M., Berdal, K. G., Brera, C., Corbisier, P., Holst-Jensen, A., Kok, E. J, et al. (2004). Detection and traceability of genetically modified organisms in the food production chain. *Food Chemistry and Toxicology, 42*, 1157–1180.
Mullis K. B., & Faloona, F. A. (1987). Specific synthesis of DNA in vitro via a polymerase-catalysed chain reaction. *Methods in Enzymology, 155*, 335–350.
Nesvold, H., Kristoffersen, A. B., Holst-Jensen, A., & Berdal, K. G. (2005). Design of a DNA chip for detection of unknown genetically modified organisms (GMOs). *Bioinformatics, 21*, 1917–1926.
Niemann, H., Kues, W. A., & Carnwath, J. W. (2005). Transgenic farm animals: Present and future. *Revue scientifique et technique (International Office of Epizootics), 24*, 285–298.
Novoselova, T. A., Meuwissen, M. P. M., & Huirne, R. B. M. (2007). Adoption of GM technology in livestock production chains: An integrating framework. *Trends in Food Science & Technology, 18*, 175–188.
Paoletti, C., Heissenberger, A., Mazzara, M., Larcher, S., Grazioli, E., Corbisier P., et al. (2006). Kernel lot distribution assessment (KeLDA): A study on the distribution of GMO in large soybean shipments, *European Food Research and Technology, 224*, 129–139.
Powell, J., & Owen, L. (2002). Reliability of food measurements: The application of proficiency testing to GMO analysis. *Accreditation and Quality Assurance, 10*, 392–402.
Ratledge, C., & Kristiansen, B. (2006). *Basic Biotechnology* (3rd ed.) Cambridge: University.

Roussel, S. A., Hardy, C. L., Hurburgh, C. R., & Rippke, G. R. (2001). Detection of Roundup Ready™ Soybeans by Near-Infrared Spectroscopy. *Applied Spectroscopy, 55*, 1425–1430.

Roy, S., & Sen, C. K. (2006). CDNA microarray screening in food safety. *Toxicology, 221*, 128–233.

Sakakibara, K., & Saito, K. (2006). Review: Genetically modified plants for the promotion of human health. *Biotechnology Letters, 28*, 1983–1991.

Sato, K., Hosokawa, K., & Maeda, M. (2007). Biosensors based on DNA-nanoparticle conjugates. *Analytical Sciences, 23*, 17–23.

Sauer, B. (2002). Chromosome manipulation by Cre-lox recombination. In N. L. Craig, R. Craigie, M. Gellert, & A. M. Lambowitz, *MobileDNA* (2nd ed., pp. 38–58), ASM Press, Washington.

Schrijver, I., Külm, M., Gardner, P. I., Pergament, E. P., & Fiddler, M. B. (2007). Comprehensive arrayed primer extension array for the detection of 59 Sequence Variants in 15 Conditions. Journal of Molecular Diagnostics, 9, 228–236.

Somma, M. (2004). Extraction and purification of DNA. In M. Querci, M. Jermini, & G. Van den Eede (Ed.). *The analysis of food samples for the presence of genetically, modified organisms*. European commission, Joint research centre. Retrieved October 16, 2007, from http://gmotraining.jrc.it.

Spoth, B., & Strauss, E. (1999). Screening for genetically modified organisms in food using Promega's Wizard resin. *Promega Notes Magazine, 73*, 23–25.

Terry, C. F., Harris N., & Parkes, H. C. (2002). Detection of genetically modified crops and their derivatives: Critical steps in sample preparation and extraction. *Journal of AOAC International, 85*(3), 768–774.

US Patent 20060185029. *Avians that produce eggs containing exogenous proteins*. Retrieved August 17, 2006, from http://www.freepatentsonline.com/20060185029.html.

Vaïtilingom M., Pijnenburg, H., Gendre, F., Brignon, P. (1999). Real-time quantitative PCR detection of genetically modified maximizer maize and roundup ready soybean in some representative foods. *Journal of Agricultural and Food Chemistry, 47*, 5261–5266.

Von Holst, Ch., Baeten, Ch. V., Berben, G., & Brambilla, G. (2004). Overview of methods for the detection of species specific proteins in feed intended for farmed animals. Retrieved September 30, 2004, from http://ec.europa.eu/food/food/biosafety/bse/bse52_en.pdf.

Wenijn, S., Siyang, S., Minnan, L., & Guangming, L. (2003). Multiplex polymerase chain reaction/membrane hybridization assay for detection of genetically modified organisms. *Journal of Biotechnology, 105*, 227–223.

Wenzl, P., Carling, J., Kudrna, D., Jaccoud, D., Huttner, E., Kleinhofs, A., & Kilian, A. (2004). Diversity Arrays Technology (DarT) for whole-genome profiling of barley Proc. Natl. Acad. Sci USA *101*, 9915–9920.

Wheeler, M. B. (2007). Agricultural application for transgeniclivestock. *Trends in Biotechnology, 25*, 204–210.

Yang, L., Guo, J., Pan, A., Zhang, H., Zhang, K., Wang, Z., et al. (2007). Event-specific quantitative detection of nine genetically modified maizes using one novel standard reference molecule. *Journal of Agricultural and Food Chemistry, 55*, 15–24

Žel, J., Cankar, K., Ravnikar, M., Camloh, M., & Gruden, K. (2006). Accreditation of GMO detection laboratories: Improving the reliability of GMO detection. *Accred Quality Assurance, 10*, 531–536.

Chapter 15
Biosensors for Detecting Pathogenic Bacteria in the Meat Industry

Evangelyn C. Alocilja

Introduction

Global meat production in 2006 increased 1.6% compared to 2005 (Food and Agriculture Organization [FAO], 2006). According to a September 2007 report by the U.S. Meat Export Federation, the U.S. beef and beef variety meat exports worldwide increased 27% in value to $1.42 billion with a volume of 425,394 metric tons (mt) while U.S. pork and pork variety meat exports were up 5% in value to $1.7 billion with a volume of 704,138 mt[1]. However, with increasing production also comes increasing product recalls, averaging 4,536 mt of meat and poultry every year since 1997 (Teratanavat & Hooker, 2004). For example, in September 2007, a major meat processing company recalled up to 9,843 mt (21.7 million pounds) of ground beef due *E. coli* O157:H7 contamination, one of the largest meat recalls in U.S. history[2]. This contamination sickened 30 people in eight states. On October 5, 2007, that company announced that it was closing its business[3]. Contamination of meat products by foodborne pathogens is increasingly a major food safety and economic concerns. Billions of dollars are lost every year in medical costs, productivity, product recalls, and jobs as a result of pathogen-contamination outbreaks. In the United States, there are up to 33 million cases of human illness each year from microbial pathogens in the food supply, with an associated cost of 2–4 billion dollars in 2006[4].

Meat products can be contaminated from a number of sources, including the environment, the animal itself, fecal-contaminated water, or from human contact during the conversion of animal meat into food products. Majority of bacterial contamination in meat is generally harmless to humans however some contaminants are

---

[1] http://www.usmef.org/TradeLibrary/News07_0912a.asp
[2] http://www.cnn.com/2007/US/09/29/meat.recall/
[3] http://www.msnbc.msn.com/id/21149977/
[4] http://www.ers.usda.gov/Data/FoodborneIllness/

E.C. Alocilja
Department of Biosystems and Agricultural Engineering, Michigan State University, 213 Farrall Hall, East Lansing, MI 48824-1323

human pathogens. *Escherichia coli* O157:H7 and *Salmonella species* are among the deadliest of the foodborne pathogens in the United States.

Escherichia coli bacteria are part of the normal flora in the intestines of all animals, including humans. Normally, *E. coli* serves a useful function in the body by suppressing the growth of harmful bacterial species and by synthesizing some vitamins. A minority of *E. coli* strains however are capable of causing human illness by various mechanisms. *E. coli* O157:H7 produces toxins that damage the lining of the intestine, cause anemia, stomach cramps and bloody diarrhea, and serious complications called hemolytic uremic syndrome (HUS) and thrombotic thrombocytopenic purpura (TTP) (Doyle, Zhao, Meng, & Zhao, 1997). In North America, HUS is the most common cause of acute kidney failure in children, who are particularly susceptible to this complication. TTP has a mortality rate of as high as 50% among the elderly (Food and Drug Administration [FDA], 2006). Recent food safety data indicates that cases of *E. coli* O157:H7 are rising in both the US and other industrialized nations (WHO, 2002).

Possible points of entry of this pathogen into the food supply chain include naturally occurring sources from wild animals and ecosystems, infected livestock, contaminated processing operations, and unsanitary food preparation practices. *E. coli* O157:H7 may be introduced into food through meat grinders, knives, cutting blocks and storage containers. Human infections with *E. coli* O157:H7 have also been traced back to individuals having direct contact with food in situations involving food handling or food preparation. *E. coli* O157:H7 has also been found in drinking water that has been contaminated by runoff from livestock farms as a result of heavy rains. Regardless of the source, *E. coli* O157:H7 has been traced to a number of food products including meat and meat products, apple juice or cider, milk, alfalfa sprouts, unpasteurized fruit juices, dry-cured salami, lettuce, game meat, and cheese curds (Doyle et al., 1997; Food and Drug Administration [FDA], 2005). The food industry is under increasing pressure to identify and control this organism on the farm, in processing facilities, and in the food supply chain.

In 2000, *E. coli* O157:H7 was the etiological agent in 69 confirmed outbreaks (twice the number in 1999) involving 1564 people in 26 states (CDC, 2001a). Of the known transmission routes, 69% were attributed to food sources, 11% to animal contact, 11% to water exposures, and 8% to person-to-person transmission (CDC, 2001a). In 2006, the associated cost of *E. coli* O157:H7 infection was $445.8 million[5].

All known strains of *Salmonella* are pathogenic with a very low infectious dose as observed in some of the foodborne outbreaks traced back to *Salmonella* contamination. *Salmonella enterica* serovar Typhimurium and *Salmonella enterica* serovar Enteritidis are the most common *Salmonella* serotypes found in the United States. According to the Centers for Disease Control and Prevention (CDC), salmonellosis is the most common foodborne illness (CDC, 2002b) with over 40,000 actual cases reported yearly in the U.S. (CDC, 2002a). Annually, approximately 1,000 deaths

---

[5] http://www.ers.usda.gov/Data/FoodborneIllness/

are caused by *Salmonella* infections in the U.S. (CDC, 2001b; Mead et al., 1999). In 2006, Salmonella poisoning cost $2.5 billion[5]. Raw meats, poultry, eggs, milk and dairy products, fish, shrimp, frog legs, yeast, coconut, sauces and salad dressing, cake mixes, cream-filled desserts and toppings, dried gelatin, peanut butter, cocoa, and chocolate are some of the foods associated with *Salmonella* infection (D'Aoust, 1997).

Typhoid and paratyphoid strains of *Salmonella* are well adapted for invasion and survival within host tissues, causing enteric fever, a serious human disease. Non-typhoid *Salmonella* causes salmonellosis, which is manifested as gastroenteritis with diarrhea, fever, and abdominal cramps. Severe infection could lead to septicemia, urinary tract infection, and even death in at-risk populations. Newborns, infants, the elderly, and immuno-compromised individuals are more susceptible to *Salmonella* infections (D'Aoust, 1997). The developing immune system in newborns and infants, the frequently weak and/or delayed immunological responses in the elderly and debilitated persons, and low gastric acid production in infants and seniors facilitate the intestinal proliferation and systemic infection of salmonellae in this susceptible population (Blaser & Newman, 1982). Moreover, in recent years, some strains of *Salmonella* have become resistant to several of the antibiotics traditionally used to treat it, in both animals and humans, making *Salmonella* infections an important health concern in both developed and developing countries. The majority of the increased incidence of resistance can be attributed to *Salmonella* Typhimurium DT104. Treatment of this infectious disease is complicated by its ability to acquire resistance to multiple antibiotics (Carlson et al., 1999). Evidence suggesting that ingestion of only a few *Salmonella* cells can develop a variety of clinical conditions (including death) is a reminder for food producers, processors, and distributors that low levels of *Salmonella* in a finished food product can lead to serious public health consequences, and undermine the reputation and economic viability of the incriminated food manufacturer.

The USDA Food Safety and Inspection Service (FSIS) has issued a "zero-tolerance" policy for *E. coli* O157:H7 and *Salmonella* species. Thus, early and rapid detection of these organisms are very important to the food industry so that appropriate measures can be taken to eliminate infection through food. The detection and identification of these foodborne pathogens in raw food materials, ready-to-eat food products, restaurants, processing and assembly lines, hospitals, ports of entry, and drinking water supplies continue to rely on conventional culturing techniques. Conventional methods involve enriching the sample and performing various media-based metabolic tests (agar plates or slants). Although these methods are highly sensitive and specific, they are elaborate, laborious, and typically require 2–7 days to obtain conclusive results (FDA, 2005).

An increased demand for rapid and high-throughput screening, especially in the clinical and pharmaceutical industries, has produced several technological developments for quickly detecting target biomolecules, including cells. Some of these emerging technologies include polymerase chain reaction (PCR), enzyme linked immunosorbent assay (ELISA), flow cytometry, chemiluminescence, molecular cantilevers, matrix-assisted laser desorption/ionization, immunomagnetic separation, artificial membranes, spectroscopy, electronic noses, and biosensors. Pathogen detection utilizing ELISA methods for determining and quantifying pathogens in food have been well established (Jiang et al., 2006; Cohn, 1998). The PCR method is extremely sensitive but requires pure sample preparation and hours of processing along with expertise in molecular biology (Meng, Zhao, Doyle, & Kresovich, 1996). Flow cytometry is another highly effective means for rapid analysis of individual cells at rates up to 1000 cells/s (McClelland & Pinder, 1994), however, it has been used almost exclusively for eukaryotic cells. Electronic noses have recently been developed for pathogen detection as well but they lack the needed specificity for zero-tolerance policy (Alocilja, Ritchie, & Grooms, 2003; Younts et al., 2003; Younts, Alocilja, Osburn, Marquie, & Grooms, 2002). These detection methods are mostly for laboratory use but cannot adequately serve the needs of health practitioners and monitoring agencies in the field. Furthermore, these systems are costly, require specialized training, have complicated processing steps in order to culture or extract the pathogen from food samples, and/or are time consuming. In comparison, a field-ready biosensor is inexpensive, easy to use, portable, and provides results in minutes. A biosensor that can detect pathogens within minutes will allow food processors to take quick corrective action when pathogens are detected, resulting in minimized product recalls and lawsuits for the industry and savings from medical costs for consumers.

Biosensors are analytical instruments possessing a bio-molecule as a receptor or reactive surface in close proximity to a transducer, which converts the binding of the target analyte with the capturing bio-molecule or receptor into a measurable signal (Muhammad-Tahir & Alocilja, 2003a; Turner & Newman, 1998). The capturing bio-molecule can be in the form of an antibody, enzyme, nucleic acid, cell, or aptamer. The transducer can be based on electrochemical (potentiometric, voltammetric, and conductometric), optical, thermal, or piezoelectric devices, field effect transistors, or surface acoustic waves. Biosensors can provide results in real-time and can be portable. These instruments often operate in a reagentless process (Muhammad-Tahir & Alocilja, 2003a, 2003b), making them user-friendly and field-ready devices. Some of the major attributes of a biosensor technology are its specificity, sensitivity, reliability, portability, real time analysis, and simplicity of operation (D'Souza, 2001; Fratamico, Strobaugh, Medina, & Gehring, 1998; Muhammad-Tahir & Alocilja, 2003a). Market potential for biosensors is huge (Alocilja & Radke, 2003; Kalorama Information, 2000). Although majority of the applications are in the medical field, recent applications are in food safety, environmental integrity, and homeland security (Pal, Alocilja, & Downes, 2007; Radke & Alocilja, 2005a, 2005b). Of particular interest is the monitoring for biological and microbial contaminants in food and water (Corry, Uilk, & Crawley, 2002; D'Souza, 2001; Ko & Grant, 2006; Komarova, Aldissi, & Bogomolova, 2005; Lazcka, Campo, Javier, Munoz, & Xavier, 2007; Mathew & Alocilja, 2005; Pal et al., 2007). Optical (DeMarco & Lim, 2002; Fratamico et al., 1998; Hoyle, 2001; Lazcka et al., 2007; Meeusen, Alocilja, & Osburn, 2005; Seo, Brackett, Hartman, & Campbell, 1999), piezoelectric (Bao, Deng, Nie, Yao & Wei, 1996), and electrochemical (Ghindilis, Atanasov, Wilkins & Wilkins, 1998;

Lazcka et al., 2007; Muhammad-Tahir & Alocilja, 2003a, 2003b; Pal et al., 2007) biosensors have been developed for the detection of pathogenic bacteria in food and water. Electrochemical biosensors, often referred to as amperometric, conductometric or impedimetric, have the advantage of being highly sensitive, rapid, inexpensive and are highly amenable towards microfabrication (Gau, Lan, Dunn, Ho, & Woo, 2001; Ghindilis et al., 1998; Muhammad-Tahir & Alocilja, 2004; Pal et al., 2007; Radke & Alocilja, 2004, 2005a, 2005b). They measure the change in electrical properties of electrode structures as cells become entrapped or immobilized on or near the electrode. For example, an amperometric immunoassay was developed for *Staphyloccoccus aureus* utilizing antibodies bound to a carbon electrode and enzyme amplifiers to amplify the detection signal, with a lower limit of detection (LLD) at 10^3 CFU/mL in 30 min (Rishpon & Ivnitski, 1997). An impedimetric biosensor for *E. coli* O157:H7 was developed with a LLD of 10^4 CFU/mL (Radke & Alocilja, 2005a, 2005b). Another impedimetric biosensor utilizing mircofluidics was able to detect 10^5 CFU/mL of *Listeria monocytogenes* in a few hours with respect to a sterile buffer (Gomez et al., 2001). A conductometric biosensor was developed for *E. coli* strains, *Salmonella species*, and *Bacillus* species in selected food matrices (Muhammad-Tahir & Alocilja, 2004; Pal et al., 2007). Using porous filter membranes, a flow-through conductometric immuno-filtration biosensor was developed for the detection of *Escherichia coli* O157:H7 in liquid media (Abdel-Hamid, Ivnitski, Atanasov, Wilkins, 1998; Muhammad-Tahir & Alocilja, 2003a). An enzyme-based amperometric biosensor for quantifying *Escherichia coli* had LLD of 10^3 CFU/mL in 60–75 min (Mittelmann, Ron, & Rishpon, 2002). An enzyme-based impedimetric biosensor was successfully used to detect the presence of glucose oxidase binding to interdigitated electrodes deposited on silicon oxide surfaces (Van Gerwen et al., 1998). A similar architecture was used to detect urea concentrations as low as $50\,\mu M$ on interdigitated electrodes with immobilized urease (Sheppard, Mears, & Guiseppi-Elie, 1995). Enzyme-based chemiluminescence technique for pathogen detection also showed very good sensitivity (Mathew, Alagesan, & Alocilja, 2004; Mathew & Alocilja, 2004).

Immunosensors

Antibodies are frequently used in biosensor development due to their high specificity attributes (Sadana, 2002). The specificity of antibodies is based on the immunological reaction involving the unique structure recognition on the antigen surface and the antibody binding site (Barbour & George, 1997). The basic structure of all antibody or immunoglobulin molecules consists of four polypeptides–two heavy chains and two light chains joined like a capital letter "Y", and linked by disulphide bonds (Sadana, 2002). The Fc fragment comprises the effector functions, such as complement activation and cell membrane receptor interaction. The Fab fragment, on the other hand, comprises the antigen binding sites. The amino acid sequence on the tips of the "Y" (antigen binding sites) varies greatly among different antibodies. This antigen binding site is composed of 110–130 amino acids, giving the antibody its specificity for binding antigen (Barbour & George, 1997). Typically, antibodies are prepared as monoclonal or polyclonal. Polyclonal antibodies are produced by multiple clones of antibody-producing molecules. Polyclonal antibodies recognize multiple epitopes or binding sites on the surface of the antigen, making them more tolerant to the variability in antigen structures. Monoclonal antibodies on the other hand, are those that are derived from a single clone antibody-producing molecule and thus, react only to a specific epitope on the antigen (CHEMICON International, 2004). Because of their high specificity, monoclonal antibodies are excellent to be used in immunoassay techniques. Studies also show that result from monoclonal antibodies are highly reproducible between experiments (CHEMICON International, 2004). Monoclonal antibodies, however, are much more vulnerable to epitope lose due to chemical treatment compared to polyclonal antibodies.

Biosensors using antibodies as receptors are often referred to as immunosensors. In immunosensor development, antibody immobilization is a vital step. The immobilization method must preserve the biological activity of the antibody and enable efficient binding. These methods can be grouped into three main categories: adsorption; immobilization via entrapment (Vikholm, 2005; Ye, Letcher, & Rand, 1997) or immobilization via cross-linking agents (Bhatia et al., 1989; Narang, Anderson, Ligler, & Burans, 1997; Nashat, Moronne, & Ferrari, 1998; Radke & Alocilja, 2005a, 2005b; Slavik, Homola, & Brynda, 2002); and self-assembled monolayer (Chen et al., 2003; Su & Li, 2004; Vikholm, 2005).

Attachment of antibodies on quartz or glass can be achieved by simple adsorption; however, the immobilized proteins suffer partial denaturation, and tend to leach or wash off the surface (Bhatia et al., 1989; Huang et al., 2004; Zhou & Muthuswamy, 2004). Also, this approach does not provide permanent attachment because the complex is weakly bound to the solid support by adsorption.

Immobilization by entrapment on a pre-coated crystal with polyethylenimine (PEI) was also reported by Lin and Tsai (Lin & Tsai, 2003; Tsai & Lin, 2005). Immobilization through Protein A coupling (Boltovets et al., 2002; Su & Li, 2005) is another reported method. Protein A is a cell wall protein, produced by most strains of *Staphylococcus aureus*. Protein A is a directed immobilization method due to its natural affinity towards the Fc region of IgG molecules. This does not block the active sites of the antibodies for analyte binding (Babacan, Pivarnik, Letcher & Rand, 2000). In the PEI–glutaraldehyde (GA) method, immobilization of antibodies is achieved via surface aldehyde groups of GA on a quartz crystal pre-coated with PEI. Glutaraldehyde is a homobifunctional cross-linker. In this method, the antibodies are randomly oriented and bound to the active surface (Babacan et al., 2000).

The self-assembled monolayer (SAM) technique offers one of the simplest ways to provide a reproducible, ultra thin, and well-ordered layer suitable for further modification with antibodies, which has potential in improving detection sensitivity, speed, and reproducibility (Chen et al., 2003; Su & Li, 2004; Vikholm, 2005).

Covalent binding is a common method of attaching an antibody to a surface due to the strong, stable linkage that is formed. Hydroxyl groups on the biosensor platform surface provide sites for covalent attachment of organic molecules. Several investigators have modified surface hydroxyl groups to provide a functionality that would react directly with antibodies (Bhatia et al., 1989; Nashat et al., 1998; Shriver-Lake et al., 1997). Coating the biosensor surface with a silane film provides a method for modifying the reactive hydroxyl groups on the surface to attach cross-linking agents. Silanes reportedly used include 4-aminobutyldimethylmethoxysilane, 4-aminobutyltriethoxysilane, 3-mercaptopropyltrimethoxysilane, mercaptomethyldimethylethoxysilane, and 3-aminopropyltriethoxysilane with cross-linkers such as glutaraldehyde, N-N-maleimidobutyryloxy succinimide ester, N-succinimidyl-3-(2-pyridyldithio) propionate, N-succinimidyl-(4-iodoacetyl) aminobenzoate, and succinimidyl 4-(p-aleimidophenyl) butyrate (Bhatia et al., 1989; Shriver-Lake et al., 1997). Silanes utilized for the modification of the biosensor surface (gold, silicon, glass) introduce amino groups on it, which in turn provide reaction sites for covalently bonding to glutaraldehyde cross-linker. Antibodies are then immobilized through Schiff base to the cross-linker. These reactive glutaraldehyde polymers may bind many residues and form multi-protein complexes (Bhatia et al., 1989). These effects are likely to interfere with protein function, lowering the antibody activity on the biosensor surface. To obviate the problem, Sportsman and Wilson (Sportsman & Wilson, 1980) coated glass with glycidoxypropylsilane and oxidized the silane to produce aldehyde groups reacting directly with antibody. This method needs fewer functionalization steps and less processing time, while providing a functionalized biosensor with a high activity of immobilized antibodies.

The optimization of bio-molecule immobilization however, is a critical issue in the performance of a biosensor. Most current immobilization methodologies are not at 100% efficiency due to the limited immobilization capacity, partial loss of bioactivity of the immobilized molecules (Bunde, Jarvi, & Rosentrerer, 1998), cross-reactivity of enzymes (Cheung, Stockton, & Rubner, 1997), nonspecific binding (analyte binding occurs at places where it should not), and interference with the transducer elements (Scheller et al., 1991). For instance, during the antibody immobilization process on the conducting polymer film, the antibody may bind at the Fc region, on both Fc and Fb regions, or one of the antigen binding sites. Lu et al. emphasizes that if immobilization occurs on the antigen-binding sites, the ability of the antibody to bind the antigen may be lost completely or at least to a high degree, and thus affecting the overall performance of the biosensor (Lu, Smyth, & O'Kennedy, 1996).

Antibody-based biosensors have been developed for pathogen analysis in various matrices. A quartz crystal microbalance (QCM)-based biosensor was used to detect *Salmonella* species in milk samples with a detection limit of 10^6 CFU/ml (Park, Kim, & Kim, 2000). *Listeria* and *Salmonella* spp. were detected with a similar detection limit by a surface plasmon resonance (SPR) biosensor (Koubova et al., 2001). QCM and SPR biosensors for different bacterial targets, such as *Pseudomonas aeruginosa, Bacillus cereus*, and *E. coli* O157:H7, were later developed by researchers (Kim, Park, & Kim, 2004; Meeusen et al., 2005; Su & Li, 2004; Su & Li, 2005; Vaughan, Carter, O'Sullivan, & Guilbault, 2003), and showed no significant improvement in detection limits. In a dipstick-type assay, Park and Durst (2000) were able to detect *E. coli* O157:H7 in food matrices at a low detection limit of about 10^3 CFU/ml without any enrichment required.

An impedance biosensor chip for detection of *E. coli* O157:H7 was developed based on the surface immobilization of affinity-purified antibodies onto indium tin oxide (ITO) electrode chips, with a detection limit of 6×10^3 cells/ml (Ruan, Yang, & Li, 2002). Shah, Chemburu, Wilkins, and Abdel-Hamid (2003) developed an amperometric immunosensor with a graphite-coated nylon membrane serving as a support for antibody immobilization and as a working electrode. This approach was used for detection of *E. coli* with a low detection limit of 40 CFU/ml. An antibody-based conductometric biosensor developed by Muhammad-Tahir and Alocilja also showed a similar detection limit of 50 CFU/ml for bacteria and 10^3 cell culture infective dose (CCID)/ml for viruses within 10 min (Muhammad-Tahir & Alocilja, 2003a, 2003b; Muhammad-Tahir & Alocilja, 2004; Muhammad-Tahir, Alocilja, & Grooms, 2005a, 2005b, 2007).

An antibody-based fiber-optic biosensor, to detect low levels of *L. monocytogenes* cells following an enrichment step, was developed using a cyanine 5-labeled antibody to generate a specific fluorescent signal. The sensitivity threshold was about 4.3×10^3 CFU/ml for a pure culture of *L. monocytogenes* grown at 37°C. Results could only be obtained after 2.5 h of sample processing. In less than 24 h, this method could detect *L. monocytogenes* in hot dog or bologna naturally contaminated or artificially inoculated with 10^1 to 10^3 CFU/g after enrichment in buffered Listeria enrichment broth (Geng, Morgan & Bhunia, 2004). In another study, a microcapillary flow injection liposome immunoanalysis system (mFILIA) was developed for the detection of heat-killed *E. coli* O157:H7. Liposomes tagged with anti-*E. coli* O157:H7 and an encapsulating fluorescent dye were used to generate fluorescence signals measured by a fluorometer. The mFILIA system successfully detected as few as 360 cells/ml with a total assay time of 45 min (Ho, Hsu, & Huang, 2004).

The ability to detect small amounts of bacterial organisms was demonstrated using micro electromechanical systems (MEMS) for the qualitative detection of specific *Salmonella enterica* strains with a functionalized silicon nitride microcantilever. Detection was achieved due to a change in the surface stress on the cantilever surface in situ upon binding of a small number of bacteria (Weeks et al., 2003). A MEMS fabricated high-density microelectrode array biosensor, developed for the detection of *E. coli* O157:H7 using a change in impedance caused by the bacteria measured over a frequency range of 100 Hz–10 MHz, was able to detect and discriminate 10^4–10^7 CFU/ml (Radke & Alocilja, 2005a, 2005b). Nanotechnology-based biosensors coupled with novel signal processing architectures are also fast emerging (Gore, Chakrabartty, Pal & Alocilja, 2006; Yang, Chakrabartty, & Alocilja, 2007; Zuo, Chakrabartty, Muhammad-Tahir, Pal, & Alocilja, 2006).

The potential use of immunosensors is due to their general applicability, specificity and selectivity of the antigen-antibody reaction, and the high sensitivity of the method. The antigen-antibody complex may be utilized in all types of sensors. However, immunosensors cannot be employed to specifically detect viable cells.

The antibodies used are selective to the epitope on the antigen. If the epitope is present on a living or dead microorganism, the antibody will capture the antigen and register a positive signal. In the case of bacterial foodborne pathogens that must be ingested by the host to cause disease, a positive result from non-viable cells may raise false alarm.

DNA Sensors

A DNA probe is a segment of nucleic acid that specifically recognizes, and hybridizes to, a nucleic acid target. The recognition is dependent upon the formation of stable hydrogen bonds between the two nucleic acid strands. This contrasts with interactions of antibody-antigen complex formation where hydrophobic, ionic, and hydrogen bonds play a role. The bonding between nucleic acids takes place at regular (nucleotide) intervals along the length of the nucleic acid duplex, whereas antibody-protein bonds occur only at a few specific sites (epitopes). The frequency of bonding is reflected in the higher association constant for a nucleic acid duplex in comparison with an antibody-protein complex, and thus indicates that highly specific and sensitive detection systems can be developed using nucleic acid probes (McGown, Joseph, Pitner, Vonk, & Linn, 1995; Skuridin, Yevdokimov, Efimov, Hall, & Turner, 1996). The specificity of nucleic acid probes relies on the ability of different nucleotides to form bonds only with an appropriate counterpart. Since the nucleic acid recognition layers are very stable, an important advantage of nucleic acid ligands as immobilized sensors is that they can easily be denatured to reverse binding and then regenerated simply by controlling buffer-ion concentrations (Graham, Leslie, & Squirrell, 1992). Most nucleic acid biosensors are based on this highly specific hybridization of complementary strands of DNA and also RNA molecules.

Besides the immobilization techniques described for antibodies, DNA probes also can be immobilized on a biosensor platform via avidin-biotin chemistry (Wang, Tombelli, Minunni, Spiriti & Mascini, 2004). Avidin is a basic glycoprotein with an isoelectric point (pI) of approximately 10, originally isolated from chicken egg white. Biotin is a naturally occurring vitamin found in every living cell (Savage et al., 1994). The most commonly used method in surface plasmon resonance (SPR)-based optical sensing is the immobilization of biotinylated DNA probes onto a layer of streptavidin. Streptavidin is a biotin-binding protein isolated from culture broth of *Streptomyces avidinii*. Streptavidin binds four moles of biotin per mole of protein, and has a pI of approximately 5–6 (Savage et al., 1994). Streptavidin is covalently linked to carboxylated dextran fixed onto the gold surface of the SPR chip on a self-assembled monolayer of 11-mercaptoundecanol (Bianchi et al., 1997; Feriotto, Borgatti, Mischiati, Bianchi, & Gambari, 2002; Gao et al., 2004; Jordan, Frutos, Thiel, & Corn, 1997; Ruan, Zeng, Varghese, & Grimes, 2004; Silin & Plant, 1997; Tombelli, Mascini, Sacco, & Turner, 2000). This method is very efficient in terms of sensitivity, selectivity, and stability of the realized sensor chip for DNA hybridization detection (Kukanskis et al., 1999; Mariotti, Minunni, & Mascini, 2002; Silin & Plant, 1997).

The detection of bacteria and other pathogens in food, drinking water, and air, based on their nucleic acid sequences, has been well explored. Biosensors have been developed to detect DNA hybridization at sub-picomolar to micromolar levels using gravimetric detection systems (Sung Hoon, In Seon, Namsoo, & Woo Yeon, 2001). A quartz crystal nanobalance system could detect DNA hybridization at 0.3 nanogram levels using frequency shift nanogravimetric measurement (Nicolini et al., 1997). Tombelli et al. (2000) developed a DNA piezoelectric biosensor for the detection of bacterial toxicity based on the detection of PCR amplified *aer* gene of *Aeromonas hydrophila*. The biosensor was applied to vegetables, environmental water, and human specimens (Tombelli et al., 2000). The biosensor was able to successfully distinguish between samples containing the pathogen and those not contaminated. Zhao et al. (2001) developed a QCM biosensor using 50 nm gold nanoparticles as the amplification probe for DNA detection in the order of 10 fM of target, which is higher than what has been ever reported using the same method. The high sensitivity was explained by the weight of the larger particles, and the larger area occupied by the larger particles that needed less target DNA for their binding (Zhao et al., 2001). Another QCM biosensor applied to the detection of *E. coli* in water in combination with PCR amplification (of the *lac* gene) was able to detect a 10 fg of genomic *E. coli* DNA (few viable *E. coli* cells in 100 ml of water) (Mo, Zhou, Lei, & Deng, 2002). When used for detection of *Hepatitis B virus*, Zhou, Liu, Hu, Wang, and Hu (2002) observed that the QCM could detect frequency shifts of DNA hybridization as a linear relationship in the range 0.02-0.14 µg/ml with a detection limit of 0.1 µg/ml, similar to the QCM biosensor developed for *Pseudomonas aeruginosa* (He & Liu, 2004). An electrochemical DNA biosensor has been developed for water quality monitoring of *E. coli* contaminants by Rodriguez and Alocilja (2005).

Optical biosensor systems developed for DNA detection exhibit higher sensitivity compared to the QCM biosensors. An automated optical biosensor system based on fluorescence excitation and detection in the evanescent field of a quartz fiber was used to detect 16-mer oligonucleotides in DNA hybridization assays. The detection limit for the hybridization with a complementary fluorescein-labeled oligonucleotide was 2×10^{-13} M (Abel, Weller, Duveneck, Ehrat, & Widmer, 1996). Another optical fiber evanescent wave DNA biosensor used a molecular beacon (MB) DNA probe that became fluorescent upon hybridization with target DNA. The detection limit of the evanescent wave biosensor with synthesized complementary DNA was 1.1 nM although testing with environmental samples was not performed (Liu & Tan, 1999). Liu et al. later developed MB-DNA biosensors with micrometer to submicrometer sizes for DNA/RNA analysis. The MB-DNA biosensor was highly selective with single base-pair mismatch identification capability, and could detect 0.3 nM and 10 nM of rat gamma-actin mRNA with a 105-µm biosensor and a submicrometer biosensor, respectively (Liu et al., 2000).

Optical biosensors targeting RNA as the analyte offer an added advantage over traditional DNA-based detection methods, i.e., viable cell detection. Baeumner, Cohen, Miksic and Min (2003) detected as few as 40 *E. coli* cells/ml in samples using a simple optical dipstick-type biosensor coupled to Nucleic Acid Sequence Based Amplification (NASBA), emphasizing the fact that only viable cells were detected, and no false positive signals were obtained from dead cells present in the sample. The detection of viable cells is important in respect to food safety, and also food and environmental sample sterilization assessments. Similarly, a biosensor for the protozoan parasite *Cryptosporidium parvum* was developed (Esch, Locascio, Tarlov, & Durst, 2001). Hartley and Baeumner (2003) developed a simple membrane strip-based biosensor for the detection of viable *Bacillus anthracis* spores. The study combined the optical detection process with a spore germination procedure as well as a nucleic acid amplification reaction to identify as little as one viable *B. anthracis* spore in 12 h. A quantitative universal biosensor was developed on the basis of olignucleotide sandwich hybridization for the rapid (30 min total assay time) and highly sensitive (1 nM) detection of specific nucleic acid sequences (Baeumner, Pretz, & Fang, 2004). The biosensor consisted of a universal (polyethersulfone) membrane, a universal dye-entrapping liposomal nanovesicle, and two oligonucleotides – a reporter and a capture probe that could hybridize specifically with the target nucleic acid sequence. Limits of detection of 1 nM per assay and dynamic ranges between 1–750 nM were obtained. While the RNA-based biosensor can be an excellent tool for detection of viable bacterial cells, inherent disadvantages to the technique include the short life span of the mRNA target, high susceptibility to contaminants and inhibitors from environmental and food samples, and need for complex detection systems.

Other biosensors targeting DNA that have been developed include high throughput PCR biosensors (Nagai, Murakami, Yokoyama, & Tamiya, 2001), carbon nanotube-based field effect transistor biosensor (Maehashi, Matsumoto, Kerman, Takamura, & Tamiya, 2004), microcantilever-based cyclic voltammetry biosensor (Zhang & Li, 2005), pulsed amperometry- (Ramanaviciene & Ramanavicius, 2004), capacitance- (Berney et al., 2000; Lee, Choi, Pio, Seo, & Lee, 2002), absorbance- (Mir & Katakis, 2005), and MEMS-based biosensors (Gau et al., 2001).

MEMS Approach to Biosensor Fabrication

Fabrication of biosensors comes in many forms and styles. One of the more recent methods is the use of the microelectromechanical systems (MEMS) approach. A MEMS-based biosensor using antibodies as capturing molecule is illustrated in this article. The design, fabrication, performance, and illustrations presented in this section are excerpted from three journal articles by Radke & Alocilja (2004, 2005a, 2005b), and where appropriate, reprinted with permission from ©IEEE.

Detection Principle

Gold-on-titanium interdigitated electrode array is used as a sensing surface on silicon chip to detect the presence of bacteria in solution. When the interdigitated array is immersed in a sample solution, the active area is exposed to the bacteria which bind to the antibodies on the biosensor surface. When this occurs, a region of 2–4 µm (size of bacteria) above the sensor surface becomes modified. Surface antibodies immobilized between the electrodes via heterobifunctional crosslinkers act as tethers, which serve the purpose of holding the bacteria in place. Different cellular concentrations of bacteria bound to the sensor surface yield different changes in electrical impedance between the electrodes. The change in impedance is proportional to the bacterial concentration. The change is due to the resistivity of the bacteria cytoplasm and the capacitance of the lipid bilayer membrane (Tien & Ottova-Leitmannova, 2000). Cell suspensions exhibit dispersion (polarization) due to cell membrane and cytoplasm biomass in the audio (α-dispersion) and radio (β-dispersion) frequency range of 10–10^7 Hz (Cuireanu, Levadoux & Goldstein, 1997; Markx & Davey, 1999). Figure 15.1 shows the impedance elements created by the immobilized bacteria as a function of the solution resistance (R_{SOL}), the capacitance representing the dielectric behavior of the solution (C_{SOL}), and the impedance of bacteria, which consists of the resistance of the cytoplasm (R_{CYT}), and the resistance and capacitance of the cell membrane (R_{BLM} and C_{BLM}, respectively). Additionally, the parasitic capacitance (C_{PAR}) represents the double-layer capacitance at the electrode surface and the capacitance created from the oxide separation of the gold electrodes and the silicon surface.

Sensor Design

The spacing of the interdigitated electrodes is based on the size of the bacteria and the percentage of total electric field strength near the sensor surface. Since the mean length and diameter of *E. coli* is 2.57 µm and 0.49 µm, respectively, with a semi-log normal distribution (Koppes, Woldringh, & Nanninga, 1978), these factors are critical in the design. The application of the biosensor is to test for bacteria in a complex food matrix hence the change in impedance due to foreign particles near the sensor surface needs to be minimized. To help with the physical design of the biosensor, a simulation of the electric field was performed on different electrode

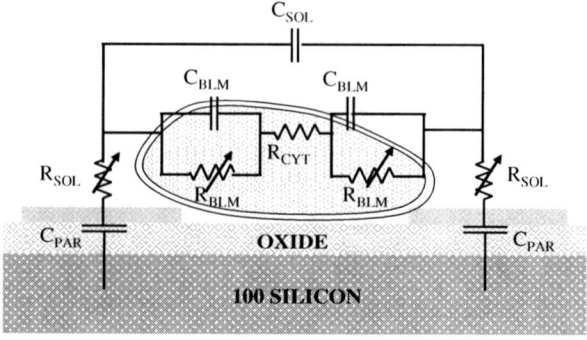

Fig. 15.1 Circuit model of the impedance of bacteria immobilized between two interdigitated electrodes (Reprinted from Radke, S., & Alocilja, E. (2004). *IEEE Sensors Journal*, 4(4), 434–440.)

widths and spacing to determine the electric field strength near the sensor surface. The electric field simulation used a relative permittivity of 60, 10 and 80 for the cell cytoplasm, cell membrane, and testing solution, respectively (Suehiro, Hamada, Noutomi, Shutou, & Hara, 2003; Wiegand, Arribas-Layton, Hillebrandt, Sackmann, & Wagner, 2002). The optimum electrode width and spacing was determined to be 3 μm and 4 μm, respectively, based on a mean *E. coli* length of 2.5 μm and a target of 90% of total electric field strength below a distance of 5 μm from the sensor surface. The active area was determined based on an estimate of the surface area required to accommodate 10^7 colony forming units (CFU) of immobilized *E. coli* cells on the sensor surface. Two electrode arrays, each measuring 1.0 mm × 0.8 mm, were selected to achieve the required active area of the sensor. Figure 15.2a shows the simulation results for an applied potential of 50 mV and interdigitated electrodes with a width of 3 μm and spacing of 4 μm. Figure 15.2b shows a cross-section of the sensor suspended in a solution with bacteria immobilized on the surface. Figure 15.2c shows the percentage of total electric field that is concentrated near the sensor surface. For example, if there is a particle floating 10 μm above the sensor surface, the effect on sensor impedance will be minimal since 98% of the electric field is below 10 μm.

Fabrication

The biosensor was fabricated from 500–550 μm thick 4-inch diameter silicon wafer which is of p-type 100 orientation. Thermal oxide layer (2 μm thick) was grown over the silicon surface to serve as an insulator between the electrodes and the silicon

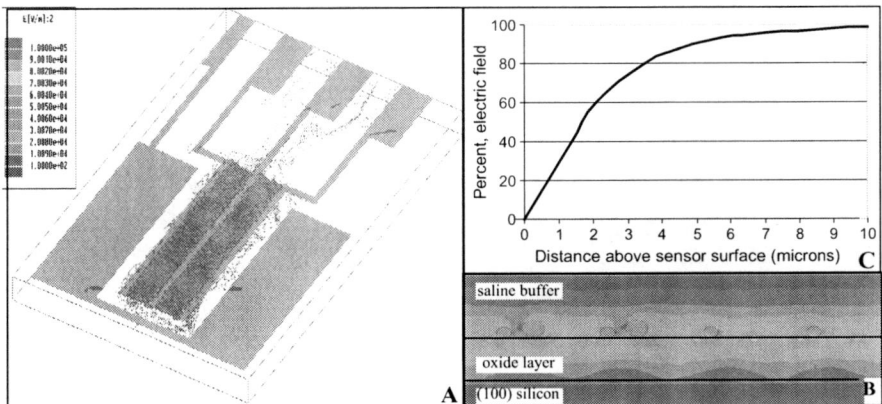

Fig. 15.2 FEA simulation of biosensor electric field magnitude: (a) isometric view of the entire biosensor; (b) cross-section of the biosensor with immobilized bacteria; (c) percentage of total electric field below a given distance above sensor surface; interdigitated electrode width of 3 μm, spacing of 4 μm (Reprinted from Radke, S., & Alocilja, E. (2004). *IEEE Sensors Journal*, 4(4), 434–440.)

substrate. Micro-electrode arrays were fabricated by patterning S1805 photo resist and depositing 5 nm of titanium (for adhesion) and topped with 50 nm of gold. Using ultrasonic activation, a lift-off process developed the interdigitated electrode array. For ease of fabrication, two small electrode arrays were fabricated to reduce the aspect ratio as compared to one large array. The sensor included an interdigitated electrode array with circuit traces connecting to bond pads. Each electrode finger had a length of 750 μm, a width of 3 μm, and an in-between spacing of 4 μm. Each sensor chip was diced to a dimension of 12 mm × 8 mm. Each chip had a total of 1700 electrodes and a large active area totaling 9.6 mm^2, as shown in Fig. 15.3.

After electrode array fabrication, antibodies were immobilized to the oxide surface of the sensor. The antibodies were attached to the surface via heterobifunctional crosslinkers using an established process (Bhatia et al., 1989). Briefly, the biosensor was cleaned and activated by immersing the sensor in nitric acid for 30 min followed by rinsing with distilled water. Silanization of the chip surface occurred after immersing the chip in a mixture of hydrochloric acid and methanol for 30 min followed with immersion in sulfuric acid. The silanizing of 3-Mercaptomethyldimethylethoxysilane (MDS) to the sensor surface occurred in dry toluene. Crosslinkers, N-y-maleimidobutyryloxy succinimide ester (GMBS) dissolved in dimethylformamide (DMF) and ethanol, were attached to the chip surface after silanization. Antibodies were added to the sensor surface to attach to the crosslinkers. The next step was to package the biosensor. Epoxy was used to bond the backside of the device to a rectangular piece of PVC for support. The bond pads were connected to a zero insertion force (ZIF) socket with leads terminating to an impedance analyzer, completing the fabrication process.

The biosensor chips were inspected in order to validate the physical properties as shown in Fig. 15.4. A metallurgical microscope was used to validate the fabrication of the electrode array for complete deposition of metal and to ensure that the high aspect ratio electrodes (250:1) were within tolerance (Fig. 15.4a). Atomic

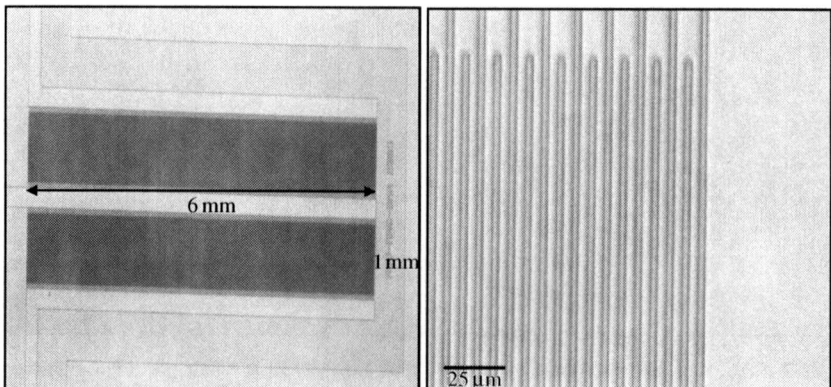

Fig. 15.3 Interdigitated gold electrode arrays: (left) two high density, high aspect ratio electrode arrays; (right) end view of one electrode array (Reprinted from Radke, S., & Alocilja, E. (2004). *IEEE Sensors Journal*, 4(4), 434–440.)

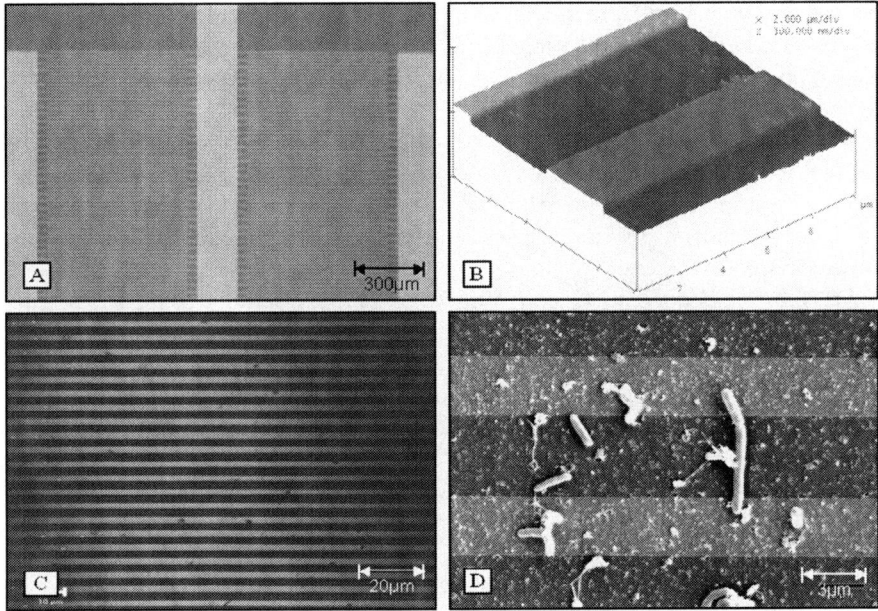

Fig. 15.4 Analysis of microfabricated device: (a) two high density, high aspect ratio electrode arrays; (b) AFM image of sensor surface; (c) CLSM image of antibodies labeled with FITC (light coloring is due to fluorescence); (d) SEM image of bacteria bound to antibodies immobilized to biosensor surface (Reprinted from Radke, S., & Alocilja, E. (2005b). *IEEE Sensors Journal*, 5(4), 744–750.)

force microscopy (AFM) was used to measure the length, width and thickness of the electrode and the spacing between opposing electrode fingers (Fig. 15.4b). Confocal laser scanning microscopy (CLSM) was used to validate antibody immobilization and determine where it occurred (dark spots, Fig. 15.4c). Antibody concentration on the surface was not determined, though a concentration of 150 μg/mL was used in the immobilization procedure. Scanning electron microscope (SEM) was used to verify binding of bacteria to the antibodies immobilized on the biosensor surface (Fig. 15.4d).

Testing

To test the biosensor performance, nonpathogenic Escherichia coli (ATCC #25922) and pathogenic *E. coli* O157:H7 (ATCC #43895) were grown in nutrient broth. A 10 μL loop of each isolate was cultured in 10 mL of nutrient broth and incubated for 24 h at 37°C to make a stock culture of each organism. Ten-fold serial dilutions of the organisms were prepared from the stock cultures through serial dilution in 0.1% of peptone water. The different concentrations were added to neutral phosphate buffer creating test samples of 20 mL with concentrations ranging from 10^0 CFU/mL to 10^7 CFU/mL. The sample concentrations were then determined by the standard plating method according to the FDA Bacteriological Analytical Manual (Food and Drug Administration [FDA], 1998). Sorbitol MacConkey agar was used to plate 100 µL of each serial dilution. The colonies were counted after 24 h of incubation at 37°C.

The biosensor was immersed into the sample for 2 min to allow antibody-antigen reaction to occur. A potential of 50 mV was applied across the electrodes and the impedance magnitude and phase angle of the biosensor was measured with an HP 4192A Impedance Analyzer. The sample testing process, including data acquisition, required 5 min.

For the nonpathogenic *E. coli*, four trials of each serially diluted concentration from 10^0 CFU/mL to 10^7 CFU/mL were tested against a sterile blank solution. For *E. coli* O157:H7, three trials of each serially diluted concentration from 10^0 CFU/mL to 10^7 CFU/mL were tested against a sterile blank solution. The means and standard deviations of the impedance magnitude and phase angle were calculated for frequencies between 10 Hz–13 MHz. The differences between means was determined and analyzed based on a one-way ANOVA and t test to a significance of 95% ($p < 0.05$).

Results

Modification of the sensor with antibodies caused an irreversible thin film of silanes to build up on the sensor surface. The repeated treatment of the oxide surface with oxysilanes resulted in a permanent change to the sensor surface even though attempts were made to clean the sensor after each use. Figure 15.5a and b show the difference between a clean sensor and a surface modified with crosslinkers and antibodies via silanization. As a result of this finding, a new sensor was used for each trial and then was sterilized and discarded after testing.

The change in impedance of the biosensor was directly proportional to the number of bacteria immobilized on the sensor surface. The impedance spectra for different concentrations of bacteria are shown in Fig. 15.6. The impedance caused

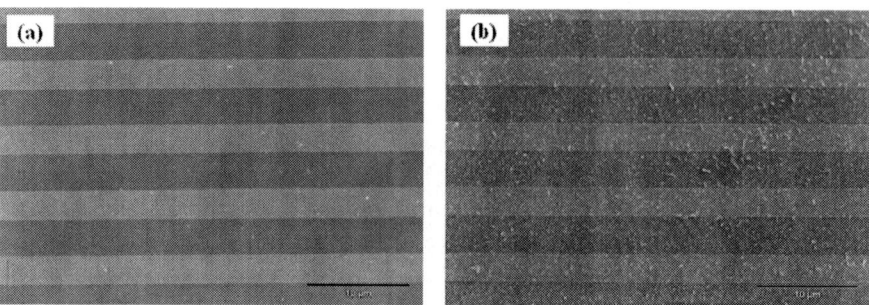

Fig. 15.5 SEM micrographs of the sensor electrodes: (a) clean surface; (b) surface with immobilized antibodies via silanization (Reprinted from Radke, S., & Alocilja, E. (2004). *IEEE Sensors Journal, 4*(4), 434–440.)

15 Biosensors for Detecting Pathogenic Bacteria in the Meat Industry

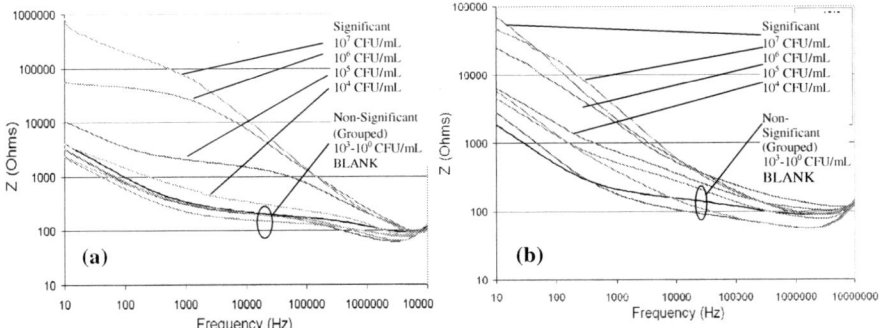

Fig. 15.6 Impedance for a frequency distribution from 10 Hz–10 MHz: (a) nonpathogenic *E. coli*; (b) *E. coli* O157:H7 (Reprinted from Radke, S., & Alocilja, E. (2004). *IEEE Sensors Journal, 4*(4), 434–440.)

by the bacteria linearly increased with the number of cells present in solution. Figure 15.7a shows that a solution with low cell concentrations result in few bacteria being immobilized on the biosensor surface while Fig. 15.7b shows that a solution with high cell concentrations results in bacteria completely covering the sensor surface.

At low frequencies, different cellular concentrations have the effect of increasing the impedance in proportion to the number of cells present in solution for concentrations greater than 10^3 CFU/mL. The polarization effect of the bacteria on the biosensor surface only began to change the impedance at a concentration of 10^4 CFU/mL and higher because of the sufficient number of bacteria required to change the impedance above the base impedance of the biosensor. At higher frequencies, the impedance became less reactive and more resistive potentially due to the relaxation of small dipole species (water molecules) at high frequencies. Also, the reactance caused by the capacitance (parasitic, the electrode double layer, and

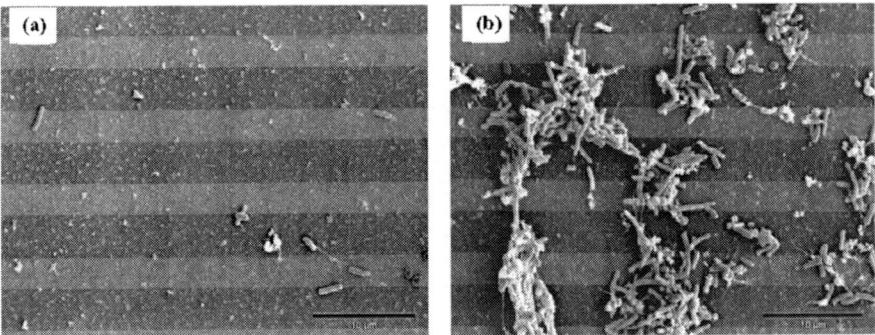

Fig. 15.7 SEM micrographs of bacteria bound to the biosensor surface: (a) sample of 10^2 CFU/mL; (b) sample of 10^6 CFU/mL (Reprinted from Radke, S., & Alocilja, E. (2004). *IEEE Sensors Journal, 4*(4), 434–440.)

capacitance due to bacteria) was minimized at high frequencies and the impedance curves converged toward the resistance of the testing solution.

Between pathogenic and nonpathogenic bacteria, it is shown that *E. coli* O157:H7 yields lower impedance than *E. coli* K12. This effect is most pronounced at low frequencies, as shown in Fig. 15.8a. For both bacteria species, there is no statistically significant detectable difference in impedance between the blank and bacteria concentrations from 10^0 to 10^3 CFU/mL at a frequency of 1 kHz. This could be due to the effect of double layer capacitance and parasitic capacitance found in the sensor, which acts independently of whether bacteria are present in solution. With increasing frequency, the presence of bacteria has a decreasing effect on overall sensor impedance resulting in a convergence of the impedance for reasons already described above. It is suspected that the pathogenic bacteria did not bind as well as the nonpathogenic strain, due to the specificity of the antibodies used for the bacterial species. Additionally, decreased sensitivity at low concentrations may be due to surface chemistry issues related to the silanization process allowing some antibodies to bind to the gold electrodes.

After measuring the impedance data against time, it was found that the impedance increased the longer the biosensor remained in solution when measured at low frequencies. Figure 15.8b shows Cole-Cole plots of the impedance for 5, 20, 35 and 65 min after insertion in the test solution. The increase in impedance is due to an increased number of bacteria cells binding to the sensor surface over time. The increase in impedance is most pronounced within the first 20 min after insertion into solution. After 35 min, the rate of cell binding on the surface slows down considerably. Given enough elapsed time, dense packing of bacteria will eventually occur on the surface. By taking impedance measurements at specific time intervals, the effect of bacteria concentration is shown.

For nonpathogenic *E. coli*, the statistical analysis shows that the biosensor has a lower detection limit of 10^5 CFU/mL with respect to the blank when using a log transformation of impedance data at a frequency of 1 kHz (Fig. 15.9). For

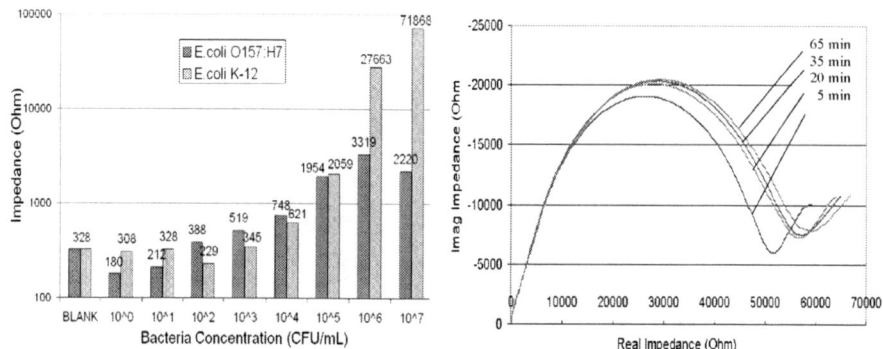

Fig. 15.8 (a) Comparison of *E. coli* and *E. coli* O157:H7 at 1 kHz frequency and (b) Cole-Cole Plot for *E. coli* at 10^6 CFU/mL (Reprinted from Radke, S., & Alocilja, E. (2004). *IEEE Sensors Journal*, 4(4), 434–440.)

Count of Bacteria Species (CFU/mL)	Mean ± SD and significance[1]	
	E. coli K-12 Impedance $(\log \Omega)^2$	*E. coli* O157:H7 Impedance $(\log \Omega)^3$
BLANK	2.42 ± 0.13 a	2.29 ± 0.32 a
2×10^0	2.45 ± 0.04 a	2.25 ± 0.05 a
2×10^1	2.51 ± 0.32 a	2.32 ± 0.10 a
2×10^2	2.32 ± 0.04 a	2.50 ± 0.33 a
5×10^3	2.52 ± 0.03 a	2.70 ± 0.16 a,b
1×10^4	2.73 ± 0.07 a,b	2.87 ± 0.02 b,c
1×10^5	3.20 ± 0.16 b	3.28 ± 0.09 c,d
1×10^6	4.40 ± 0.16 c	3.50 ± 0.16 d
1×10^7	4.67 ± 0.31 c	3.34 ± 0.09 c,d

[1] Means with same letter are not significantly different ($p > 0.05$)
[2] Log transform of K-12 impedance data from a frequency of 1 kHz
[3] Log transform of O157:H7 impedance data from a frequency of 1 kHz

Fig. 15.9 Statistical significance of mean differences between concentrations (Reprinted from Radke, S., & Alocilja, E. (2004). *IEEE Sensors Journal, 4*(4), 434–440.)

pathogenic *E. coli* O157:H7, a lower detection limit of 10^4 CFU/mL is obtained with respect to the blank.

Concluding Comments

Biosensors are emerging as potential alternatives to rapid screening and detection of microbial contaminants in meat and meat products. They are certainly useful for field, on-the-floor type of devices because they are small, portable, and inexpensive. Many of them are disposable, hence regeneration time is eliminated. They are useful tools to support alongside laboratory-based microbiological methods that are proven to be sensitive, specific, and reliable. They are expected to play a major role in controlling, preventing, and responding to pathogen contamination in the food supply in general and meat industry in particular.

Acknowledgements I would like to acknowledge the publications and thesis manuscripts of my graduate students over the years, particular those by Stephen Radke, Zarini Muhammad-Tahir, and Finny Mathew, whose works have been highly instrumental in writing this chapter. I would also like to acknowledge all my funding sources without which the experimental works would not have been possible.

References

Abdel-Hamid, Ivnitski, D., Atanasov, P., Wilkins, E. (1998). Flow-through immunofiltration assay system for rapid detection of *E. coli* O157:H7, *Biosensors and Bioelectronics, 14*, 309–316.
Abel, A. P., Weller, M. G., Duveneck, G. L., Ehrat, M., & Widmer, H. M. (1996). Fiber-optic evanescent wave biosensor for the detection of oligonucleotides. *Analytical Chemistry, 68*, 2905–2912.

Alocilja, E. C., & Radke, S. (2003). Market analysis of biosensors for food safety. *Biosensors and Bioelectronics Journal, 18*(5–6), 841–846.

Alocilja, E. C., Ritchie, N., & Grooms, D. (2003). Protocol development using an electronic nose for differentiating *E. coli* strains. *IEEE Sensors Journal, 3*(6), 801–805.

Babacan, S., Pivarnik, P., Letcher, S., & Rand, A. G. (2000). Evaluation of antibody immobilization methods for piezoelectric biosensor application. *Biosensors and Bioelectronics, 15*, 615–621.

Baeumner, A. J., Cohen, R. N., Miksic, V., & Min, J. (2003). RNA biosensor for the rapid detection of viable *Escherichia coli* in drinking water. *Biosensors and Bioelectronics, 18*, 405–413.

Baeumner, A. J., Pretz, J., & Fang, S. (2004). A universal nucleic acid sequence biosensor with nanomolar detection limits. *Analytical Chemistry, 76*, 888–894.

Bao, L., Deng, L., Nie, L., Yao, S., & Wei, W. (1996). Determination of microorganisms with a quartz crystal microbalance sensor. *Analytical Chemical Acta, 319*, 97–101.

Barbour, W. M., & George, T. (1997). Genetic and immunologic techniques for detecting foodborne pathogens and toxins. In T. J. Montville (Ed.), *Food microbiology: Fundamentals and frontiers* (pp. 30–65). Washington, DC: ASM Press.

Berney, H., West, J., Haefele, E., Alderman, J., Lane, W., & Collins, J. K. (2000). A DNA diagnostic biosensor: Development, characterisation and performance. *Sensors and Actuators B: Chemical, 68*, 100–108.

Bhatia, S. K., Shriver-Lake, L. C., Prior, K. J., Georger, J. H., Calvert, J. M., Bredehorst, R., et al. (1989). Use of thiol-terminal silanes and heterobifunctional crosslinkers for immobilization of antibodies on silica surfaces. *Analytical Biochemistry, 178*, 408–413.

Bianchi, N., Rutigliano, C., Tomasetti, M., Feriotto, G., Zorzato, F., & Gambari, R. (1997). Biosensor technology and surface plasmon resonance for real-time detection of HIV-1 genomic sequences amplified by polymerase chain reaction. *Clinical and Diagnostic Virology, 8*, 199–208.

Blaser, M. J., & Newman, L. S. (1982). A review of human salmonellosis I. Infective dose. *Reviews of Infectious Diseases, 4*, 1096–1106.

Boltovets, P. M., Boyko, V. R., Kostikov, I. Y., Dyachenko, N. S., Snopok, B. A., & Shirshov, Y. M. (2002). Simple method for plant virus detection: Effect of antibody immobilization technique. *Journal of Virological Methods, 105*, 141–146.

Bunde, R. L., Jarvi, E. J., & Rosentrerer, J. J. (1998). Piezoelectric quartz crystal biosensor. *Talanta, 46*, 1223–1229.

Carlson, S. A., Bolton, L. F., Briggs, C. E., Hurd, H. S., Sharma, V. K., Fedorka Cray, P. J., et al. (1999). Detection of multiresistant *Salmonella typhimurium* DT104 using multiplex and fluorogenic PCR. *Molecular and Cellular Probes, 13*, 213–222.

CDC. (2001a). Outbreaks caused by Shiga toxin-producing Escherichia coli-Summary of 2000 Surveillance Data. *Centers for Disease Control and Prevention*. From http://www.cdc.gov/foodborneoutbreaks/ecoli/2000_summaryLetter.pdf.

CDC. (2001b). Salmonellosis. From http://www.cdc.gov/ncidod/dbmd/diseaseinfo/salmonellosis_g.htm.

CDC. (2002a). Notice to readers: Final 2001 reports of notifiable diseases. *Morbidity and Mortality Weekly Report, 51*, 710.

CDC. (2002b). Preliminary FoodNet data on the incidence of foodborne illnesses – selected sites, United States, 2001. *Morbidity and Mortality Weekly Report, 51*, 325–329.

CDC. (2002c). Report on the decline of foodborne illness. *Centers for Disease Control and Prevention*. From http://www.cdc.gov/foodborne/publications/201-nelson_2004.pdf.

CHEMICON International. (2004). Introduction to antibodies. CHEMICON International, Inc. From http://www.chemicon.com/resource/ANT101/a1.asp.

Chen, Z. Z., Wang, K. M., Yang, X. H., Huang, S. S., Huang, H. M., Li, D., et al. (2003). Determination of hepatitis B surface antigen by surface plasmon resonance biosensor. *Acta Chimica Sinica, 61*, 137–140.

Cheung, J. H., Stockton, W. B., & Rubner, M. F. (1997). Molecular-level processing of conjugated polymers: Layer-by-layer manipulation of polyaniline via electrostatic interactions. *Macromolecules, 30*, 2712–2716.

Cohn, G. E. (1998). Systems and technologies for clinical diagnostics and drug discovery. *SPIE Proceedings, 3259*, 11–17.
Corry, B., Uilk, J., & Crawley, C. (2003). *Analytica Chemica Acta, 496*, 103–116.
Cuireanu, M., Levadoux, W., & Goldstein, S. (1997). Electrical impedance studies on a culture of a newly discovered strain of *Steptomyces*. *Enzyme and Microbial Technology, 21*, 441–449.
D'Aoust, J. Y. (1997). Salmonella species. In T. J. Montville (Ed.), *Food microbiology: Fundamentals and frontiers* (pp. 138–139). Washington, DC: ASM.
DeMarco, D., & Lim, D. (2002). Detection of *Escherichia coli* O157:H7 in 10- and 25-gram ground beef samples with an evanescent-wave biosensor with silica and polystyrene waveguides. *Journal of Food Protection, 65*, 596–602.
Doyle, M. P., Zhao, T., Meng, J., & Zhao, S. (1997). Escherichia coli O157:H7. *Food microbiology fundamentals and frontiers*. Washington, D.C: American Society for Microbiology.
D'Souza, S. F. (2001). Microbial biosensors (Review). *Biosensors and Bioelectronics, 16*, 337–353.
Esch, M. B., Locascio, L. E., Tarlov, M. J., & Durst, R. A. (2001). Detection of viable *Cryptosporidium parvum* using DNA-modified liposomes in a microfluidic chip. *Analytical Chemistry, 73*, 2952–2958.
FAO-Food and Agriculture Organization. (2006). Global market analysis. *Food Outlook No. 1*. Retrieved June 2006, from FTP://FTP.FAO.ORG/DOCREP/FAO/009/J7927E/J7927E00.PDF.
FDA-Food and Drug Administration. (1998). *Bacteriological analytical manual*. Food and Drug Administration, Gaithersburg, MD.
FDA-Food and Drug Administration. (2005). *Bacteriological analytical manual*. Rockville, MD, USA: Food and Drug Administration. From http://www.cfsan.fda.gov/~ebam/bam-toc.html.
FDA-Food and Drug Administration. (2006). *Foodborne pathogenic microorganisms and natural toxins handbook: The "Bad Bug Book"*. FDA-CFSAN. From http://www.cfsan.fda.gov/~mow/intro.html.
Feriotto, G., Borgatti, M., Mischiati, C., Bianchi, N., & Gambari, R. (2002). Biosensor technology and surface plasmon resonance for real-time detection of genetically modified roundup ready soybean gene sequences. *Journal of Agricultural and Food Chemistry, 50*, 955–962.
Fratamico, P. M., Strobaugh, T. P., Medina, M. B., & Gehring, A. G. (1998). Detection of *Escherichia coli* O157:H7 using a surface plasmon resonance biosensor. *Biotechnology Techniques, 12* (7), 571–576.
Gao, Z. X., Fang, Y. J., Ren, J., Ning, B., Zhu, H. Z., & He, Y. H. (2004). Studies on biotin-avidin indirect conjugated technology for a piezoelectric DNA sensor. *International Journal of Environmental Analytical Chemistry, 84*, 599–606.
Gau, J. -J., Lan, E. H., Dunn, B., Ho, C. -M., & Woo, J. C. S. (2001). A MEMS based amperometric detector for *E. coli* bacteria using self-assembled monolayers. *Biosensors and Bioelectronics, 16*, 745–755.
Geng, T., Morgan, M. T., & Bhunia, A. K. (2004). Detection of low levels of *Listeria monocytogenes* cells by using a fiber-optic immunosensor. *Applied and Environmental Microbiology, 70*, 6138–6146.
Ghindilis, A., Atanasov, P., Wilkins, M., & Wilkins, E. (1998). Immunosensors: Electrochemical sensing and other engineering approaches. *Biosensors and Bioelectronics, 13*, 113–131.
Gomez, R., Bashir, R., Sarikaya, A., Ladisch, M., Sturgis, J., Robinson, J., et al. (2001). Microfluidic biochip for impedance spectroscopy of biological species. *Biomedical Microdevices, 3*, 201–209.
Gore, A., Chakrabartty, S., Pal, S., & Alocilja, E. C. (2006). A multi-channel femtoampere-sensitivity potentiostat array for biosensing applications. *IEEE Transactions on Circuits and Systems, 53*(11), 2357–2363.
Graham, C. R., Leslie, D., & Squirrell, D. J. (1992). Gene probe assays on a fibre-optic evanescent wave biosensor. *Biosensors and Bioelectronics, 7*, 487–493.
Hartley, H. A., & Baeumner, A. J. (2003). Biosensor for the specific detection of a single viable B. anthracis spore. *Analytical Bioanalytical Chemistry, 376*, 319–327.

He, F. J., & Liu, S. Q. (2004). Detection of *P. aeruginosa* using nano-structured electrode-separated piezoelectric DNA biosensor. *Talanta, 62*, 271–277.

Ho, J. A., Hsu, H. W., & Huang, M. R. (2004). Liposome-based microcapillary immunosensor for detection of *Escherichia coli* O157:H7. *Analytical Biochemistry, 330*, 342–349.

Hoyle, B. (2001). High-tech biosensor speeds bacteria detection. *ASM News, 67*, 434–435.

Huang, T. S., Tzeng, Y., Liu, Y. K., Chen, Y. K., Walker, K. R., Guntupalli, R., et al. (2004). Immobilization of antibodies and bacterial binding on nanodiamond and carbon nanotubes for biosensor applications. *Diamond and Related Materials, 13*, 1098–1102.

Jiang, H., Adams, C., Graziano, N., Roberson, A., McGuire, M., & Khiari, D. (2006). Enzyme-linked immunosorbent analysis (ELISA) of atrazine in raw and finished drinking water. *Environmental Engineering Science, 23* (2), 357–366.

Jordan, C. E., Frutos, A. G., Thiel, A. J., & Corn, R. M. (1997). Surface plasmon resonance imaging measurements of DNA hybridization adsorption and streptavidin/DNA multilayer formation at chemically modified gold surfaces. *Analytical Chemistry, 69*, 4939–4947.

Kalorama Information. (2000). MarketResearch.com, New York, p. 313.

Kim, N., Park, I. S., & Kim, D. K. (2004). Characteristics of a label-free piezoelectric immunosensor detecting *Pseudomonas aeruginosa*. *Sensors and Actuators B-Chemical, 100*, 432–438.

Ko, S., & Grant, S. A. (2006). A novel FRET-based optical fiber biosensor for rapid detection of *Salmonella typhimurium*. *Biosensors and Bioelectronics, 21*, 1283–1290.

Komarova, E., Aldissi, M., & Bogomolova, A. (2005). Direct electrochemical sensor for fast reagent-free DNA detection. *Biosensors and Bioelectronics, 21*, 182–189.

Koppes, L., Woldringh, C., & Nanninga, N. (1978). Size variations and correlation of different cell cycle events in slow-growing *Escherichia coli*. *Journal of Bacteriology, 134*, 423–433.

Koubova, V., Brynda, E., Karasova, L., Skvor, J., Homola, J., Dostalek, J., et al. (2001). Detection of foodborne pathogens using surface plasmon resonance biosensors. *Sensors and Actuators B-Chemical, 74*, 100–105.

Kukanskis, K., Elkind, J., Melendez, J., Murphy, T., Miller, G., & Garner, H. (1999). Detection of DNA hybridization using the TISPR-1 surface plasmon resonance biosensor. *Analytical Biochemistry, 274*, 7–17.

Lazcka, O., Campo, F., Javier, D., Munoz, F., & Xavier, F. (2007). Pathogen detection: A perspective of traditional methods and biosensors. *Biosensors and Bioelectronics, 22*, 1205–1217.

Lee, J. S., Choi, Y.-K., Pio, M., Seo, J., & Lee, L. P. (2002). Nanogap capacitors for label free DNA analysis. *BioMEMS and Bionanotechnology, 729*, 185–190.

Lin, H. C., & Tsai, W. C. (2003). Piezoelectric crystal immunosensor for the detection of staphylococcal enterotoxin B. *Biosensors and Bioelectronics, 18*, 1479–1483.

Liu, X., Farmerie, W., Schuster, S., & Tan, W. (2000). Molecular beacons for DNA biosensors with micrometer to submicrometer dimensions. *Analytical Biochemistry, 283*, 56–63.

Liu, X., & Tan, W. (1999). A fiber-optic evanescent wave DNA biosensor based on novel molecular beacons. *Analytical Chemistry, 71*, 5054–5059.

Lu, B., Smyth, M. R., and O'Kennedy, R. (1996). Oriented immobilization of antibodies and its applications in immunoassays and immunosensors. *Analyst, 121*, 29R–32R.

Maehashi, K., Matsumoto, K., Kerman, K., Takamura, Y., & Tamiya, E. (2004). Ultrasensitive detection of DNA hybridization using carbon nanotube field-effect transistors. *Japanese Journal of Applied Physics Part 2-Letters & Express Letters, 43*, L1558–L1560.

Mariotti, E., Minunni, M., & Mascini, M. (2002). Surface plasmon resonance biosensor for genetically modified organisms detection. *Analytica Chimica Acta, 453*, 165–172.

Markx, G., & Davey, C. (1999). The dielectric properties of biological cells at radiofrequencies: Applications in biotechnology. *Enzyme and Microbial Technology, 25*, 161–171.

Mathew, F., Alagesan, D., & Alocilja, E. C. (2004). Chemiluminescence detection of *Escherichia coli* in fresh produce obtained from different sources. *Luminescence Journal, 19*, 193–198.

Mathew, F., & Alocilja, E. C. (2004). Enzyme-based detection of *Escherichia coli*. *Transactions of the ASAE, 47* (1), 357–362.

Mathew, F., & Alocilja, E. C. (2005). Porous silicon-based biosensor for pathogen detection. *Biosensors and Bioelectronics Journal, 20* (8),1656–1661.

McClelland, R., & Pinder, A. (1994). Detection of Salmonella typhimurium in dairy products with flow cytometry and monoclonal antibodies. *Applied Environmental Microbiology A, 60,* 4255–4262.

McGown, L. B., Joseph, M. J., Pitner, J. B., Vonk, G. P., & Linn, C. P. (1995). The nucleic-acid ligand – a new tool for molecular recognition. *Analytical Chemistry, 67,* A663–A668.

Mead, P. S., Slutsker, L., Dietz, V., McGaig, L., Bresee, J., Shapiro, C., et al. (1999). Food-related illnesses and death in the United States. *Emerging Infectious Disease, 5,* 607–625.

Meeusen, C., Alocilja, E. C., & Osburn, W. (2005). Detection of *E. coli* O157:H7 using a miniaturized surface plasmon resonance biosensor. *Transactions of the ASAE, 48*(6), 2409–2416.

Meng, J., Zhao, S., Doyle, M., & Kresovich, S. (1996). Polymerase chain reaction for detection *E. coli* O157:H7. *International Journal of Food Microbiology, 32,* 103–113.

Mir, M., & Katakis, I. (2005). Towards a fast-responding, label-free electrochemical DNA biosensor. *Analytical and Bioanalytical Chemistry, 381,* 1033–1035.

Mittelmann, A. S., Ron, E. Z., & Rishpon, J. (2002). Amperometric quantification of total coliforms and specific detection of *Escherichia coli*. *Analytical Chemistry, 74*(4), 903–907.

Mo, X. T., Zhou, Y. P., Lei, H., & Deng, L. (2002). Microbalance-DNA probe method for the detection of specific bacteria in water. *Enzyme and Microbial Technology, 30,* 583–589.

Muhammad-Tahir, Z., & Alocilja, E. (2004). A disposable biosensor for pathogen detection in fresh produce samples. *Biosystems Engineering, 88,* 145–151.

Muhammad-Tahir, Z., & Alocilja, E. C. (2003a). A conductimetric biosensor for biosecurity. *Biosensors and Bioelectronics, 18,* 813–819.

Muhammad-Tahir, Z., & Alocilja, E. C. (2003b). Fabrication of a disposable biosensor for *Escherichia coli* O157:H7 detection. *IEEE Sensors Journal, 3,* 345–351.

Muhammad-Tahir, Z., Alocilja, E. C., & Grooms, D. L. (2005a). Polyaniline synthesis and its biosensor application. *Biosensors and Bioelectronics, 20,* 1690–1695.

Muhammad-Tahir, Z., Alocilja, E. C., & Grooms, D. L. (2005b). Rapid detection of *Bovine Viral Diarrhea Virus* as surrogate of bioterrorism agents. *IEEE Sensors Journal, 5,* 757–762.

Muhammad-Tahir, Z., Alocilja, E. C., & Grooms, D. L. (2007). Indium tin oxide-polyaniline biosensor: Fabrication and characterization. *Sensors Journal, 7,* 1123–1140.

Nagai, H., Murakami, Y., Yokoyama, K., & Tamiya, E. (2001). High-throughput PCR in silicon based microchamber array. *Biosensors and Bioelectronics, 16,* 1015–1019.

Narang, U., Anderson, G. P., Ligler, F. S., & Burans, J. (1997). Fiber optic-based biosensor for ricin. *Biosensors and Bioelectronics, 12,* 937–945.

Nashat, A. H., Moronne, M., & Ferrari, M. (1998). Detection of functional groups and antibodies on microfabricated surfaces by confocal microscopy. *Biotechnology and Bioengineering, 60,* 137–146.

Nicolini, C., Erokhin, V., Facci, P., Guerzoni, S., Ross, A., & Paschkevitsch, P. (1997). Quartz balance DNA sensor. *Biosensors and Bioelectronics, 12,* 613–618.

Pal, S., Alocilja, E. C., & Downes, F. P. (2007). Nanowire labeled direct-charge transfer biosensor for detecting bacillus species. *Biosensors & Bioelectronics Journal, 22,* 2329–2336.

Park, I. S., Kim, W. Y., & Kim, N. (2000). Operational characteristics of an antibody-immobilized QCM system detecting Salmonella spp. *Biosensors and Bioelectronics, 15,* 167–172.

Park, S., & Durst, R. A. (2000). Immunoliposome sandwich assay for the detection of Escherichia coli O157:H7. *Analytical Biochemistry, 280,* 151–158.

Radke, S., & Alocilja, E. (2004). Design and fabrication of an impedimetric biosensor. *IEEE Sensors Journal, 4,* 434–440.

Radke, S., & Alocilja, E. C. (2005b). A microfabricated biosensor for detecting foodborne bioterrorism agents. *IEEE Sensors Journal, 5* (4), 744–750.

Radke, S. M., & Alocilja, E. C. (2005a). A high density microelectrode array biosensor for detection of *E. coli* O157:H7. *Biosensors and Bioelectronics SPEC ISS, 20,* 1662–1667.

Ramanaviciene, A., & Ramanavicius, A. (2004). Pulsed amperometric detection of DNA with an ssDNA/polypyrrole-modified electrode. *Analytical and Bioanalytical Chemistry, 379*, 287–293.

Rishpon, J., & Ivnitski, D. (1997). An amperometric enzyme-channeling immunosensor. *Biosensors and Bioelectronics, 12*, 195–204.

Rodriguez, M., & Alocilja, E. (2005). Embedded DNA-polypyrrole biosensor for rapid detection of *Escherichia coli*. *IEEE Sensors Journal, 5*, 733–736.

Ruan, C. M., Yang, L. J., & Li, Y. B. (2002). Immunobiosensor chips for detection of *Escherichia coli* O157: H7 using electrochemical impedance spectroscopy. *Analytical Chemistry, 74*, 4814–4820.

Ruan, C. M., Zeng, K. F., Varghese, O. K., & Grimes, C. A. (2004). A staphylococcal enterotoxin B magnetoelastic immunosensor. *Biosensors and Bioelectronics, 20*, 585–591.

Sadana, A. 2002. *Engineering biosensors: Kinetic and design application*. London: Academic Press.

Savage, M. D., Mattson, G., Desai, S., Nielander, G. W., Morgensen, S., & Conklin, E. J. (1994). *Avidin-biotin chemistry: A handbook*. Rockford, IL: Pierce Chemical Company.

Scheller, F. W., Hintscher, R., Pfeiffer, P., Schubert, F., Riedel, K., & Kindervater, R. (1991). Biosensors: Fundamentals, applications and trends. *Sensors and Actuators B, 4*, 197–206.

Seo, K. H., Brackett, R. E., Hartman, N. F., & Campbell, D. P. (1999). Development of a rapid response biosensor for detection of *Salmonella* Typhimurium. *Journal of Food Protection, 62*, 431–437.

Shah, J., Chemburu, S., Wilkins, E., & Abdel-Hamid, I. (2003). Rapid amperometric immunoassay for *Escherichia coli* based on graphite coated nylon membranes. *Electroanalysis, 15*, 1809–1814.

Sheppard, N. F., Mears, D. J., & Guiseppi-Elie, A. (1995). Model of an immobilized enzyme conductimetric urea biosensor. *Biosensors and Bioelectronics, 11*, 967–979.

Shriver-Lake, L. C., Donner, B., Edelstein, R., Breslin, K., Bhatia, S. K., & Ligler, F. S. (1997). Antibody immobilization using heterobifunctional crosslinkers. *Biosensors and Bioelectronics, 12*, 1101–1106.

Silin, V., & Plant, A. (1997). Biotechnological applications of surface plasmon resonance. *Trends in Biotechnology, 15*, 353–359.

Skuridin, S. G., Yevdokimov, Y. M., Efimov, V. S., Hall, J. M., & Turner, A. P. F. (1996). A new approach for creating double-stranded DNA biosensors. *Biosensors and Bioelectronics, 11*, 903–911.

Slavik, R., Homola, J., & Brynda, E. (2002). A miniature fiber optic surface plasmon resonance sensor for fast detection of Staphylococcal enterotoxin B. *Biosensors and Bioelectronics, 17*, 591–595.

Sportsman, J. R., & Wilson, G. S. (1980). Chromatographic properties of silica-immobilized antibodies. *Analytical Chemistry, 52*, 2013–2018.

Su, X. L., & Li, Y. (2005). Surface plasmon resonance and quartz crystal microbalance immunosensors for detection of *Escherichia coli* O157 : H7. *Transactions of the ASAE, 48*, 405–413.

Su, X. L., & Li, Y. B. (2004). A self-assembled monolayer-based piezoelectric immunosensor for rapid detection of *Escherichia coli* O157 : H7. *Biosensors and Bioelectronics, 19*, 563–574.

Suehiro, J., Hamada, R., Noutomi, D., Shutou, M., & Hara, M. (2003). Selective detection of viable bacteria using dielectrophoretic impedance measurement method. *Journal of Electrostatics, 57*, 157–168.

Sung Hoon, R., In Seon, P., Namsoo, K., & Woo Yeon, K. (2001). Hybridization of *Salmonella* spp.-specific nucleic acids immobilized on a quartz crystal microbalance. *Food Science and Biotechnology, 10*, 663–667.

Teratanavat, R., & Hooker, N. H. (2004). Understanding the characteristics of US meat and poultry recalls: 1994–2002. *Food Control, 5*, 359–367.

Tien, H., & Ottova-Leitmannova, A. (2000). *Membrane biophysics as viewed from experimental bilayer lipid membranes*. Amsterdam: Elsevier.

Tombelli, S., Mascini, M., Sacco, C., & Turner, A. P. F. (2000). A DNA piezoelectric biosensor assay coupled with a polymerase chain reaction for bacterial toxicity determination in environmental samples. *Analytica Chimica Acta, 418*, 1–9.

Tsai, W. C., & Lin, I. C. (2005). Development of a piezoelectric immunosensor for the detection of alpha-fetoprotein. *Sensors and Actuators B-Chemical, 106*, 455–460.

Turner, A. P., & Newman, J. D. (1998). An introduction to biosensor. In T. W. Gateshead (Ed.), *Biosensor for Food Analysis* (pp. 13–27). UK: Athaenaeum Press Ltd.

Van Gerwen, P., Laureyn, W., Laureys, W., Huyberechts, G., Op De Beeck, M., Baert, K., et al. (1998). Nanoscaled interdigitated electrode arrays for biochemical sensors. *Sensors and Actuators B, 49*, 73–80.

Vaughan, R. D., Carter, R. M., O'Sullivan, C. K., & Guilbault, G. G. (2003). A quartz crystal microbalance (QCM) sensor for the detection of *Bacillus cereus*. *Analytical Letters, 36*, 731–747.

Vikholm, I. (2005). Self-assembly of antibody fragments and polymers onto gold for immunosensing. *Sensors and Actuators B-Chemical, 106*, 311–316.

Wang, R. H., Tombelli, S., Minunni, M., Spiriti, M. M., & Mascini, M. (2004). Immobilisation of DNA probes for the development of SPR-based sensing. *Biosensors and Bioelectronics, 20*, 967–974.

Weeks, B. L., Camarero, J., Noy, A., Miller, A. E., Stanker, L., & De Yoreo, J. J. (2003). A microcantilever-based pathogen detector. *Scanning, 25*, 297–299.

WHO. (2002). *Terrorist threats to food: Guidance for establishing and strengthening prevention and response systems*. Geneva, Switzerland: World Health Organization Food Safety Dept.

Wiegand, G., Arribas-Layton, N., Hillebrandt, H., Sackmann, E., & Wagner, P. (2002). Electrical properties of supported bilayer membranes. *Journal of Physical Chemistry B, 106*, 4245–4254.

Yang, L., Chakrabartty, S., & Alocilja, E. C. (2007). Fundamental building blocks for molecular bio-wire based forward-error correcting biosensors. *Nanotechnology Journal, 18*, 424017 (6pp).

Ye, J. M., Letcher, S. V., & Rand, A. G. (1997). Piezoelectric biosensor for detection of *Salmonella* Typhimurium. *Journal of Food Science, 62*, 1067.

Younts, S., Alocilja, E. C., Osburn, W. N., Marquie, S., & Grooms, D. L. (2002). Differentiation of *Escherichia coli* O157:H7 from non-O157:H7 *E. coli* serotypes using a sensor-based, computer-controlled detection system. *Transactions of the ASAE, 44*, 1681–1685.

Younts, S. Alocilja, E., Osburn, W., Marquie, S., Gray, J., & Grooms, D. (2003). Experimental use of a gas sensor-based instrument for differentiation of *E. coli* O157:H7 from non-O157:H7 *E. coli* field isolates. *Journal of Food Protection, 66*, 1455–1458.

Zhang, Z. X., & Li, M. Q. (2005). Electrostatic microcantilever array biosensor and its application in DNA detection. *Progress in Biochemistry and Biophysics, 32*, 314–317.

Zhao, H. Q., Lin, L., Li, J. R., Tang, J. A., Duan, M. X., & Jiang, L. (2001). DNA biosensor with high sensitivity amplified by gold nanoparticles. *Journal of Nanoparticle Research, 3*, 321–323.

Zhou, A. H., & Muthuswamy, J. (2004). Acoustic biosensor for monitoring antibody immobilization and neurotransmitter GABA in real-time. *Sensors and Actuators B-Chemical, 101*, 8–19.

Zhou, X. D., Liu, L. J., Hu, M., Wang, L. L., & Hu, J. M. (2002). Detection of *Hepatitis B virus* by piezoelectric biosensor. *Journal of Pharmaceutical and Biomedical Analysis, 27*, 341–345.

Zuo, Y., Chakrabartty, S., Muhammad-Tahir, Z., Pal, S., & Alocilja, E. C. (2006). Spatio-temporal processing for multichannel biosensors using support vector machines. *IEEE Sensors Journal, 6*, 1644–1651.

Chapter 16
Immunology-Based Techniques for the Detection of Veterinary Drug Residues in Foods

Milagro Reig and Fidel Toldrá

Introduction

Veterinary drugs are used in farm animals, via the feed or the drinking water, to prevent the outbreak of diseases or even for the treatment of diseases. However, the growth of animals may be promoted through the use of hormones and antibiotics. Depending on the type of residue and the application and washing conditions, these substances or its metabolites may remain in meat and other foods of animal origin and may cause adverse effects on consumers' health. This is the main reason why its use is strictly regulated or even banned (case of the European Union) in different countries. Antibiotics typically used for growth promotion include chloramphenicol, nitrofurans, and enrofloxacin but others like sulphonamides, macrolides etc. may also be used (Reig & Toldrá, 2007). An irreversible type of bone marrow depression that might lead to aplastic anaemia may be caused by chloramphenicol (Mottier et al., 2003), while some allergic reactions as well as emergence of drug-resistant bacteria may be caused by enrofloxacin (Cinquina, Longo, Anastasi, Giannetti, & Cozzani, 2003). One of the major metabolites of nitrofurans is furazolidone that has been reported to have mutagenic and carcinogenic properties (Guo, Chou, & Liao, 2003). There is also some evidence on the toxicity of sulphonamides on the thyroid gland (Pecorelli, Bibi, Fioroni, & Galarini, 2004). Other important concern about the presence of antibiotics residues in animal foodstuffs is related to the selection of resistant bacteria in the gastrointestinal tract and disruption of the colonization barrier of the resident intestinal microflora (Cerniglia & Kotarski, 2005). Intestinal flora may vary depending on the diet and thus is subject to large variations in the proportion of major bacterial species (Moore & Moore, 1995). The presence of antibiotics in meat may alter the intestinal microflora that is essential for human physiology, food digestion, and metabolism of nutrients (Chadwick, George, & Claxton, 1992; Vollard & Clasener, 1994) contributing to the development of antibiotic resistance in the indigenous microflora and impairing colonization

---

F. Toldrá
Instituto de Agroquímica y Tecnología de Alimentos (CSIC), PO Box 73, 46100 Burjassot (Valencia), Spain

resistance, that can increase its susceptibility to infection by pathogenic microorganisms (Cerniglia & Kotarski, 1999).

Thus, it is important to control the presence of veterinary drug residues in meat and other animal foodstuffs in order to ensure consumer protection which is regulated by the different countries, for instance the Council Directive 96/23/EC (EC, 1996) in the EU where national monitoring programmes and sampling procedures have been set up. These regulations usually imply the analysis of a wide variety of residues in a large number of samples making it necessary to have available screening techniques (Bergweff, 2005; Reig & Toldrá, 2008a). Immunoassays are relatively simple to use and rapid and offer important additional advantages like high specificity and sensibility (Reig & Toldrá, 2008b). This depends on the specificity and affinity of the antibody and the form used to quantify the antibody-antigen interaction. The application of immunoassays and other related techniques for the detection of veterinary drug residues, including growth promoter agents, are described in this chapter.

Molecular Recognition

Usual clean-up of complex samples like meat are based on solid phase extraction technique which is very fast and economic even though it lacks enough selectivity. Immunoaffinity extraction is very selective but expensive. So, methods based on molecular recognition mechanisms for clean up of samples have been developed in recent years. Molecular imprinted polymers (MIPs) have shown promising results for the isolation of low amounts of residues as those found in meat. These are cross-linked polymers prepared in the presence of a template molecule that can be a specific analyte or drug. When this template is removed, the polymer offers a binding site complementary to the template structure or closely related analogues. MIPs have better stability than antibodies because they can support high temperatures, larger pH ranges and a wide variety of organic solvents. This extraction procedure is also known as molecularly imprinted solid phase extraction (MISPE) (Baggiani, Anfossi, & Giovannoli, 2007; Widstrand et al., 2004). Figure 16.1 shows a typical procedure. The choice of the appropriate molecule as template is the critical factor for a reliable analysis (Stolker, Zoonties, & Van Ginkel, 1998). The extracted residues are then analyzed by LC-MS and have shown good quantitative results for chloramphenicol (Boyd et al., 2007), cimaterol, ractopamine, clenproperol, clenbuterol, brombuterol, mabuterol, mapenterol and isoxsurine but not for salbutamol and terbutaline (Berggren, Bayoudh, Sherrington, & Ensing, 2000; Kootstra et al., 2005; Stubbings et al., 2005). MISPE can be used on-line, either packaged in a cartridge for the extraction of a particular analyte in a complex sample or in a 96-well extraction plate. Another possibility is to use an ultra filtration device followed by centrifugation and collection of the pellet. The analyte can be separated from the antibody with a mixture of methanol and acetic acid (Johansson & Hellenas, 2004).

16 Detection of Veterinary Drug Residues in Foods

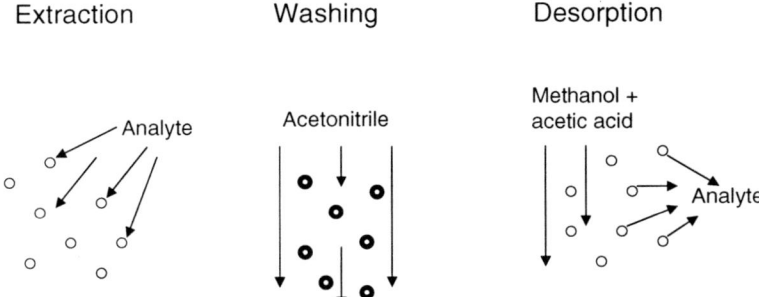

Fig. 16.1 Stages in the extraction of a particular analyte from a sample. (1) Extraction, where the analyte binds to the molecular imprinted polymer. (2) Washing, where the interfering substances are eluted while the analyte is retained in the polymer surface. (3) Desorption, where the analyte is desorbed and recovered

Immunoafinity Chromatography

This type of chromatography is intended for the purification of a particular residue (analyte) and is based on the interaction between the antigen and the antibody which is very specific. A scheme of the procedure is shown in Fig. 16.2. The columns are packaged with a solid matrix, usually a gel like sepharose 4B, having a bound antibody which is specific for the target analyte. The extract, containing the target analyte, is injected into the column and a buffer is eluted through the column for rinsing.

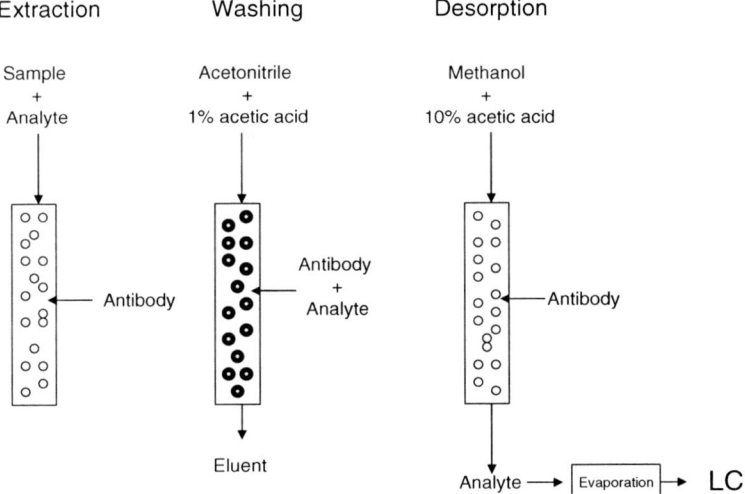

Fig. 16.2 Stages in the immunoaffinity chromatography purification of a particular analyte. 1) Extraction, where the analyte binds to the antibody immobilized to the packaging. 2) Washing, where the interfering substances are eluted while the analyte is retained in the packaging. 3) Desorption, where the analyte is free from its bind to the antibody and recovered

The analyte remains in the column retained by the antibody bound to the matrix. The rest of the extract is eluted. Then, the eluting buffer is changed to an antibody-antigen dissociating buffer so that the target analyte is eluted and recovered in high concentration (see Fig. 16.2). For example, zearalenone present in feed is extracted with acetonitrile-water and sodium chloride. Once extracted, the extract is diluted and applied to an immunoaffinity column that was eluted with water or phosphate buffer for washing; the zearalenone is then eluted with methanol, evaporated and dissolved in the mobile phase for its injection into the LC chromatograph (Campbell & Armstrong, 2007). In another procedure for zeranol, the column contains monoclonal antibodies raised against zeranol coupled to CNBr-activated Sepharose 4B (Zhang et al., 2006). The dry reagent time-resolved fluoroimmunoassay (TR-FIA) applied to zeranol analysis has been enhanced (LOD 0.16 ng/g) by using previous purification of the sample by immunoaffinity chromatography (Tuomola, Cooper, & Lahdenpera, 2002). Immunoaffinity columns prepared by coupling anti-avermectin polyclonal antibody with CNBr-activated Sepharose 4B gave a good clean-up of the sample for the detection of avermectins (He, Hou, Jiang, & Shen, 2005).

Immunoafinity chromatographic columns are highly specific for each type of analyte and are very useful to reduce the detection limit. They are only limited by potential interferences mainly due to potential cross-reactions by other residues of the sample with the antibody (Godfrey, 1998). Another disadvantage of these columns is the high cost and they can only be re-used a limited number of times.

Immunoassay Kits

Immunological methods are very specific for a given residue because they are based on the interaction between the antigen and antibody. A good number of immunoassays have been developed in recent years for the detection of veterinary drugs residues in foods. These methods are based on enzyme-linked immunosorbent assays (ELISA), enzyme immunoassay (EIA), radio immunoassay (RIA), arrays and chips (biosensors). These methods are effective for the specific detection of the analyte and the binding between the antibody and the target analyte but are subject to interferences by unwanted cross reactions, or unspecific binding by substances exceeding the analyte and present in the sample. In other cases, a high background signal may mask the assay. The detection when using ELISA or EIA kits is easily observed by a color change proportional to the amount of target analyte present in the sample. A similar change of color is the basis for dipsticks, that consist of an antibody immobilized to the end of a plastic stick (Link, Weber, & Fussenegger, 2007). The stick is successively transferred, with the inclusion of intermediate rinsings, from the sample to the conjugate and into the substrate (Levieux, 2007). In some cases, luminescence or fluorescence detectors may increase the sensibility (Roda et al., 2003; Zhang, Zhang, Shi, Eremin, & Shen, 2006). In the case of RIA, the detection is based on the measurement of the radioactivity of the immunological complex (Elliot et al., 1998; Samarajeewa, Wei, Huang, & Marshall, 1991). New enzyme immunoassays are reported in literature in

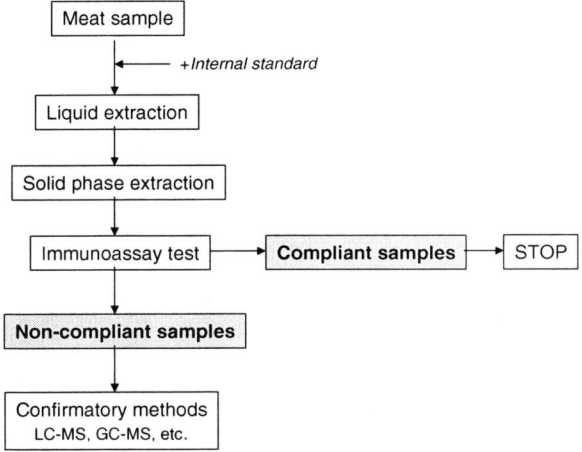

Fig. 16.3 Example of a typical standard procedure for routinary control of residues in meat samples

recent years (see Table 16.1). A scheme of the general procedure for the analysis of a meat sample is shown in Fig. 16.3.

A large number of ELISA kits, with measurement based on color development, are commercially available. These kits are intended for the detection of agonists, esteroids, estrogens, antibiotics, etc. Several examples of commercial available ELISA test kits for the detection of different residues and its respective detection limits are shown in Table 16.2. In general, the limits of detection depend on the previous extraction and clean-up of the sample (Cooper, Delahaut, Fodey, & Elliott, 2004; De Wasch et al., 2001; Gaudin, Cadieu, & Maris, 2003). ELISA test kits are very sensitive, especially for certain substances, and can give semiquantitative results. It must be taken into account that this is a screening technique. Some false positives may arise from interferences by other substances, for instance, cross-reactions with substances that have a structural resemblance to the assayed substance or, less often, changes in pH or composition of the buffers. In case of uncertainty, samples must be subjected to confirmatory analysis (i.e. gas chromatography or HPLC coupled to mass spectrometry) for further confirmation of the suspected compound.

Several inter-laboratory tests have been performed to check and compare different kits from different suppliers for specific residues. So, a study for the detection of chloramphenicol in pigs was carried out in 2002 involving 20 European laboratories that received 8 coded frozen muscle samples with different levels of chloramphenicol. The results were good since the false compliant results were 0% and the false non-compliant results were 10% (Gaudin et al., 2003). Another wide range study for the detection of five antimicrobial promoters (zinc bacitracin, spiramycin, tylosin, virginiamycin and olaquindox) in feedstuffs with 15 European laboratories representing government agencies and private industry also revealed good results with a 1.9% false non-compliant rate and 0% false compliant rate (Situ, Grutters, van Wichen, & Elliott, 2006). These results confirmed that this type of

Table 16.1 Limits of detection or quantitation (CCβ) of ELISA test kits developed for different residues

Type of residue	Group	Foodstuff	Detection limit*	Reference
Erythromycin	Antibiotic	Bovine meat	0.4 ng/mL	Draisci et al., 2001
Tylosin	"	"	4 ng/mL	"
Oxytetracycline	"	Chicken meat	< EU MRL	Dewalsch et al., 2001
Chlortetracycline	"	"	"	"
Doxyckine	"	"	"	"
Tetracycline	"	"	"	"
Bacitracin	Antibiotic	Feed	1 μg/g	Situ & Elliott, 2005
Tylosin	"	"	"	"
Spiramycin	"	"	"	"
Virginiamycin	"	"	"	"
Olaquindox	"	"	"	"
Sulphachlorpyridazine	"	Meat	100 ng/g	Wang, Wang, Duan, & Kennedy, 2006
Tetracycline	"	Pig plasma	10 ng/mL	Lee, Lee, Ryu, Lee, & Cho, 2001
Tylosine	"	Water	0.1 ng/mL	Kumar, Thompson, Singh, Chander, & Gupta, 2004
Tetracycline	"	"	0.05 ng/mL	"
Chloramphenicol	"	Chicken muscle	6 ng/L	Zhang et al., 2006
Diethylestilbestrol	Estrogen	Chicken meat	0.07 ng/mL	Xu, Chu, et al., 2006a
Hexoestrol	"	Pork meat	0.07 ng/mL	Xu, Chu, et al., 2006b
Avermectins	Insecticidal	Bovine liver	1.06 ng/mL	Shi et al., 2006
Medroxyprogesterone acetate	Steroid	Meat	0.096 ng/g	Chifang, Chuanlai, Zhengyu, Xiaogang, & Liying, 2006
Semicarbazide	Nitrofuran	Chicken meat	CCβ = 0.25 ng/g	Cooper et al., 2007a
Dimetridazole	Nitroimidazoles	Chicken muscle	CCβ = 2 ng/g	Huet et al., 2005
Metronidazole	"	"	CCβ = 10 ng/g	"
Ronidazole	"	"	CCβ = 20 ng/g	"
Hydroxydimetridazole	"	"	CCβ = 20 ng/g	"
Ipronidazole	"	"	CCβ = 40 ng/g	"
Azaperol	Sedative	Pork kidney	5 ng/g	Cooper, Samsonova, Plumpton, Elliott, & Kennedy, 2007b
Azaperone	"	"	15 ng/g	"
Carazolol	"	"	5 ng/g	"
Acepromazine	"	"	5 ng/g	"
Chlorpromazine	"	"	20 ng/g	"
Propionylpromazine	"	"	5 ng/g	"

*Limit of detection except CCβ when indicated.

Table 16.2 Examples of commercial available ELISA test kits for different residues

Type of residue	Group*	Main substance to detect	Detection limit (ng/mL)
Estrogens	A 1	Diethylestylbestrol	0.2
Esteroids	A 3	Trenbolone	0.5
Resorcylic acid lactones	A 4	Zeranol	0.25
B-agonists	A 5	Clenbuterol	0.3
Antibiotics	A 6	Chloramphenicol	0.5
Antibiotics	B 1	Sulphonamides	0.5
Antibiotics	B 1	Tylosine	5.0
Antibiotics	B 1	Gentamicine	1.5
Corticoids	B 2f	Dexamethasone	2.5

*According to European Council Directive 96/23/EC (EC, 1996).

kits could be used for screening purposes. In other cases, immunoassays are not so effective due to cross reactions. For instance, usual immunoassay kits for zeranol give cross-reactions with toxins formed by *Fusarium* spp leading to false positive results. A study for the detection of zeranol and the toxin α-zearalenol in bovine urine was carried out in 4 EU National Reference Laboratories. In this study, the assay was a dry reagent time-resolved fluoroimmunoassay (TR-FIA) and no false positive results were obtained (Cooper et al., 2003). This type of method solved the cross-reaction problem (Wang & Wang, 2007). Using previous purification of the sample by immunoaffinity chromatography, a limit for the detection of zeranol of 0.16 ng/g could be achieved (Tuomola et al., 2002).

Thus, ELISA test kits offer important advantages like the possibility to analyze a large number of samples per kit, relatively short time to response, its high specificity and good sensitivity when compared to conventional detection methods (Hahnau & Jülicher, 1996; Toldrá & Reig, 2006). Another advantage of these ready-to-use test kits is that they can be used within the food-processing facility without the need to transport the samples to the laboratory (Paraf & Sarradin, 1996). However, ELISA test kits are unsuitable for multi-residue analysis and give cross-reactions with structurally related compounds. In addition, ELISA kits have experienced a large increase in the costs making it restrictive for a large extended use.

Recent research includes the development of polyclonal antibodies able to detect 3-amino-2-oxazolidinone (AOZ), which is a stable metabolite of the nitrofuran antibiotic furazolidone (Cooper, Delahaut, Fodey, & Elliott, 2004b). A generic dipstick was also developed for the detection of antibiotics. It consists of membrane strips coated with streptavidin and immobilized biotinylated operator DNA. When the sample contains antibiotics, they trigger the dose-dependent release of the capture DNA-biosensor interaction and the color developed can be determined at 450 nm. (Link et al., 2007). An immunochromatographic lateral-flow test strip was developed for the detection of sulfonamides in chicken muscle. It was in the competitive reaction format and the sensitivity for sulfonamides (sulfamonomethoxine, sulfamethoxydiazine, sulfadimethoxine, and sulfadiazine), that only required 15 minutes, was as low as 10 ng/mL (Wang et al., 2007).

Biosensors

The rapid developments in biological and chemical technology have prompted the miniaturization for integrated, high-throughput and high-sensitivity analysis. These new biosensors have been applied for the rapid detection of veterinary drugs in foods of animal origin. The basis consists of the interaction of an immobilized antibody on the surface of a transducer that interacts with the analyte in the sample, in most cases with no need for sample clean-up (Wang, Wang, Duan, & Kennedy, 2006). The changes at the surface are converted into a measurable signal (optical or electronical detection) (De Wasch et al., 2001, Draisci et al., 2001). An advantage of biosensors is the possibility for simultaneous detection, at a time, of multiple veterinary drugs residues in a sample (Kumar, Thompson, Singh, Chander, & Gupta, 2004). The surface plasmon resonance (SPR) measures variations in the refractive index of the solution adjacent to the metal surface (Cooper, 2004; Hauguey & Baxter, 2006). The immobilized antibody on the surface causes a shift in the resonance angle but when a high molecular antigen binds to the antibody, it causes a further change (Gründig, & Renneberg, 2002). This biosensor, which is commercially available as a fully automated system with the name Biacore^{TM}, allows real time quantification of the antibody-antigen interaction. Important advantages are simplicity, high precision, speed (few minutes per sample) and automation. Furthermore, they provide a wide variability of target analytes and sensibility depending on the signal to noise ratio. So, it has been successfully applied to the detection of different veterinary drugs residues like ractopamine (Thompson et al., 2008), nitroimidazoles (Connolly et al., 2007; Cooper, Samsonova, Plumpton, Elliott, & Kennedy, 2007a; Situ & Elliott, 2005), clenbuterol in urine at a limit of detection of 0.27 ng/mL (Haughey et al., 2001), flumequine in broiler muscle at a CCα of 500 ng/g (Haasnoot, Gerçek, Cazemier, & Nielen, 2007), chloramphenicol in poultry with a CCα of 0.005 ng/g (Ferguson et al., 2005) and chloramphenicol glucuronide in kidney at CCα_i0.1 ng/g (Ashwin et al., 2005) and sulphonamides (sulfametazine and sulfadiazine) in pork meat at a LOD of 0.1 ng/g (Crooks, Baxter, O'Connor, & Elliot, 1998). Twenty sulfonamide residues were detected with a SPR biosensor in porcine muscle at a limit of detection of 16.9 ng/g (McGrath, Baxter, Ferguson, Haughey, & Bjurling, 2005). Even a specific antibody mutant was developed for the SPR biosensor towards a multi-sulfonamide detection (Bienenmann-Ploum, Korpimaki, Haasnoot, & Kohen, 2005). SPR sensor system was also successfully tested in a slaughterhouse under cold and humid conditions (Baxter, O'Connor, Haughey, Crooks, & Elliott, 1999). Another SPR biosensor, where the extract is injected over the surfaces of two chip flow cells, has been developed for the detection of thiamphenicol, florenicol, florefenicol amine and chloramphenicol in shrimps. The limits of quantitation (CCβ) were 0.1, 0.2, 250 and 0.5 ppb, respectively (Dumont, Huet, Traynor, Elliott, & Delahaut, 2006). Different kits (Qflex® kit) with specific antibodies against a wide variety of residues are commercially available.

Other biosensors are based on the use of biochip arrays or microarrays. Small molecule microarrays use small molecules as probes immobilized on a variety of surfaces (Zuo and Ye, 2006). The target residues are conjugated with a carrier protein and printed on modified glass slides. These residues are still able to interact with the respective antibodies in solution or in the sample. The system is incubated for antigen-antibody reaction that is detected by using fluorescence labelled secondary antibody. This biosensor has been successfully applied to the detection of clenbuterol, chloramphenicol and tylosin (Peng & Bang-Ce, 2006). These chips are specific for a certain number of residues like chloramphenicol in meat (Gaudin et al., 2003), nitroimidazoles in chicken (Huet et al., 2005) as well as clenbuterol and the hormone analogues ethinylestradiol and trenbolone (Johansson & Hellenas, 2001).

A broad spectrum protein biosensor was developed for the assay of tetracycline, streptogramin and macrolide antibiotics (Weber et al., 2005). This biosensor is based on the interaction of protein sensor with specific DNA fragments chemically linked to a solid surface. This interaction gives a particular response that is converted into a colorimetric readout for each class of antibiotics and the color intensity is proportional to the antibiotic concentration.

References

Ashwin, H. M., Stead, S. L., Taylor, J. C., Startin, J. R., Richmond, S. F., Homer, V., et al. (2005). Development and validation of screening and confirmatory methods for the detection of chloramphenicol and chloramphenicol glucuronide using SPR biosensor and liquid chromatography-tandem mass spectrometry. *Analytica Chimica Acta, 529*, 103–108.

Baggiani, C., Anfossi, L., & Giovannoli, C. (2007). Solid phase extraction of food contaminants using molecular imprinted polymers. *Analytica Chimica Acta, 591*, 29–39.

Baxter, G. A., O'Connor, M. C., Haughey, S. A., Crooks, S. R. H., & Elliott, C. T. (1999). Evaluation of an immunobiosensor for the on-site testing of veterinary drug residues at an abattoir. Screening for sulfamethazine in pigs. *Analyst, 124*, 1315–1318.

Berggren, C., Bayoudh, S., Sherrington, D., & Ensing, K. (2000). Use of molecularly imprinted solid-phase extraction for the selective clean-up of clenbuterol from calf urine. *Journal of Chromatography A, 889*, 105–110.

Bergwerff, A. A. (2005). Rapid assays for detection of residues of veterinary drugs. In A. van Amerongen, D. Barug, & M. Lauwars (Eds.), *Rapid methods for biological and chemical contaminants in food and feed* (pp. 259–292). Wageningen: Academic Publishers.

Bienemann-Ploum, M., Korpimaki, T., Haasnoot, W., & Kohen, F. (2005). Comparison of multi sulfonamide biosensor immunoassays. *Analytica Chimica Acta, 529*, 115–122.

Boyd, B., Bjork, H., Billing, J., Shimelis, O., Axelsson, S., Leonora, M., et al. (2007). Development of an improved method for trace analysis of chloramphenicol using molecularly imprinted polymers. *Journal of Chromatography A, 1174*, 63–71.

Campbell, H. M., & Armstrong, J. F. (2007). Determination of zearalenone in cereal grains, animal feed, and feed ingredients using immunoaffinity column chromatography and liquid chromatography: Interlaboratory study. *Journal of AOAC International, 90*, 1610–1622.

Cerniglia, C. E., & Kotarski, S. (1999). Evaluation of veterinary drug residues in food for their potential to affect human intestinal microflora. *Regulatory Toxicology and Pharmacology 29*, 238–261.

Cerniglia, C. E., & Kotarski, S. (2005). Approaches in the safety evaluations of veterinary antimicrobial agents in food to determine the effects on the human intestinal microflora. *Journal of Veterinary Pharmacology and Therapy, 28*, 3–20.

Chadwick, R. W., George, S. E., & Claxton, L. D. (1992). Role of gastrointestinal mucosa and microflora in the bioactivation of dietary and environmental mutagens or carcinogens. *Drug Metabolism Reviews, 2*, 425–492.

Chifang, P., Chuanlai, X., Zhengyu, J., Xiaogang, C., & Liying, W. (2006). Determination of anabolic steroid residues (medroxyprogesterone acetate) in pork by ELISA and comparison with liquid chromatography tandem mass spectrometry. *Journal of Food Science, 71*, C044–C050.

Cinquina, A. L. Longo, F., Anastasi, G., Giannetti, L., & Cozzani, R. (2003). Validation of a high-performance liquid chromatography method for the determination of oxytetracycline, tetracycline, chlortetracycline and doxycycline in bovine milk and muscle. *Journal of Chromatography A, 987*, 227–233.

Connolly, L., Thompson, C. S., Haughey, S. A., Traynor, I. M., Tittlemeier, S., & Elliot, C. T. (2007). The development of a multi.nitorimidazole residue analysis assay by optical biosensor via a proof of concept project to develop and assess a prototype test kit. *Analytica Chimica Acta, 598*, 155–161.

Cooper, J., Delahaut, P., Fodey, T. L., & Elliott, C. T. (2004). Development of a rapid screening test for veterinary sedatives and the beta-blocker carazolol in porcine kidney by ELISA. *Analyst, 129*, 169–174.

Cooper, K. M., Caddell, A., Elliott, C. T., & Kennedy, D. G. (2004). Production and characterisation of polyclonal antibodies to a derivative of 3-amino-2-oxazolidinone, a metabolite of the nitrofuran furazolidone. *Analytica Chimica Acta, 520*, 79–86.

Cooper, K. M., Ribeiro, L., Alves, P., Vozikis, V., Tsitsamis, S., Alfredssonk, G., et al. (2003). Interlaboratory ring test of time-resolved fluoroimmunoassays for zeranol and a-zearalenol and comparison with zeranol test kits. *Food Additives and Contaminants, 20*, 804–812.

Cooper, K. M., Samsonova, J. V., Plumpton, L., Elliott, C. T., & Kennedy, D. G. (2007a). Enzyme immunoassay for semicarbazide—The nitrofuran metabolite and food contaminant. *Analytica Chimica Acta, 592*, 64–71.

Cooper, K. M., Samsonova, J. V., Plumpton, L., Elliott, C. T., & Kennedy, D. G. (2007b). Enzyme immunoassay for semicarbazide—The nitrofuran metabolite and food contaminant. *Analytica Chimica Acta, 592*, 64–71.

Crooks, S. R. H., Baxter, G. A., O'Connor, M. C., & Elliot, C. T. (1998). Immunobiosensor – an alternative to enzyme immunoassay screening for residues of two sulfonamides in pigs. *Analyst, 123*, 2755–2757.

De Wasch, K., Okerman, L., Croubels, S., De Brabander, H., Van Hoof, J., & De Backer, P. (2001). Detection of residues of tetracycline antibiotics in pork and chicken meat: Correlation between results of screening and confirmatory tests. *Analyst, 123*, 2737–2741.

Draisci, R. delli Quadri, F., Achene, L., Volpe, G., Palleschi, L., & Palleschi, G. (2001). A new electrochemical enzyme'linked immunosorbent assay for the screening of macrolide antibiotic residues in bovine meat. *Analyst, 126*, 1942–1946.

Dumont, V., Huet, A. C., Traynor, I., Elliott, C., & Delahaut, P. (2006). A surface plasmon resonance biosensor assay for the simultaneous determination of thiamphenicol, florefenicol, florefenicol amine and chloramphenicol residues in shrimps. *Analytica Chimica Acta, 567*, 179–183.

EC. (1996). Council Directive 96/23/EEC of 29 April 1996 on measures to monitor certain substances and residues thereof in live animals and animal products. *Official Journal of the European Community L, 125*, 10.

Elliott, C. T., Baxter, G. A., Hewitt, S. A., Arts, C. J. M., van Baak, M., Hellenas, K. E., et al. (1998). Use of biosensors for rapid drug residue analysis without sample deconjugation or clean-up: a possible way forward. *Analyst, 123*, 2469–2473.

Ferguson, J., Baxter, A., Young, P., Kennedy, G., Elliott, C., Weigel, S., et al. (2005). Detection of chloramphenicol and chloramphenicol glucuronide residues in poultry muscle, honey, prawn and milk using a surface plasmon resonance biosensor and Qflex® kit chloramphenicol. *Analytica Chimica Acta, 529*, 109–113.

Gaudin, V., Cadieu, N., & Maris, P. (2003). Inter-laboratory studies for the evaluation of ELISA kits for the detection of chloramphenicol residues in milk and muscle. *Food and Agricultural Immunology, 15*, 143–157.

Godfrey, M. A. J. (1998). Immunoafinity extraction in veterinary residue analysis-a regulatory viewpoint. *Analyst, 123*, 2501–2506.

Gründig, B., & Renneberg, R. (2002). Chemical and biochemical sensors. In A. Katerkamp, B. Gründig, & R. Renneberg (Eds.), *Ullmann's Encyclopedia of Industrial Chemistry* (pp. 87–98). Verlag: Wiley-VCH.

Guo, J. J., Chou, H. N., & Liao, I. C. (2003). Disposition of 3-(4-cyano-2-oxobutylidene amino)-2-oxazolidone, a cyano-metabolite of furazolidone, in furazolidone-treated grouper. *Food Additives and Contaminants, 20*, 229–236.

Haasnoot, W., Gerçek, H., Cazemier, G., & Nielen, M. W. F. (2007). Biosensor immunoassay for flumequine in broiler serum and muscle. *Analytica Chimica Acta, 586*, 312–318.

Hahnau, S., & Jülicher, B. (1996). Evaluation of commercially available ELISA test kits for the detection of clenbuterol and other β-agonists. *Food Additives and Contaminants, 13*, 259–274.

Haughey, S. A., & Baxter, C. A. (2006). Biosensor screening for veterinary drug residues in foodstuffs. *Journal of AOAC International, 89*, 862–867.

He, J. H., Hou, X. L., Jiang, H. Y., & Shen, J. Z. (2005). Multiresidue analysis of avermectins in bovine liver by immunoaffinity column cleanup procedure and liquid chromatography with fluorescence detector. *Journal of AOAC International, 88*, 1099–1103.

Haughey, S. A., Baxter, G. A., Elliot, C. T., Persson, B., Jonson, C., & Bjurling, P. (2001). Determination of clenbuterol residues in bovine urine by optical immunobiosensor assay. *Journal of AOAC International, 84*, 1025–1030.

Huet, A. C. Mortier, L., Daeseleire, E., Fodey, T., Elliott, C. T., & Delahaut, P. (2005). Development of an ELISA screening test for nitroimidazoles in egg and chicken muscle *Analytica Chimica Acta, 534*, 157–162.

Johansson, M. A., & Hellenas, K. E. (2001). Sensor chip preparation and assay construction for immunobiosensor determination of beta-agonists and hormones. *Analyst, 126*, 1721–1727.

Johansson, M. A., & Hellenas, K. E. (2004). Immunobiosensor determination of b-agonists in urine using integrated immunofiltration clean-up. *International Journal of Food Science and Technology, 39*, 891–898.

Kootstra, P. R. Kuijpers, C. J. P. F., Wubs, K. L., van Doorn, D., Sterk, S. S., van Ginkel, L. A., et al. (2005). The analysis of beta-agonists in bovine muscle using molecular imprinted polymers with ion trap LCMS screening. *Anaytica Chimica Acta, 529*, 75–81.

Kumar, K., Thompson, A., Singh, A. K., Chander, Y., & Gupta, S. C. (2004). Enzyme-linked immunosorbent assay for ultratrace determination of antibiotics in aqueous samples. *Journal of Environmental Quality, 33*, 250–256.

Lee, H. J., Lee, M. H., Ryu, P. D., Lee, H., & Cho, M. H. (2001). Enzyme-linked immunosorbent assay for screening the plasma residues of tetracycline antibiotics in pigs. *Jounal of Veterinary Medicine, 63*, 553–556.

Levieux, D. (2007). Immunodiagnosctic technology and its applications. In L. M. L. Nollet, & F. Toldrá (Eds.), *Advances in Food Diagnostics* (pp. 211–227). Ames, Iowa: Blackwell Publishing.

Link, N., Weber, W., & Fussenegger, M. (2007). A novel generic dipstick-based technology for rapid and precise detection of tetracycline, streptogramin and macrolide antibiotics in food samples. *Journal of Biotechnology, 128*, 668–680.

McGrath, T., Baxter, A., Ferguson, J., Haughey, S., & Bjurling, P. (2005). Multi sulfonamide screening in porcine muscle using a surface plasmon resonance biosensor. *Analytica Chimica Acta, 529*, 123–127.

Moore, W. E. C. & Moore, L. H. (1995). Intestinal floras of populations that have risk of colon cancer. *Applied and Environmental Microbiology, 61*, 3202-3207.

Mottier, P., et al. (2003). Determination of the antibiotic chloramphenicol in meat and seafood products by liquid chromatography–electrospray ionization tandem mass spectrometry. *Journal of Chromatography A, 994*, 75–84.

Paraf, A., & Sarradin, P. (1996). Immunochemistry in food control. *Recent Research Developments in Nutrition, 1*, 99–114.

Pecorelli, I., Bibi, R., Fioroni, L., & Galarini, R. (2004). Validation of a confirmatory method for the determination of sulphonamides in muscle according to the European Union regulation 2002/657/EC. *Journal of Chromatography A, 1032*, 23–29.

Peng, Z., & Bang-Ce, Y. (2006). Small molecule microarrays for drug residue detection in foodstuffs. *Journal of Agricultural and Food Chemistry, 54*, 6978–6983.

Reig, M., & Toldrá, F. (2007). Chemical origin toxic compounds. In F. Toldrá, Y. H. Hui, I. Astiasarán, W. K. Nip, J. G. Sebranek, E. T. F. Silveira, et al. (Eds.), *Handbook of fermented meat and poultry* (pp. 469–475). Ames, Iowa: Blackwell Publishing.

Reig, M., & Toldrá, F. (2008a). Veterinary drug residues in meat: Concerns and rapid methods for detection. *Meat Science, 78*, 60–67.

Reig, M., & Toldrá, F. (2008b). Growth promoters. In L. M. L. Nollet, & F. Toldrá (Eds.), *Handbook of muscle foods analysis*. Boca Raton, Fl: CRC Press (in press).

Roda, A., Manetta, A. C., Portanti, O., Mirasoli, M., Guardigli, M., Pasini, P., et al. (2003). A rapid and sensitive 384-well microtitre format chemiluminiscent enzyme immunoassay for 19.nortestosterone. *Luminescence, 18*, 72–78.

Samarajeewa, U., Wei, C. I., Huang, T. S., & Marshall, M. R. (1991). Application of immunoassay in the food industry. *Critical Reviews in Food Science and Nutrition, 29*, 403–434.

Shi, W. M., He, J. H., Jiang, H. Y., Hou, X. L., Yang, J. H., & Shen, J. Z. (2006). Determination of multiresidue of avermectins in bovine liver by an indirect competitive ELISA. *Journal of Agricultural and Food Chemsitry, 54*, 6143–6146.

Situ, C., & Elliott, C. T. (2005). Simultaneous and rapid detection of five banned antibiotic growth promoters by immunoassay. *Analytica Chimica Acta, 529*, 89–96.

Situ, C., Grutters, E., van Wichen, P., & Elliott, C. T. (2006). A collaborative trial to evaluate the performance of a multi-antibiotic enzyme-linked immunosorbent assay for screening five banned antimicrobial growth promoters in animal feedingstuffs. *Analytica Chimica Acta, 561*, 62–68.

Stolker, A. A. M., Zoonties, P. W., & Van Ginkel, L. A. (1998). The use of supercritical fluid extraction for the determination of steroids in animal tissues. *Analyst, 123*, 2671–2676.

Stubbings, G., Tarbin, J., Cooper, A., Sharman, M., Bigwood, T., & Robb, P. (2005). A multi-residue cation-exchange clean up procedure for basic drugs in produce of animal origin. *Analyitica Chimica Acta, 547*, 262–268.

Thompson, C. S., Haughey, S. A., Traynor, I. M., Fodey, T. L., Elliot, C. T., Antignac, J. P., et al. (2008). Effective monitoring of ractopamine residues in samples of animal origin by SPR biosensor and mass spectrometry. *Analytica Chimica Acta, 608*, 217–225.

Toldrá, F., & Reig, M. (2006). Methods for rapid detection of chemical and veterinary drug residues in animal foods. *Trends Food Science and Technology, 17*, 482–489.

Tuomola, M., Cooper, K. M., & Lahdenpera, S. (2002). A specificity-emhanced time-resolved fluoroimmunoassay for zeranol employing the dry reagent all-in-one-well principle. *Analyst, 127*, 83–86.

Vollard, E. J., & Clasener, H. A. L. (1994). Colonization resistance. *Antimicrobial Agents and Chemotherapy, 38*, 409–414.

Wang, S., Wang, Z. L., Duan, Z. J., & Kennedy, I. (2006). Analysis of sulphonamide residues in edible animal products: A review. *Food Additives and Contaminants, 23*, 362–384.

Wang, S., & Wang, X. H. (2007). Analytical methods for the determination of zeranol residues in animal products: A review. *Food Additives and Contaminants, 24*, 573–582.

Wang, X. L., Li, K., Shi, D. S., Xiong, N., Jin, X., Yi, J. D., et al. (2007). Development of an immunochromatographic lateral-flow test strip for rapid detection of sulfonamides in eggs and chicken muscles. *Journal of Agricultural and Food Chemistry, 55*, 2072–2078.

Weber, C. C., Link, N., Fux, C., Zisch, A. H., Weber, W., & Fussenegger, M. (2005). Broad-spectrum protein biosensors for class-specific detection of antibiotics. *Biotechnology and Bioengineering, 89*, 9–17.

Widstrand, C. Larsson, F., Fiori, M., Civitareale, C., Mirante, S., & Brambilla, G. (2004). Evaluation of MISPE for the multi-residue extraction of β-agonists from calves urine. *Journal of Chromatography B, 804*, 85–91.

Xu, C. L., Chu, X. G., Peng, C. F., Liu, L. Q., Wang, L. Y., & Jin, Z. (2006). Comparison of enzyme-linked immunosorbent assay with liquid chromatography-tandem mass spectrometry for the determination of diethylstilbesterol residues in chicken and liver tissues. *Biomedical Chromatography, 20*, 1956–1064.

Xu, C. L., Peng, C. F., Liu, L. Q., Wang, L. Y., Jin, Z. Y., & Chu, X. G. (2006). Determination of hexoestrol residues in animal tissues based on enzyme-linked immunosorbent assay and comparison with liquid chromatography-tandem mass spectrometry. *Journal of Pharmaceutical and Biomedical Analysis, 41*, 1029–1036.

Zhang, S. X., Zhang, Z., Shi, W. M., Eremin, S. A., & Shen, J. Z. (2006). Development of a chemiluminescent ELISA for determining chloramphenicol in chicken muscle. *Journal of Agricultural and Food Chemistry, 54*, 5718–5722.

Zhang, W., Wang, H. H., Wang, J. P., Li, X. W., Jiang, H. Y., & Shen, J. Z. (2006). Multiresidue determination of zeranol and related compounds in bovine muscle by gas chromatography/mass spectrometry with immunoaffinity cleanup. *Journal of AOAC International,l 89*, 1677–1681.

Zuo, P., & Ye, B. C. (2006). Small molecule microarrays for drug residue detection in foodstuffs. *Journal of Agricultural and Food Chemistry, 54*, 6978–6983.

Chapter 17
Antimicrobial Activity of Bacteriocins and Their Applications

Eleftherios H. Drosinos, Marios Mataragas, and Spiros Paramithiotis

Introduction

Bacteriocins are peptides or proteins that exert an antimicrobial action against a range of microorganisms. Their production can be related to the antagonism within a certain ecological niche, as the producer strain, being itself immune to its action, generally gains a competitive advantage. Many Gram-positive and Gram-negative microorganisms have been found to produce bacteriocins. The former, and especially the ones produced by lactic acid bacteria, has been the field of intensive research during the last decades mainly due to their properties that account for their suitability in food preservation and the benefits arising from that, and secondarily due to the broader inhibitory spectrum compared to the ones produced by Gram-negative microorganisms. Bacteriocins are ribosomally synthesized peptides whose responsible genes are frequently associated with mobilisable elements on the chromosome such as in association with transposons or on plasmids (Deegan, Cotter, Hill, & Ross, 2006). This property is one of the main differences that distinguish them from antibiotics (Cleveland, Montville, Nes, & Chikindas, 2001). The latter are considered being secondary metabolites and although many of them comprise amino acids, they are enzymatically rather than ribosomally synthesized.

Classification and Biochemical Properties

Bacteriocin classification has been a field of extensive debate. The original classification scheme has been proposed by Klaenhammer (1993); it was then modified by Nes et al. (1996), van Belkum and Stiles (2000) and Nes and Holo (2000) and new propositions were recently made by Cotter, Hill, and Ross (2005) and Franz, van Belkum, Holzapfel, Abriouel, and Galvez (2007), with the latter referring mainly to enterococcal bacteriocins, since many of them appear as atypical

---

E.H. Drosinos
Laboratory of Food Quality Control and Hygiene, Department of Food Science and Technology, Agricultural University of Athens, 75, Iera Odos Street, Votanicos, GR-118 55, Athens, Greece

and structurally distinct from the previously described general classes (Table 17.1). Moreover, the inclusion of a fifth class consisting of ribosomally synthesized non-modified head-to-tail-ligated cyclic antibacterial peptides has been proposed by Kemperman et al. (2003), a suggestion that has met with controversial reactions since these bacteriocins could fit into the second class (Cotter et al., 2005). It is, though, generally accepted that on the basis of their structure as well as biochemical properties, bacteriocins are divided into three clusters: lantibiotics, non-lantibiotics and non-lantibiotics heat-labile large molecules. Since the vast majority of the bacteriocins produced by lactic acid bacteria belong to the first two classes, a huge amount of data currently exists concerning their production and excresion, immunity, mode of action and factors that affect their production or activity *in vitro* or *in situ*.

Lantibiotics are characterized by the presence of thioether cross-links termed lanthionines or methyl-lanthionines in addition to the amino acids 2,3-didehydroalanine and (Z)-2,3-didehydrobutyrine. The lantibiotic biosynthetic genes are clustered and located on the bacterial chromosome or on large plasmids. The products of the biosynthetic genes have been classified into structural, modification, processing, transport, immunity and regulation proteins. The precursor peptide that is initially produced contains an amino-terminal leader sequence of 23–59 amino acids and a structural region that is further modified in order to introduce several structural motifs. Several roles have been assigned to the leader peptide; among them is a contribution to the cell's immunity by retaining the lantibiotic inactive within the cell, as a transport aid, and as a recognition sequence for the biosynthetic enzymes responsible for production of the active lantibiotic (Chatterjee, Paul, Xie, & van der Donk, 2005). After the modification of the precursor peptide, the leader peptide must be cleaved off and translocated across the cytoplasmic membrane. This is performed either by a serine-type protease or an ATP-binding cassette protein that has also been found to be responsible for its secretion. As far as the mode of action of lantibiotics is concerned, they seem to be far more complex than initially suspected. The antimicrobial activity involves one or more mechanisms that may include disruption of cell wall biosynthesis, inhibition of spore outgrowth, and pore formation that may or may not be aided by prior docking on cellular targets (Chatterjee et al., 2005; Nagao et al., 2006; Patton & van der Donk, 2005; Xie & van der Donk, 2004).

Encoding of Class II bacteriocins' protein set, share many similarities with respect to lantibiotics. All genes are organised on plasmids, on the chromosome or both and consist of a structural gene and a cognate immunity gene. It is not rare that these two genes are clustered together with another group of genes encoding dedicated proteins for the export of the bacteriocin from the cell. Thus, class II bacteriocins seem to require a minimum of four genes encoding for the bacteriocin prepeptide, cognate immunity protein, a dedicated cell-membrane-associated, ATP-binding cassette (ABC) transport protein and a membrane-bound accessory protein (van Belkum and Stiles, 2000). They are, thus, synthesized as precursor bacteriocins with little or no antagonistic activity. The formation of the active peptide requires the cleavage of the N-terminus extension following the export from the cell. The

17 Antimicrobial Activity of Bacteriocins and Applications

Table 17.1 Proposals for the classification of bacteriocins

Class	Klaenhammer (1993)	Nes et al. (1996)	van Belkum and Stiles (2000)	Nes and Holo (2000)	Cotter et al. (2005)	Franz et al. (2007)
I	Lantibiotics; small membrane-active peptides (<5 kDa) containing lanthionine, β-methyl lantionine and dehydrated residues	Lantibiotics	Lantibiotics	Lantibiotics	Lantibiotics	Lantibiotic enterocins
II	Small heat-stable, non-lanthionine containing membrane-active peptides (<10 kDa)	Small heat-stable non-lantibiotics	Non-lantibiotics	Nonmodified heat stable bacteriocins	Nonlantibiotics	small nonlantibiotic peptides
IIa	*Listeria*-active peptides with YGNGVXC motif near N-terminus and GG leader peptide	Pediocin-like bacteriocins with strong antilisterial effect	Cystibiotics with two disulfite bridges resulting from four cysteine residues	Pediocin-like bacteriocins (*Listeria*-active)	–	Enterocins of the pediocin family
IIb	Poration complexes consisting of two peptides with GG leader peptide	Two peptide bacteriocins	Cystibiotics with one disulfite bridge resulting from two cysteine residues in the N-terminal section of the peptide	Two-peptide bacteriocins	–	Enterocins synthesized without a leader peptide

Table 17.1 (Continued)

Class	Klaenhammer (1993)	Nes et al. (1996)	van Belkum and Stiles (2000)	Nes and Holo (2000)	Cotter et al. (2005)	Franz et al. (2007)
IIc	Thiol-avtivated peptides requiring reduced cysteine residues for activity and GG leader peptide	sec-depended secreted bacteriocins	Cystibiotics with one disulfite bridge that spans the N- and C-terminal sections of the peptide	Other bacteriocins	—	Other linear nonpediocin-like enterocins
IId	—	—	Thiolbiotics with one or no cysteine residues	—	—	—
IIe	—	—	Two–peptide bacteriocins	—	—	—
IIf	—	—	Atypical bacteriocins	—	—	—
III	Large heat–labile proteins (>30 kDa)	Large heat–labile proteins (>30 kDa)	Large heat–labile proteins	Large heat–labile bacteriocins	Bacteriolysins (large, heat–labile murein hydrolases)	Cyclic enterocins
IV	Complex bacteriocins composed of protein plus one or more chemical moieties (lipid, carbohydrate) required for activity	—	—	—	—	Large enterocins

principal target of these is the cytoplasmic membrane through permealization and concomitant leakage, without, though, conclusive experimental evidence for the existence or necessity of a bacteriocin receptor (Nes and Holo, 2000).

In Vitro and In Situ Bacteriocin Production

Typical spontaneous sausage fermentation is driven by lactic acid bacteria followed by members of the Micrococcaceae family and enterococci which are very frequently isolated in high numbers, suggesting a possible contribution to the fermentation process. Many researchers have studied their biochemical properties and members of nearly all genera have been found to produce bacteriocins (Table 17.2). The antimicrobial range varies from narrow to broad and from species within the same genus to a variety of Gram-positive and Gram-negative microorganisms, including several pathogens.

Table 17.2 Some bacteriocinogenic strains isolated from meat products

Species	Reference
Lc. lactis (strains BB24, WNC 20)	Rodriguez et al. (1995); Noonpakdee et al. (2003)
Lb. sakei (strains L45, 148, CTC494, CTC372, Lb706, Lb674)	Mortvedt et al. (1991); Sobrino et al. (1992); Aymerich et al. (2000b); Schillinger and Luecke (1989); Holck et al. (1994)
Pd. acidilactici (strains HA-6111-2, HA-5692-3, L50, H)	Albano et al. (2007a); Cintas et al. (1995); Motlagh et al. (1992)
Lb. curvatus (strains L442, LTH 1174)	Mataragas et al. (2003b); Messens et al. (2003)
Cb. divergens (strain 750)	Holck et al. (1996)
Cb. piscicola (strains LV17, JG126)	Ahn and Stiles (1990); Jack et al. (1996)
Ent. faecium (strains L1, P13, L50, T136, CTC492)	Lyon et al. (1995); Cintas et al. (1997); Cintas et al. (1998); Casaus et al. (1997); Aymerich et al. (1996)
Lb. plantarum (strains CTC305, UG1, 35d)	Aymerich et al. (2000b); Enan et al. (1996); Messi et al. (2001)
Leu. mesenteroides (strains E131, L124)	Drosinos et al. (2006); Mataragas et al. (2003b)
Leu. carnosum (strains OZ, 4010)	Osmanagaoglu (2007); Budde et al. (2003)
Leu. gelidum (strain A-UAL187)	Harding and Shaw (1990)
Pd. parvulus (strain 133)	Schneider et al. (2006)
Pd. pentosaceus	Albano et al. (2007b)
St. warneri (strains FM10, FM20 and FM30)	Prema et al. (2006)
St. xylosus (strain 1E)	Villani et al. (1997)

Lc.: Lactococcus; Lb.: Lactobacillus; Pd.: Pediococcus; Cb.: Carnobacterium; Ent.: Enterococcus; Leu.: Leuconostoc; St.: Staphylococcus

Factors Affecting In Vitro Bacteriocin Production

Once a bacteriocinogenic strain is isolated and the genetic and phenotypic characterization of the bacteriocin has been completed, optimization of the bacteriocin production in synthetic media under strictly defined conditions and secondarily an *in situ* application consist the next step of research efforts. Since foods are complex systems with several factors influencing microbial growth and metabolism, these factors need to be tested on the performance of the bacteriocinogenic strains (Hugas, 1998). In order to evaluate the potential use of bacteriocinogenic LAB strains in the food systems, parameters known to influence growth and bacteriocin production should be considered. Such factors include the following:

Microbial Strain

A bacteriocin may be produced by more than one microbial strain and/or species. Yang and Ray (1994) found that nisin and leuconocin Lcm1 production substantially varied between the different strains producing these two bacteriocins, whereas pediocin AcH did not show such variation.

Physiological State of the Bacteriocinogenic Strain

Bacteriocins are considered primary metabolites, because their production starts early (early exponential phase) and it is completed by the end of the exponential growth phase or at the early stages of the stationary phase such as, in the case of, lactocin S, sakacin P, mesenterocin Y, curvaticin L442 and a bacteriocin produced by *Leuconostoc mesenteroides* L124 (Drosinos et al., 2005b; Drosinos, Mataragas, & Metaxopoulos, 2006; Mataragas et al., 2003b; Mataragas, Drosinos, Tsakalidou, & Metaxopoulos, 2004; Mortvedt-Abildgaard et al., 1995; Xiraphi et al., 2006). However, there are bacteriocins with maximum concentration at the early stages of the exponential growth phase e.g. a bacteriocin produced by *Leu. gelidum* UAL 187 (Stiles & Hastings, 1991) or in the middle of the same phase (helveticin J) (Joerger & Klaenhammer, 1986).

Growth Medium

Type and level of nutrients (sugars, nitrogen, phosphorus, magnesium, calcium, etc.) influence the bacteriocin production. LAB possesses a limited number of catabolic pathways providing energy for cell synthesis and the energy demanding bacteriocin production. Therefore, conditions favoring cell growth will, probably, improve bacteriocin production (growth-associated process). However, an extremely rich environment will not necessarily lead to a gain of bacteriocin production. Several studies have shown that the type and level of sugars (e.g. glucose, sucrose, fructose, etc.) used as carbon sources substantially affect the bacteriocin synthesis (Barcena, Sineriz, Gonzalez de Llano, Rodriguez, & Suarez, 1998; Biswas, Ray, Johnson, &

Ray, 1991; Drosinos et al., 2005a, 2006; Mataragas et al., 2004; Matsusaki, Endo, Sonomoto, & Ishikazi, 1996). Bacteriocin production was stimulated at high glucose concentrations (Drosinos et al., 2005b), because sometimes the stress situation created by the high glucose content (higher osmotic pressures) causes better bacteriocin production (De Vuyst & Vandamme, 1994). *Leu. mesenteroides* E131 in the presence of fructose, instead of glucose, displayed better bacteriocin production (Drosinos et al., 2005b, 2006). This happens because fructose, as an energy source, is utilized more effectively by leuconostocs under the same pH and temperature conditions. Leuconostocs reduce some of the fructose to mannitol leading to energy gain as the cells produce ATP during this process and accordingly more ATP per time unit (Drosinos et al., 2005b). Nitrogen also plays a significant role. Kim, Hall, and Dunn (1997) observed that the maximum concentration of nisin is increased by the increase of the contained nitrogen due to increased supply of essential amino acids for cell formation and bacteriocin production. Type of nitrogen source has an effect on bacteriocin production (Mataragas et al., 2004). Various nitrogen sources with different number of amino acids along the chain of the contained peptides gave different results. Nitrogen sources (e.g. yeast extract) with a relatively large proportion of free amino acids and short peptides (2 or 3 amino acids) favored bacteriocin production than sources with long peptides (8 amino acids) (Mataragas et al., 2004) due to energy saving (Aasen, Moretro, Katla, Axelsson, & Storro, 2000; Benthin, Schulze, Nielsen, & Villadsen, 1994). However, an optimal ratio between carbon and nitrogen source is required to ensure satisfactory bacteriocin production because when carbon or nitrogen is in excess the others become a limiting factor for bacteriocin synthesis (Mataragas et al., 2004). Presence of various elements such as phosphorus, magnesium and calcium influence bacteriocin production but this action is strain specific (Parente & Ricciardi, 1999). Tween 80 (polyoxyethylene sorbitan mono-oleate) favors bacteriocin production in some cases (Daba, Lacroix, Huang, & Simard, 1993; Matsusaki et al., 1996; Parente & Hill, 1992); however, enhancement in bacteriocin activity, in the presence of Tween 80 seems to be related with decreased adsorption of the bacteriocin molecules onto the cellular membrane (Joosten & Nunez, 1995).

Environmental Conditions: pH and Temperature

pH and temperature conditions constitute two of the most critical factors which significantly affect bacteriocin production by bacteriocinogenic strains. Frequently, maximum bacteriocin concentration is achieved at pH (usually within the range 5.5–6.0) and/or temperature values slightly lower than those optimal for growth (De Vuyst, Callewaert, & Crabbe, 1996; Drosinos et al., 2005a, 2006; Kaiser & Montville, 1993; Mataragas, Drosinos, & Metaxopoulos, 2003a; Meghrous, Huot, Quittelier, & Petitdemange, 1992; Parente & Ricciardi, 1994). Only some bacteriocins display better production at lower pH values (5.0) (Barcena et al., 1998; Biswas et al., 1991; Mortvedt-Abildgaard et al., 1995; Yang and Ray, 1994). Bacteriocin production is enhanced at lower values of environmental factors than those of optimum, probably because the whole bacteriocin synthesis process is favored by a relatively lower growth rates. Increased growth rates do not necessarily mean better bacteriocin production (De Vuyst et al., 1996; Mataragas et al., 2003a). Lower growth rates at sub-optimal pH or temperature values indicate potentially better utilization of energy and essential metabolites (Aasen et al., 2000; Moretro, Aasen, Storro, & Axelsson, 2000). At higher temperatures or pH values bacteriocin production is lower because energy needs for maintenance purposes are higher when temperature or pH increases (Drosinos et al., 2005b; Leroy & De Vuyst, 1999). Maintenance operations such as turnover of macromolecules (DNA, RNA) and maintenance of the potential along the membrane of cells are growth dependant and faster growth rates mean more energy is required for maintenance (Nielsen, Nikolajsen, & Villadsen, 1991). However, there are cases of better bacteriocin production at temperatures close to optimum for growth (Daba et al., 1993; Drosinos et al., 2005b, 2006; Matsusaki et al., 1996; Meghrous et al., 1992; Lejeune, Callewaert, Crabbe, & De Vuyst, 1998; Parente & Ricciardi, 1994). Finally, certain LAB such as *Leu. mesenteroides* subsp. *mesenteroides* FR52, produce more than one bacteriocin requiring different environmental conditions. Maximum concentration of mesenterocin 52A is observed at pH 5.5 and temperature 20°C and of mesenterocin 52B at pH 5.0 and temperature 25°C (Krier, Revol-Junelles, & Germain, 1998).

Predictive modeling has gained increased attention and mathematical models have been developed to describe the growth and bacteriocin production of the bacteriocin-producing strains added to foods as starter or protective cultures (Leroy & De Vuyst, 2003; Messens, Neysens, Vasieleghem, Vanderhoeven, & De Vuyst, 2002; Messens et al., 2003). Modeling contributes to the determination of how environmental factors affect the growth and bacteriocin production (Leroy & De Vuyst, 2003; Leroy, Verluyten, Messens, & De Vuyst, 2002). Experimental and mathematical calculation of the various parameters of growth and bacteriocin production are displayed in Table 17.3. For more detailed information on the analytical procedure of the differential equations readers should refer to the literature listed in the Table 17.3 and McKellar (1997).

One method that is usually followed for the optimization of bacteriocin production is that of varying one factor at a time. The other factors are kept constant. As it can be seen, this technique is laborious and requires a lot of time in case of several factors under study. For this reason, statistical experimental designs have been developed to evaluate the influence of substrate composition on growth and bacteriocin production (Dominguez, Bimani, Caldera-Olivera, & Brandelli, 2007; Rollini & Manzoni, 2005).

Factors Related to Food Ingredients and Additives

Apart from the above mentioned parameters, the *in vitro* study of bacteriocin production and activity has often implicated various ingredients used for the production of fermented sausages. Hugas, Garriga, Pascual, Aymerich, and Monfort (2002) studied the effect of various ingredients, used in the manufacture of fermented sausages, on the antilisterial performance of the bacteriocinogenic *Lactobacillus*

17 Antimicrobial Activity of Bacteriocins and Applications

Table 17.3 Experimental and mathematical estimation of the parameters of growth and bacteriocin production

Parameter/Unit	Equation	Symbol/Unit	Comment	Reference
Experimental estimation				
Maximum specific growth rate (μ_{max}) (h^{-1})	$dX/dt = \mu_{max} \times X$	t, time (h) X, biomass concentration (g CDM l^{-1})	Calculated by linear regression from plots of ln$X$ vs t	De Vuyst et al. (1996)
Bacteriocin production rate (k_b) [×10^6 AU (g CDM)$^{-1}$]	$B - B_0 = k_b \times (X - X_0)$	B, bacteriocin activity (AU l^{-1}) B_0, initial bacteriocin activity (AU l^{-1}) X_0, initial biomass concentration (g CDM l^{-1})	Calculated by linear regression from plots of $B - B_0$ vs $X - X_0$	Parente & Ricciardi (1994); De Vuyst et al.(1996)
Specific rate of bacteriocin degradation (k_{inact}) (h^{-1})	$dB/dt = -k_{inact} \times B$	X', CDM concentration at which bacteriocin production stops	Calculated by linear regression from plots of ln$B$ vs t, when $X > X'$.	Lejeune et al. (1998)
Cell yield ($Y_{X/S}$) [g CDM (g sugar)$^{-1}$]	$X - X_0 = -Y_{X/S} \times (S - S_0)$	S, residual sugar concentration (g l^{-1}) S_0, initial sugar concentration (g l^{-1})	Calculated by linear regression from plots of $X - X_0$ vs S-S_0	De Vuyst et al. (1996)
Maintenance coefficient (m_S) [g sugar (g CDM)$^{-1}$ h^{-1}]	$S - S_0 = -m_S \times (X - X_0) \times t$	X_{max}, maximum attainable biomass concentration	Calculated by linear regression from plots of $S - S_0$ vs $X - X_0$, when $X = X_{max}$	Parente & Ricciardi (1994); Lejeune et al. (1998)
Mathematical estimation				
Cell growth (μ_{max} and X_{max}) (h^{-1} and g CDM l^{-1}, respectively)	$dX/dt = \mu_{max} \times (1 - X/X_{max}) \times X$	see previous	-	Lejeune et al. (1998); Leroy & De Vuyst (1999)

Table 17.3 (continued)

Parameter/Unit	Equation	Symbol/Unit	Comment	Reference
Bacteriocin production (k_b and k_{inact}) (see previous)	$dB/dt = k_b \times \frac{dX}{dt} - k_{inact} \times X \times B$	see previous	-	Lejeune et al. (1998); Leroy & De Vuyst (1999)
Glucose consumption ($Y_{X/S}$ and m_S) (see previous)	$dS/dt = -Y_{X/S} \times \frac{dX}{dt} - m_S \times X$	see previous	-	Lejeune et al. (1998); Leroy & De Vuyst (1999)
Lactic and acetic acid production ($Y_{L/S}$ and $Y_{A/S}$) [g lactic or acetic acid (g sugar)$^{-1}$]	$dL/dt = -Y_{L/S} \times \frac{dS}{dt}$	L and A, lactate and acetate concentration, respectively (g l^{-1}) $Y_{L/S}$ and $Y_{A/S}$, yield coefficients for the conversion of sugar into lactic and acetic acid, respectively	For acetate: dL and $Y_{L/S}$ are substituted by dA and $Y_{A/S}$, respectively	Lejeune et al. (1998); Leroy & De Vuyst (1999)

CDM, cell dry mass

sakei CTC494 *in vitro* and in model fermented sausages. The influence of sausage ingredients on sakacin K production was variable; sodium chloride mixed with nitrate and nitrite decreased total bacteriocin production, whereas sodium chloride and pepper, while stimulating growth of the strain, did not affect bacteriocin production. Sodium chloride protected *Listeria monocytogenes* from the action of the bacteriocin, but black pepper and sodium nitrite enhanced sakacin K activity *in vitro* and in sausages, surpassing the protective effect of sodium chloride and further inhibiting the growth of *L. monocytogenes*. The same effect observed with black pepper and nitrite was found when manganese was used as a sausage ingredient instead of pepper, suggesting that one of the active factors in pepper was manganese. Verluyten et al. (Verluyten, Leroy, & de Vuyst, 2004; Verluyten, Messens, & De Vuyst, 2003, 2004) studied the effect of sodium chloride, sodium nitrite as well as the effect of different spices relevant for the production of fermented sausages on curvacin A, a listericidal bacteriocin produced by *Lb. curvatus* LTH 1174, a strain isolated from fermented sausage. The strain was highly sensitive to nitrite; even a concentration of 10 ppm of sodium nitrite inhibited its growth and both the volumetric and specific bacteriocin production. Both cell growth and bacteriocin activity were affected by changes in sodium chloride concentration. It clearly slowed down the growth of *Lb. curvatus* LTH 1174, but more importantly, it had a detrimental effect on specific curvacin A production and hence on overall bacteriocin activity. Even a low salt concentration (2%, wt/vol) decreased bacteriocin production, while growth was unaffected at this concentration. The inhibitory effect of NaCl was mainly due to its role as an aw-reducing agent. Further, it was clear that salt interfered with bacteriocin induction. Additionally, when 6% (wt/vol) sodium chloride was added, the minimum biomass concentration necessary to start the production of curvacin A was 0.90 g (cell dry mass) per liter. Addition of a cell-free culture supernatant or a protein solution as a source of induction factor resulted in an increase in the maximum attainable bacteriocin activity. As far as the effect of the spices were concerned, pepper, nutmeg, rosemary, mace, and garlic all decreased the maximum specific growth rate, while paprika was the only spice that increased it. The effect on the lag phase was minor except for nutmeg and especially for garlic, which increased it, yet garlic was stimulatory for biomass production. The maximum attainable biomass concentration was severely decreased by the addition of 0.40% (wt/vol) nutmeg, while 0.35% (wt/vol) garlic or 0.80% (wt/vol) white pepper increased biomass concentration. Nutmeg decreased both growth and bacteriocin production considerably. Garlic was the only spice enhancing specific bacteriocin production, resulting in a higher bacteriocin activity in the cell-free culture supernatant. Finally, lactic acid production was stimulated by the addition of pepper, and this was not due to the manganese present. Addition of spices to the sausage mixture is clearly a factor that will influence the effectiveness of bacteriocinogenic starter cultures in fermented-sausage manufacturing. Aymerich, Artigas, Garriga, Monfort, and Hugas (2000a) studied the effect of sausage ingredients and additives on Enterocin A and B production by *Enterococcus faecium* strain CTC492, and concluded that it was significantly inhibited by sodium chloride, nitrate and pepper.

In Situ *Study of Bacteriocin Antimicrobial Activity*

The *in situ* application of bacteriocins has been a field of intensive study. Its effectiveness against pathogens has been studied in raw, cooked or fermented meat products. However, such an application might face several limitations occurring due to various reasons, the most common being the relatively narrow spectrum of action, the possible inactivation in food systems by proteases or as a result of their interactions with food constituents. Additionally, the emergence of resistant mutants, that has been demonstrated (Rekhif, Atrih, & Lefebvre, 1994), might lead to a loss of antimicrobial activity of bacteriocins against sensitive strains during food storage.

Currently, there are two major approaches that are commonly used. The first one requires the inoculation of the commodity with a bacteriocinogenic strain. This strain can either fulfill all the criteria in order to be used as a starter culture, or it can be used as an adjunct culture provided that it can not only tolerate the action of the starter cultures, but can proliferate and produce bacteriocin as well. Alternatively, the bacteriocin itself, after being purified to some degree, can be added as an ingredient in the food matrix or immobilized on a proper agent, such as silicon surfaces, soy- and corn- based films, cellulose, cellophane and several others. (Daeschel, Mcguire, & Almakhla, 1992; Guerra, Macias, Agrasar, & Castro 2005; Luchansky & Call, 2004; Natrajan & Sheldon, 2000; Siragusa, Cutter, & Willett, 1999).

Since the first attempts for meat and meat products biopreservation (Schillinger & Luecke, 1987) many interesting studies have taken place, referring to a wide range of products and thus extensive data currently exists. Both approaches are currently used in fermented meat products, although the addition of the bacteriocinogenic strain is preferred due to legal and economic implications

Nisin, produced by *Lactococcus lactis*, was the first and most prominent member of lantibiotics, that was discovered just before penicillin, and has been used extensively as a food preservative. Despite the promising results that are published regularly, concerning a variety of products such as minced raw buffalo meat (Pawar, Malik, Bhilegaonkar, & Barbuddhe, 2000), beef products (Cutter & Siragusa, 1994; Millette, Le Tien, Smoragiewicz, & Lacroix, 2007), raw and cooked pork (Fang & Lin, 1994; Murray & Richard, 1997), and sucuks (Hampikyan & Ugur, 2007), it seems to be a common belief that, compared to dairy products, its effectiveness is reduced due to its low solubility, uneven distribution and lack of stability. Moreover the required dose to be effective is uneconomical and exceeding the acceptable daily intake (Hugas, 1998).

It has been previously shown that members of nearly all genera present in spontaneous sausage fermentation have been found to produce bacteriocins. Concequently, a wide range of bacteriocinogenic microorganisms or their bacteriocins have been tested for their capacity to contribute to the safety of a variety of products.

The efficacy of *Pediococcus acidilactici* strains PAC 1.0, H and PO2 to control growth of *L. monocytogenes* as well as *Pd. acidilactici* PA-2 against *Escherichia coli* O157:H7 and *L. monocytogenes* during the manufacture of dry sausage has been studied by Foegeding, Thomas, Pilkington, and Klaenhammer (1992), Luchansky et al. (1992) and Lahti, Johansson, Honkanen-Buzalski, Hill, and Nurmi (2001).

In the first case, the contribution of pediocin to the inactivation of *L. monocytogenes* strains was confirmed, while in the second case, the incorporation of a bacteriocinogenic strain of *Pd. acidilactici* (PA-2) as a starter culture had no effect on the numbers of *E. coli* O157:H7, regardless of the inoculum level, and decreased the population of *L. monocytogenes* but only when it was present in high numbers (5.05–5.41 log cfu g^{-1}). Recently, the antilisterial activity of two strains of *Pd. pentosaceus* was verified during the production of 'alheiras' sausage (Albano et al., 2007b), as well as the ability of a bacteriocin produced by a *Pd. acidilactici* strain to reduce the numbers of *L. monocytogenes* inoculated on the surface of raw meat and stored at 4°C or 15°C, in addition to the bacteriostatic effect against *Clostridium perfringens* has been demonstrated (Nieto-Lozano, Reguera-Useros, del Pelaez-Martinez, & Hardisson de la Torr, 2006).

The effect of the bacteriocinogenic strain *Staphylococcus xylosus* 1E has been studied by Villani et al. (1997). It was able to reduce the viable counts of *L. monocytogenes* in Naples-type sausages after 21 days of maturation. Moreover, no *L. monocytogenes* were recovered after 75 days in sausages inoculated with *St. xylosus* 1E, while the pathogen was still present at this time in control sausages in which *L. monocytogenes* was challenged.

As far as the *Carnobacterium* sp. are concerned, addition of piscicolin 126, a bacteriocin produced by *Carnobacterium piscicola* JG126 to a devilled ham paste test food system inhibited the growth of *L. monocytogenes* for at least 14 days (Jack et al., 1996).

Ananou, Maqueda, Martinez-Bueno, Galvez, and Valdivia (2005) studied the anti-staphylococcal activity of bacteriocin AS-48. The latter was added either as a semi-purified form or by *in situ* production via the incorporation of the producer strain *Ent. faecalis* A-48-32 in the starter culture. The producer strain managed to develop well and the best result was achieved with a bacteriocinogenic strain inoculum of 10^7 CFU/g. The effectiveness of the enterocin CCM 4231 in controlling *L. monocytogenes* contamination in dry fermented Hornad salami was verified by Laukova, Czikkova, Laczkova, and Turek (1999).

Campanini, Pedrazzoni, Barbuti, and Baidini (1993) observed that the addition of starter cultures prevented growth of *L. monocytogenes* and the pathogen was detected after enrichment. Listeria appeared to be absent only in salami inoculated with the bacteriocinogenic *Lb. plantarum* MCS strain. Vermeiren, Devlieghere, Vandekinderen, and Debevere (2006) studied two lactic acid bacteria, a non-bacteriocinogenic *Lb. sakei* subsp. *carnosus* strain 10A and lactocin S producing *Lb. sakei* strain 148 for their usefulness as a protective culture in the biopreservation of cooked meat products using a cooked ham model inoculated with a cocktail of three *L. monocytogenes* strains. The application of the non-bacteriocinogenic strain at 10^6 cfu/g was capable to limit the growth of *L. monocytogenes* to <1 $\log^{10}$ cfu/g during 27 days at 7°C, whilst an application level of 10^5 cfu/g failed to prevent the growth of the pathogen to acceptable levels. Lowering the temperature to 4°C or switching from vacuum packaging to modified atmosphere packaging did not influence the ability of the strain to dominate. A combination of strain 10A and 4°C or a MAP containing 50% CO_2 completely inhibited the growth of the pathogen. On the other hand the bacteriocinogenic strain failed to demonstrate an antagonistic effect towards *L. monocytogenes*. Mataragas et al. (2003b) studied the antagonistic activity of *Lb. curvatus* strain L442 and *Leu. mesenteroides* L124 against different strains of *L. monocytogenes* and *L. innocua* in vacuum- or modified atmosphere-packaged sliced cooked cured pork shoulder at 4°C. Inoculation of the meat products with both lactic acid bacteria resulted in a decrease of the listeriae population, whereas addition of the bacteriocins seemed to be more effective as the listeriae population was reduced to below enumeration limit regardless of the type of packaging. Dicks, Mellett, and Hoffman (2004) applied two bacteriocinogenic strains as starter cultures in the production of Ostrich meat salami. *Lb. plantarum* strain 423 was abble to produce plantaricin 423 while *Lb. curvatus* strain DF126 produced curvacin DF126. Neither of the two bacteriocins inhibited the growth of *Micrococcus* sp. MC50 and did not have any inhibitory effect on either of the producer strains. Curvacin DF126 and plantaricin 423 inhibited the growth of *L. monocytogenes* in salami. However, after 15 h of fermentation, the viable count of *L. monocytogenes* LM1 increased, probably due to a decrease in activity of the bacteriocins and/or the development of resistant bacterial cells. Ghalfi, Benkerroum, Doguiet, Bensaid, & Thonart (2007) examined the effectiveness of a combination of cell-adsorbed bacteriocin (CAB) of a *Lactobacillus curvatus* strain with oregano or savory essential oil to control *L. monocytogenes* in pork meat. Addition of oregano or savory essential oil exhibited a synergistic effect with CAB to control *L. monocytogenes* in pork meat during storage at 4°C.

Vignolo, Fadda, de Kairuz, Holgado, de, and Oliver (1998) inoculated meat slurry with *L. monocytogenes* and applied different levels of curing additives (NaCl, $NaNO_2$, ascorbic acid, alginate meat binder, sodium lactate) in order to study their influence on the inhibitory effect of lactocin 705 at 20°C. The results obtained, indicated that the use of lactocin 705 to control *L. monocytogenes* was less effective in the presence of curing ingredients. Benkerroum et al. (2005) combined lyophilized bacteriocinogenic strains (*Lc. lactis* subsp. *lactis* LMG21206 and *Lb. curvatus* LBPE) with a non-inhibitory to *L. monocytogenes* commercial starter culture in order to study their *in vivo* effectiveness to control *L. monocytogenes* in dry-fermented sausages. The meat batter was experimentally contaminated with a mixture of four different strains of *L. monocytogenes*. The results showed that *L. monocytogenes* did not grow in any of the contaminated batches, but no significant decrease was observed either in the positive control (no added starter culture) or in samples fermented with the bacteriocinogenic starter culture during the fermentation period and up to 15 days of drying.

Bacteriocins in the Context of the Hurdle Theory

The effectiveness of bacteriocins in various food commodities has been widely studied and its contribution to their safety can hardly be questioned. It is though a fact that the food matrix along with several other specific processing parameters affects their efficacy on a variable degree, leading some times to a minimization of its action. Thus, in recent years, application of bacteriocins as part of hurdle-technology has gained attention. It has been shown that several bacteriocins might exhibit additive or synergistic effects when used in combination with other antimicrobial agents or processes, including chemicals, packaging, thermal or non-thermal treatments.

The combination of bacteriocins with several chemical substances such as curing agents, organic acids, chelating agents and the phenolic compounds of essential oils, has been studied both *in vitro* and *in situ*. Curing agents have a variable effect on bacteriocin activity. The negative effect of sodium chloride, that has already been mentioned, can be attributed to ionic interactions between the bacteriocin molecules and charged groups, involved in bacteriocin binding to target cells, induced conformational changes to bacteriocin molecules or even to changes in the cell envelope of the target organisms (Bhunia, Johnson, Ray, & Kalchayanand, 1991; Jydegaard, Gravesen, & Knøchel, 2000; Lee, Iwata, & Oyagi, 1993). On the other hand, nitrite addition enhanced bacteriocin activity against target microorganisms. For example, nisin activity was enhanced in presence of nitrite against clostridial endospores outgrowth (Taylor, Somer, & Kruger, 1985) and *L. monocytogenes* (Gill & Holley, 2003). Similarly, the activity of enterocin EJ97, together with nitrite, was amplified against *L. monocytogenes*, *Bacillus coagulans*, *B. macroides/maroccanus*, and AS-48 against *B. cereus* (Abriouel, Maqueda, Galvez, Martinez-Bueno, & Valdivia, 2002; Garcia et al., 2003; Garcia et al., 2004a, Garcia et al., 2004b. Organic acids and chelating agents also increase bacteriocin activity, the former via facilitated translocation through the cell wall or through the increase of their solubility and concomitantly facilitating their diffusion in the food matrix and the latter by allowing bacteriocins to easily reach the cytoplasmic membrane of Gram-negative bacteria. Concerning the synergistic effect with the active components of essential oils that has been exhibited using a variety of combinations between bacteriocins and the phenolic compounds (Grande et al., 2007), their impact on the organoleptic properties of the final product should be of concern when applying in a food matrix.

The combined effect of bacteriocins with vacuum or modified-atmosphere packaging has also been studied to some extent for various foodstuffs including a variety of meat products. Apart from the individual action that these may exhibit, and their combined effect, it had been shown that carbon dioxide act synergistically with nisin by enhancing its permealisation through the cytoplasmic membrane of *L. monocytogenes* (Nilsson et al., 2000).

Thermal treatment that is quite often used in food processing might have positive and negative effects, from a microbiological point of view. The former refer to the reduction of the viable cell counts, whereas the surviving ones are sublethally injured, while the latter to the possible activation of bacterial endospores. Several studies have confirmed the synergistic action of bacteriocins with thermal treatment, in both a further reduction of viable cell counts and protection against proliferation of the endospores that might even lead and in some cases to a reduction of the thermal treatment intensity (Ananou, Valdivia, Martínez Bueno, Galvez, & Maqueda, 2004; Grande et al., 2006). Moreover, due to the damage undergone by the heat treatment, an induced sensitivity to Gram-negative bacteria seemed to take place, against several bacteriocins (Ananou, Galvez, Martinez-Bueno, Maqueda, and Valdivia, 2005; Bakes, Kitis, Quattlebaum, & Barefoot, 2004; Boziaris, Humpheson, & Adams, 1998) that otherwise would have been ineffective.

During food processing, several treatments take place that do not necessary involve heat exposure. Irradiated, pulsed electric or magnetic fields and high hydrostatic pressure are processes that can be used in a variety of foodstuffs aiming at the reduction of viable cell counts. Several reports on the combined effect of these processes with bacteriocins are currently available and finely reviewed by Galvez, Abriouel, Lopez, and Ben Omar (2007)

The utilization of bacteriocins in meat preservation as natural antimicrobials has attracted the interest in current research. The antimicrobial action of that implicit determinant in the meat microbial ecosystem, however, has to be quantified by the recent developments of predictive modelling in foods. In addition, multidisciplinary efforts of researchers from the fields of molecular biology, microbial ecology, food safety, meat and meat products technology should be applied, as a holistic approach to bacteriocin application in meat. The extent of the application of bacteriocins as a hurdle and the success to this end will be the outcome of these coordinated efforts.

References

Aasen, I. M., Moretro, T., Katla, T., Axelsson, L., & Storro, I. (2000). Influence of complex nutrients, temperature and pH on bacteriocin production by *Lactobacillus sakei* CCUG 42687. *Applied Microbiology and Biotechnology*, 53, 159–166.

Abriouel, H., Maqueda, M., Galvez, A., Martinez-Bueno, M., & Valdivia, E. (2002). Inhibition of bacterial growth, enterotoxin production, and spore outgrowth in strains of *Bacillus cereus* by bacteriocin AS-48. *Applied and Environmental Microbiology*, 68, 1473–1477.

Ahn, C., & Stiles, M. E. (1990). Plasmid-associated bacteriocin production by a strain of Carnobacterium piscicola from meat. *Applied and Environmental Microbiology*, 56, 2503–2510.

Albano, H., Todorov, S. D., van Reenen, C. A., Hogg, T., Dicks, L. M. T., & Texixeira, P. (2007a). Characterization of two bacteriocins produced by Pediococcus acidilactici isolated from 'Alheira', a fermented sausage traditionally produced in Portugal. *International Journal of Food Microbiology*, 116, 239–247.

Albano, H., Oliveira, M., Aroso, R., Cubero, N., Hogg, T., & Teixeira, P. (2007b). Antilisterial activity of lactic acid bacteria isolated from 'Alheiras' (traditional Portuguese fermented sausages): In situ assays. *Meat Science*, 76, 796–800.

Ananou, S., Galvez, A., Martinez-Bueno, M., Maqueda, M., & Valdivia, E. (2005). Synergistic effect of enterocin AS-48 in combination with outer membrane permeabilizing treatments against *Escherichia coli* O157:H7. *Journal of Applied Microbiology*, 99, 1364–1372.

Ananou, S., Maqueda, M., Martinez-Bueno, M., Galvez, A., & Valdivia, E. (2005). Control of *Staphylococcus aureus* in sausages by enterocin AS-48. *Meat Science*, 71, 549–556

Ananou, S., Valdivia, E., Martínez Bueno, M., Galvez, A., & Maqueda, M. (2004). Effect of combined physico-chemical preservatives on enterocin AS-48 activity against the enterotoxigenic *Staphylococcus aureus* CECT 976 strain. *Journal of Applied Microbiology*, 97, 48–56.

Aymerich, T., Holo, H., Havarstein, L.S., Hugas, M., Garriga, M., & Nes, I. F. (1996). Biochemical and genetic characterization of enterocin A from Enterococcus faecium, a new antilisterial bacteriocin in the pediocin family of bacteriocins. *Applied and Environmental Microbiology*, 62, 1676–1682.

Aymerich, T., Artigas, M. G., Garriga, M., Monfort, J. M., & Hugas, M. (2000a). Effect of sausage ingredients and additives on the production of enterocins A and B by *Enterococcus faecium* CTC492.Optimization of in vitro production and anti-listerial effect in dry.fermented sausages. *Journal of Applied Microbiology, 88*, 686–694.

Aymerich, M. T., Garriga, M., Monford, J. M., Nes, I., & Hugas, M. (2000b). Bacteriocin-producing lactobacilli in Spanish-style fermented sausages: characterization of bacteriocins. *Food Microbiology, 17*, 33–45.

Bakes, S. H., Kitis, F. Y. E., Quattlebaum, R. G., & Barefoot, S. F. (2004). Sensitization of Gram-negative and Gram-positive bacteria to jenseniin G by sublethal injury. *Journal of Food Protection, 67*, 1009–1013.

Barcena, B. J. M., Sineriz, F., Gonzalez de Llano, D., Rodriguez, A., & Suarez, J. E. (1998). Chemo-stat production of plantaricin C by *Lactobacillus plantarum* LL41. *Applied and Environmental Microbiology, 57*, 3512–3514.

Benkerroum, N., Daoudi, A., Hamraoui, T., Ghalfi, H., Thiry, C., Duroy, M., Evrart, P., Roblain, D., & Thonart, P. (2005). Lyophilized preparations of bacteriocinogenic *Lactobacillus curvatus* and *Lactococcus lactis* subsp. *lactis* as potential protective adjuncts to control *Listeria monocytogenes* in dry-fermented sausages. *Journal of Applied Microbiology, 98*, 56–63.

Benthin, S., Schulze, U., Nielsen, J., & Villadsen, J. (1994). Growth energetics of *Lactococcus cremoris* FD1 during energy-, carbon-, and nitrogen-limitation in steady state and transient cultures. *Chemical Engineering Science, 49*, 589–609.

Bhunia, A. K., Johnson, M. C., Ray, B., & Kalchayanand, N. (1991). Mode of action of pediocin AcH from *Pediococcus acidilactici* H on sensitive bacterial strains. *Journal of Applied Bacteriology, 70*, 25–33.

Biswas, S. R., Ray, P., Johnson, M. C., & Ray, B. (1991). Influence of growth conditions on the production of a bacteriocin, pediocin AcH, by *Pediococcus acidilactici* H. *Applied and Environmental Microbiology, 57*, 1265–1267.

Boziaris, I. S., Humpheson, I., & Adams, M. R. (1998). Effect of nisin on heat injury and inactivation of *Salmonella enteritidis* PT4. *International Journal of Food Microbiology, 43*, 7–13.

Budde, B. B., Hornbaek, T., Jacobsen, T., Barkholt, V., & Koch, A. G. (2003). Leuconostoc carnosum 4010 has the potential for use as a protective culture for vacuum-packed meats: Culture isolation, bacteriocin identification, and meat application experiments. *International Journal of Food Microbiology, 83*, 171–184.

Campanini, M., Pedrazzoni, I., Barbuti, S., & Baidini, P. (1993). Behaviour of *Listeria monocytogenes* during the maturation of naturally and artificially contaminated salami: Effect of lactic-acid bacteria starter cultures *International Journal of Food Microbiology, 21*, 169–175.

Casaus, P., Nilsen, T., Cintas, L. M., Nes, I. F., Hernandez, P. E., & Holo, H. (1997). Enterocin B, a new bacteriocin from Enterococcus faecium T136 which can act synergistically with enterocin A. *Microbiology, 143*, 2287–2294.

Chatterjee, C., Paul, M., Xie, L., & van der Donk, W. A. (2005). Biosynthesis and mode of action of Lantibiotics. *Chemistry Reviews, 105*, 633–683.

Cintas, L. M., Rodriguez, J. M., Fernandez, M. F., Sletten, K., Nes, I. F., Hernandez, P. E., & Holo, H. (1995). Isolation and Characterization of Pediocin L50, a new bacteriocin from Pediococcus acidilactici with a broad inhibitory spectrum. *Applied and Environmental Microbiology, 61*, 263–2648.

Cintas, L. M., Casaus, P., Havarstein, L. S., Hernandez, P. E., & Nes, I. F. (1997). Biochemical and genetic characterization of enterocin P, a novel sec-dependent bacteriocin from Enterococcus faecium P13 with a broad antimicrobial spectrum. *Applied and Environmental Microbiology, 63*, 4321–4330.

Cintas, L. M., Casaus, P., Holo, H., Hernandez, P. E., Nes, I. F., & Havarstein, L. S. (1998). Enterocins L50A and L50B, two novel bacteriocins from Enterococcus faecium L50, and related to staphylococcal hemolysins. *Journal of Bacteriology, 180*, 1988–1994.

Cleveland, J., Montville, T. J., Nes, I. F., & Chikindas, M. L. (2001). Bacteriocins: Safe, natural antimicrobials for food preservation. *International Journal of Food Microbiology, 71*, 1–20.

Cotter, P. D., Hill, C., & Ross, R. P. (2005). Bacteriocins: Developing innate immunity for food. *Nature Reviews Microbiology, 3*, 777–788.

Cutter, C. N., & Siragusa, G. R. (1994). Decontamination of beef carcass tissue with nisin using a pilot scale model carcass washer. *Food Microbiology, 11*, 481–489.

Daba, H., Lacroix, C., Huang, J., & Simard, R. (1993). Influence of growth conditions on production and activity of mesenterocin 5 by a strain of *Leuconostoc mesenteroides*. *Applied Microbiology and Biotechnology, 39*, 166–173.

Daeschel, M. A., Mcguire, J., & Almakhla, H. (1992). Antimicrobial activity of nisin adsorbed to hydrophilic and hydrophobic silicon surfaces. *Journal of Food Protection, 55*, 731–735.

De Vuyst, L., Callewaert, R., & Crabbe, K. (1996). Primary metabolite kinetics of bacteriocin biosynthesis by *Lactobacillus amylovorus* and evidence for stimulation of bacteriocin under unfavourable growth conditions. *Microbiology, 142*, 817–827.

De Vuyst, L., & Vandamme, E. J. (1994). *Bacteriocins of lactic acid bacteria: Microbiology, genetics and applications*. London: Blackie Academic and Professional.

Deegan, L. H., Cotter, P. D., Hill, C., & Ross, P. (2006). Bacteriocins: Biological tools for biopreservation and shelf-life extension. *International Dairy Journal, 16*, 1058–1071.

Dicks, L. M. T., Mellett, F. D., & Hoffman, L. C. (2004). Use of bacteriocin-producing starter cultures of *Lactobacillus plantarum* and *Lactobacillus curvatus* in production of ostrich meat salami. *Meat Science, 66*, 703–708.

Dominguez, A. P. M., Bimani, D., Caldera-Olivera, F., & Brandelli, A. (2007). Cerein 8 production in soybean protein using response surface methodology. *Biochemical Engineering Journal, 35*, 238–243.

Drosinos, E. H., Mataragas, M., & Metaxopoulos, J. (2005a). Biopreservation: A new direction towards food safety. In A. P. Riley (Ed.), *New developments in food policy, control and research* (pp. 31–64). New York: Nova Science Publishers, Inc.

Drosinos, E. H., Mataragas, M., & Metaxopoulos, J. (2006). Modeling of growth and bacteriocin production by *Leuconostoc mesenteroides* E131. *Meat Science, 74*, 690–696.

Drosinos, E. H., Mataragas, M., Nasis, P., Galiotou, M., & Metaxopoulos, J. (2005b). Growth and bacteriocin production kinetics of *Leuconostoc mesenteroides* E131. *Journal of Applied Microbiology, 99*, 1314–1323.

Enan, G., El-Essawy, A. A., Uyttendaele, M., & Debevere, J. (1996). Antibacterial activity of Lactobacillus plantarum UG1 isolated from dry sausage: Characterization production and bactericidal action of plantaricin UG1. *International Journal of Food Microbiology, 30*, 189–215.

Fang, T. J., & Lin, L. W. (1994). Growth of *Listeria monocytogenes* and *Pseudomonas fragi* on cooked pork in a modified atmosphere packaging/nisin combination. *Journal of Food Protection, 57*, 479–485.

Foegeding, P. M., Thomas, A. B., Pilkington, D. H., & Klaenhammer, T. R. (1992). Enhanced control of *Listeria monocytogenes* by in situ-produced pediocin during dry fermented sausage productiont. *Applied and Environmental Microbiology, 58*, 884–890.

Franz, C. M. A. P., van Belkum, M. J., Holzapfel, W. H., Abriouel, H., & Galvez, A. (2007). Diversity of enterococcal bacteriocins and their grouping in a new classification scheme. *FEMS Microbiology Reviews, 31*, 293–310.

Galvez, A., Abriouel, H., Lopez, R. L., & Ben Omar, N. (2007). Bacteriocin-based strategies for food biopreservation. *International Journal of Food Microbiology* (in press) doi:10.1016/j.ijfoodmicro.2007.06.001.

Garcia, M. T., Ben Omar, N., Lucas, R., Perez-Pulido, R., Castro, A., Grande, M. J., Martinez-Canamero, M., & Galvez, A. (2003). Antimicrobial activity of enterocin EJ97 on *Bacillus coagulans* CECT 12. *Food Microbiology, 20*, 533–536.

Garcia, M. T., Lucas, R., Abriouel, H., Ben Omar, N., Perez, R., Grande, M. J., Martinez-Canamero, M., & Galvez, A. (2004a). Antimicrobial activity of enterocin EJ97 against '*Bacillus macroides/Bacillus maroccanus*' isolated from zucchini purée. *Journal of Applied Microbiology, 97*, 731–737.

Garcia, M. T., Martinez Canamero, M., Lucas, R., Ben Omar, N., Perez Pulido, R., & Galvez, A. (2004b). Inhibition of *Listeria monocytogenes* by enterocin EJ97 produced by *Enterococcus faecalis* EJ97. *International Journal of Food Microbiology, 90*, 161–170.

Ghalfi, H., Benkerroum, N., Doguiet, D. D. K., Bensaid, M., & Thonart, P. (2007). Effectiveness of cell-adsorbed bacteriocin produced by *Lactobacillus curvatus* CWBI-B28 and selected essential oils to control *Listeria monocytogenes* in pork meat during cold storage. *Letters in Applied Microbiology, 44*, 268–273.

Gill, A. O., & Holley, R. A. (2003). Interactive inhibition of meat spoilage and pathogenic bacteria by lysozyme, nisin and EDTA in the presence of nitrite and sodium chloride at 24°C. *International Journal of Food Microbiology, 80*, 251–259.

Grande, Ma. J., Lucas, R., Abriouel, H., Valdivia, E., Ben Omar, N., Maqueda, M., Martinez-Bueno, M., Martinez-Canamero, M., & Galvez, A. (2006). Inhibition of toxicogenic *Bacillus cereus* in rice-based foods by enterocin AS-48. *International Journal of Food Microbiology, 106*, 185–194.

Grande, Ma. J., Lucas, R., Abriouel, H., Valdivia, E., Ben Omar, N., Maqueda, M., Martinez-Canamero, M., & Galvez, A. (2007). Treatment of vegetable sauces with enterocin AS-48 alone or in combination with phenolic compounds to inhibit proliferation of *Staphylococcus aureus*. *Journal of Food Protection, 70*, 405–411.

Guerra, N. P., Macias, C. L., Agrasar, A. T., & Castro, L. P. (2005). Development of a bioactive packaging cellophane using Nisaplin as biopreservative agent. *Letters in Applied Microbiology, 40*, 106–1610.

Hampikyan, H., & Ugur, M. (2007). The effect of nisin on *L. monocytogenes* in Turkish fermented sausages (sucuks). *Meat Science, 76*, 327–332.

Harding, C. D., & Saw, B. G. (1990). Antimicrobial activity of Leuconostoc gelidum against closely related species and Listeria monocytogenes. *Journal of Applied Bacteriology, 69*, 648–654.

Holck, A. L., Axelsson, L., Huhne, K., & Krockel, L. (1994). Purification and cloning of sakacin 674, a bacteriocin from Lactobacillus sake Lb674. *FEMS Microbiology Letters, 115*, 143–150.

Holck, A., Axelsson, L., & Schillinger, U. (1996). Divergicin 750, a novel bacteriocin produced by Carnobacterium divergens 750. *FEMS Microbiology Letters, 136*, 163–168.

Hugas, M. (1998). Bacteriocinogenic lactic acid bacteria for the biopreservation of meat and meat products. *Meat Science, 49*, S139–S150.

Hugas, M., Garriga, M., Pascual, M., Aymerich, M. T., & Monfort, J. M. (2002). Enhancement of sakacin K activity against *Listeria monocytogenes* in fermented sausages with pepper or manganese as ingredients. *Food Microbiology, 19*, 519–528.

Jack, R. W., Wan, J., Gordon, J., Harmark, K., Davidson, B. E., Hillier, A. J., Wettenhall, R. E. H., Hickey, M. W., & Coventry, M. J. (1996). Characterization of the chemical and antimicrobial properties of piscicolin 126, a bacteriocin produced by Carnobacterium piscicola JG126. *Applied and Environmental Microbiology, 62*, 2897–2903.

Joerger, M. C., & Klaenhammer, T. R. (1986). Characterization and purification of helveticin J and evidence for a chromosomally determined, bacteriocin produced by *Lactobacillus helveticus* 481. *Journal of Bacteriology, 167*, 439–446.

Joosten, H. M. L. J., & Nunez, M. (1995). Adsorption of nisin and enterocin 4 to polypropylene and glass surface and its prevention by tween 80. *Letters in Applied Microbiology, 21*, 389–392.

Jydegaard, A.-M., Gravesen, A., & Knøchel, S. (2000). Growth condition-related response of *Listeria monocytogenes* 412 to bacteriocin inactivation. *Letters in Applied Microbiology, 31*, 68–72.

Kaiser, A. L., & Montville, T. J. (1993). The influence of pH and growth rate on the production of the bacteriocin, bavaricin MN, in batch and continuous fermentations. *Journal of Applied Bacteriology, 75*, 536–540.

Kemperman, R., Kuipers, A., Karsens, H., Nauta, A., Kuipers, O., & Kok, J. (2003). Identification and characterization of two novel clostridial bacteriocins, Circularin A and Closticin 574. *Applied and Environmental Microbiology, 69*, 1589–1597.

Kim, W. S., Hall, R. J., & Dunn, N. W. (1997). The effect of nisin concentration and nutrient depletion on nisin production of *Lactobacillus lactis*. *Applied Microbiology and Biotechnology, 50*, 429–433.

Klaenhammer, T. R. (1993). Genetics of bacteriocins produced by lactic acid bacteria. *FEMS Microbiology Reviews, 12*, 39–86.

Krier, F., Revol-Junelles, A. M., & Germain, P. (1998). Influence of temperature and pH production of two bacteriocins by *Leuconostoc mesenteroides* subsp. *mesenteroides* FR52 during batch fermentation. *Applied Microbiology and Biotechnology, 50*, 359–363.

Lahti, E., Johansson, T., Honkanen-Buzalski, T., Hill, P., & Nurmi, E. (2001). Survival and detection of *Escherichia coli* O157:H7 and *Listeria monocytogenes* during the manufacture of dry sausage using two different starter cultures. *Food Microbiology, 18*, 75–85.

Laukova, A., Czikkova, S., Laczkova, S., & Turek, P. (1999). Use of enterocin CCM 4231 to control *Listeria monocytogenes* in experimentally contaminated dry fermented Hornad salami. *International Journal of Food Microbiology, 52*, 115–119.

Lee, S., Iwata, T., & Oyagi, H. (1993). Effects of salts on conformational change of basic amphipathic peptides from β-structure to α-helix in the presence of phospholipid liposomes and their channel-forming ability. *Biochimica et Biophysica Acta, 1151*, 75–82.

Lejeune, R., Callewaert, R., Crabbe, K., & De Vuyst, L. (1998). Modelling the growth and bacteriocin production by *Lactobacillus amylovorus* DCE 471 in batch cultivation. *Journal of Applied Microbiology, 84*, 159–168.

Leroy, F., & De Vuyst, L. (1999). Temperature and pH conditions that prevail during the fermentation of sausages are optimal for the production of the antilisterial bacteriocin sakacin K. *Applied and Environmental Microbiology, 65*, 974–981.

Leroy, F., & De Vuyst, L. (2003). A combined model to predict the functionality of the bacteriocin-producing *Lactobacillus sakei* strain CTC 494. *Applied and Environmental Microbiology, 69*, 1093–1099.

Leroy, F., Verluyten, J., Messens, W., & De Vuyst, L. (2002). Modelling contributes to the understanding of the different behaviour of bacteriocin-producing strains in a meat environment. *International Dairy Journal, 12*, 247–253.

Luchansky, J. B., & Call, J. E. (2004). Evaluation of nisin-coated cellulose casings for the control of *Listeria monocytogenes* inoculated onto the surface of commercially prepared frankfurters. *Journal of Food Protection, 67*, 1017–1021.

Luchansky, J. B., Glass, K. A., Harsono, K. D., Degnan, A. J., Faith, N. G., Cauvin, B., Baccus-Taylor, G., Arihara, K., Bater, B., Maurer, A. J., & Cassens, R. G. (1992). Genomic analysis of pediococcus starter cultures used to control *Listeria monocytogenes* in Turkey summer sausage. *Applied and Environmental Microbiology, 58*, 3053–3059.

Lyon, W. J., Olson, D. G., & Murano, E. A. (1995). Isolation and purification of enterocin EL1, a bacteriocin produced by a strain of Enterococcus faecium. *Journal of Food Protection, 58*, 890–898.

Mataragas, M., Drosinos, E. H., & Metaxopoulos, J. (2003a). Antagonistic activity of lactic acid bacteria against *Listeria monocytogenes* in sliced cooked cured pork shoulder stored under vacuum or modified atmosphere at $4\pm2°C$. *Food Microbiology, 20*, 259–265.

Mataragas, M., Drosinos, E. H., Tsakalidou, E., & Metaxopoulos, J. (2004). Influence of nutrients on growth and bacteriocin production by *Leuconostoc mesenteroides* L124 and *Lactobacillus curvatus* L442. [International Journal of General and Molecular Microbiology] *Antonie van Leeuwenhoek, 85*, 191–198.

Mataragas, M., Metaxopoulos, J., Galiotou, M., & Drosinos, E. H. (2003b). Influence of pH and temperature on growth and bacteriocin production by *Leuconostoc mesenteroides* L124 and *Lactobacillus curvatus* L442. *Meat Science, 64*, 265–271.

Matsusaki, H., Endo, N., Sonomoto, K., & Ishikazi, A. (1996). Lantibiotic nisin Z fermentative production by *Lactobacillus lactis* IO-1: Relationship between production of the lantibiotic and lactate and cell growth. *Applied Microbiology and Biotechnology, 45*, 36–40.

McKellar, R. C. (1997). A heterogeneous population model for the analysis of bacterial growth kinetics. *International Journal of Food Microbiology, 36*, 179–186.

Meghrous, J., Huot, E., Quittelier, M., & Petitdemange, H. (1992). Regulation of nisin biosynthesis by continuous cultures and by resting cells of *Lactococcus lactis* subsp. *lactis*. *Research in Microbiology, 143*, 879–890.

Messens, W., Neysens, P., Vansieleghem, W., Vanderhoeven, J., & De Vuyst, L. (2002). Modeling growth and bacteriocin production by *Lactobacillus amylovorus* DCE 471 in response to temperature and pH values used for sourdough fermentations. *Applied and Environmental Microbiology, 68*, 1431–1435.

Messens, W., Verluyten, J., Leroy, F., & De Vuyst, L. (2003). Modeling growth and bacteriocin production by *Lactobacillus curvatus* LTH 1174 in response to temperature and pH values used for European sausage fermentation processes. *International Journal of Food Microbiology, 81*, 41–52.

Messi, P., Bondi, M., Sabia, C., Battini, R., & Manicardi G. (2001). Detection and preliminary characterization of a bacteriocin (plantaricin 35d) produced by a Lactobacillus plantarum strain. *International Journal of Food Microbiology, 64*, 193–198.

Millette, M., Le Tien, C., Smoragiewicz, W., & Lacroix, M. (2007). Inhibition of *Staphylococcus aureus* on beef by nisin-containing modified alginate films and beads. *Food Control, 18*, 878–884.

Moretro, T., Aasen, I. M., Storro, I., & Axelsson, L. (2000). Production of sakacin P by *Lactobacillus sakei* in a completely defined medium. *Journal of Applied Microbiology, 88*, 536–545.

Mortvedt, C. I., Nissen-Meyer, J., Sletten, K., & Nes I. F. (1991). Purification and amino acid sequence of lactocin S, a bacteriocin produced by Lactobacillus sake L45. *Applied and Environmental Microbiology, 57*, 1829–1834.

Mortvedt-Abildgaard, C., Nissen-Meyer, J., Jelle, B., Grenov, B., Skaugen, M., & Nes, I. F. (1995). Production and pH-dependent bactericidal activity of lactocin S, a lantibiotic from *Lactobacillus sake*. *Applied and Environmental Microbiology, 61*, 175–179.

Motlagh, A. M., Bhunia, A. K., Szostek, F., Hansen, T. R., Johnson, M. C., & Ray B. (1992). Nucleotide and amino acid sequence of pap-gene (pediocin AcH production) in Pediococcus acidilactici H. *Letters in Applied Microbiology, 15*, 45–48.

Murray, M., & Richard, J. A. (1997). Comparative study of the antilisterial activity of nisin A and pediocin AcH in fresh ground pork stored aerobically at 5°C. *Journal of Food Protection, 60*, 1534–1540.

Nagao, J. -I., Asaduzzaman, S. M., Aso, Y., Okuda, K. -I., Nakayama, J., & Sonomoto, K. (2006). Lantibiotics: Insight and foresight for new paradigm. *Journal of Bioscience and Bioengineering, 102*, 139–149.

Natrajan, N., & Sheldon, B. W. (2000). Efficacy of nisin-coated polymer films to inactivate *Salmonella typhimurium* on fresh broiler skin. *Journal of Food Protection, 63*, 1189–1196.

Nes, I. F., Diep, D. B., Havarstein, L. S., Brurberg, M. B., Eijsink, V., & Holo, H. (1996). Biosynthesis of bacteriocins in lactic acid bacteria. *Antonie van Leeuwenhoek, 70*, 113–128.

Nes, I. F., & Holo, H. (2000). Class II antimicrobial peptides from lactic acid bacteria. *Biopolymers (Peptide Science), 55*, 50–61.

Nielsen, J., Nikolajsen, K., & Villadsen, J. (1991). Structured modelling of a microbial system II. Experimental verification of a structured lactic acid fermentation model. *Biotechnology and Bioengineering, 38*, 11–23.

Nieto-Lozano, J. C., Reguera-Useros, J. I., Pelaez-Martinez, M., del, C., & Hardisson de la Torr, A. (2006). Effect of a bacteriocin produced by *Pediococcus acidilactici* against *Listeria monocytogenes* and *Clostridium perfringens* on Spanish raw meat. *Meat Science, 72*, 57–61.

Nilsson, L., Chen, Y., Chikindas, M. L., Huss, H. H., Gram, L., & Montville, T. J. (2000). Carbon dioxide and nisin act synergistically on *Listeria monocytogenes*. *Applied and Environmental Microbiology, 66*, 769–774.

Noonpakdee, W., Santivarngkna, C., Jumriangrit, P., Sonomoto, K., & Panyim, S. (2003). Isolation of nisin-producing Lactococcus lactis WNC 20 strain from nham, a traditional Thai fermented sausage. *International Journal of Food Microbiology, 81*, 137–145.

Osmanagaoglu, O. (2007). Detection and characterization of Leucocin OZ, a new anti-listerial bacteriocin produced by Leuconostoc carnosum with a broad spectrum of activity. *Food Control, 18*, 118–123.

Parente, E., & Hill, C. (1992). A comparison of factors affecting the production of two bacteriocins from lactic acid bacteria. *Journal of Applied Bacteriology, 73*, 290–298.

Parente, E., & Ricciardi, A. (1994). Influence of pH on the production of enterocin 1146 during batch fermentation. *Letters in Applied Microbiology, 19*, 12–15.
Parente, E., & Ricciardi, A. (1999). Production, recovery and purification of bacteriocins from lactic acid bacteria. *Applied Microbiology and Biotechnology, 52*, 628–638.
Patton, G. C., & van der Donk, W. A. (2005). New developments in lantibiotic biosynthesis and mode of action. *Current Opinion in Microbiology, 8*, 543–551.
Pawar, D. D., Malik, S. V. S., Bhilegaonkar, K. N., & Barbuddhe, S. B. (2000). Effect of nisin and its combination with sodium chloride on the survival of *Listeria monocytogenes* added to raw buffalo meat mince. *Meat Science, 56*, 215–219.
Prema, P., Bharathy, S., Palavesam, A., Sivasubramanian, M., & Immanuel G. (2006). Detection, purification and efficacy of warnerin produced by Staphylococcus warneri. *World Journal of Microbiology and Biotechnology, 22*, 865–872.
Rekhif, N., Atrih, A., & Lefebvre, G. (1994). Selection and properties of spontaneous mutants of *Listeria monocytogenes* ATTC 15313 resistant to different bacteriocins produced by lactic acid bacteria strains. *Current Microbiology, 28*, 237–242.
Rodriguez, J. M., Cintas, L. M., Casaus, P., Horn, N., Dodd, H. M., Hernandez, P. E., & Gasson, M. J. (1995). Isolation of nisin-producing Lactococcus lactis strains from dry fermented sausages. *Journal of Applied Bacteriology, 78*, 109–115.
Rollini, M., & Manzoni, M. (2005). Influence of different fermentation parameters on glutathione volumetric productivity by *Saccharomyces cerevisiae*. *Process Biochemistry, 41*, 1501–1505.
Schneider, R., Fernandez, F. J., Aquilar, M. B., Guerrero-Legarreta, I., Alpuche-Solis, A., & Ponce-Alquicira, E. (2006). Partial characterization of a class IIa pediocin produced by Pediococcus parvulus 133 strain isolated from meat (Mexical 'chorizo'). *Food Control, 17*, 909–915.
Schillinger, U., & Luecke, F. K. (1987). Lactic acid bacteria on vacuum packaged meat and their influence on shelf life. *Fleischwirtschaft, 67*, 1244–1248.
Schillinger, U., & Luecke, F. K. (1989). Antibacterial activity of Lactobacillus sake isolated from meat. *Applied and Environmental Microbiology, 55*, 1901–1906.
Siragusa, G. R., Cutter, C. N., & Willett, J. L. (1999). Incorporation of bacteriocin in plastic retains activity and inhibits surface growth of bacteria on meat. *Food Microbiology, 61*, 229–235.
Sobrino, O. J., Rodriguez, J. M., Moreira, W. L., Cintas, L. M., Fernandez, M. F., Sanz, B., & Hernandez, P.E. (1992). Sakacin M, a bacteriocin-like substance from Lactobacillus sake 148. *International Journal of Food Microbiology, 16*, 215–225.
Stiles, M. E., & Hastings, J. W. (1991). Bacteriocin production by lactic acid bacteria: Potential for use in meat preservation. *Trends in Food Science and Technology, 2*, 247–251.
Taylor, J. I., Somer, E. B., & Kruger, L. A. (1985). Antibotulinal effectiveness of nisin-nitrite combinations in culture medium and chicken frankfurter emulsions. *Journal of Food Protection, 48*, 234–249.
van Belkum, M. J., & Stiles, M. E. (2000). Nonlantibiotic antimicrobial peptides from lactic acid bacteria. *Natural Product Reports, 17*, 323–365.
Verluyten, J., Leroy, F., & de Vuyst, L. (2004). Effects of different spices used in production of fermented sausages on growth of and curvacin A production by *Lactobacillus curvatus* LTH 1174. *Applied and Environmental Microbiology, 70*, 4807–4813.
Verluyten, J., Messens, W., & De Vuyst, L. (2003). The curing agent sodium nitrite, used in the production of fermented sausages, is less inhibiting to the bacteriocin-producing meat starter culture *Lactobacillus curvatus* LTH 1174 under anaerobic conditions. *Applied and Environmental Microbiology, 69*, 3833–3839.
Verluyten, J., Messens, W., & De Vuyst, L. (2004). Sodium chloride reduces production of curvacin A, a bacteriocin produced by *Lactobacillus curvatus* strain LTH 1174, originating from fermented sausage. *Applied and Environmental Microbiology, 70*, 2271–2278.
Vermeiren, L., Devlieghere, F., Vandekinderen, I., & Debevere, J. (2006). The interaction of the non-bacteriocinogenic *Lactobacillus sakei* 10A and lactocin S producing *Lactobacillus sakei* 148 towards *Listeria monocytogenes* on a model cooked ham. *Food Microbiology, 23*, 511–518.

Vignolo, G., Fadda, S., de Kairuz, M. N., Holgado, A. P., de, R., & Oliver, G. (1998). Effects of curing additives on the control of *Listeria monocytogenes* by lactocin 705 in meat slurry. *Food Microbiology, 15,* 259–264.

Villani, F., Sannino, L., Moschetti, G., Mauriello, G., Pepe, O., Amodio-Cocchieri R., & Coppola, S. (1997). Partial characterization of an antagonistic substance produced by Staphylococcus xylosus 1E and determination of the effectiveness of the producer strain to inhibit Listeria monocytogenes in Italian sausages. *Food Microbiology, 14,* 555–566.

Xie, L., & van der Donk, W. A. (2004). Post-translational modifications during lantibiotic biosynthesis. *Current Opinion in Chemical Biology, 8,* 498–507.

Xiraphi, N., Georgalaki, M., Van Driessche, G., Devreese, B., Van Beeumen, J., Tsakalidou, E., Metaxopoulos, J., & Drosinos, E. H. (2006). Purification and characterization of curvaticin L442, a bacteriocin produced by *Lactobacillus curvatus* L442. [International Journal of General and Molecular Microbiology] *Antonie van Leeuwenhoek, 89,* 19–26.

Yand, R., & Ray, B. (1994). Factors influencing production of bacteriocins by lactic acid bacteria. *Food Microbiology, 11,* 281–291.

Chapter 18
Bioprotective Cultures

Graciela Vignolo, Silvina Fadda, and Patricia Castellano

Introduction

Although the globalization and liberalization of world food trade offers many benefits and opportunities, it also presents new risks. As food is distributed from the original points of production, processing, and packaging to locations thousands of kilometers away, there is an increased risk of cross-border transmission of infectious agents and exposure of consumers to new hazards. In the last few years, food safety concerns have increased their importance due to the dramatic impact that they have on public health. Despite the recent progress in food biotechnology, the meat industry is still under scrutiny from consumers and media due to recent food safety crises; a series of food scandals erupted involving meat and meat products, which impacted prominently on consumer confidence. The outbreaks of bovine spongiform encephalitis (BSE), foot-and-mouth disease and the spread of avian influenza, among other factors, have triggered a sudden lack of consumer confidence in meat and meat products leading to a dramatic fall in demand. Outbreaks of gastroenteritis have been repeatedly associated with the consumption of raw meat and raw meat products. In 2003, over 120,000 disease episodes caused by foodborne pathogens were reported in Germany, *Salmonella* being the most frequently involved pathogen (Bremer et al., 2005). Awareness of the consequences of the meat borne pathogen *Escherichia coli* O157:H7 has increased in the general public opinion making this organism a household name in the 21st century (Ransom et al., 2003). Recently, the preliminary FooNet data on the incidence of infection with pathogens transmitted through food in the USA showed a decline in infections caused by *Campylobacter, Listeria, Shigella* and *Yersinia*; however, those caused by Shiga toxin-producing *Escherichia coli* 057 and *Salmonella* did not decrease significantly, while *Vibrio* infections increased (Centers for Disease Control and Prevention [CDC], 2007).

Undoubtedly, the major threat to food safety is the emergence of "new" pathogens. The recent role of *Listeria monocytogenes, E. coli* O157:H7, *Campylobacter jejuni, Y. enterocolitica* and *Vibrio parahemolyticus* as foodborne

G. Vignolo
Centro de Referencia para Lactobacilos (CERELA), CONICET. Chacabuco 145, T4000ILC Tucumán, Argentina

microorganisms has been related to the increase in outbreaks when compared to traditional food pathogens (Church, 2004; Elmi, 2004). Modern life conditions, with some of them related to or being the result of globalization, ensure that factors responsible for disease emergence are more strongly prevalent than ever. Even when the categorization of these factors is somewhat arbitrary, some of the main ones are ecological changes, intensive agriculture, anomalies in the climate, human demographical changes and behavior, travel and commerce, technology and industry, microbial adaptation and change and the breakdown of public health measures. Further on, concerning pathogens, their most striking feature is their diversity and selection for drug resistance, suggesting that infections will continue to emerge and will probably increase, which emphasizes the urgent need for effective surveillance and control (Morse, 2004, 2007). Additionally, considering that most of the meat consumed in developed countries is produced in developing countries, the existent gap between them in monitoring and surveillance systems for quality and safety of foods can have a significant impact on foodborne illnesses (Cahill & Jouve, 2004).

On these basis, the need for solutions regarding the hygienic quality of food is stated. Consumers are increasingly demanding food that is free from pathogens, with minimal processing and fewer preservatives and additives, but keeping its sensorial quality. As a response to these conflicting demands, current trends in the food industry include the investigation of alternative inhibitors for use in foods. Biopreservation has gained increasing attention as a means of naturally controlling the shelf life and safety of foods. The use of bioprotective cultures to ensure the hygienic quality of food is a promising tool. Even though it was highlighted by Holzapfel, Geisen, & Schillinger, (1995), it should be considered only as an additional measure to good manufacturing, processing, storage and distribution practices. Some microorganisms, commonly associated with meats, have shown antagonism towards pathogenic and spoilage bacteria. Lactic acid bacteria (LAB) have a major potential for use in biopreservation because they are safe to consume (GRAS status) and during storage they naturally dominate the micro-biota in many foods. During the last 25 years knowledge of LAB antimicrobial peptides and bacteriocins, has dramatically improved; however, their application has not met with equal success. Their somewhat limited host range application, the food composition factors affecting bacteriocin effectiveness and the restrictive legislation concerning food additives are among the reasons that might explain the lack of industrial applications of bacteriocins. From a regulatory perspective, nisin, a bacteriocin produced by *Lactococcus lactis*, is the only bacteriocin approved by the FDA for use in more than 50 countries. It has gained widespread application in the food industry and it is also the most extensively studied bacteriocin. Thus, an alternative approach to introduce bacteriocins in food is the use of live LAB cultures that produce bacteriocins *in situ* in the food. In this case, the bacteriocin-producing strains are either substituted for the whole starter culture, a part of it or are subsequently applied to the food as a bioprotective culture to improve the safety of the product. In this contribution, the use of antagonistic microorganisms to inhibit and/or inactivate pathogens and spoilage micro-biota in meat and meat products is discussed, with particular reference to the use of bacteriocin-producing LAB to kill pathogens and/or contaminants as well as spoilage LAB present in these products.

Microbiology of Meat and Meat Products

The characteristic microbial populations that develop in meat and meat products are the result of the effect of the prevailing environmental conditions on the growth of the types of microbes initially present in the raw materials or introduced by cross-contamination or processing. The intrinsic and extrinsic factors governing microbial growth will determine the type and number of bacteria present in meat, these factors being predominantly chemical (concentration and availability of nutrients, pH, redox potential, buffering capacity, a_w, meat structure) and related to storage and processing conditions, respectively. The extrinsic factors are often manipulated to extend the shelf life of meat products with temperature and oxygen availability being the major parameters. Hygiene during slaughter and dressing of carcasses together with prompt and adequate cooling are of major importance for meat quality and safety.

Raw Meat

The microbiology of meat carcasses is highly dependent on the conditions under which animals were reared, slaughtered and processed; the condition of animals at slaughter, contamination spread during the process and time-temperature of storage are important factors that will determine the microbiological quality of meat (Chang, Mills, & Cutter, 2003; Fegan, Vanderlinde, Higgs, & Desmarchelier, 2004; Savell, Mueller, & Baird, 2005). Regional and seasonal differences in the prevalence of pathogens on the hides of cattle presented for harvest at commercial beef processing plants were also suggested; hide data reflecting the regional prevalence while carcass data being indicative of differences in slaughter practices (Rivera-Betancourt et al., 2004). After carcasses dressing, the micro-biota comprises a mixture of mesophiles and psychrotrophs, which are gradually selected during chilling of meat; mesophiles growth will no longer occur and a psychrotrophic micro-biota will develop. Since most pathogens are mesophiles, meat obtained in good hygienic conditions would not be expected to pose sanitary risk. Carcass decontamination strategies using sanitizing solutions have been successfully applied for pathogen reduction in beef carcass, especially when combined with pre-chill treatments (Castillo, Lucia, Mercado, & Acuff, 2001; Kanellos & Burriel, 2005; Ransom et al., 2003).

Salmonellae are typically intestinal pathogens and represent an important organism of public health significance. High levels of *Salmonella* in the meat may arise for animal production practices at the rearing stage as well as cross-contamination after slaughter either at the abattoir or at the retail level (Beach, Murano, & Acuff, 2002).

Salmonella prevalence in beef carcasses at different stages through the beef meat production as well as trimmings was reported to be in the range of 3–12% (McEvoy, Doherty, Sheridan, Blair, & McDowell, 2003), while a lower detection impact (0–0.2% of beef carcasses) was obtained after a recent national survey in Australia (Phillips, Jordan, Morris, Jenson, & Sumner, 2006). In pig slaughterhouses, the occurrence of *Salmonella* showed high variability with average values between 1.7% and 12% for pork carcasses while a higher incidence in environmental samples was reported (Hald, Wingstrand, Swanenburg, von Altrock, & Thorberg, 2003; Pala & Sevilla, 2004; Rodriguez, Pangloli, Richards, Mount, & Draughton, 2006). Food-poisoning *Staphylococci* are also widely distributed, with meat contamination being generally associated with highly manual handled foods and for the most part the aetiologic agent is *Staphylococcus aureus* and its related heat stable enterotoxins (Balaban & Rasooly, 2000). Although the microbiological quality of beef is highly dependent on slaughterhouses, dressing operations and hygiene of processing lines, the hands of workers and environmental sites associated with the evisceration process were the principal sources (Desmarchelier, Higgs, Mills, Sullivan, & Vanderlinde, 1999; Shale, Jacoby, & Plaatjies, 2006). The presence of *Staphylococcus* in bio-aerosols from red-meat abattoirs as well as bovine mastitis in dairy cows may also constitute a risk of foodborne pathogen contamination (Pitkala, Haveri, Pyorala, Myllys, & Honkanen-Buzalski, 2004; Shale, Lues, Venter, & Buys, 2006). Pig carcasses are an important source of contamination with *S. aureus* mainly due to the sequential steps of slaughter (Pala & Sevilla, 2004; Spescha, Stephan, & Zweifel, 2006). Time and temperature abuse of a food product contaminated with enterotoxigenic *S. aureus* can result in enterotoxin formation; its occurrence in foods of animal origin was reported to be extremely variable (Holeckova et al., 2002).

Enterohemorrhagic shiga toxin-producing *E. coli* (STEC) emerged as a foodborne pathogen more significantly than other well-known ones because of the severe consequences of infection, its low infection dose, its unusual acid tolerance and its apparently special but inexplicable association with ruminants used for food. This pathogen causes serious complications in humans such as hemorrhagic colitis and hemolytic uremic syndrome (Griffin & Tauxe, 1991). The increased prevalence of *E. coli* O157:H7 in foods may be associated with the consolidation of the beef industry into fewer but larger production and processing units. Reports from France, the UK, the USA, Canada, Japan, Australia, and Argentina confirmed that cattle are the major reservoir of STEC pathogenic for humans with large prevalence rates (0.2–27.8%) and a high incidence of hemolytic uremic syndrome (Hussein & Bollinger, 2005; Omisakin, MacRae, Ogden, & Strachan, 2003). In particular, young beef steers from the beef-producing areas in Argentina were reported to be important reservoirs of STEC strains (Meichtri et al., 2004), especially the hides and feces of animals at slaughter. Even when mishandling practices in processing plants are highly responsible for contamination of meat and meat products (Normano et al., 2004), intervention technologies for reducing *E. coli* O157:H7 on beef carcasses, cuts and trimmings proved to be effective in beef packaging/processing industry (Marshall, Niebuhr, Acuff, Lucia, & Dickson, 2005; Ransom et al., 2003). On the other hand, *Listeria monocytogenes* has continued to raise food safety concerns, especially with respect to ready-to-eat (RTE) products. Listeriosis, caused by this pathogen, is a significant public-health concern and as a result of its clinical severity and high mortality rates, world-wide foodborne outbreaks in which meat products were implicated have occurred (Okutani, Okada, Yamamoto, & Igimi, 2004; Vaillant et al., 2005). Due to its ubiquitous character, *L. monocytogenes* can grow at temperatures ranging from 1 to 45°C, at a pH between 4.6–9.6, in the presence of high salt concentrations, and can survive attached to the processing equipment by biofilm formation (Autio et al., 1999). A higher incidence of *Listeria* spp. on ground meat than on carcasses or boneless meat cuts has been reported; its prevalence being mainly due to cross-contamination increasing from the farm to the manufacture plants (Lin et al., 2006; Thevenot, Dernburg, & Vernozy-Rozand, 2006). The slaughterhouse environment constitutes the primary source of *Listeria*- contaminated meat products while human contact probably accounts for a portion of this contamination (Marzoca, Marucci, Sica, & Alvarez, 2004; Peccio, Autio, Korkeala, Rosmini, & Trevisani, 2003). High average values for *Listeria* spp and *L. monocytogenes* (73% and 75%, respectively) were reported in frozen beef (Hassan, Purwati, Radu, Rahim, & Rusul, 2001). Although many risk assessment strategies have been developed to control *L. monocytogenes* in foods (Chen, Ross, Scott, & Gombas, 2003; ILSI Research Foundation; Risk Science Institute, 2005; Salvat & Fravalo, 2004), data suggest that this pathogen cannot be completely eliminated from the environment and food products.

Species within the genus *Campylobacter* and *Yersinia* have also emerged as pathogens of human public health concern (Centers for Disease Control and Prevention [CDC], 2002). *Campylobacter* is the most common cause of human foodborne illness in the US because of its high degree of virulence and its widespread prevalence in foods of animal origin. This pathogen has been isolated from beef at retail sale, indicating that beef can serve as a potential vehicle for its transmission to humans. *C. jejuni* can be transferred from hides to meat during slaughter and dressing of beef carcasses (Inglis, Kalischuk, Busz, & Kastelic, 2005. Even after high initial carcass prevalence, chilling showed to be an efficient critical control point to eliminate *Campylobacter* from carcass surfaces. *Y. enterocolitica* is also frequently associated with pigs and pork products and can be transmitted to humans by consumption of raw, undercooked or re-contaminated processed meats. The prevalence of *Y. enterocolitica* in pig herds has been reported to range between 40 to 65%; tonsils and oral cavity being important reservoirs (Bhaduri, Wesley, & Bush, 2005; Gurtler, Alter, Kasimir, Linnebur, & Fehlhaber, 2005). Other pathogens of human health concern that may be present but remain undetected in slaughtered animals involve streptococci, clostridia and corinebacteria. Among these, *Clostridium perfringes* is a leading cause of bacterial foodborne illness in countries where consumption of meat and poultry is high. Meat animals are subjected to a number of clostridial diseases and these bacteria may be present in carcasses. The consumption of red meats was implicated in infectious intestinal outbreaks in which *C. perfringes* was the most frequently reported organism (Smerdon, Adak, O'Brien, Gillespie, & Reacher, 2001).

After slaughter and dressing of carcasses, bacterial growth will be depend on the storage conditions. During storage, environmental factors such as temperature, gaseous atmosphere and meat pH will allow certain bacteria to grow. Cold storage of meat will decrease bacterial growth; only 10% of the bacteria initially present being able to grow at refrigeration temperatures. Oxygen restriction by the use of vacuum or modified atmospheres will drastically reduce the presence of *Pseudomonas* and bacterial flora will be gradually selected towards CO_2-tolerant organisms. Under these conditions the dominating microorganisms include: *Brochothrix thermosphacta, Enterobacteriaceae* and LAB. A pattern of succession was reported during a 16-week storage of vacuum-packaged beef at −1.5°C between strains of *Carnobacterium, Lactobacillus, Leuconostoc* and *Pediococcus* (Jones, 2004). In addition, when microbial diversity by direct molecular approach was studied on vacuum-packaged Argentinean raw beef, *Lactobacillus sakei, Lactobacillus curvatus, Leuconostoc gelidum* and *Leuconostoc carnosum* were determined as the dominant LAB species (Fontana, Cocconcelli, & Vignolo, 2006). Nevertheless, *B. thermosphacta* was also found to be a numerical significant component of the micro-biota of meat and meat products stored under these conditions (Fontana et al., 2006; Sakala et al., 2002). LAB and *B. thermosphacta* significantly influence the quality of meat and meat products, both being associated with their spoilage. The effect of storage temperature and gas permeability of the packaging film on the growth of LAB and *B. thermosphacta* also determines the quantities and the type of microorganisms growing in these conditions (Cayré, Garro, & Vignolo, 2005).

Meat Products

Fermented dry sausages are known as "shelf stable products". This term refers to those products that do not require refrigeration or freezing for safety and acceptable organo-leptic characteristics since most often they are stored at room temperature. Different species of LAB and Gram-positive coagulase-negative cocci (GCC) are the microorganisms primarily responsible for sausage fermentation. Molds and yeast also play an important role by bringing about characteristic surface appearance and flavors. It is well known that LAB, in particular lactobacilli play an important role in meat fermentation processes. They contribute to the hygienic and sensory quality of meat products mainly through carbohydrates and protein catabolism resulting in pH reduction, production of antimicrobial agents such as organic acids, inhibitory peptides or bacteriocins and generation of flavor compounds (Demeyer, 2003; Lücke, 2000; Talon, Leroy-Sétrin, & Fadda, 2003). On the other hand, GCC participate in color development and stabilization through a nitrate reductase activity that leads to the formation of nitroso-myoglobin. In addition, their antioxidant potential due to catalase activity, aroma formation as well as removal of excess nitrate in the meat batter will contribute to achieve a stable fermented product (Tjener, Stahnke, Andersen, & Martinussen, 2004). In spontaneously fermented European sausages, homo-fermentative lactobacilli constitute the predominant flora throughout ripening. Recently, studies carried out on small-scale processing facilities confirmed, on the basis of their genetic fingerprints, that *Lb. sakei* and *Lb. curvatus* were the predominant species isolated from artisanal fermented sausages (Fontana, Cocconcelli, & Vignolo, 2005; Rantsiou et al., 2005). The commercial starter cultures in Europe are generally made up of a balanced mixture of LAB (*Lactobacillus, Pediococcus*) and GCC (*Staphylococcus, Kocuria*). In Mediterranean countries, yeasts (*Debaryomices, Candida*) and moulds (*Penicillium*) are also used to inoculate the surface of sausages.

Apart from LAB, GCC, molds, and yeasts involved in sausage fermentation, beef and pork meat as the major components of dry cured sausages regularly contain pathogenic bacteria and are often implicated in the spread of foodborne diseases. Raw dry sausage materials (meat and casings) are the principal vehicles for pathogens and contaminating microorganisms. Food-poisoning *S. aureus, Salmonellae* and *C. perfringens* have been traditionally implicated in fermented dry sausage (Holley, Lammerding, & Tittiger, 1988; Wen & McClane, 2004). On the other hand, the emergent pathogens within the genera *Campylobacter* and *Yersinia* as well as *L. monocytogenes* and shiga toxin-producing *E. coli* have also been involved in outbreaks caused by fermented sausage (Centers for Disease Control and Prevention [CDC], 1995; Normano et al., 2004; Thevenot et al., 2006; Tilden et al., 1996). Particularly, since there is no epidemiological evidence for the involvement of fermented sausages in recent outbreaks of listeriosis, it has been recommended that up to 100 cells of *L. monocytogens* per gram of this product can be tolerated (ICMSF, 2002).

In the last 20 years, as a consequence of increasing consumer demands, the search by the meat industry, for innovative solutions to market requirements, resulted in a wide spectrum of new different RTE meat based products. Processed meat that refers to products subjected to a technological process such as heat treatment or a decrease in water activity, involves whole meat pieces as well as comminuted RTE meat products. The microbiological status of processed meat products is highly dependent on raw materials, hygiene and handling procedures. Cooked meat and poultry products are generally either pre-packaged in a permeable over-wrapping film or cut from a bulk block and may become contaminated after production or processing by retailers and caterers. Even when plant sanitation strategies and processing practices are adequate and the product has been correctly handled after cooking, they are often insufficient to prevent contamination with *L. monocytogenes, S. aureus, Salmoneallae, C. perfringes* and STEC. They are likely to be present as a result of post-processing contamination principally due to poor cooling procedures and mishandling practices. Even when most processed meat products undergo post-packaging thermal pasteurization, contamination at the end of shelf life can occur. In particular, *L. monocytogenes* contamination typically develops from post-thermal process mishandling, because this pathogen is normally killed by the thermal process used for RTE meats. Considering that *L. monocytogenes* is more heat resistant than *Salmonella* and *E. coli* O157:H7, pasteurization treatments for *L. monocytogenes* may have additional benefits by reducing the levels of other pathogens on processed meats. However, the presence of *S. aureus* (Ingham et al., 2005); *E. coli* O157:H7 (Sagoo, Little, Allen, Williamson, & Grant, 2007); *C. perfringes* (Kalinowski, Tompkin, Bodnaruk, & Pruett, 2003), and *L. monocytogenes* (Gombas, Chen, Clavero, & Scott, 2003; Sagoo et al., 2007) have been reported in cooked RTE meat products.

Shelf Stability and Hurdle Presence in Meat and Meat Products

The microbial safety and stability as well as the nutritional and sensory quality of meat and meat products are based on the application of combined preservative factors or "hurdles". The hurdle technology, derived from the understanding of the hurdle effect (Leistner, 1997, 2000), refers to the deliberate combination of existing (temperature, a_w, Eh, preservatives) and novel preservation techniques (gas packaging, bacteriocins) in order to establish a series of more selective preservative factors (hurdles) that spoilage and pathogenic microorganisms should not be able to overcome. The extrinsic factors often manipulated to extend the shelf life of meat are mainly concerned with storage and processing conditions and can be considered as a set of hurdles applied for its stabilization. Since temperature is the most important factor affecting the micro-biota of meat, cooling of carcasses will be the first hurdle that spoilage bacteria have to overcome during meat conditioning. Nowadays, meat trade throughout the world is carried out in vacuum packed controlled atmosphere packs, the evolution of the raw meat trade relying upon low temperature and modified packaging as the most important hurdles. On the other hand, the range of minimally processed RTE meat products available in the marketplace continues to increase. Furthermore, a number of foods that are packaged under anaerobic conditions rely on refrigeration as the sole defense against pathogen growth. Fermented meat products (salami and salami-type sausage) become stable and safe through a sequence of hurdles, some of which are specifically included ($NaCl$, $NaNO_2/NaNO_3$, ascorbate) while others are indirectly created in the stuffed mix (low Eh, antagonistic substances, low water activity). The effective hurdles, inhibitory to pathogens, present in fresh meat, fermented meat as well as in RTE meat products are shown in Table 18.1.

Bioprotective Cultures

The preservation of foods using their natural and controlled micro-biota and/or their antimicrobial metabolites has been termed bioprotection or biopreservation to differentiate it from artificial (chemical) preservation (Stiles, 1996). Antagonistic cultures added to foods to inhibit pathogens and/or extend the shelf life while changing the sensory properties as little as possible are called protective cultures (Lücke, 2000). The main objective of this concept is storage life extension as well as the enhancement of food safety. LAB have a major potential for use in biopreservation because they are safe for consumption (GRAS), and during storage

Table 18.1 Effective hurdles inhibitory to pathogens present in fresh meat, fermented meat and RTE meat products

	Hurdles		
	Fresh meat	Fermented meat products	RTE meat products
Brochothrix thermosphacta	Bioprotection	LAB (bacteriocins); pH < 5.8	Bioprotection; HHP/bacteriocins
Campylobacter jejuni	Chilling/Sanitizers/OAS; LTF; VP/MAP	LAB (acid); NaCl; $NaNO_2$	Pasteurization; HHP/bacteriocins; LST; VP/MAP
Clostridium perfringens	Chilling/Sanitizers; Bioprotection	LAB (acid, bacteriocins)	Pasteurization; Bioprotection; LST
E. coli O157:H7	Chilling/Sanitizers/OAS; LAB (acid, bacteriocins/EDTA); VP/MAP	LAB (acid); $a_w < 0.80$; pH < 5.0	Pasteurization; HHP/bacteriocins; LST; VP/MAP
LAB spoilage	Bioprotection	Bioprotection	Pasteurization; Bioprotection; HHP/bacteriocins
L. monocytogenes	Chilling/Sanitizers; Bioprotection; VP/MAP	LAB (acid, bacteriocins); $a_w < 0.90$; spices; NaCl	Pasteurization; LST; HHP; OAS; Bioprotection; VP/MAP
Salmonella	Chilling/Sanitizers/OAS; VP/MAP; LAB (bacteriocins/EDTA)	pH < 5.0; $a_w < 0.95$; $NaCl/NaNO_2$; HHP	Pasteurization; HHP/bacteriocins; LST; VP/MAP
S. aureus	Chilling/Sanitizers; Bioprotection; VP/MAP	pH < 5.1; $a_w < 0.86$; LAB (bacteriocins)	Pasteurization; LST; VP/MAP; LAB (bacteriocins)
Yersinia enterocolitica	Chilling/Sanitizers/OAS; VP/MAP	LAB (acid)	Pasteurization; HHP/bacteriocins; LST; VP/MAP

LST: Low storage temperature; LTF: Long term freezing; HHP: High hydrostatic pressure; VP: Vacuum packaging; MAP: Modified atmosphere packaging; LAB: Lactic acid bacteria; OAS: organic acid solutions

they naturally dominate the micro-biota of many foods. In addition, antimicrobial peptides produced by LAB can be easily broken down by digestive proteases so as not to disturb the gut micro-biota. LAB can exert a bioprotective or inhibitory effect against other microorganisms as a result of competition for nutrients and/or of the production of bacteriocins or other antagonistic compounds such as organic acids, hydrogen peroxide and enzymes. A distinction can be made between starter cultures and protective cultures in which metabolic activity (acid production, protein hydrolysis) and antimicrobial action constitute the major objectives, respectively. Food processors face a major challenge with consumers demanding safe foods with a long shelf life, but also expressing a preference for minimally processed products, less severely damaged by heat and freezing and not containing chemical preservatives. Bacteriocins constitute an attractive option that could provide at least part of the solution; they also have implications for the development of desirable flora in fermented foods by conferring a rudimentary form of innate immunity to foodstuffs and helping processors extend their control over the food flora long after manufacture (Cotter, Hill, & Ross, 2005). Bacteriocins can be introduced into food in at least three different ways: (i) by *in situ* produced bacterial cultures; (ii) by direct addition of purified or semi-purified bacteriocins to food or (iii) as an ingredient based on a fermentate of a bacteriocin-producing strain.

Bacteriocins Produced by LAB

The preservative ability of LAB in foods is attributed to the production of antimicrobial metabolites including organic acids and bacteriocins. Acid production is a common feature among LAB; however, not all LAB can produce antimicrobial peptides during growth. The production of these peptides seems to be a common phenotype among LAB since numerous bacteriocins have been isolated over the last three decades, varying in size from small (<3 kDa), post-translationally heavily modified peptides to large heat labile proteins. Bacteriocins produced by LAB are a heterogeneous group of peptides and proteins. The last proposed revised classification scheme divides bacteriocins into two categories: the lanthionine-containing lantibiotics (class I) and the non-lanthionine-containing bacteriocins (class II), while moving the large, heat-labile murein hydrolases (formerly class III bacteriocins) to a separate designation called bacteriolysins (Cotter et al., 2005). These small peptides produced by LAB are secreted into the microbial environment where they may cause the inhibition of spoilage bacteria and foodborne pathogens. Despite the dramatic increase in the number of novel bacteriocins discovered in the past two decades, biochemical and genetic characterization has only been carried out on some of them. Most bacteriocins identified to date and produced by LAB belong to classes I and II. Although they are generally heat resistant, they can be inactivated by proteolytic enzymes in foods and due to their hydrophobic character, can be bound by fats and phospholipids. The bacteriocins of interest to the meat industry are class II bacteriocins, which involve class IIa pediocin-like or *Listeria*-active and class IIb two-peptide bacteriocins. A bacteriocinogenic culture should be well adapted to grow in the product, survive throughout its storage, compete with the relatively high indigenous microbial loads of raw meat, actively inhibit pathogenic and spoilage bacteria, and not alter the sensory properties of meat except under temperature-abuse conditions. Moreover, the use of bioprotective cultures in refrigerated vacuum-packaged meat and meat products would be more feasible from an economic point of view, and without legal restrictions, compared to the addition of the bacteriocins produced. The use of bacteriocinogenic LAB may be a better choice for introducing bacteriocins into meats because they can provide a source of bacteriocins over a longer period of time.

Bioprotection Applied to Raw Meat

LAB growth on meat is considered as a "hidden" fermentation, since due to the low carbohydrate content and the strong buffering capacity of the meat these bacteria do not produce a dramatic change in sensory characteristics. Since meat cannot be pasteurized prior to the addition of LAB cultures, a biopreservative culture in the meat must compete with the natural micro-biota present. LAB that grow at refrigeration temperatures are the natural psychrotrophic species that prevail in meat ecosystems. Decontamination of carcasses with organic acid solutions has been studied without apparent success (Castillo et al., 2001; Kanellos & Burriel, 2005); whereas decontamination of meat surfaces involving a mixture of nisin (500 IU/ml) and lactic acid applied on beef carcasses at specific points during slaughter resulted in a significant reduction in bacterial populations (total coliforms and *E. coli*) by 2 log units (Barboza de Martínez, Ferrer, & Salas, 2002).

Systems for retail meat distribution are principally based on vacuum-packaging of meat cuts using low gas permeability films and refrigeration. Under these conditions, CO_2 concentration in the pack is critical for microbial inhibition (Phillips, 1996). For maximum antimicrobial effect, the storage temperature of a modified atmosphere packaged (MAP) product should be kept as low as possible, since a dramatic decrease in CO_2 solubility is produced when temperature increases. Under these conditions, the emergent psychrotrophic pathogens such as *L. monocytogenes* and the Gram-negative *Y. enterocolitica* and *Aeromonas hydroophila* have been observed to grow (Church & Parsons, 1995). An additional hurdle to reduce the risk of *L. monocytogenes* would be the use of competitive-bacteriocin producing cultures referred to as bioprotective cultures.

Lactobacillus curvatus CRL705 in Vacuum-Packaged Fresh Beef Stored at Refrigeration Temperatures

Lb. curvatus CRL705 isolated from artisanal Argentinean sausages produces lactocin 705 and lactocin AL705; a two-peptide (class IIb) and a pediocin-like (class IIa) bacteriocin, respectively. Lactocin 705 is a bacteriocin whose activity depends on the complementary action of two peptides (lac705α and lac705β) of 33 amino acid residues each. The genetic characterization as well as its mechanism of action against target microorganisms have been reported (Castellano, Raya, & Vignolo, 2003; Castellano, Vignolo, Farías, Arrondo, & Chehín, 2007; Cuozzo, Castellano, Sesma, Vignolo, & Raya, 2003). Lactocin 705 was found to be active against closely related LAB and *B. thermosphacta* while lactocin AL705 is an anti-*Listeria* bacteriocin (Castellano, Holzapfel, & Vignolo, 2004; Castellano & Vignolo, 2006). From an application point of view, the use of bacteriocinogenic LAB strains as bioprotective cultures for commercial meat preservation would be more feasible, since nisin is currently the only bacteriocin approved for use in foods. Consequently, the effect of the bacteriocin-producing strain *Lb. curvatus* CRL705 as well as the produced bacteriocins lactocin 705 and lactocin AL705 were first evaluated on the control of *Listeria innocua* 7 and the deteriorative strain *Lb. sakei* CRL1424 (isolated from spoiled vacuum-packaged raw meat) in a meat slurry during storage under vacuum at chill temperatures. Results indicated that the bioprotective culture *Lb. curvatus* CRL705 was able to grow in the meat system, to compete with the indigenous LAB micro-biota, retaining its inhibitory effect against *Listeria* at low temperatures (Fig. 18.1a and b), (Castellano et al., 2004). The inhibitory effect of *Lb. curvatus* CRL705 as a biopreservative culture and of its bacteriocins against *L. innocua*, *B. thermosphacta* and indigenous LAB was compared in vacuum-packaged bovine meat discs stored at 2°C. When the growth of psychrophilic micro-biota, naturally present in the meat, was evaluated in the absence and presence of lactocin 705, final cell counts of 5×10^6 CFU/cm^2 and 5×10^4 CFU/cm^2 after 36 days were obtained respectively, showing a 2 log reduction on the meat (Fig. 18.2a). On the other hand, *Lb. curvatus* CRL705 used as a bioprotective culture as well as lactocin 705 were shown to be similarly effective in preventing the growth of *L. innocua* in raw meat discs throughout the storage period (Fig 18.2b), while *B. thermosphacta* experienced a reduction of 1.5 log cycles after the addition of either the protective culture or lactocin 705 (Fig. 18.2c). A similar listeriostatic effect was reported by Katla et al. (2002), when comparing the anti-*Listeria* activity of sakacin P and the sakacin P-producing *Lb. sakei* strain on chicken cold cuts. Moreover, Hugas, Pages, Garriga, & Monfort, (1998), working

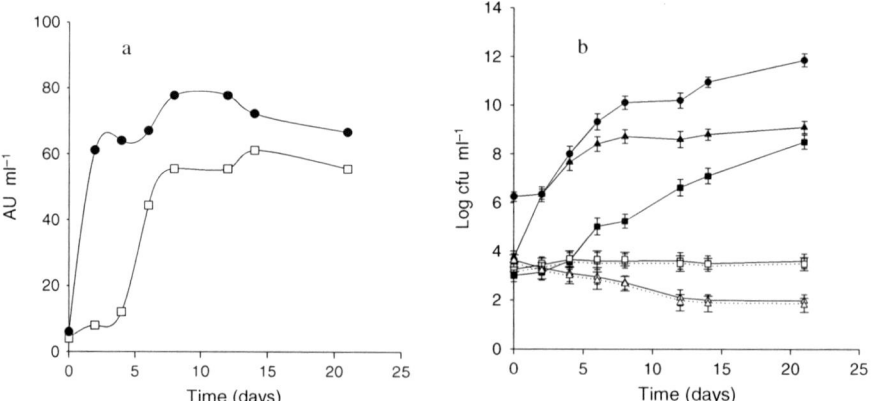

Fig. 18.1 Bacteriocin activity and growth of *Lb. sakei* CRL1424 and *L. innocua* 7 in mixed culture with *Lb. curvatus* CRL705 at 4°C in meat slurry. (a) Bacteriocin activity in AU ml^{-1} of supernatant of *Lb. curvatus* CRL705 against *L. innocua* 7 (□) and *Lb. plantarum* CRL691 (●). (b) Growth of *Lb. curvatus* CRL705 (●), *Lb. sakei* CRL1424 (▲) and *L. innocua* 7 (■) alone, and *Lb. sakei* CRL1424 (△) and *L. innocua* 7 co-inoculate with *Lb. curvatus* CRL705 (□). Dashed lines indicate growth of *Lb. sakei* CRL1424 and *L. innocua* 7 in combined cultures (*Lb. sakei* CRL1424 + + *L. innocua* 7 CRL705)

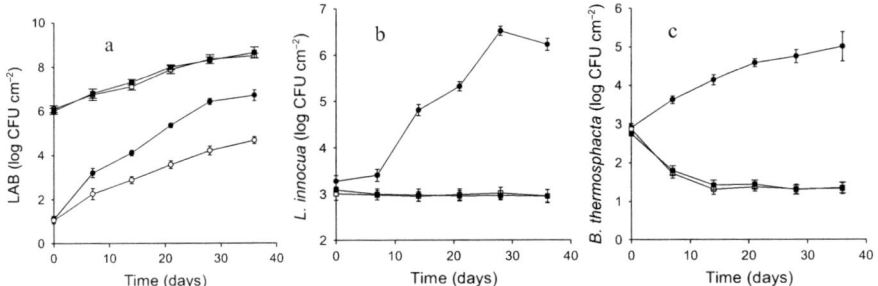

Fig. 18.2 Growth of LAB, *L. innocua* and *B. thermosphacta* on meat discs incubated at 2°C for 36 days. (a) control (●); lactocin 705 addition (○); *Lb. curvatus* CRL705 growth in the presence/absence of spectinomycin (□)/(■). (b) control (●); addition of *Lb. curvatus* CRL705 (■) and lactocin AL705 (□). (c) control (●); addition of *Lb. curvatus* CRL705 (■) and lactocin 705 (□). The numbers represent duplicates of treated and control samples

on vacuum-packaged fresh meat products also found that *Lb. sakei* CTC494 and the bacteriocin produced, sakacin K, exerted a listeriostatic effect. On the other hand, Jacobsen, Budde, & Koch, (2003) showed that the live cells of *Lc. carnosum* 4010 were more effective than leucocins alone for growth inhibition of *L. monocytogenes* in sliced meat products. In contrast, the addition of *Enterococcus faecium* CTC492 to fermented sausages and hamburgers failed to exert any positive anti-listerial effect compared to the batches treated with enterocins A and B, this being attributed to the high inhibition of the producer strain by refrigeration and sausage ingredients (Aymerich, Garriga, et al., 2000). In addition, *B. thermosphacta* population was also reduced by *Leuconostoc mesenteroides* L124 and *Lb. curvatus* L442 or their bacteriocins when they were added to cooked cured meat products under vacuum or MAP conditions at 4°C. The complete inhibition of *B. thermosphacta* on vacuum-packaged fresh beef treated with nisin was also reported by Tu & Mustapha (2002). Inoculation of slices of beef *M. semitendinous* with the bacteriocin-like inhibitory substance producer, *Lb. sakei* CECT4808, provided an additional hurdle to improve the refrigerated storage life without causing any sensorial alteration (Katikou, Ambrosiadis, Georgantelis, Koidis, & Georgakis, 2005). The control of indigenous LAB in meat without compromising its microbiological safety would provide shelf life extension. Since parts of the naturally growing LAB are suppressed, the addition of a selected bioprotective culture could be an effective way of controlling undesirable organo-leptic changes in vacuum-packaged meat.

Bioprotection Applied to Fermented Meat Products

Even when dry fermented sausage processing conditions, curing additives, and the presence of LAB starter cultures act as effective hurdles for pathogen control, they are not sufficient to prevent the survival of *L. monocytogenes* and STEC during the manufacturing process. The use of bacteriocinogenic strains as starter cultures to reduce the risk of *L. monocytogenes* has been suggested as an extra hurdle to ensure product safety and quality. Besides, strategies to overcome the barrier presented by the outer membrane in Gram-negative bacteria such as the addition of chelating agents could improve bacteriocin efficiency (Belfiore, Castellano, & Vignolo, 2007). At present only nisin and pediocin PA-1/AcH are widely used in foods. Nisin is used mostly as Nisaplin (Danisco), a preparation containing 2.5% nisin. Pediocin PA1 for food biopreservation has also been commercially exploited as ALTA 2431 (Quest), based on a fermentate from a pediocin PA-1-producing strain of *Pediococcus acidilactici*; its use being covered by several US and European patents. Among class Ia bacteriocins, nisin has not been quite successful in meat products because of its low solubility, uneven distribution and lack of stability (Aymerich, Hugas, & Monfort, 1998).

Even when lactococcal bacteriocins seem to be not particularly adapted to sausage technology, the use of lacticin 3147-producing strains as starter cultures in the production of salami showed that the bacteriocin was produced throughout manufacture and that the final overall characteristics were similar to the control. The protective ability of this lantibiotic was demonstrated by the significant reduction in *L. innocua* and *S. aureus* in spiked sausages (Scannell, Ross, Hill, & Arendt, 2000). The effect of *in situ* production of a bacteriocin by *Lactococcus lactis* M on *L. monocytogenes* during fermentation and storage of "merguez" sausage showed a greater reduction of this pathogen in samples fermented with the bacteriocinogenic strain (Benkerroum, Daoudi, & Kamal, 2003). Nevertheless, nisin producers isolated from fermented meat products seem to have some ecological advantages that may improve their effectiveness in these products. Among bacteriocinogenic strains producing class IIa bacteriocins, *Lb. sakei*, sakacin producers, have been assayed in *in situ* experiments, in which they demonstrated a strong activity against *Listeria*. Particularly, in fermented dry sausages, inoculation with the bacteriocinogenic strains *Lb. sakei* CTC494 and *Lb. curvatus* LTH1174 producer of sakacin K and curvacin A respectively, showed a reduction in *L. ivanovii* by more than 1-4 log cycles compared to a non-bacteriocinogenic control (Hugas, Garriga, Aymerich, & Monfort, 2002). In addition, the inhibition of *L. monocytogenes* in ostrich meat salami by using bacteriocin-producing starter cultures of *Lb. plantarum* and *Lb. curvatus* strains was also reported (Dicks, Mellet, & Hoffman, 2004). On the contrary, the inoculation in sausages as a starter culture of the sakacin K producer, *Lb. sakei* LTH673, isolated from meat failed to lead the fermentation, although it reduced the number of *Listeria* compared to the non-bacteriocinogenic starter (Hugas, Neumeyer, Pagés, Garriga, & Hammes, 1997). Pediocin-producing strains of *P. acidilactici* were used to reduce the counts of *L. monocytogenes* by 1–2.5 logs more than a non-pediocin producing strain during summer sausage fermentation (Foegeding, Thomas, Pilkington, & Klaenhammer, 1992). Moreover, when the survival of *L. monocytogenes* was compared in German, American and Italian-style fermented sausages using pediococci as a starter culture, higher reductions in *Listeria* numbers were observed in the American–style sausages after fermentation and drying. The higher fermentation temperature used in American-style sausages (25–35°C) may have a positive influence on the production of bacteriocins by *P. acidilactici*.

Although the use of bacteriocin-producing enterococci in cheese manufacture is well documented, information on their potential as biopreservatives in the meat industry is scarce. Bacteriocin-producing *E. faecium* strains have been included in the starter cultures in Spanish-style dry fermented sausage to effectively inhibit *Listeria* (Callewaert, Hugas, & De Vuyst, 2000) The bacteriocin-producing *Enterococcus casseliflavus* IM416K1, isolated from Italian "cacciatore" sausage, showed a high effectiveness in *L. monocytogenes* elimination while a listeriostatic effect was obtained when enterocin CCM4231, produced by a strain of *E. faecium*, was added to the sausage batter during the manufacture of Hornád salami, a dry fermented traditional Slovakian product (Lauková, Czikková, Laczková, & Turek, 1999; & Sabia, de Niederhausern, Messi, Manicardi, & Bondi, 2003). In contrast, *E. faecium* CTC492, an enterocin A and B producer, was unable to exert any positive anti-listerial effect on fermented sausages compared to the batches treated with enterocins A and B; this being attributed to a high inhibition of the producer strain by refrigeration and sausage ingredients (Aymerich, Artigas, Garriga, Monfort, & Hugas, 2000). Since enterococci are hardly competitive in the early stages of the fermentation in reducing listerial counts, bacteriocin-producing enterococcal strains may be used as adjunct cultures during sausage manufacture.

The use of a bacteriocinogenic strain as a new functional starter culture may offer food safety advantages, without risks for human health due to toxicological side effects. It represents a safe way of natural preservation that has probably occurred in fermented foods for centuries. Moreover, bacteriocin production in fermented sausages does not generally lead to organo-leptic imperfections, as has been shown in sensorial studies with different strains (Coffey et al., 1998). However, some disadvantages have to be considered related to the lower *in situ* bacteriocin activity compared to the one expected from *in vitro* experiments; this being due to low production, genetic instability, inability for uniform distribution throughout the product, low solubility, inactivation by meat proteases, resistance of the target strain, and interference by meat components (Leroy, Verluyten, & De Vuyst, 2006). The use of strains well adapted to the sausage environment, preferably sausage isolates as well as strain compatibility (including GCC) must be considered for optimal performance and bacteriocin production. In this context, it has been shown that the sausage isolates *Lb. sakei* CTC494, *Lb. curvatus* LTH1174 and L442 optimally produce bacteriocins under conditions of pH and temperature that prevail during European sausage fermentation (Leroy et al., 2006). Moreover, a bioprotective culture must be selected according to the specific formulation and technology used in the fermentation. The performance of bacteriocinogenic LAB as starters or co-cultures will depend on the processing technology (pediococci are more suitable than lactobacilli in fast fermentation at high temperatures), the presence of nitrate or nitrate/nitrite, NaCl and spices (Hugas et al., 1998; Verluyten, Leroy, & De Vuyst, 2004; Vignolo, Fadda, Kairuz, Ruiz Holgado, & Oliver, 1998).

Bioprotection Applied to Ready-to-Eat Meat Products

The microbiological status of processed meat products is highly dependent on raw materials and hygiene of the handling procedures. Even when these products are heated to a temperature between 65–75°C and most vegetative cells are killed, post-heat treatment re-contamination will determine their shelf life. Cooked meat and poultry products are generally either pre-packaged in a permeable over-wrapping film or cut from a bulk block and may become contaminated during production or processing at home or food establishments. The low salt content, pH around 6.0 and water activity higher that 0.95 constitute weak hurdles to prevent the growth of organisms associated with post-processing contamination. Even when processing conditions are adequate and the product has been correctly handled after processing, *S. aureus*, *L. monocytogenes*, *Salmonellae* and shiga toxin-producing *E. coli* may be implicated in RTE contamination. Since products such as frankfurters and cooked ham are massively consumed, effective hurdle technologies to inhibit the growth of these pathogens should be implemented.

During the last 20 years, a number of studies have focused on the biopreservation of RTE meat products with LAB, mainly to control the presence of *L. monocytogenes* which has become a major concern for the meat processing industry. Cooked meat products offer no protection against *L. monocytogenes* outgrowth if contamination occurs after post-processing during slicing and/or packaging. This pathogen will grow exponentially even at refrigeration temperatures reaching high numbers within days, the control of *L. monocytogenes* being difficult due to its ability to grow at a wide temperature range (3 to 45°C), pH tolerance (4.8 to 9.6), high NaCl levels (8–9%) and sodium nitrite (< 1000 ppm). The use of meat borne LAB as protective cultures for the biopreservation of cooked meat products has been extensively reported. Effective control of *L. monocytogenes* after refrigerated storage of different RTE meat products was obtained when selected strains of *Lactobacillus* and *P. acidilactici*, previously isolated from the same products, were directly applied on their surface (Amézquita & Brashears, 2002; Vermerien, Devlieghere, & Debevere, 2004). Similarly, the addition of *Lb. sakei* TH1 as a bioprotective culture to cooked, sliced and vacuum-packaged ham and servelat sausage resulted in the inhibition of *L. monocytogenes* at 8 and 4°C during 28 days of storage (Bredholt, Nesbakken, & Holck, 2001). In addition, when the live cells of the bacteriocin producing *Lc. carnosum* 4010 were applied by spray to cooked and sliced pork saveloy, it was also reported to be effective in the control of *L. monocytogenes* during 4 weeks of refrigerated storage (Jacobsen et al., 2003). The anti-listerial effect of bioprotective cultures and/or their bacteriocins will depend on the product composition and the presence of competitive microflora. The addition of enterocins A and B produced by *E. faecium* CTC492 (4,800 AU/g) to cooked ham and pork liver paté to inhibit *L. innocua* showed an effective reduction in its numbers; this being higher in paté than in cooked ham. Without competitive microflora and favorable physico-chemical conditions, *Listeria* growth will be promoted, but the higher efficiency of enterocins in the fatter product (paté) is consistent with the hydrophobic nature of bacteriocins (Aymerich, Garriga, et al., 2000).

Even when studies on biopreservation in cooked meat products have mainly focused on the inhibition of food pathogens such as *L. monocytogenes*, the presence of spoilage LAB as well as *B. thermosphacta* is of concern because of the high economic losses. Product handling after cooking plus slicing prior to packaging re-contaminates them with about 0.5–2 log CFU/g of total bacteria, mainly LAB. When stored under vacuum-packaging or MAP and refrigeration conditions, LAB will dominate the spoilage process. The metabolic activity of LAB results in the spoilage appearing as sour, off-flavors, off-odors, milky exudates, slime production, swelling of the package through gas production and discoloration such as greening (Samelis & Georgiadou, 2000). The spoilage microflora of vacuum packaged cooked meat products as well as salty dry products mainly consists of *Lactobacillus, Leuconostoc, Carnobacterium* and *Weissella* species and *B. thermosphacta* (Koort et al., 2006; Samelis, Björkroth, Kakouri, & Rementzis, 2006; Samelis & Georgiadou, 2000). LAB that are homo-fermentative, salt tolerant, psychrotrophic, adapted to meat substrates and have a good potential to be used for bioprotection of RTE products. The inoculation of vacuum-packaged sliced cooked ham and frankfurters with the bioprotective culture *Lactobacillus alimentarius* FloraCarn L2, at refrigeration temperatures, showed 1 week and 19 days extension in their shelf life respectively (Kotzekidou & Bloukas, 1998). When the bacteriocins producing *Lc. mesenteroides* L124 and *Lb. curvatus* L422 were used as bioprotective cultures in sliced cooked cured pork shoulder and frankfurter-type sausage, a reduction in the numbers of *B. thermosphacta* and enterococci was obtained thus extending their shelf life (Metaxopoulos, Mataragas, & Drosinos, 2002). Specifically, the inhibitory effect of lactocin S producing *Lb. sakei* 148 on hetero-fermentative LAB present in a cooked model ham was demonstrated by Vermeiren et al. (2004). The use of *E. faecium* CTC492 and *Lb. sakei* CTC494 as bioprotective cultures prevented ropiness due to *Lb. sakei* CTC476 until 21 days of storage at 8°C in sliced cooked meat products (Hugas et al., 2002).

Application of bioprotective cultures to RTE products will depend on every particular technology. Specifically, for cooked ham, inoculation with the bacteriocinogenic strain in the bulk mass during the tumbling procedure (5–6°C, during 24 h) will ensure its growth and bacteriocin production. Since bacteriocins are thermoresistant they would remain active in the cooked product mass after pasteurization and during storage, preventing contaminant and/or pathogen growth. On the other hand, during frankfurter processing bacteriocins may be added to meat batter before cooking, included in the casing as a folding additive or sprayed on the surface during post-processing and packing. The anti-listerial effect of bacteriocins was found to be higher when applied in the meat batter of the frankfurter-type sausage (Hugas et al., 2002).

Strategies to Improve Bioprotective Effect

Even when many bacteriocinogenic strains inhibited *L. monocytogenes in vitro* and in meat systems, most studies on the *in situ* effect were carried out on single strains of the pathogen or used high levels of contamination. Significant variations in the sensitivity of the *Listeria* strains to the same bacteriocin as well as the emergence of spontaneous resistant mutants to individual bacteriocins were reported (Castellano, Farías, Holzapfel, & Vignolo, 2001). For improved control of target microorganisms and bacteriocin-resistant strains, a combined use of bacteriocins was proposed. The use of nisin (class Ia), lactocin 705 (class IIb) and enterocin CRL35 (class IIa) combination against different *L. monocytogenes* strains in meat slurry showed no viable counts after incubation (Vignolo et al., 2000). The simultaneous addition of a nisin-curvaticin 13 combination also led to the absence of viable *Listeria* cells in the broth, preventing the re-growth of bacteriocin resistant mutants (Bouttefroy & Milliere, 2000). However, when the direct addition of enterocin A, sakacin K (produced by *E. faecium* CTC492 and *Lb. sakei* CTC494, respectively), Nisaplin (nisin) and ALTA 2351 (pediocin) to frankfurter sausages meat batter was carried out, only the batch with enterocin and sakacin showed significant *Listeria* reduction numbers (7.6 log cycles during 60 days of storage at 3.5°C). No significant differences were observed when sakacin, enterocin or nisin were applied separately or with ALTA (Hugas et al., 2002). It is generally assumed that a mixture containing more than one bacteriocin will be bactericidal to more cells in a sensitive population, since cells resistant to one bacteriocin would be killed by another.

Combinations of selected and those that are generally recognized as safe preservative antimicrobial solutions applied to processed meat surface may also provide a safeguard to contamination by *L. monocytogenes*. In this context, lactic, acetic, benzoic, propionic and sorbic acids or their salts in combination with bacteriocins were reported to be highly inhibitory. Post-processing sequential treatment of dipping solutions of Nisaplin (5000 IU/ml) plus lactic acid, reduced *L. monocytogenes* to undetectable levels in vacuum-packaged bologna ham stored at 10°C when compared to nisin applied alone (Geornaras et al., 2005). In order to find hurdles to avoid the risk of pathogen growth during storage and to reduce the levels of chemicals applied, new food preservation technologies complementary and alternative to traditional techniques have been assayed in combination with bacteriocins. The effectiveness of high hydrostatic pressure combined with nisin and lactate salts to inhibit the growth of *L. monocytogenes* and *Salmonella* in sliced cooked ham was demonstrated. It seems that the injured survivors after pressure became sensitive to nisin resulting in a viability loss (Aymerich, Jofre, Garriga, & Hugas, 2005). On the other hand, combining pediocin (ALTA2341) with post-packaging thermal pasteurization showed an effective treatment combination for improved control of *L. monocytogenes* on frankfurters (Chen, Sebranek, Dickson, & Mendonca, 2004). Moreover, a combination of the non-bacteriocinogenic *Lb. sakei* subsp. *carnosus* 10A and lactocin S producing *Lb. sakei* 148 with MAP containing 50% CO_2, completely inhibited the growth of *L. monocytogenes* on a model cooked ham (Vermeiren, Devlieghere, Vandekinderen, & Devebere, 2006). Food-grade chelators such as EDTA in combination with bacteriocins were also assayed as part of the hurdle concept to overcome Gram-negative resistance to bacteriocins. Vacuum-packaged and refrigerated fresh beef treated with nisin combined with EDTA increased its shelf life by inhibiting the growth *E. coli* as well as *B. thermosphacta* (Tu & Mustapha, 2002).

Influence of the Bioprotective Culture on Meat Sensorial Changes

The use of selected LAB strains as bioprotective cultures to extend the shelf life of chill stored vacuum-packaged meat and meat products implies the growth and acid production of the bacteriocinogenic strain. As stated above, LAB induces typical sensory changes such as souring, gas, SH_2 and slime thus resulting in costly losses to the industry. Organo-leptic defects related to lactic acid production would be perceived by consumers as sour or cheesy off odors and/or off tastes and even a discoloration process may occur due to meat protein denaturation. When the protective culture *Lb. curvatus* CRL705 applied on raw meat discs did not result in significant pH changes after incubation for 36 days at 2°C compared to the control, this indicated that no organo-leptic defects related to pH decrease were produced by the bioprotective culture (Castellano & Vignolo, 2006). This small pH decline may be attributed to the buffering capacity of the meat and the potential contribution of LAB to the breakdown of meat proteins and peptides, this being responsible for the generation of basic compounds which partially neutralize the acid produced and prevent an excessive acidification of the meat. In addition to a mild acidification, a bioprotective culture used to extend the shelf life of fresh meat should neither be a gas-forming nor an exo-polysaccharide producer to avoid bulging in the packages due to gas accumulation and slime formation on meat surfaces. To fulfill industrial requirements, a protective culture should be easy to culture, easy to apply to the meat and with a reliable and reproducible result.

Conclusions

Nowadays, consumers are greatly concerned about the relationship between food and health. The demand for reduced use of additives and processing seems contradictory for a market demanding safer and tastier foods. These market demands have placed the food industry under pressure to search for innovative solutions. Meat and meat products, in particular, face the challenge of overcoming the lack of consumer confidence that emerged during the last decade due to BSE and foodborne outbreaks caused by emerging pathogens. Novel insights into the metabolism of LAB offer perspectives for the application of a new generation of functional cultures. New starter LAB cultures with an industrially important functionality are being developed, mainly contributing to microbial safety or offering one or more sensorial, technological, nutritional or health advantages. The use of competitive bacteriocinogenic LAB as bioprotective cultures may well provide at least part of the solution. However, the application of biopreservation technology to meat and meat products should be considered only as an additional hurdle to good manufacturing processing and distribution practices.

References

Amézquita, A., & Brashears, M. (2002). Competitive inhibition of *Listeria monocytogenes* in ready-to-eat products by lactic acid bacteria. *Journal of Food Protection, 65*, 316–325.

Autio, T., Hielm, S., Miettinen, M., Sjöberg, A., Aamisalo, K., Björkroth, J., et al. (1999). Sources of *Listeria monocytogenes* contamination in a cold-smoked rainbow trout processing plant detected by pulsed-field gel electrophoresis typing. *Applied Environmental Microbiology, 65*, 150–155.

Aymerich, M. T., Artigas, M. G., Garriga, M., Monfort, J. & Hugas, M. (2000). Effect of sausage ingredients and additives on the production of enterocins A and B by *Enterococcus faecium* CTC492. Optimization of in vitro production and anti-listerial effect in dry fermented sausages. *Journal of Applied Microbiology, 88*, 686–694.

Aymerich, M. T., Garriga, M., Ylla, J., Vallier, J., Monfort, J., & Hugas, M. (2000). Applications of enterocins as biopreservatives against *Listeria innocua* in meat products. *Journal of Food Protection, 63*, 721–726.

Aymerich, M. T., Hugas, M., & Monfort, J. M. (1998). Review: Bacteriocinogenic lactic acid bacteria associated with meat products. *Food Science and Technology International, 4*, 141–158.

Aymerich, M. T., Jofre, A., Garriga, M., & Hugas, M. (2005). Inhibition of *Listeria monocytogenes* and *Salmonella* by natural antimicrobials and high hydrostatic pressure in sliced cooked ham. *Journal of Food Protection, 68*, 173–177.

Balaban, N., & Rasooly, A. (2000). Staphylococcal enterotoxins. *International Journal of Food Microbiology, 61*, 1–10.

Barboza de Martínez, Y., Ferrer, K., & Salas, E. (2002). Combined effects of lactic acid and nisin solution in reducing levels of microbiological contamination in red meat carcasses. *Journal of Food Protection, 65*, 1780–1783.

Beach, J. C., Murano, E.-A., & Acuff, G. R. (2002). Prevalence of *Salmonella* and *Campylobacter* in beef from transport to slaughter. *Journal of Food Protection, 65*, 1687–1693.

Belfiore, C., Castellano, P., & Vignolo, G. (2007). Reduction of *Escherichia coli* population following treatment with bacteriocins from lactic acid bacteria and chelators. *Food Microbiology, 24*, 223–229.

Benkerroum, N., Daoudi, A., & Kamal, M. (2003). Behaviour of *Listeria monocytogenes* in raw sausages (merguez) in presence of a bacteriocins-producing lactococcal strain as a protective culture. *Meat Science, 63*, 479–484.

Bhaduri, S., Wesley, I., & Bush, E. (2005). Prevalence of pathogenic *Yersinia enterocolitica* strains in pigs in the United States. *Applied and Environmental Microbiology, 71*, 7117–7121.

Bouttefroy, A., & Milliere, J. B. (2000). Nisin-curvaticin 13 combinations for avoiding the regrowth of bacteria resistant cell of *Listeria monocytogenes* ATCC 15313. *International Journal of Food Microbiology, 62*, 65–75.

Bredholt, S., Nesbakken, T., & Holck, A. (2001). Industrial application of an antilisterial strain of *Lactobacillus sakei* as a protective culture and its effect on the sensory acceptability of cooked, sliced, vacuum-packaged meats. *International Journal of Food Microbiology, 66*, 191–196.

Bremer, V., Bocter, N., Rehmet, S., Klein, G., Breuer, T., & Ammon, A. (2005). Consumption, knowledge, and handling of raw meat: A representative cross-sectional survey in Germany, March 2001. *Journal of Food Protection, 68*, 785–789.

Cahill, S. M., & Jouve, J.-L. (2004). Microbiological risk assessment in developing countries. *Journal of Food Protection, 67*, 2016–2023.

Callewaert, R., Hugas, M., & De Vuyst, L. (2000). Competitiveness and bacteriocin production of Enterococci in the production of Spanish-style dry fermented sausages. *International Journal of Food Microbiology, 57*, 33–42.

Castellano, P., Farías, M. E., Holzapfel, W., & Vignolo, G. (2001). Sensitivity variations of *Listeria* strains to the bacteriocins lactocin 705, enterocin CRL35 and nisin. *Biotechnology Letters, 23*, 605–608.

Castellano, P., Holzapfel, W., & Vignolo, G. (2004). The control of *Listeria innocua* and *Lactobacillus sakei* in broth and meat slurry with the bacteriocinogenic strain *Lactobacillus casei* CRL705. *Food Microbiology, 21*, 291–298.

Castellano, P., Raya, R., & Vignolo, G. (2003). Mode of action of the two-component bacteriocin lactocin 705. *International Journal of Food Microbiology, 85*, 35–43.

Castellano, P., & Vignolo, G. (2006). Inhibition of *Listeria innocua* and *Brochothrix thermosphacta* in vacuum-packaged meat by the addition of bacteriocinogenic *Lactobacillus curvatus* CRL705 and its bacteriocins. *Letters in Applied Microbiology, 43*, 194–199.

Castellano, P., Vignolo, G., Farías, R., Arrondo, J., & Chehín, R. (2007). Molecular view by Fourier transform infrared spectroscopy of the relationship between lactocin 705 and membranes: Speculations on antimicrobial mechanism. *Applied of Environmental Microbiology, 73*, 415–420.

Castillo, A., Lucia, L., Mercado, I., & Acuff, G. (2001). In-plant evaluation of a lactic acid treatment for reduction of bacteria on chilled beef carcasses. *Journal of Food Protection, 64*, 738–740.

Cayré, M. E., Garro, O., & Vignolo, G. (2005). Effect of storage temperature and gas permeability of packaging film on the growth of lactic acid bacteria and *Brochothrix thermosphacta* in cooked meat emulsions. *Food Microbiology, 22*, 505–512.

CDC, Centers for Disease Control and Prevention (1995). *Escherichia coli* O157:H7 outbreak linked to commercially distributed dry cured salami-Washington and California. *Morbidity and Mortality Weekly Report, 44*, 157–160.

CDC, Centers for Disease Control and Prevention (2002). Incidence of foodborne illness: Preliminary FoodNet data on the incidence of foodborne illness-selected sites, United States. *Morbidity and Mortality Weekly Report, 51*, 325–329.

CDC, Centers for Disease Control and Prevention (2007). Preliminary FoodNet data on the incidence of infection with pathogens transmitted commonly through food – 10 States, 2006. *Morbidity and Mortality Weekly Report, 56*, 336–339.

Chang, V., Mills, E., & Cutter, C. (2003). Reduction of bacteria on pork carcasses associated with chilling method. *Journal of Food Protection, 66*, 1019–1024.

Chen, C., Sebranek, J., Dickson, J., & Mendonca, A. (2004). Combining pediocin (ALTA2341) with postpackaging pasteurization for control of *Listeria monocytogenes* frankfurters. *Journal of Food Protection, 67*, 1855–1865.

Chen, Y., Ross, W., Scott, V., & Gombas, D. (2003). *Listeria monocytogenes*: Low levels equal low risk. *Journal of Food Protection, 66*, 570–577.

Church, D. (2004). Major factors affecting the emergence and re-emergence of infectious diseases. *Clinics in Laboratory Medicine, 24*, 559–586.

Church, I., & Parsons, A. (1995). Modified atmosphere packaging technology: A review. *Journal of Science and Food Agriculture, 67*, 143–152.

Coffey, A., Ryan, M., Ross, R., Hill, C., Arendt, E., & Schwarz, G. (1998). Use of a broad-host-range bacteriocins-producing *Lactococcus lactis* transconjugant as an alternative starter for salami manufacture. *International Journal of Food Microbiology, 43*, 231–235.

Cotter, P., Hill, C., & Ross, P. (2005). Bacteriocins: Developing innate immunity for foods. *Nature Reviews Microbiology, 3*, 777–788.

Cuozzo, S., Castellano, P., Sesma, F., Vignolo. G., & Raya, R. (2003). Differential roles of two component peptides of lactocin 705 in antimicrobial activity. *Current Microbiology, 46*, 180–183.

Demeyer, D. (2003). Meat fermentation: Principles and applications. In Y. Hui, L. Goddik, A. Hansen, J. Josephsen, W. Nip, P. Stanfield, et al. (Eds.), *Handbook of Food and Beverage Fermentation Technology* (pp. 353–367). New York: Marcel Dekker.

Desmarchelier, P., Higgs, G., Mills, L., Sullivan, A., & Vanderlinde, P. (1999). Incidence of coagulase positive *Staphylococcus* on beef carcasses in three Australian abattoirs. *International Journal of Food Microbiology, 47*, 221–229.

Dicks, L., Mellet, F., & Hoffman, L. (2004). Use of bacteriocin-producing starter culture of *Lactobacillus plantarum* and *Lactobacillus curvatus* in production of ostrich meat salami. *Meat Science, 66*, 703–708.

Elmi, M. (2004). Food safety: Current situation, unaddressed issues and the emerging priorities. *Eastern Mediterranean Health Journal, 10,* 794 800.

Fegan, N., Vanderlinde, P., Higgs, G., & Desmarchelier, P. (2004). The prevalence and concentration of *Escherichia coli* O157 in feces of cattle from different production systems at slaughter. *Journal of Applied Microbiology, 97,* 362–370.

Foegeding, P., Thomas, A., Pilkington, D., & Klaenhammer, T. (1992). Enhanced control of *Listeria monocytogenes* by in situ-produced pediocin during dry fermented sausage production. *Applied & Environmental Microbiology, 58,* 3053–3059.

Fontana, C., Cocconcelli, P., & Vignolo, G. (2005). Monitoring the bacterial population dynamics during fermentation of artisanal Argentinean sausages. *International Journal of Food Microbiology, 103,* 131–142.

Fontana, C., Cocconcelli, P., & Vignolo, G. (2006). Direct molecular approach to monitoring bacterial colonization on vacuum-packaged beef. *Applied Environmental Microbiology, 72,* 5618–5622.

Geornaras, I., Belk, K., Scanga, J., Kendall, P., Smite, G., & Sofos, J. (2005). Postprocessing antimicrobial treatments to control *L. monocytogenes* in commercial vacuum-packaged bologna ham stored at 10°C. *Journal of Food Protection, 68,* 991–998.

Gombas, D., Chen, Y., Clavero, R., & Scott, V. (2003). Survey of *Listeria monocytogenes* in ready-to-eat foods. *Journal of Food Protection, 66,* 559–560.

Griffin, P., & Tauxe, R. (1991). The epidemiology of infections caused by *Escherichia coli* O157:H7, other enterohemorrhagic *E. coli* and the associated haemolytic uremic syndrome. *Epidemiology Reviews, 13,* 60–98.

Gurtler, M., Alter, T., Kasimir, S., Linnebur, M., & Fehlhaber, K., (2005). Prevalence of *Yersinia enterocolítica* in fattening pigs. *Journal of Food Protection, 68,* 850–854.

Hald, T., Wingstrand, A., Swanenburg, M., von Altrock, A., & Thorberg, B. (2003). The occurrence and epidemiology of *Salmonella* in European pig slaughterhouses. *Epidemiology Infection, 131,* 1187–1203.

Hassan, Z., Purwati, E., Radu, S., Rahim, R., & Rusul, G. (2001). Prevalence of *Listeria* spp and *Listeria monocytogenes* in meat and fermented fish in Malaysia. *Southeast Asian Journal of Tropical Medicine Public Health, 32,* 402–407.

Holeckova, B., Holoda, E., Fotta, M., Kalinacova, V., Gondol', J., & Grolmus, J. (2002). Occurrence of enterotoxigenic *Staphylococcus aureus* in food. *Annals of Agricultural and Environmental Medicine, 9,* 179–182.

Holley, R., Lammerding, A., & Tittiger, F. (1988). Microbiological safety of traditional and starter-mediated processes for the manufacture of Italian dry sausage. *International Journal of Food Microbiology, 7,* 49–62.

Holzapfel, W., Geisen, R., & Schillinger, U. (1995). Biological preservation of foods with reference to protective cultures, bacteriocins and food-grade enzymes. *International Journal of Food Microbiology, 24,* 343–362.

Hugas, M., Garriga, M., Aymerich, M., & Monfort, J., (2002). Bacterial cultures and metabolites for the enhancement of safety and quality. In F. Toldrá (Ed.), *Research advances in the quality of meat and meat products* (pp. 225–247). India: Research Signpost.

Hugas, M., Neumeyer, B., Pagés, F., Garriga, M., & Hammes, W. (1997). Comparison of bacteriocins-producing lactobacilli on *Listeria* growth in fermented sausages. *Fleischwirtschaft, 76,* 649–652.

Hugas, M., Pages, F., Garriga, M., & Monfort, J. (1998). Application of the bacteriocinogenic *Lactobacillus sakei* CTC494 to prevent growth of *Listeria* in fresh and cooked meat products packed with different atmospheres. *Food Microbiology, 15,* 639–650.

Hussein, H., & Bollinger, L. (2005). Prevalence of Shiga toxin-producing *Escherichia coli* in beef cattle. *Journal of Food Protection, 68,* 2224–2241.

ICMSF International Commission on Mirobiological Specifications for Foods. (2002). *Microorganisms in foods 7: Microbiological testing in food safety management* (pp. 285–312). New York: Kluwer Academic/Plenum Publishers.

ILSI Research Foundation; Risk Science Institute. (2005). Achieving continuous improvement in reductions in foodborne listeriosis-a risk-based approach. *Journal of Food Protection, 68*, 1932–1994.
Ingham, S., Engel, R., Fanslau, M., Schoeller, E., Searls, G., Buege, D., et al. (2005). Fate of *Staphylococcus aureus* on vacuum-packaged ready-to-eat meat products stored at 21 degrees C. *Journal Food Protection, 68*, 1911–1915.
Inglis, G., Kalischuk, L., Busz, H., & Kastelic, J. (2005). Colonization of cattle intestines by *Campylobacter jejuni* and *Campylobacter lanienae*. *Applied and Environmental Microbiology, 71*, 5145–5153.
Jacobsen, T., Budde, B., & Koch, A. (2003). Application of *Leuconostoc carnosum* for biopreservation of cooked meat products. *Journal of Applied Microbiology, 95*, 242–249.
Jones, R. J. (2004). Observations on the succession dynamics of lactic acid bacteria populations in chill-stored vacuum-packaged beef. *International Journal of Food Microbiology, 90*, 273–282.
Kalinowski, R., Tompkin, R., Bodnaruk, P., & Pruett, W. (2003). Impact of cooking, and subsequent refrigeration on the growth or survival of *Clostridium perfringes* in cooked meat and poultry products. *Journal of Food Protection, 66*, 1227–1232.
Kanellos, T., & Burriel, A. (2005). The bactericidal effects of lactic acid and trisodium phosphate on *Salmonella enteritidis* serotype pt4, total viable counts and counts of *Enterobacteriaceae*. *Food Protection Trends, 25*, 346–350.
Katikou, P., Ambrosiadis, I., Georgantelis, D., Koidis, P., & Georgakis, S. (2005). Effect of *Lactobacillus*-protective cultures with bacteriocins-like inhibitory substances' producing ability on microbiological, chemical and sensory changes during storage of refrigerated vacuum-packaged sliced beef. *Journal of Applied Microbiology, 99*, 1303–1313.
Katla, T., Møretrø, T., Sveen, I., Aasen, I., Axelsson, L., & Rørvik, L. (2002). Inhibition of *Listeria monocytogenes* in chicken cold cuts by addition of sakacin P and sakacin P-producing *Lactobacillus sakei*. *Journal of Applied Microbiology, 93*, 191–196.
Koort, J., Coenye, T., Santos, E., Molinero, C., Jaime, I., Rovira, J., et al. (2006). Diversity of *Weissella viridescens* strains associated with "Morcilla de Burgos". *International Journal of Food Microbiology, 109*, 164–168.
Kotzekidou, P., & Bloukas, J. (1998). Microbial and sensory changes in vacuum-packed frankfurter-type sausages by *Lactobacillus alimentarius* and fate of inoculated *Salmonella enteritidis*. *Food Microbiology 15*, 101–111.
Lauková, A., Czikková, S., Laczková, S., & Turek, P., (1999). Use of enterocin CCM4231 to control *Listeria monocytogenes* in experimentally contaminated dry fermented Hornád salami. *International Journal of Food Microbiology, 52*, 115–119.
Leistner, L. (1997). Microbial stability and safety of healthy meat, poultry and fish products. In A. M. Pearson, & T. R. Dutson (Eds.), *Production and processing of healthy meat, poultry and fish products* (pp. 347–360). London: Blackie Academic and Professional.
Leistner, L. (2000). Basic aspects of food preservation by hurdle technology. *International Journal of Food Microbiology, 55*, 181–186.
Leroy, F., Verluyten, J., & De Vuyst, L. (2006). Functional meat starter cultures for improved sausage fermentation. *International Journal of Food Microbiology, 106*, 270–285.
Lin, C., Takeuchi, K., Zhang, L., Dohm, C., Meyer, J., Hall, P., et al. (2006). Cross-contamination between processing equipment and deli meats by *Listeria monocytogenes*. *Journal of Food Protection, 69*, 71–79.
Lücke, F. -K. (2000). Utilization of microbes to process and preserve meat. *Meat Science, 56*, 105–115.
Marshall, K., Niebuhr, S., Acuff, G., Lucia, L., & Dickson, J. (2005). Identification of *Escherichia coli* O157:H7 meat processing indicators for fresh meat through comparison of the effects of selected antimicrobial interventions. *Journal of Food Protection, 68*, 2580–2586.
Marzoca, M., Marucci, P., Sica, M., & Alvarez, E. (2004). *Listeria* monocytogenes detection in different food products and environmental samples of supermarkets of Bahia Blanca city (Argentina). *Revista Argentina de Microbiología, 36*, 179–181.

McEvoy, J., Doherty, A., Sheridan, J., Blair, I., & McDowell, D. (2003). The prevalence of *Salmonella* spp. in bovine faecal, rumen and carcass samples at a commercial abattoir. *Journal of Applied Microbiology, 94*, 693–700.

Meichtri, L., Miliwebsky, E., Gioffre, A., Chinen, I., Baschkier, A., Chillemi, G., et al. (2004). Shiga toxin-producing *Escherichia coli* in healthy young beef steers from Argentina: Prevalence and virulence properties. *International Journal of Food Microbiology, 96*, 189–198.

Metaxopoulus, J., Mataragas, M., & Drosinos, E. (2002). Microbial interaction in cooked cured meat products under vacuum or modified atmosphere at 4°C. *Journal of Applied Microbiology, 93*, 363–373.

Morse, S. S. (2004). Factors and determinants of disease emergence. *Revue Scientifique et Technique, 23*, 443–451.

Morse, S. S. (2007).Global infectious disease surveillance and health intelligence. *Health Affairs, 26*, 1069–1077.

Normano, G., Parisi, A., Dambrosio, A., Quaglia, N., Montagna, D., Chiocco, D., et al. (2004). Typing of *Escherichia coli* O157 strains isolated from fresh sausage. *Food Microbiology, 21*, 79–82.

Okutani, A., Okada, Y., Yamamoto, S., & Igimi, S. (2004). Nationwide survey of human *Listeria monocytogenes* infection in Japan. *Epidemiology Infection, 132*, 769–772.

Omisakin, F., MacRae, M., Ogden, I., & Strachan, N. (2003). Concentration and prevalence of *Escherichia coli* O157 in cattle feces at slaughter. *Applied and Environmental Microbiology, 69*, 2444–2447.

Pala, T., & Sevilla, A. (2004). Microbial contamination of carcasses, meat and equipment from an Iberian pork cutting plant. *Journal of Food Protection, 67*, 1624–1629.

Peccio, A., Autio, T., Korkeala, H., Rosmini, R., & Trevisani, M. (2003). *Listeria monocytogenes* occurrence and characterization in meat-producing plants. *Letters in Applied Microbiology, 37*, 234–238.

Phillips, C. A. (1996). Modified atmosphere packaging and its effects on the microbiological quality and safety of produce. *International Journal of Food Science and Technology, 31*, 463–479.

Phillips, D., Jordan, D., Morris, S., Jenson, I., & Sumner, J. (2006). A national survey of the microbiological quality of beef carcasses and frozen boneless beef in Australia. *Journal of Food Protection, 69*, 1113–1117.

Pitkala, A., Haveri, M., Pyorala, S., Myllys, V., & Honkanen-Buzalski, T. (2004). Bovine mastitis in Finland 2001-prevalence, distribution of bacteria, and antimicrobial resistance. *Journal of Dairy Science, 87*, 2433–2441.

Ransom, J., Belk, K., Sofos, J., Stopforth, J., Sacanga, J., & Smith, G. (2003). Comparison of intervention technologies for reducing *Escherichia coli* O157:H7 on beef cuts and trimmings. *Food Protection Trends, 23*, 24–34.

Rantsiou, K., Drosinos, E., Gialitaki, M., Urso, R., Krommer, J., Gasparik-Reichardt, J., et al. (2005). Molecular characterization of *Lactobacillus* species isolated from naturally fermented sausages produced in Greece, Hungary and Italy. *Food Microbiology, 22*, 19–28.

Rivera-Betancourt, M., Shackelford, S., Arthur, T., Westmoreland, K., Bellinger, G., Rossman, M., et al. (2004). Prevalence of *Escherichia coli* O157:H7, *Listeria monocytogenes*, and *Salmonella* in two geographically distant commercial beef processing plants in the United States. *Journal of Food Protection, 67*, 295–302.

Rodriguez, A., Pangloli, P., Richards, H., Mount, J., & Draughton, F. (2006). Prevalence of *Salmonella* in diverse environmental farm samples. *Journal of Food Protection, 69*, 2576–2580.

Sabia, C., de Niederhausern, S., Messi, P., Manicardi, G., & Bondi, M. (2003). Bacteriocins-producing *Enterococcus casseliflavus* IM416K1, a natural antagonist for control of *Listeria monocytogenes* in Italian sausages ("cacciatore"). *International Journal of Food Microbiology, 87*, 173–179.

Sagoo, S., Little, C., Allen, G., Williamson, K., & Grant, K. (2007). Microbiological safety of retail vacuum-packed and modified-atmosphere-packed cooked meats at end of shelf life. *Journal of Food Protection, 70*, 943–951.

Sakala, R., Hayashidani, H., Kato, Y., Hirata, T., Makino, Y., Fukushima, A., et al. (2002). Change in the composition of the microflora on vacuum-packaged beef during chiller storage. *International Journal of Food Microbiology, 74*, 87–99.

Salvat, G., & Fravalo, G. (2004). Risk assessment strategies for Europe: Integrated safety strategy or final product control: Example of *Listeria monocytogenes* in processed products from pork meat industry. *Deutshe Tierärztliche Wochenschrift, 111*, 331–334.

Samelis, J., & Georgiadou, K. G. (2000). The microbial association of Greek taverna sausage stored at 4 and 10°C in air, vacuum or 100% carbon dioxide, and its spoilage potential. *Journal of Applied Microbiology, 88*, 58–68.

Samelis, J., Björkroth, J., Kakouri, A., & Rementzis, J. (2006). *Leuconostoc carnosum* associated with spoilage of refrigerated whole cooked hams in Greece. *Journal of Food Protection, 69*, 2268–2273.

Savell, J., Mueller, S., & Baird, B. (2005). The chilling of carcasses – a review. *Meat Science, 70*, 449–459.

Scannell, A., Ross, R., Hill, C., & Arendt, E. (2000). An effective lacticin biopreservative in fresh pork sausage. *Journal of Food Protection, 63*, 370–375.

Shale, K., Jacoby, A., & Plaatjies, Z. (2006). The impact of extrinsic sources on selected indicator organisms in a typical deboning room. *International Journal of Environmental Health Research, 16*, 263–272.

Shale, K., Lues, J., Venter, P., & Buys, E. (2006). The distribution of staphylococci in bioaerosols from red-meat abattoirs. *Journal of Environmental Health, 69*, 25–32.

Smerdon, W., Adak, G., O'Brien, S., Gillespie, I., & Reacher, M. (2001). General outbreaks of infectious intestinal disease linked with red meat, England and Wales, 1992–1999. *Communicable Disease and Public Health, 4*, 259–267.

Spescha, C., Stephan, R., & Zweifel, C. (2006). Microbiological contamination of pig carcasses at different stages of slaughter in two European Union-approved abattoirs. *Journal of Food Protection, 69*, 2568–2575.

Stiles, M. E. (1996). Biopreservation by lactic acid bacteria. *Antoine van Leeuwenhoek, 70*, 331–345.

Talon, R., Leroy-Sétrin, S., & Fadda, S. (2003). Dry fermented sausages. In Y. Hui, L. Goddik, A. Hansen, J. Josephsen, W. Nip, P. Stanfield, et al. (Eds.), *Handbook of food and beverage fermentation technology* (pp. 397–416). New York: Marcel Dekker.

Thevenot, D., Dernburg, A., & Vernozy-Rozand, C. (2006). An updated review of *Listeria monocytogenes* in the pork meat industry and its products. *Journal of Applied Microbiology, 101*, 7–17.

Tilden, J., Young, W., McNamara, A., Custer, C., Boesel, B., Lambert-Fair, M., et al. (1996). A new route of transmission for *Escherichia coli*: Infection from dry fermented salami. *American Journal Public Health, 86*(8 Pt 1), 1142–1145.

Tjener, K., Stahnke, L., Andersen, L., & Martinussen, J. (2004). Growth and production of volatiles by *Staphylococcus carnosus* in dry sausages: Influence of inoculation level and ripening time. *Meat Science, 67*, 447–452.

Tu, L., & Mustapha, A. (2002). Reduction of *Brochothrix thermosphacta* and *Salmonella* serotype *tyyphimurium* on vacuum-packaged fresh beef treated with nisin and nisin combined with EDTA. *Journal of Food Science, 67*, 302–306.

Vaillant, V., de Valk, H., Baron, E., Ancelle, T., Colin, P., Delmas, M., et al. (2005). Foodborne infections in France. *Foodborne Pathogens and Disease, 2*, 221–232.

Verluyten, J., Leroy, F., & De Vuyst, L. (2004). Effects of different spices used in the production of fermented sausages on growth of and curvacin A production by *Lactobacillus curvatus* LTH1174. *Applied and Environmental Microbiology, 70*, 4807–4813.

Vermerien, L., Devlieghere, F., & Debevere, J. (2004). Evaluation of meatborne lactic acid bacteria as protective cultures for the biopreservation of cooked meat products. *International Journal of Food Microbiology, 96*, 149–164.

Vermeiren, L., Devlieghere, F., Vandekinderen, I., & Devebere, J. (2006). The interaction of the non-bacteriocinogenic *Lactobacillus sakei* 10A and lactocina S producing *Lactobacillus sakei* 148 towards *Listeria monocytogenes* on a model cooked ham. *Food Microbiology, 23*, 511–518.

Vignolo, G., Fadda, S., Kairuz, M., Ruiz Holgado, A., & Oliver, G. (1998). Effects of curing additives on the control of *Listeria monocytogenes* by lactocin 705 in meat slurry. *Food Microbiology, 15*, 259–264.

Vignolo, G., Palacios, J., Farías, M., Sesma, F., Schillinger, U., Holzapfel, W. et al. (2000). Combined effect of bacteriocins on the survival of various *Listeria* especies in broth and meat system. *Current Microbiology, 41*, 410–416.

Wen, Q., & McClane, B. (2004). Detection of enterotoxigenic *Clostridium perfringes* type A isolates in American retail foods. *Applied and Environmental Microbiology, 70*, 2685–2691.

Chapter 19
Smart Packaging Technologies and Their Application in Conventional Meat Packaging Systems

Michael N. O'Grady and Joseph P. Kerry

Introduction

Preservative packaging of meat and meat products should maintain acceptable appearance, odour and flavour and should delay the onset of microbial spoilage. Typically fresh red meats are placed on trays and over-wrapped with an oxygen permeable film or alternatively, meats are stored in modified atmosphere packages (MAP) containing high levels of oxygen and carbon dioxide (80% O_2:20% CO_2) (Georgala & Davidson, 1970). Cooked meats are usually stored in 70% N_2:30% CO_2 (Smiddy, Papkovsky, & Kerry, 2002). The function of oxygen in MAP is to maintain acceptable fresh meat colour and carbon dioxide inhibits the growth of spoilage bacteria (Seideman & Durland, 1984). Nitrogen is used as an inert filler gas either to reduce the proportions of the other gases or to maintain the pack shape (Bell & Bourke, 1996). High oxygen levels promote the oxidation of muscle lipids over time with deleterious effects on fresh meat colour and quality (O'Grady et al., 1998). In cooked meat products (e.g. cured ham) low residual levels of oxygen promote pigment denaturation which imposes a dull greyness to the meat surface (Møller, Jensen, Olsen, Skibsted, & Bertelsen, 2000). Commercially, this problem is overcome with the use of an oxygen scavenger. Oxygen scavengers are examples of entities described as 'active packaging components'.

Smart packaging is a broad term encompassing a range of new packaging concepts, most of which can be placed in one of the two principle categories: active packaging and intelligent packaging. Active packaging refers to the incorporation of certain additives into packaging systems (loose within the pack, attached to the inside of packaging material or incorporated into the packaging material) with the aim of maintaining or extending product quality and shelf life. Packaging may be termed active when it performs some desired role in food preservation other than providing an inert barrier to external conditions (Hutton, 2003). Active packaging has been defined as packaging which 'changes the condition of the packed food to extend shelf-life or to improve safety or sensory properties, while maintaining the

---

J.P. Kerry
Department of Food and Nutritional Sciences, University College Cork, National University of Ireland, Western Road, Cork, Ireland

quality of packaged food' (Ahvenainen, 2003). Intelligent packaging systems are designed to monitor specific attributes of the food, or environment within the pack and provide information about the quality or safety of the product. A wide range of technologies have already been developed and commercialised to take advantage of opportunities presented by active and intelligent packaging systems.

The aim of this review is to examine the current use of active packaging and intelligent packaging technologies with meat and meat products. New or developing technologies will also be evaluated and assessed for their potential use in future meat packaging applications.

Active/Intelligent – Smart Packaging Technologies

Oxygen Scavengers

Elevated oxygen levels in food packages may significantly reduce the shelf life of meat due to colour changes, the onset of lipid oxidation, microbial growth and nutritional losses. Oxygen absorbing systems provide an alternative to vacuum and gas flushing technologies as a means of improving product quality and shelf life (Ozdemir & Floros, 2004). Modified atmosphere packaging or vacuum packaging technologies do not always facilitate complete removal of oxygen. The use of an oxygen scavenger, which absorbs residual oxygen after packaging, minimises quality changes in oxygen sensitive foods (Vermeiren, Devlieghere, Van Beest, de Kruijf, & Debevere, 1999). Existing oxygen scavenging technologies utilise one or more of the following concepts: iron powder oxidation, ascorbic acid oxidation, photosensitive dye oxidation, enzymatic oxidation (e.g. glucose oxidase and alcohol oxidase), unsaturated fatty acids (e.g. oleic or linolenic acid) rice extract or immobilised yeast on a solid substrate (Floros, Dock, & Han, 1997). More comprehensive information and details relating to oxygen scavengers can be obtained from other reviews (Floros et al., 1997; Vermeiren et al., 1999). Structurally, the oxygen scavenging component of a package can take the form of a sachet (Fig. 19.1a), label (Fig. 19.1b), film (incorporation of scavenging agent into the packaging film) (Fig. 19.2), card, closure liner or concentrate (Suppakul, Miltz, Sonneveld, & Bigger, 2003). Commercially available oxygen scavengers are predominantly based on the principle of iron oxidation (Smith, Ramaswamy, & Simpson, 1990).

Comprehensive details on a variety of commercially available oxygen scavengers are presented by Suppakul et al. (2003). Ageless® (Mitsubishi Gas Chemical Co., Japan) is the most common oxygen scavenging system based on iron oxidation (Fig. 19.1). These scavengers are designed to reduce oxygen levels to less than 0.1%. Additional examples of oxygen absorbing sachets include ATCO® (Emco Packaging Systems, UK; Standa Industrie, France), FreshPax® (Multisorb Technologies Inc., USA) and Oxysorb® (Pillsbury Co., USA). Kerry, O'Grady, & Hogan (2006) reviewed a number of studies from scientific literature where oxygen scavengers (Ageless® and FreshPax®) were used to prevent discoloration in fresh beef and

19 Smart Packaging Technologies

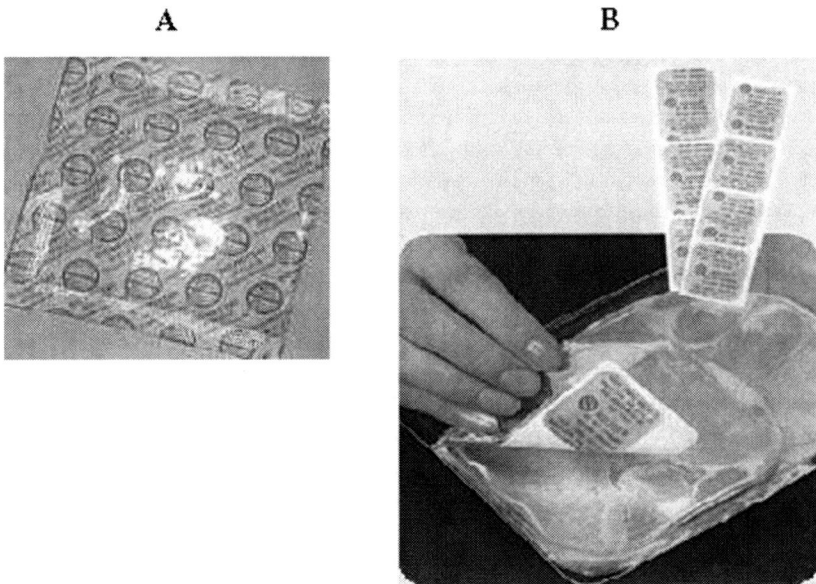

Reproduced with permission from Mitsubishi Gas Chemical Co.

Fig. 19.1 Ageless® sachet (a) and label (b) (Mitsubishi Gas Chemical Co., Japan)

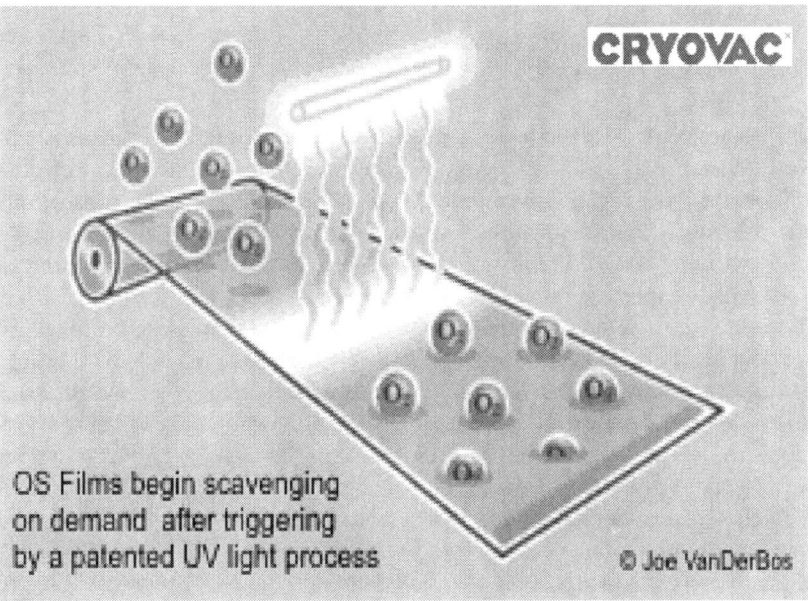

Fig. 19.2 Light-activated oxygen scavenging films (Cryovac® OS Systems, Sealed Air Corporation, USA)

enhance pork and pork product quality. While many active packaging technologies are still developmental, oxygen scavengers are widely used commercially in pre-packed cooked meat products. Emco Packaging Systems, specialists in active and intelligent packaging, are a UK manufacturer and distributor for ATCO® DE 10S self-adhesive oxygen absorbing labels. Emco supply ATCO® labels for use in pre-packed sliced cooked meats, especially hams, to meat processors in Ireland, throughout the UK and in Europe. While labels used in sliced cooked meat packages scavenge between 10 and 20 cc's of oxygen, Emco have recently launched larger oxygen scavenging labels onto the market (ATCO® 100 OS and 200 OS), which scavenge between 100 and 200 cc's oxygen, for use in larger capacity packaging applications.

Cryovac® OS Systems (Cryovac Div., Sealed Air Corporation, USA) oxygen scavenging technology incorporates a polymer-based oxygen scavenger into the packaging film. The UV light-activated Cryovac® OS2000^{TM} oxygen scavenging film (Fig. 19.2) is composed of an oxygen scavenger layer extruded into a multi-layer film and can reduce headspace oxygen levels from 1% to ppm levels in 4–10 days, comparable with oxygen scavenging sachets. These films have applications in a wide variety of food products including dried or smoked meat products and processed meats (Butler, 2002). A similar UV light-activated oxygen scavenging polymer ZERO$_2^{TM}$, developed by CSIRO, Div. of Food Science Australia in collaboration with VisyPak Food Packaging, Visy Industries, Australia, forms a layer in a multi-layer package structure and has many applications including reduced discoloration of sliced meats.

Carbon Dioxide Scavengers and Emitters

Within a packaging environment, carbon dioxide functions to suppress microbial growth. Therefore a carbon dioxide generating system can be viewed as a technique complimentary to oxygen scavenging (Suppakul et al., 2003). The permeability of carbon dioxide is 3 to 5 times higher than that of oxygen in most plastic films; therefore, it must be continuously produced to maintain the desired concentration within the package (Ozdemir & Floros, 2004). High carbon dioxide levels (10–80%) are desirable for foods such as meat and poultry in order to inhibit microbial growth and extend shelf life. Removal of oxygen from the package creates a partial vacuum which may result in the collapse of flexible packaging. Also, when a package is flushed with a mixture of gases including carbon dioxide, the carbon dioxide dissolves in the product creating a partial vacuum. In such cases, the simultaneous release of carbon dioxide from inserted sachets which consume oxygen is desirable. Such systems are based on either ferrous carbonate or a mixture of ascorbic acid and sodium bicarbonate (Rooney, 1995). Examples of commercially available dual action combined carbon dioxide generators/oxygen scavengers are Ageless® G (Mitsubishi Gas Chemical Co, Japan) and FreshPax® M (Multisorb Technologies Inc, USA). Carbon dioxide emitting sachets or labels can also be used alone. The

Verifrais™ package, manufactured by SARL Codimer (Paris, France) has been used to extend the shelf life of fresh meats. This innovative package consists of a standard modified atmosphere packaging tray, but has a perforated false bottom under which a porous sachet containing sodium bicarbonate/ascorbate is positioned. When juice exudates from the packaged meat drips onto the sachet, carbon dioxide is emitted, thus replacing any carbon dioxide absorbed by the meat and preventing package collapse.

The inhibition of spoilage bacteria utilizing active packaging technology may reduce bacterial competition and thus permit growth and toxin production by non-proteolytic *C. botulinum* or the growth of other pathogenic bacteria (Sivertsvik, 2003). Lövenklev et al. (2004) reported that while a high concentration of carbon dioxide decreased the growth rate of non-proteolytic *C. botulinum* type B, the expression and production of the toxin was greatly increased which means the risk of botulism may also be increased, rather than being reduced, if used in modified atmosphere packaging systems. Research into the safety risks associated with the use of carbon dioxide in packaging systems is necessary.

Carbon dioxide absorbers (sachets) consisting of either calcium hydroxide and sodium hydroxide, or potassium hydroxide, calcium oxide and silica gel, may be used to remove carbon dioxide during storage in order to prevent bursting of the package. Possible applications include their use in packs of dehydrated poultry products and beef jerky (Ahvenainen, 2003).

Moisture Control

The main purpose of liquid water control is to lower the water activity of the product, thereby suppressing microbial growth (Vermeiren et al., 1999). Temperature cycling of high water activity foods has led to the use of plastics with an anti-fog additive that lowers the interfacial tension between the condensate and the film. This contributes to the transparency of the film and enables the customer to clearly see the packaged food (Rooney, 1995), although it does not affect the amount of liquid water present inside the package. Several companies manufacture drip absorbent sheets or pads such as Cryovac® Dri-Loc® (Sealed Air Corporation, USA), Thermarite® or Peaksorb® (Australia), Toppan™ (Japan) and Fresh-R-Pax™ (Maxwell Chase Technologies, LLC, USA) for liquid control in high water activity foods such as meat and poultry. These systems consist of a super absorbent polymer located between two layers of a micro porous or non-woven polymer. Such sheets are used as drip-absorbing pads placed under whole chickens or chicken cuts (Suppakul et al., 2003).

Antimicrobial Packaging

Fresh meat is a highly perishable food product which, unless correctly stored, processed, packaged and distributed, spoils quickly and becomes hazardous due to microbial growth and the subsequent risk of food borne illness. Antimicrobial packaging is a promising form of active packaging especially for meat products. Since microbial contamination of meat products occurs primarily at the surface, due to post processing handling, attempts have been made to improve safety and to delay spoilage by the use of antibacterial sprays or dips. Limitations of such antibacterials include neutralisation of compounds on contact with the meat surface or diffusion of compounds from the surface into the meat mass. Incorporation of bactericidal agents into meat formulations may result in partial inactivation of the active compounds by meat constituents and therefore exert a limited effect on the surface microflora (Quintavalla & Vicini, 2002). Antimicrobial food packaging materials have to extend the lag phase and reduce the growth phase of microorganisms, in order to extend shelf life and to maintain product quality and safety (Han, 2000). Comprehensive reviews on antimicrobial food packaging have been published previously (Appendini & Hotchkiss, 2002; Suppakul et al., 2003) and more recently by Coma (2008). To confer antimicrobial activity, antimicrobial agents may be coated, incorporated, immobilised, or surface modified onto package materials (Suppakul et al., 2003). A comprehensive list of antimicrobial agents for use in antimicrobial films, containers and utensils is presented in a review by Suppakul et al. (2003). The classes of antimicrobials listed range from acid anhydride, alcohol, bacteriocins, chelators, enzymes, organic acids and polysaccharides. Examples of commercial antimicrobial materials in the form of concentrates (e.g. AgION^{TM}, AgION Technologies LLC, USA), extracts (Nisaplin® (Nisin), Integrated Ingredients, USA) and films (Microgard^{TM} Rhone-Poulenc, USA) were also presented. Antimicrobial packages have had relatively few commercial successes except in Japan where Ag-substituted zeolite is the most common antimicrobial agent incorporated into plastics. Ag-ions inhibit a range of metabolic enzymes and have strong antimicrobial activity (Vermeiren et al., 1999). Antimicrobial films can be classified into two types: those that contain an antimicrobial agent which migrates to the surface of the food and those which are effective against surface growth of microorganisms without migration. The antibacterial activity of antimicrobial agents that are coated, directly incorporated or immobilised onto packaging films in meat and meat products has been reviewed by Coma (2008) and Kerry et al. (2006).

Sensors

In modified atmosphere packaged meats, the headspace concentrations of oxygen and carbon dioxide are useful indicators of the meat product quality. The headspace profiles of oxygen and carbon dioxide change over time as a function of factors such as product type, respiration, packaging material, pack size, volume ratios, storage conditions and package integrity. An optical sensor approach to monitor gas phases in modified atmosphere packaged products offers an alternative to the conventional methods such as GC, GC/MS and portable gas analysers.

Many intelligent packaging concepts involve the use of sensors and indicators. A sensor is defined as a device used to detect, locate or quantify energy or matter, giving a signal for the detection or measurement of a physical or chemical property to which the device responds (Kress-Rogers, 1998b). A sensor device must provide continuous output of a signal. The majority of sensors are composed of a receptor and a transducer. In the receptor, physical or chemical information is transformed into a form of energy which may be measured by the transducer. The transducer is a device capable of transforming the energy carrying the physical or chemical information about the sample into a useful analytical signal. Transducers with potential use in meat packaging systems include electrical, optical, thermal or chemical signal domains (Kerry et al., 2006). Development of a potential sensor for rapid quantification of chemical or physical indicators of food quality is known as the 'marker approach' (Kress-Rogers, 2001). Intelligent packaging incorporating gas sensor technology provides a means by which the determination of indicator headspace gases gives information on meat quality and package integrity rapidly and inexpensively.

The use of sensors in the meat industry is limited; however, significant practical steps towards more widespread use have been made (Kerry & Papkovsky, 2002). High development and production costs, strict industry specifications, safety considerations and relatively limited demand from the meat industry and consumers are considered as the main obstacles to commercial use. Greater pressure on food manufacturers to guarantee safety, quality and traceability is likely to promote future establishment of commercial sensor technology in food packaging.

Gas Sensors

Gas sensors are devices which respond reversibly and quantitatively to gaseous analytes by changing the physical parameters of the sensor. Systems currently available for gas detection include amperometric oxygen sensors, potentiometric carbon dioxide sensors, metal oxide semiconductor field effect transistors, organic conducting polymers and piezo-electric crystal sensors (Kress-Rogers, 1998a). Conventional systems for oxygen sensors based on electrochemical methods have a number of limitations (Trettnak, Gruber, Reiniger, & Klimant, 1995) including consumption of analyte (oxygen), cross-sensitivity to carbon dioxide and hydrogen sulphide and fouling of sensor membranes. They also involve destructive analysis of packages.

In recent years, a number of instruments and materials for optical oxygen sensing have been reported (Papkovsky, Ponomarev, Trettnak, & O'Leary, 1995; Trettnak et al., 1995). Such sensors are usually comprised of a solid-state material and operate on the principle of luminescence quenching or absorbance changes caused by direct contact with the analyte. They are chemically inert, do not consume analytes, and provide a non-invasive technique for gas analysis through translucent materials.

Approaches to opto-chemical sensing include: a fluorescence-based system using a pH sensitive indicator (Wolfbeis, Weis, Leiner, & Ziegler, 1988), absorption-based colorimetric sensing realised through a visual indicator (Mills, Qing Chang, & McMurray, 1992), and an energy transfer approach using phase fluorimetric detection (Neurater, Klimant, & Wolfbeis, 1999). The latter allows for the possibility of combining oxygen and carbon dioxide measurements in a single sensor through compatibility with previously developed oxygen sensing technology. Most carbon dioxide sensors, however, have been developed for biomedical applications and the use of existing carbon dioxide sensors in food packaging applications is currently not feasible (Kerry & Papkovsky, 2002).

Fluorescence-Based Oxygen Sensors

Fluorescence-based oxygen sensors represent the most promising systems to date for remote measurement of headspace gases in packaged meat products. A number of disposable oxygen sensing prototypes have been developed which may be produced at low cost and provide rapid determination of oxygen concentration (Kerry & Papkovsky, 2002). The active component of a fluorescence-based oxygen sensor usually consists of a long-delay fluorescent or phosphorescent dye encapsulated in a solid polymer matrix. The dye-polymer coating is applied as a thin film coating on a suitable solid support. Molecular oxygen, present in the packaging headspace, penetrates the sensitive coating through simple diffusion and quenches the luminescence by a dynamic, i.e. collisional, mechanism. Oxygen is quantified by measuring changes in luminescence parameters against a pre-determined calibration. The process is reversible and clean; neither the dye nor the oxygen is consumed in the photochemical reactions involved, no by-products are generated and the whole cycle can be repeated.

Materials for oxygen sensors must meet strict sensitivity and working performance requirements in order to meet the suitability requirements for commercial intelligent packaging applications. They must also have fluorescence characteristics suited to the construction of simple measuring devices. Fluorescence and phosphorescence dyes with lifetimes in the micro-second range are best suited to oxygen sensing in food packaging. Other necessary features include suitable intensity, well resolved excitation and emission long wave bands and good photo-stability characteristics of the indicator dye. Such features allow sensor compatibility with simple opto-electronic measuring devices (LEDs, photodiodes etc), minimise interference by scattering and sample fluorescence and allow long-term operation without recalibration (Papkovsky et al., 1995). Materials using fluorescent complexes of ruthenium, phosphorescent palladium(II)- and platinum(II)-porphyrin complexes and related structures have shown considerable promise as oxygen sensors (Papkovsky et al., 1991; Papkovsky et al., 1995).

The combination of indicator dye and encapsulating polymer medium determines the sensitivity and effective working range of such sensors. For the purpose of food packaging applications, dyes with relatively long emission lifetimes ($\sim$ 40–500 µs) such as Pt-porphyrins combined with polystyrene as polymer matrix appear to offer greatest potential (Papkovsky, Papovskaia, Smyth, Kerry, & Ogurtsov, 2000). Other polymers with good gas-barrier properties such as polyamide, polyethylene terephthalate and PVC are not suitable for oxygen sensing as oxygen quenching is slow in such media. The use of plasticized polymers is also unsuitable due to toxicity concerns associated with potential plasticizer migration. Information regarding sensor fabrication and criteria to be considered for commercial sensor uptake, in food packaging, is reviewed by Kerry et al. (2006) and Kerry & Papkovsky (2002).

The ability of fluorescence-based oxygen sensors to accurately determine oxygen levels has been demonstrated in packaged beef, poultry and pork products (Fitzgerald et al., 2001; Papkovsky, Smiddy, Papkovskaia, & Kerry, 2002; Smiddy, Fitzgerald, et al., 2002; Smiddy, Papkovskaia, Papkovsky, & Kerry, 2002; Smiddy, Papkovskaia, et al., 2002;). The development of oxygen sensors is indicative of a move towards commercialisation of indicator-based intelligent meat packaging systems. It has been estimated that each sensor should cost less than €0.01 to produce (Kerry & Papkovsky, 2002) and impact minimally on packaged meat production costs.

Biosensors

The recently developed biosensor technologies represent an additional area with potential for application in intelligent meat packaging systems. Biosensors are compact analytical devices which detect, record and transmit information pertaining to biological reactions (Yam, Takhistov, & Miltz, 2005). They consist of a bio-receptor specific to a target analyte and a transducer to convert biological signals to a quantifiable electrical response. Bio-receptors are organic materials such as enzymes, antigens, microbes, hormones and nucleic acids. Transducers may be electrochemical, optical, or calorimetric and are system dependent. Intelligent packaging systems incorporating biosensors have the potential for extreme specificity and reliability. Market analysis of pathogen detection and safety systems for the food packaging industry suggests that biosensors offer considerable promise for future growth (Alocilja & Radke, 2003).

Toxin Guard[TM] (Toxin Alert, Ontario, Canada), a commercially available biosensor, is a patented diagnostic system incorporating antibodies printed on polyethylene-based plastic packaging capable of detecting target pathogens such as *Salmonella sp.*, *Campylobacter sp.*, *Escherichia coli 0517* and *Listeria sp.* (Bodenhammer, 2002; Bodenhammer, Jakowski, & Davies, 2004). When food packaging comes into contact with targeted bacteria, a positive test, indicated by a dramatic visual signal, alerts the consumer or retailer. Toxinguard[TM] can be targeted to detect freshness degradation, as well as the presence of specific food hazards such as pesticides or indicators of genetic modification.

Bioett AB (Lund, Sweden) has developed a system based on a biosensor for temperature monitoring. The Bioett System monitors the accumulated effect of temperature on products over time. The system consists of a chip-less RF circuit with a built-in biosensor which can be read with a handheld scanner at various points in the supply chain. Information is stored in a database and can be used to analyze the cold chain and validate that the agreed temperature has been maintained. A Time Temperature Biosensor (TTB), attached to 5 kg cases of frozen meat balls, is activated at source. The biosensor registers the accumulated temperatures that the product has been exposed to and this information can be used to optimize and monitor the cold-chain distribution system. A scanner can read the biosensor via radio waves (radio frequency) and also uniquely identifies the goods using a barcode system. The scanner also incorporates a software defined radio subsystem which can also be used for reading of RFID tags. Such systems give an insight into products that are likely to become mainstream in years to come.

Indicators

Indicators may be defined as substances which indicate the presence, absence or concentration of another substance, or the degree of reaction between two or more substances by means of a characteristic change, especially in colour. By contrast with sensors, indicators are not composed of receptor and transducer components and communicate information directly through a visual change (Kerry et al., 2006). A number of commercially available indicators are available for use with packaged meats and meat products.

Integrity Indicators

Non-invasive indicator systems provide qualitative or semi-quantitative information through visual colorimetric changes or through comparison with standard references. The majority of indicators have been developed to test package integrity. The most common cause of integrity damage in flexible plastic packages is associated with leaking seals (Hurme, 2003). Attachment of a leak indicator or sensor to a package ensures package integrity through the distribution chain.

A number of studies on package integrity in modified atmosphere packaged meat products (Ahvenainen, Eilamo, & Hurme, 1997; Eilamo, Ahvenainen, Hurme, Heiniö, & Mattila-Sandholm, 1995; Randell et al., 1995; Smolander, Hurme, & Ahvenainen, 1997) have established critical leak sizes and associated quality deterioration. Standard (destructive) manual methods for package integrity and leak testing are both laborious and can test only limited numbers of packs (Hurme, 2003). Currently available non-destructive detection systems have disadvantages such as requirements for specialised equipment, slow sampling time and an inability to detect leakages that are penetrable by pathogens (Hurme & Ahvenainen, 1998; Stauffer, 1988). Much of the research into integrity detection has focused on visual oxygen indicators for modified atmosphere packaged products. Many visual oxygen indicators have been patented and are based mainly on the use of redox dyes (Davies & Gardner, 1996; Krumhar & Karel, 1992; Mattila-Sandholm, Ahvenainen, Hurme, & Järvi-Kääriänen, 1995; Yoshikawa, Nawata, Goto, & Fujii, 1987). Testing and validation of such devices in modified atmosphere packaged minced steaks and minced meat pizzas have been reported (Ahvenainen et al., 1997; Eilamo et al., 1995). Disadvantages of such devices include high sensitivity (colour change as a result of low oxygen concentrations (0.1%)) and reversibility (leak-induced increases in oxygen consumed through subsequent microbial growth).

19 Smart Packaging Technologies

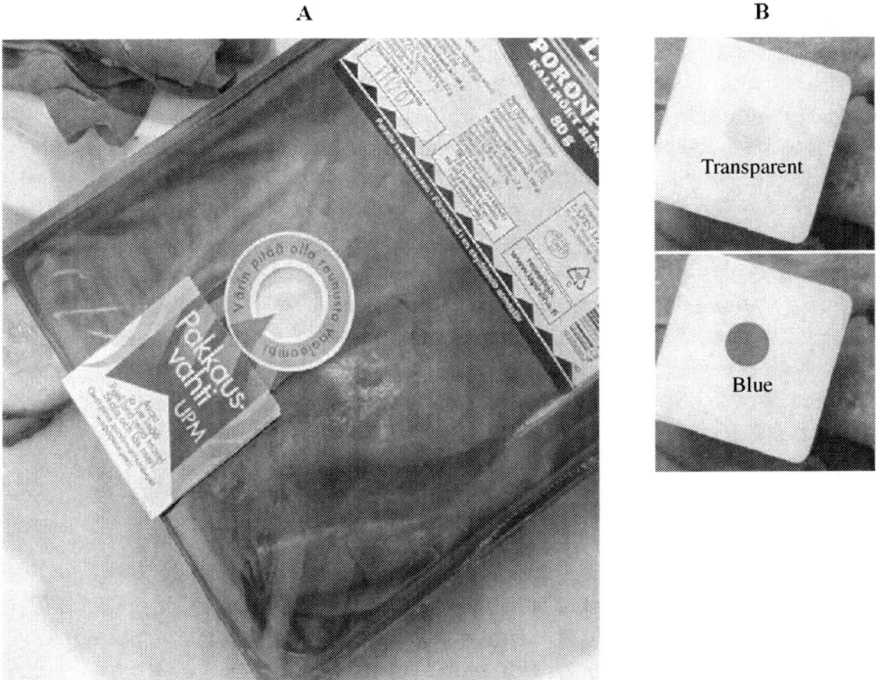

Reproduced with permission from UPM.

Fig. 19.3 The UPM Shelf Life Guard (UPM, Helsinki, Finland)

A commercial example, the UPM Shelf Life Guard (UPM, Helsinki, Finland) indicator label monitors the integrity of modified atmosphere packages (Fig. 19.3a). The label is attached to the inside of the packaging so that it can be viewed through the packages transparent outer shell. The presence of air in modified atmosphere packs reduces meat product shelf life. The label contains a redox dye, held between laminated layers, which react with oxygen. In modified atmosphere packages containing carbon dioxide, nitrogen or a mixture of both, the dye remains transparent. A colour change from transparent to blue (Fig. 19.3b) indicates that air has replaced the gases within the modified atmosphere pack i.e., the package is no longer intact and has been damaged, or a leak has occurred. The color change of the packaging label allows the consumer to make a personal assessment of the product based on his/her own sensory findings and consume the product well ahead of the best before date. UPM Shelf Life Guard has been successfully tested by Lapin Liha, a Finnish company who incorporated the labels into their packaged reindeer meat products in Tampere University Hospital in Finalnd and in Wigren 'siskonmakkara' traditional Finnish sausages. UPM also manufacture Freshness Guard where the indicator reacts to growing levels of nitrogen compounds in poultry or fish products stored in vacuum or modified atmosphere packages. Both UPM Shelf Life Guard and the Freshness Guard adhere to the European Union regulations (EU 1935/2004) with regard to food packaging.

An indicator system, specifically designed for modified atmosphere packaged foods, containing both an oxygen sensitive dye and an oxygen absorbing component exemplifies active and intelligent packaging in a single system (Mattila-Sandholm, Ahvenainen, Hurme, & Järvi-Kääriänen, 1998). A number of companies have manufactured oxygen indicators, the main application of which has been confirmation of the function of oxygen absorbers. Trade names for such devices include Vitalon®, Samso-Checker® and Ageless Eye® (Fig. 19.4). The Ageless Eye® is an in-package monitor which indicates the presence of oxygen at a glance.

A visual carbon dioxide indicator system consisting of calcium hydroxide (carbon dioxide absorber) and a redox indicator dye incorporated in polypropylene resin was described by Hong and Park (2000) and may be applicable to certain meat packaging applications.

Freshness Indicators

Freshness indicators provide direct product quality information resulting from microbial growth or chemical changes within a food product. Microbiological quality may be determined through reactions between indicators placed within the package and microbial growth metabolites (Smolander, 2003). The number of practical concepts of intelligent package indicators for freshness detection is limited; however, potential exists for the development of freshness indicators based on established knowledge of quality indicating metabolites. The improved detection of biochemical changes during storage and spoilage of foods (Dainty, 1996; Nychas, Drosinos, & Board, 1998) provide the basis by which freshness indicators may be

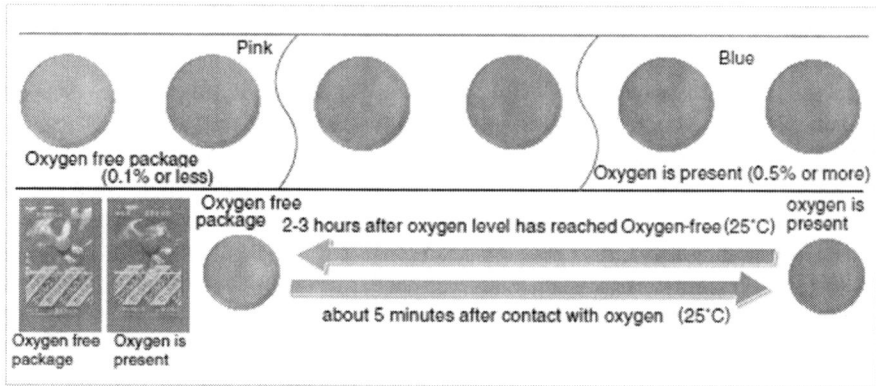

Reproduced with permission from Mitsubishi Gas Chemical Co.

Fig. 19.4 Ageless-Eye® oxygen indicator used for oxygen detection (Mitsubishi Gas Chemical Co., Japan)

developed based on target metabolites associated with microbiologically-induced deterioration.

The formation of different potential indicator metabolites in meat products is dependent on the interaction between product type, associated spoilage flora, storage conditions and the packaging system.

A number of marker metabolites associated with muscle food products exist upon which indicator development may be based. Organic acids such as n-butyrate, L-lactic acid, D-lactate and acetic acid change concentration during storage and offer potential as indicator metabolites for a number of meat products (Shu, Håkanson, & Mattiason, 1993). Colour based pH indicators offer potential for use as indicators of these microbial metabolites. Ethanol is an important indicator of fermentative metabolism of lactic acid bacteria. Randell et al. (1995) reported an increase in the ethanol concentration of anaerobic modified atmosphere packaged marinated chicken as a function of storage time.

Biogenic amines, formed from the decarboxylation of amino acids, are an indicator of bacterial growth and spoilage. Given the toxicological concerns associated with these compounds and their lack of impact on sensory meat quality, the development of effective amine indicators would prove advantageous. Food Quality Sensor International (FQSI, Inc., Lexington, MA USA) has developed a revolutionary new smart sensor label which senses spoilage in fresh meat and poultry products (Fig. 19.5). The label functions by detecting volatile biogenic amines and is unaffected by modified atmosphere packaging gases. The SensorQ^{TM} stick-on sensor label is applied by the meat packer to the inside wrap of meat and poultry packages to provide the consumer with a clear indication of product freshness. When the inside of the quality 'Q' on the label is tangerine orange, the product is fresh (Fig. 19.5a). When bacterial growth inside the package reaches a critical level, the

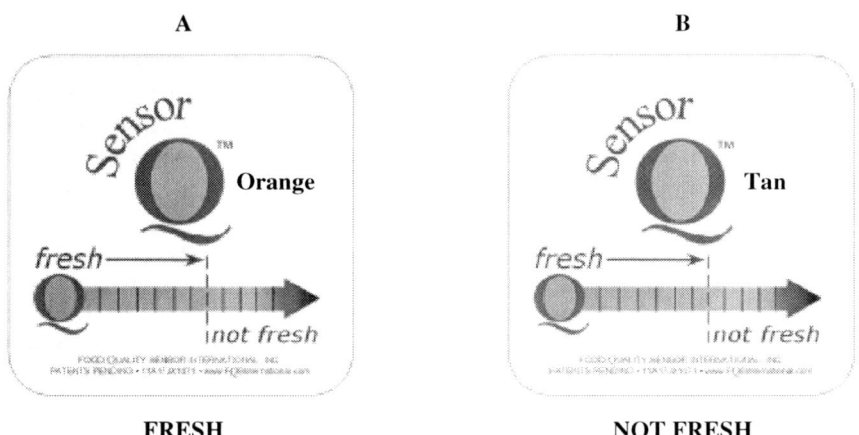

Reproduced with permission from Food Quality Sensor International, Inc.

Fig. 19.5 SensorQ^{TM} Smart Sensor Label (Food Quality Sensor International (FQSI), Inc., USA)

orange colour turns to tan which indicates spoilage (Fig. 19.5b). The SensorQ label is made of food grade materials and is economical, costing less than 1% of the total value of the average package of meat or poultry it labels. Market research indicated that over 95% of consumers would opt to purchase a product with a freshness indicator attached. FSQI are currently in the final stages of testing and validating SensorQ and expect to launch the product in late 2008.

Hydrogen sulphide, a breakdown product of cysteine, with intense off-flavours and low threshold levels, is produced during the spoilage of meat and poultry by a number of bacterial species. It forms a green pigment when bound to myoglobin and this pigment formed the basis for the development of an agarose-immobilised, myoglobin-based freshness indicator in un-marinated broiler pieces (Smolander et al., 2002). The indicator was not affected by the presence of nitrogen or carbon dioxide.

Freshness indicators based on broad spectrum colour changes have a number of disadvantages which need to be resolved before widespread commercial uptake is likely. A lack of specificity means that colour changes indicating contamination can occur in products free from any significant sensory or quality deterioration. The presence of certain target metabolites is not necessarily an indication of poor quality. More exact correlations need to be established between target metabolite, product type and organo-leptic quality and safety. The possibilities of false-negatives are likely to dissuade producers from adopting indicators unless specific indication of actual spoilage can be guaranteed (Kerry et al., 2006).

Time-Temperature Indicators

A time-temperature indicator (or integrator) (TTI) is defined as a device (small tag or label) used to show a measurable, time-temperature dependent change that reflects the full or partial temperature history of a food product to which it is attached (Taoukis & Labuza, 1989). Operation of TTIs is based on mechanical, chemical, electrochemical, enzymatic or microbiological change, usually expressed as a visible response in the form of a mechanical deformation, colour development or colour movement (Taoukis & Labuza, 2003). Therefore, the visible response gives a cumulative indication of the storage temperature to which the TTI has been exposed. TTIs are classified as either partial history or full history indicators, depending on their response mechanism.

Effective TTIs are required to indicate clear, continuous, irreversible reaction to changes in temperature. TTIs should also be small, reliable, low cost, easily integrated into food packaging, have a long pre- and post-activation shelf life and be unaffected by ambient conditions other than temperature. TTIs should also be flexible to a range of temperatures, robust, pose no toxicological or safety hazard and convey information in a clear manner.

A large number of TTI types have been developed and patented, the principles and applications of which have been reviewed previously (Taoukis & Labuza, 2003). TTIs currently commercially available include a number of diffusion, enzymatic and polymer-based systems.

Diffusion-Based TTIs

The 3M Monitor Mark® (3M Company, St. Paul, Minnesota, USA) is a non-reversible indicator, dependent on the diffusion of a coloured fatty acid ester along a porous wick made of high quality blotting paper. The measurable response is the distance of the advancing diffusion front from the origin. The useful range of temperatures and the response life of the TTI are determined by the type and concentration of ester. FreshnessCheck® (3M Company) incorporates a visco-elastic material which migrates into a diffusively light-reflective porous matrix at a temperature dependent rate. This results in a progressive change in light transmission of the porous matrix and provides a visual response.

Enzymatic TTIs

The VITSAB® TTI (VITSAB A.B., Malmö, Sweden) is based on a pH induced colour change resulting from controlled enzymatic hydrolysis of a lipid substrate. The indicator consists of two separate compartments containing an aqueous solution of lipolytic enzymes, the lipid substrate suspended in an aqueous medium and a pH indicator mix. Different enzyme–substrate combinations are available to give a variety of response lives and temperature dependencies. Activation of the TTI is brought about by mechanical breakage of the seal separating the two compartments. Hydrolysis of the substrate decreases the pH and results in a colour change from dark green to bright yellow. Visual evaluation of the colour change is made by reference to a five-point colour scale. CheckPoint® labels are the latest TTIs developed by VITSAB, which comprise a label type designed to create a better subjective reading response for users and offer direct application to poultry and ground beef products.

The Cryolog (Gentilly, France) (eO)® adhesive TTI label is in the form of a small gel pad, shaped like the petals of a flower, which changes from green (good) to red (not good). The pH induced colour change is due to a microbial growth within the gel itself. The TRACEO® (Cryolog) transparent adhesive label is designed for use on refrigerated products and placed over the barcode. The colour of the transparent adhesive label changes from colourless to red when the product is no longer fit for consumption.

Polymer-Based TTIs

The Fresh-Check® TTI (TEMPTIME Corporation, NJ, USA) is a self-adhesive device specifically formulated to match the shelf life of food products to which it is affixed (Fig. 19.6). The reactive centre circle portion of the Fresh-Check® label, based on solid-state polymerization of substituted monomers, darkens irreversibly in response to specific time and temperature. The colour change is faster at higher temperatures and slower at lower temperatures. When the centre circle is lighter than the reference colour, the product is safe to use assuming it is within its expiration date. When the centre circle has the same colour as the reference colour, the endpoint

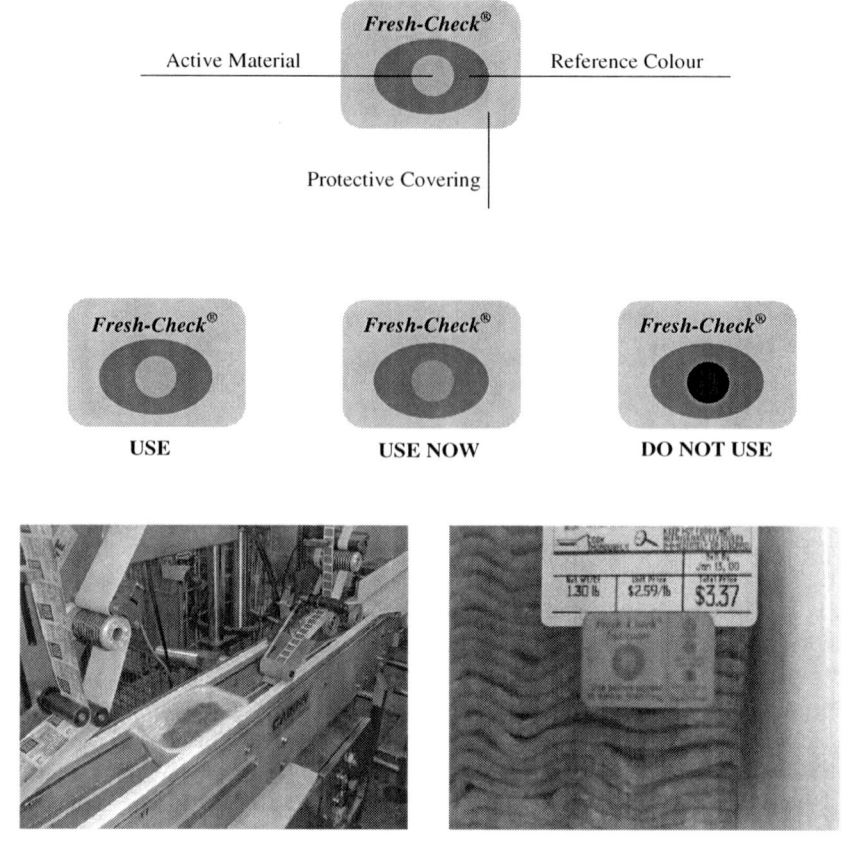

Reproduced with permission from TempTime Corporation.

Fig. 19.6 The Fresh-Check indicator label (TEMPTIME Corporation, NJ, USA)

has been reached and the product should not be used. When the centre circle is darker than the reference colour, the endpoint has been exceeded and the product should not be used. Fresh-Check indicators act as a warning device to show if the product has been exposed to excessive heat at any time during storage, distribution and through final use. Warning is through a visual indicator that a time temperature profile, as specified by the manufacturer has been exceeded. Commercially, Fresh-Check® labels are currently used in modified atmosphere packaged ground beef (Fig. 19.6).

The OnVu^{TM} TTI labels (Ciba Specialty Chemicals Inc., Switzerland) are based on organic pigments which change colour with time at rates determined by temperature (Fig. 19.7). The TTI label consists of a heart shaped 'apple' motif containing an inner heart shape. The image is stable until activated by UV light from an LED lamp, when the inner heart changes to a deep blue colour. A filter is then added over

19 Smart Packaging Technologies

Reproduced with permission from Ciba Specilaity Chemicals.

Fig. 19.7 The OnVu^{TM} time-temperature indicator (Ciba Speciality Chemicals Inc., Switzerland and Freshpoint Holdings SA, Switzerland)

the label to prevent it from being recharged. The blue inner heart changes to white as a function of time and temperature (Fig. 19.7). The system can be applied as a label or printed directly onto the package.

The Food Sentinel System™ from SIRA Technologies, USA uses a modified barcode containing a proprietary thermo-chromic printing ink which is printed in a non-scannable colour. On encountering product abuse, the thermo-chromic ink changes to an irreversible deep magenta colour which is visible and detectable during scanning (Fig. 19.8). Therefore it is possible to add a temperature and shelf life monitor to any product barcode, thus preventing the sale of contaminated food and archiving of the incident.

A number of TTI systems have their histories read by RFID readers (see Sect. 'Radio Frequency Identification') rather than by visual means. These include Bioett®, Timestrip®, KSW Microtec® and TempTime®. The use of electronic readers allows storage of information and subsequent downloading to local networks and databases.

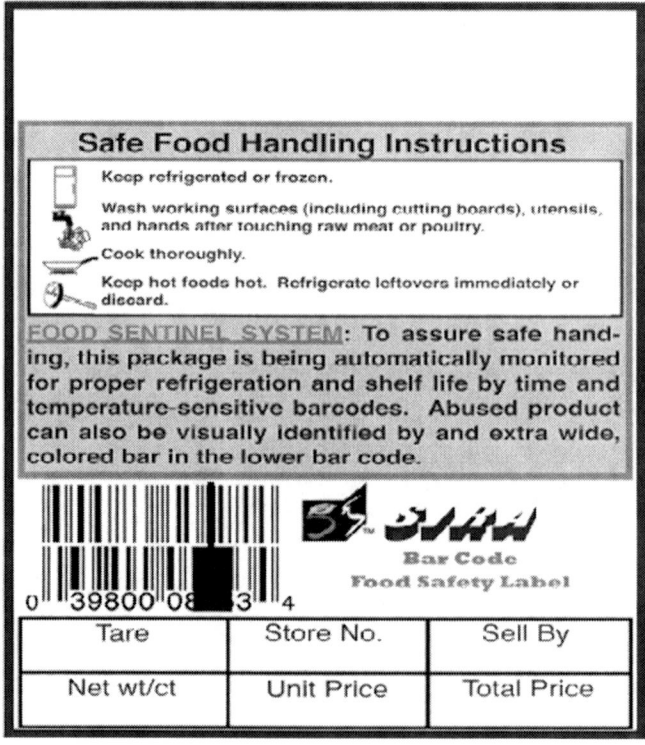

Reproduced with permission from Sira Technologies.

Fig. 19.8 The Food Sentinel System™ (Sira Technologies, CA, USA)

The use of TTIs, applied to packaged food products, has been reported previously (Riva, Piergiovanni, & Schiraldi, 2001; Shimoni, Anderson, & Labuza, 2001; Welt, Sage, & Berger, 2003). In frozen pork, Yoon, Lee, Kim, Kim, & Park (1994) demonstrated a positive correlation between oxidative stability and TTI colour change using a phospholipid/phospholipase-based TTI. Smolander, Alakomi, Ritvanen, Vainionpää, and Ahvenainen (2004) determined the applicability of VITSAB®, Fresh-Check® and 3M Monitor® TTIs for monitoring the quality of modified atmosphere packaged broiler cuts at different temperatures. Further details on the suitability for, and projected uptake of TTIs by the food industry are reviewed by Kerry et al. (2006).

Radio Frequency Identification

Radio Frequency Identification (RFID) is an electronic, information-based form of intelligent packaging. RFID uses tags to capture, store and transmit accurate real-time information about assets (cattle, containers, bins etc) to a user's information system (Townsend & Mennecke, 2008). RFID is one of the many automatic-identification technologies (a group which includes barcodes) and offers a number of potential benefits to the meat production, distribution and retail chain. These include traceability, inventory management, labour saving costs, security and promotion of quality and safety (Mousavi, Sarhadi, Lenk, & Fawcett, 2002).

RFID systems include two basic components: the interrogator (transmitter and receiver), and the transponder (the tag itself) (Townsend & Mennecke, 2008). While tags are relatively simple, better inventory information than barcode or human entry systems can be gained through tracking software. RFID tags have an advantage over barcoding, as the tags can be embedded within a container or package without adversely affecting the data. RFID tags also provide a non-contact, non-line-of-sight ability to gather real-time data and can penetrate non-metallic materials including bio-matter (Mennecke & Townsend, 2005). Tags can hold simple information (such as identification numbers) for tracking or can carry more complex information (with storage capacity at present up to about 1MB) such as temperature and relative humidity data, nutritional information, cooking instructions. Read-only and read/write tags are also available depending on the requirements of the application in question.

Tags are classified into two types: active tags function with battery power, broadcast a signal to the RFID reader and operate at a distance of up to 50 metres. Passive tags have a shorter reading range (up to about 5 metres) and are powered by the energy supplied by the reader. Common RFID frequencies range from low ($\sim$ 125 KHz) to UHF (850–900 MHz) and microwave frequencies ($\sim$ 2.45 GHz). Low frequency tags are cheaper, use less power and are better able to penetrate non-metallic objects (Townsend & Mennecke, 2008).

At present, Canada and Australia have mandated live-animal traceability from birth to slaughter using RFID-based systems. The key to individual animal traceability lies in the ability to transfer animal information sequentially and accurately to subparts of the animal during production. RFID-based tracking systems provide

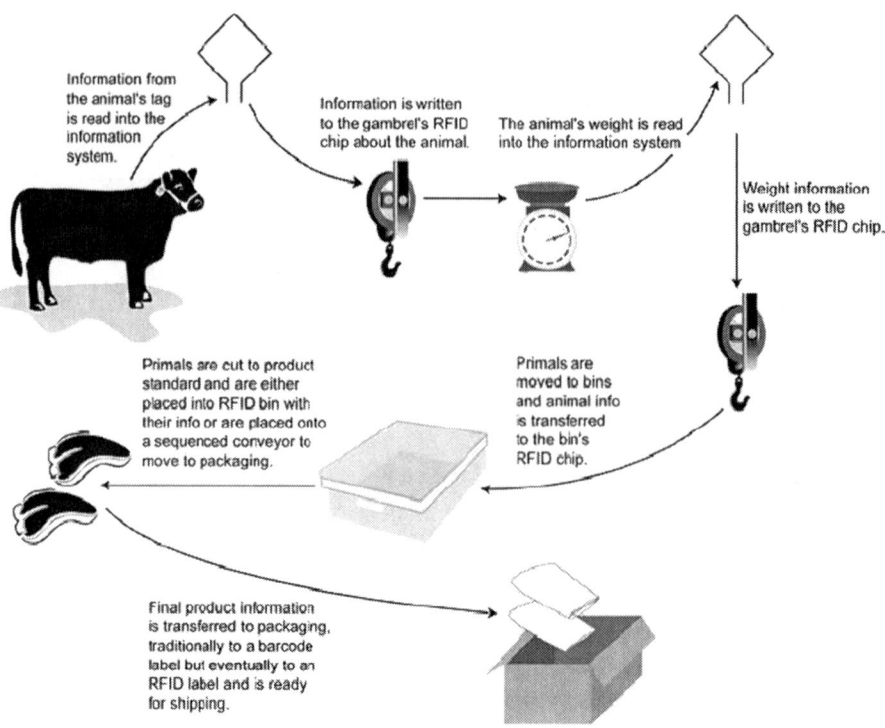

Reproduced with permission from CABI.

Fig. 19.9 Illustration of a fundamental RFID tracking process (Townsend & Mennecke, 2008)

an automated method of contributing significantly to that information exchange (Mennecke & Townsend, 2005). Currently, individually RFID tagged meat products are not available to the consumer, although the use of RFID tagging of meat cuts has been extended to the pig processing industry (Dalehead Foods, Cambridge, UK), where tracking occurs from the individual pig to its subsequent primal pieces i.e. hams. Although the purpose of this tracking scheme is for quality control, employee accountability and precision cutting, and does not extend beyond the cutting room floor or provide information about the individual animal with the final product, it does exemplify the developing use of RFID technology within the meat industry. An illustration of a fundamental RFID tracking process is outlined in Fig. 19.9. Future aspects and costs associated with RFID technology are reviewed by Kerry et al. (2006).

Convenience Smart Packaging Applications in Meat Products

Changes in consumer type, attitudes and preferences, has increased the popularity of convenience-orientated ready-prepared meals. As a result, the demand for food packaging has increased prompting further development and growth in smart packaging technologies. In response to demands and growth in the 'Heat n' Serve' convenience food category, the Kepak Group (Ireland) developed the 'Global Cuisine' range (Fig. 19.10) consisting of pre-cooked meat joints (beef, pork, chicken and turkey) combined with natural gravy and vacuum skin packaged in pre-formed trays. Cryovac Darfresh® vacuum skin packaging technology enables products to be cooked, shipped, stored, displayed, sold, re-heated and served all in the same package. From a manufacturing perspective, vacuum skin packaging results in fewer processing steps and eliminates secondary product handling. Products can be cooked (re-heated) by the consumer in a microwave in approximately 7 minutes, and have a shelf life of up to 21 days when stored between 0 and 4°C. The package does not require ventilation holes to be punctured in the package before microwave heating. During product heating, the vacuum-skin film forms a bubble (Fig. 19.11a–d), trapping moisture and flavour, subsequently self-vents and relaxes over the product. The 'stay cool' side handles reduce the risk of burns as the tray is removed from the microwave and eliminates handling messy hot pouches associated with current 'Heat n' Serve' microwavable meals. The package contains an easy-open feature which enables consumers to peel away the cover film and serve after heating.

©*Food Packaging Group, University College Cork (UUC), Ireland.*

Fig. 19.10 Global Cuisine 'Heat n' Serve' range (Kepak Group, Ireland)

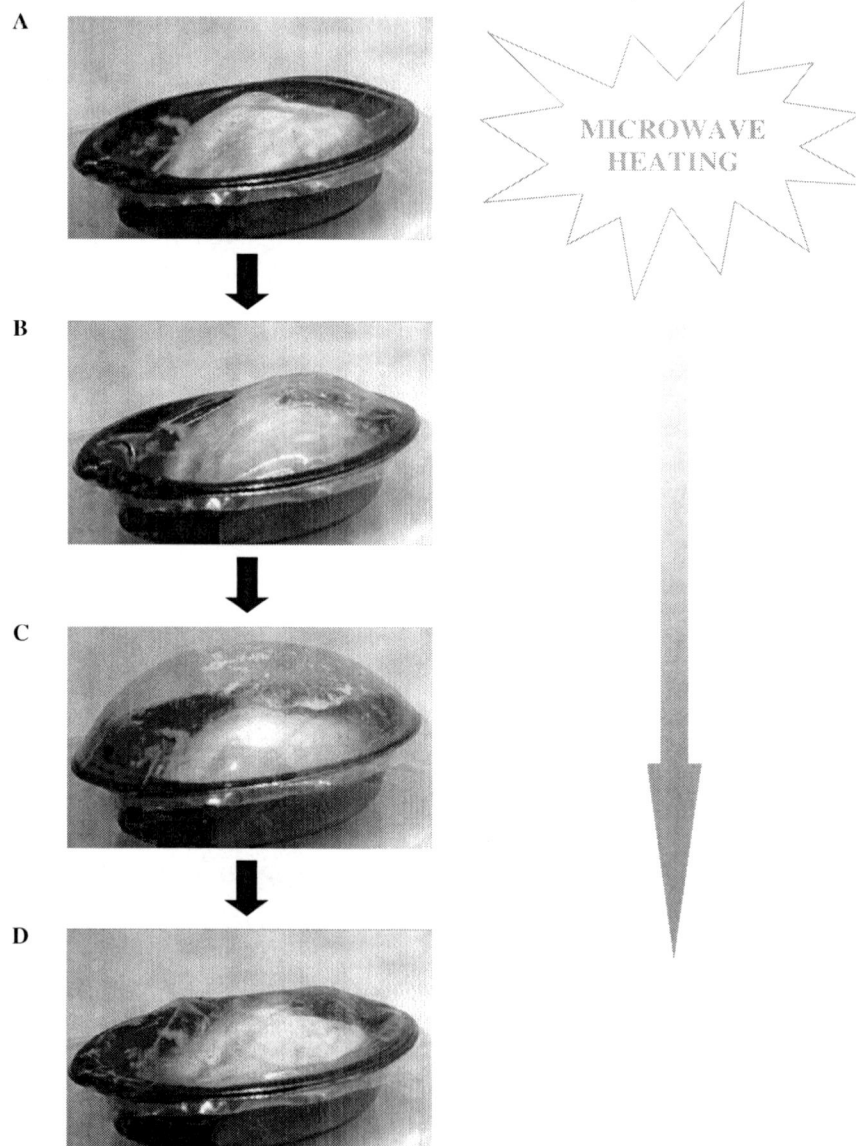

©*Food Packaging Group, UCC.*

Fig. 19.11 Schematic representation (a–d) of the bubble formed by the vacuum-skin film during microwave re-heating of a Global Cuisine 'Heat n' Serve' meat product

©*Food Packaging Group, UCC.*

Fig. 19.12 Ready meal packaging (Marks and Spencer plc) containing a SmartCode for use with a Smart Oven

Smart cooking is a new cooking innovation combining the cooking capabilities of a convection oven with microwave and grill cooking. The Smart cooking process is made possible through the innovation of Smart ovens (e.g. Samsung BCE 1197) which read special SmartCodes (2Dimentional barcodes). The SmartCode is scanned by the oven scanner and the Smart oven converts the code into a cooking instruction. Every SmartCode (Fig. 19.12) contains a unique set of instructions which provide the Smart oven with the correct temperature, microwave power and time to cook the food to perfection. SmartCodes cook meals to the same quality each time delivering consistent oven cooked results.

Conclusions

Interest in the use of smart packaging (active and intelligent) technologies has increased in recent years. The most highly developed and widely used form of active packaging is oxygen absorption or scavenging systems. Additional active packaging technologies include carbon dioxide scavengers and emitters, moisture control agents and antimicrobial packaging technologies. Intelligent packaging technologies are systems which monitor the condition of packaged foods to give information regarding the quality of the packaged food during transport and storage. Sensor technology, indicators (integrity, freshness and time-temperature (TTI) indicators) and radio frequency identification (RFID) all have potential applications in smart packaging of meat and meat products; however, further research is necessary. The increased demand for ready prepared meals will serve to further advance smart packaging research and technologies, for use in the convenience food sector. The recognition of the benefits of smart packaging by the meat industry, development of economically viable packaging systems and increased consumer acceptance are necessary for commercial realization of smart packaging technologies for use with meat and meat products in the future.

References

Ahvenainen, R. (2003). Active and intelligent packaging: An introduction. In R. Ahvenainen (Ed.), *Novel food packaging techniques* (pp. 5–21). Cambridge, UK: Woodhead Publishing Ltd.

Ahvenainen, R., Eilamo, M., & Hurme, E. (1997). Detection of improper sealing and quality deterioration of modified-atmosphere-packed pizza by a colour indicator. *Food Control, 8*, 177–184.

Alocilja, E. C., & Radke, S. M. (2003). Market analysis of biosensors for food safety. *Biosensors and Bioelectronics, 18*, 841–846.

Appendini, P., & Hotchkiss, J. H. (2002). Review of antimicrobial food packaging. *Innovative Food Science and Emerging Technologies, 3*, 113–126.

Bell, R. G. & Bourke, B. J. (1996). Recent developments in packaging of meat and meat products. *Proceedings of the international developments in process efficiency and quality in the meat industry* (pp. 99–119). Dublin Castle, Ireland.

Bodenhammer, W. T. (2002). Method and apparatus for selective biological material detection. *US Patent 6376204*.

Bodenhammer, W. T., Jakowski, G., & Davies, E. (2004). Surface binding of an immunoglobulin to a flexible polymer using a water soluble varnish matrix. *US Patent 6692973*.

Butler, B. L. (2002). Cryovac® OS2000™ Polymeric oxygen scavenging systems. *Presented at Worldpak 2002*. From http://www.sealedair.com/library/articles/article-os2000.html.

Coma, V. (2008). Bioactive packaging technologies for extended shelf life of meat-based products. *Meat Science, 78*, 90–103.

Dainty, R. H. (1996). Chemical/biochemical detection of spoilage. *International Journal of Food Microbiology, 33*, 19–33.

Davies, E. S., & Gardner, C. D. (1996). Oxygen indicating composition. *British Patent 2298273*.

Eilamo, M., Ahvenainen, R., Hurme, E., Heiniö, R. L., & Mattila-Sandholm, T. (1995). The effect of package leakage on the shelf life of modified atmosphere packed minced meat steaks and its detection. *Lebensmittel-Wissenschaft und Technologie, 28*, 62–71.

Fitzgerald, M., Papkovsky, D. B., Kerry, J. P., O'Sullivan, C. K., Buckley, D. J., & Guilbault, G. G. (2001). Nondestructive monitoring of oxygen profiles in packaged foods using phase-fluorimetric oxygen sensor. *Journal of Food Science, 66*, 105–110.

Floros, J. D., Dock, L. L., & Han, J. H. (1997). Active packaging technologies and applications. *Food Cosmetics and Drug Packaging, 20*, 10–17.

Georgala, D. L., & Davidson, C. L. (1970). Food package. *British Patent 1199998*.

Han, J. H. (2000). Antimicrobial food packaging. *Food Technology, 54*, 56–65.

Hong, S. I., & Park, W. S. (2000). Use of color indicators as an active packaging system for evaluating kimchi fermentation. *Journal of Food Engineering, 46*, 67–72.

Hurme, E. (2003). Detecting leaks in modified atmosphere packaging. In R. Ahvenainen (Ed.), *Novel food packaging techniques* (pp. 276–286). Cambridge, UK: Woodhead Publishing Ltd.

Hurme, E., & Ahvenainen, R. (1998). A nondestructive leak detection method for flexible food packages using hydrogen as a tracer gas. *Journal of Food Protection, 61*, 1165–1169.

Hutton, T. (2003). Food packaging: An introduction. *Key topics in food science and technology – number 7* (p. 108). Chipping Campden, Gloucestershire, UK: Campden and Chorleywood Food Research Association Group.

Kerry, J. P., O'Grady, M. N., & Hogan, S. A. (2006). Past, current and potential utilization of active and intelligent packaging systems for meat and muscle-based products: A review. *Meat Science, 74*, 113–130.

Kerry J. P., & Papkovsky, D. B. (2002). Development and use of non-destructive, continuous assessment, chemical oxygen sensors in packs containing oxygen sensitive foodstuffs. *Research Advances in Food Science, 3*, 121–140.

Kress-Rogers, E. (1998a). Chemosensors, biosensors and immunosensors. In E. Kress-Rodgers (Ed.), *Instrumentation and sensors for the food industry* (pp. 581–669). Cambridge, UK: Woodhead Publishing Ltd.

Kress-Rogers, E. (1998b). Terms in instrumentation and sensors technology. In E. Kress-Rodgers (Ed.), *Instrumentation and sensors for the food industry* (pp. 673–691). Cambridge, UK: Woodhead Publishing Ltd.

Kress-Rogers, E. (2001). Instrumentation for food quality assurance. In E. Kress-Rodgers, & C. J. B. Brimelow (Eds.), *Instrumentation and sensors for the food industry* (2nd ed., pp. 581–669). Cambridge, UK: Woodhead Publishing Ltd.

Krumhar, K. C., & Karel, M. (1992). Visual indicator system. *US Patent 5096813*.

Lövenklev, M., Artin, I., Hagberg, O., Borch, E., Holst, E., & Rådström, P. (2004). Quantitative interaction effects of carbon dioxide, sodium chloride, and sodium nitrite on neurotoxin gene expression in nonproteolytic *Clostridium botulinum* type B. *Applied and Environmental Microbiology, 70*, 2928–2934.

Mattila-Sandholm, T., Ahvenainen, R., Hurme, E., & Järvi-Kääriänen, T. (1995). Leakage indicator. *Finnish Patent 94802*.

Mattila-Sandholm, T., Ahvenainen, R., Hurme, E., & Järvi-Kääriänen, T. (1998). Oxygen sensitive colour indicator for detecting leaks in gas-protected food packages. *European Patent EP 0666977*.

Mennecke, B., & Townsend, A. (2005). Radio frequency identification tagging as a mechanism of creating a viable producer's brand in the cattle industry. *MATRIC (Midwest Agribusiness research and Information Center) Research Paper 05-MRP 8* (http://www.matric.iastate.edu).

Mills, A., Qing Chang, Q., & McMurray, N. (1992). Equilibrium studies on colorimetric plastic film sensors for carbon dioxide. *Analytical Chemistry, 64*, 1383–1389.

Møller, J. K. S., Jensen, J. S., Olsen, M. B., Skibsted, L. S., & Bertelsen, G. (2000). Effect of residual oxygen on colour stability during chill storage of sliced, pasteurised ham packaged in modified atmosphere. *Meat Science, 54*, 399–405.

Mousavi, A., Sarhadi, M., Lenk, A., & Fawcett, S. (2002). Tracking and traceability in the meat processing industry: A solution. *British Food Journal, 104*, 7–19.

Neurater, G., Klimant, I., & Wolfbeis, O. S. (1999). Microsecond lifetime-based optical carbon dioxide sensor using luminescence resonance energy transfer. *Analytica Chimica Acta, 382*, 67–75.

Nychas, G. E., Drosinos, E. H., & Board, R. G. (1998). Chemical changes in stored meat. In A. Davies, & R. G. Board (Eds.), *The microbiology of meat and poultry* (pp. 288–326). London: Blackie Academic & Professional.

O'Grady, M. N., Monahan, F. J., Bailey, J., Allen, P., Buckley, D. J., & Keane, M. G. (1998). Colour-stabilising effect of muscle vitamin E in minced beef stored in high oxygen packs. *Meat Science, 50*, 73–80.

Ozdemir, M., & Floros, J. D. (2004). Active food packaging technologies. *Critical Reviews in Food Science and Nutrition, 44*, 185–193.

Papkovsky, D. B., Olah, J., Troyanovsky, I. V., Sadovsky, N. A., Rumyantseva, V. D., Mironov, A. F., et al. (1991). Phosphorescent polymer films for optical oxygen sensors. *Biosensors and Bioelectronics, 7*, 199–206.

Papkovsky, D. B., Papovskaia, N., Smyth, A., Kerry, J. P., & Ogurtsov, V. I. (2000). Phosphorescent sensor approach for a non-destructive measurement of oxygen in packaged foods. *Analytical Letters, 33*, 1755–1777.

Papkovsky, D. B., Ponomarev, G. V., Trettnak, W., & O'Leary, P. (1995). Phosphorescent complexes of porphyrin-ketones: Optical properties and application to oxygen sensing. *Analytical Chemistry, 67*, 4112–4117.

Papkovsky, D. B., Smiddy, M. A., Papkovskaia, N. Y., & Kerry, J. P. (2002). Nondestructive measurement of oxygen in modified atmosphere packaged hams using a phase-fluorimetric sensor system. *Journal of Food Science, 67*, 3164–3169.

Quintavalla, S., & Vicini, L. (2002). Antimicrobial food packaging in meat industry. *Meat Science 62*, 373–380.

Randell, K., Ahvenainen, R., Latva-Kala, K., Hurme, E., Mattila-Sandholm, T., & Hyvönen, L. (1995). Modified atmosphere-packed marinated chicken breast and rainbow trout quality as affected by package leakage. *Journal of Food Science, 60*, 667–672, 684.

Riva, M., Piergiovanni, L., & Schiraldi, A. (2001). Performance of time-temperature indicators in the study of temperature exposure of packaged fresh foods. *Packaging Technology and Science, 14*, 1–9.

Rooney, M. L. (1995). Active packaging in polymer films. In M. L. Rooney (Ed.), *Active food packaging* (pp. 74–110). Glasgow: Blackie Academic and Professional.

Seideman, S. C., & Durland, P. R. (1984). The utilization of modified atmosphere packaging fro fresh meat: A review. *Journal of Food Quality, 6*, 239–252.

Shimoni, E., Anderson, E. M., & Labuza, T. P. (2001). Reliability of time temperature indicators under temperature abuse. *Journal of Food Science, 66*, 1337–1340.

Shu, H. C., Håkanson, E. H., & Mattiason, B. (1993). D-lactic acid in pork as a freshness indicator monitored by immobilized D-lactate dehydrogenase using sequential injection analysis. *Analytica Chimica Acta, 283*, 727–737.

Sivertsvik, M. (2003). Active packaging in practice: Fish. In R. Ahvenainen (Ed.), *Novel food packaging techniques* (pp. 384–400). Cambridge, UK: Woodhead Publishing Ltd.

Smiddy, M., Fitzgerald, M., Kerry, J. P., Papkovsky, D. B., O'Sullivan, C. K., & Guilbault, G. G. (2002). Use of oxygen sensors to non-destructively measure the oxygen content in modified atmosphere and vacuum packed beef: Impact of oxygen content on lipid oxidation. *Meat Science, 61*, 285–290.

Smiddy, M., Papkovskaia, N., Papkovsky, D. B., & Kerry, J. P. (2002). Use of oxygen sensors for the non-destructive measurement of the oxygen content in modified atmosphere and vacuum packs of cooked chicken patties; impact of oxygen content on lipid oxidation. *Food Research International, 35*, 577–584.

Smiddy, M., Papkovsky, D. B., & Kerry, J. P. (2002). Evaluation of oxygen content in commercial modified atmosphere packs (MAP) of processed cooked meats. *Food Research International, 35*, 571–575.

Smith, J. P., Ramaswamy, H. S., & Simpson, B. K. (1990). Developments in food packaging technology. Part II. Storage aspects. *Trends in Food Science and Technology, 1*, 111–118.

Smolander, M. (2003). The use of freshness indicators in packaging. In R. Ahvenainen (Ed.), *Novel food packaging techniques* (pp. 128–143). Cambridge, UK: Woodhead Publishing Ltd.

Smolander, M., Alakomi, H.-L., Ritvanen, T., Vainionpää, J., & Ahvenainen, R. (2004). Monitoring of the quality of modified atmosphere packaged broiler cuts stored in different temperature conditions. A. Time-temperature indicators as quality-indicating tools. *Food Control, 15*, 217–229.

Smolander, M., Hurme, E., & Ahvenainen, R. (1997). Leak indicators for modified-atmosphere packages. *Trends in Food Science & Technology, 8*, 101–106.

Smolander, M., Hurme, E., Latva-Kala, K., Luoma, T., Alakomi, H. L., & Ahvenainen, R. (2002). Myoglobin-based indicators for the evaluation of freshness of unmarinated broiler cuts. *Innovative Food Science and Emerging Technologies, 3*, 279–288.

Stauffer, T. (1988). Non-destructive in-line detection of leaks in food and beverage packages – an analysis of methods. *Journal of Packaging Technology, 2,* 147–149.

Suppakul, P., Miltz, J., Sonneveld, K., & Bigger, S. W. (2003). Active packaging technologies with an emphasis on antimicrobial packaging and its applications. *Journal of Food Science, 68,* 408–420.

Taoukis, P. S., & Labuza, T. P. (1989). Applicability of time temperature indicators as shelf life monitors of food products. *Journal of Food Science, 54,* 783–788.

Taoukis, P. S., & Labuza, T. P. (2003). Time-temperature indicators (TTIs). In R. Ahvenainen (Ed.), *Novel food packaging techniques* (pp. 103–126). Cambridge, UK: Woodhead Publishing Ltd.

Townsend, A., & Mennecke, B. (2008). Application of radio frequency identification (RFID) in meat production: Two case studies. *CAB Reviews: Perspectives in Agriculture, Veterinary Science, Nutrition and Natural Resources, 3,* 1–10.

Trettnak, W., Gruber, W., Reiniger, F., & Klimant, I. (1995). Recent progress in optical sensor instrumentation. *Sensors and Actuators B, 29,* 219–225.

Vermeiren, L., Devlieghere, F., Van Beest, M., de Kruijf, N., & Debevere, J. (1999). Developments in the active packaging of foods. *Trends in Food Science and Technology, 10,* 77–86.

Welt, B. A., Sage, D. S., & Berger, K. L. (2003). Performance specification of time-temperature integrators designed to protect against botulism in refrigerated fresh foods. *Journal of Food Science, 68,* 2–9.

Wolfbeis, O. S., Weis, L. J., Leiner, M. J. P., & Ziegler, W. E. (1988). Fibre-optic fluorosensor for oxygen and carbon dioxide. *Analytical Chemistry, 60,* 2028–2030.

Yam, K. L., Takhistov, P. T., & Miltz, J. (2005). Intelligent packaging: Concepts and applications. *Journal of Food Science, 70,* 1–10.

Yoon, S. H., Lee, C. H., Kim, D. Y., Kim, J. W., & Park, K. H. (1994). Time-temperature indicator using a phospholipids-phospholipase system and application to storage of frozen pork. *Journal of Food Science, 59,* 490–493.

Yoshikawa, Y., Nawata, T., Goto, M., & Fujii, Y. (1987). Oxygen indicator. *US Patent 4169811.*

Chapter 20
Meat Safety and Regulatory Aspects in the European Union

Ron H. Dwinger, Thomas E. Golden, Maija Hatakka, and Thierry Chalus

Introduction to Meat Safety

Meat safety is concerned with chemical, physical and biological aspects. With regard to the chemical aspects, residues and contaminants should be kept at as low a level as possible, but should certainly not exceed the maximum levels laid down in Community legislation. To prevent residues and contaminants in meat, it is essential to follow good agricultural practice, which involves feeding and management requirements, observing the correct withdrawal period, following treatment of animals, with veterinary medicines, a strict selection of raw materials, correct use of pesticides on grassland, preventing the access of animals to toxins or environmental contaminants, etc.

With regard to the physical aspects it is important that the slaughterhouse and processing industry implements a HACCP programme. By implementing such a programme physical hazards will have to be addressed. Consequently, a metal detector is a regular piece of equipment in the meat processing industry.

With regard to the microbiological aspects meat inspection has focused traditionally on the detection of the major zoonotic diseases occurring in domestic animals, such as tuberculosis, cysticercosis, glanders, etc. In order to detect these diseases it was necessary to palpate and incise various parts of each slaughtered animal. However, these diseases either have been largely eradicated from herds kept under modern management conditions or do not occur in the majority of the very young and generally healthy animals slaughtered nowadays. Moreover, it has been shown that meat inspection is, in some specific cases, not the most sensitive way to detect infestation. For example, it has been shown that in a sero-epidemiological study of *Taenia saginata* cysticercosis in cattle presented for slaughter in Belgium, the prevalence of bovine cysticercosis was more than 10 times higher with the antigen detection ELISA than by classical meat inspection (Dorny et al., 2000). Furthermore, micro-organisms that are of increasing zoonotic importance in modern animal husbandry systems, like salmonella and campylobacter, are readily transmitted

---

R. H. Dwinger
VWA, Room 9A12, Postbus 19506, 2500 CM The Hague, The Netherlands

from one carcase to the next by various manipulations required to be performed during the traditional meat inspection procedures. Modern meat inspection should be based on risk assessment and should prevent cross-contamination in the slaughter hall (Berends, 1998). In addition, meat inspection can be improved by imposing stricter hygienic measures at the farm level and by requiring the farm operator to send relevant management and health information to the slaughterhouse for those animals that are to be slaughtered in the next 24 hours. These principles have been introduced in the new hygiene related legislation introduced in the EU in recent years. The legislation will be explained in detail and related legislative aspects will be mentioned.

Background to Regulatory Aspects in the EU

Following several disease outbreaks and food contamination scares in Europe in recent years, the Commission adopted the White Paper on Food Safety in 2000. This ambitious programme contains a number of recommendations aimed at increasing food safety, improving the traceability of food products and regaining consumer confidence in the food industry. To this end a package of proposals for new legislation on food and feed has been prepared with the following characteristics: to complete and update the legislation; to improve official controls and ensure their efficient implementation.

General Food Law

The objective of the general food law (Regulation (EC) No. 178/2002[1] of the European Parliament and of the Council) is to create a framework ensuring universal consistency between all food end feed related legislation. It lays down guiding principles and establishes common definitions, such as for food business operator, retail, risk and hazard. It aims at bringing together various related aspects of Community legislation by including not only safety aspects but also protection of consumer interests, by covering all foods and by including both national and Community levels. Furthermore, the Regulation puts the overall responsibility for producing safe food on the food business operator. It requires the food business operator to have a system in place enabling them to identify the supplier(s) of the raw materials and the immediate customer(s) of their products in order to ensure traceability. Other issues that are covered within this Regulation are the principles of risk analysis, withdrawal of food from the market by the food business operator if safety is at stake and the precautionary principle, which enables the adoption of provisional risk management measures as long as scientific uncertainty persists. Finally, it lays down the principles and requirements for the rapid alert system for food and feed

---

[1] OJ L 31, 1.2.2002, p. 1

(RASFF) and for the establishment of the European Food Safety Authority (EFSA). The RASFF is a network involving Member States, the Commission and EFSA. Whenever a member of the network has any information relating to the existence of a serious direct or indirect risk to human health, this information is immediately notified to the Commission under the RASFF. The Commission immediately transmits the information to the other members of the network, which should take corrective measures, if applicable. EFSA, based in Parma, Italy, is in charge of risk assessment and related risk communication. On the basis of the risk assessment, the legislators and decision makers within the Commission and the Member States will develop and implement risk management measures and will communicate these measures accordingly.

Legislation on Hygiene

The "hygiene package" consists of a total of five legislative parts, of which four were adopted in April 2004, and provided the Member States and the stakeholders with a preparatory period of 18 months before becoming applicable with effect from 1 January 2006. The hygiene package puts the responsibility for producing safe food on the food business operator, while the competent authority of the Member State verifies correct implementation of the new rules. Production should be based on good hygienic practice and the HACCP principles and products are subject to microbiological criteria and temperature limits. The legislative texts deal with a variety of food types and cover the entire food chain ("from stable to table"). Two of the Regulations apply directly to food business operators:

- Regulation (EC) No 852/2004 of the European Parliament and of the Council on the hygiene of foodstuffs[2]
- Regulation (EC) No 853/2004 of the European Parliament and of the Council laying down specific hygiene rules for food of animal origin[3]

Regulation (EC) No 852/2004

The Regulation lays down general hygiene requirements to be respected by food businesses at all stages of the food chain including primary production. The Regulation does not apply to small quantities of primary products supplied directly by the producer to the final consumer or to local retail establishments that directly supply the final consumer. Examples of such products are vegetables, fruits, eggs and raw milk or products collected in the wild such as mushrooms and berries.

The Regulation requires all food business operators to put in place, implement and maintain a permanent procedure based on the Hazard Analysis and Critical

---

[2] OJ L 139, 30.4.2004, p. 1, corrected by OJ L 226, 25.6.2004, p. 3.
[3] OJ L 139, 30.4.2004, p. 55, corrected by OJ L 226, 25.6.2004, p. 22.

Control Point (HACCP) principles with the exception of those involved in primary production. Food hygiene is the result of the implementation by food businesses of pre-requisite requirements (such as concerning infrastructure and equipment, pest control, water quality, personal hygiene, etc.) and procedures based on the HACCP principles. The pre-requisite requirements provide the foundation for effective HACCP implementation and should be in place before a HACCP based procedure is established. The pre-requisite requirements to be respected are laid down in an Annex to the Regulation. The Regulation allows the HACCP based procedures to be implemented with flexibility so as to ensure that they can be applied in all situations. Guides to good practice for hygiene and for the application of the HACCP principles developed by the food business sectors themselves, either at national or at Community level, should help businesses to implement HACCP-based procedures tailored to the characteristics of their production.

In addition, the Regulation requires food businesses to be registered with the competent authority, this being a simple procedure, whereby the competent authority is informed about the address of the establishment and the activity carried out.

Regulation (EC) No 853/2004

The Regulation is more specific than the previous one by laying down the hygiene requirements to be respected by food businesses handling food of animal origin such as meat, live bivalve molluscs, fishery products, raw milk and dairy products, eggs and egg products, frogs' legs and snails, collagen and gelatine at all stages of the food chain. The Regulation does not apply to retail, which for food hygiene purposes means all activities involving direct sale or supply of food of animal origin to the final consumer. In such cases Regulation (EC) 852/2004 will apply. Establishments handling foods of animal origin for which the Regulation lays down requirements in an Annex, must be approved. Approval procedures involve an on-site visit by the competent authority to verify if the establishment fulfils all the requirements concerning infrastructure, equipment and hygiene. Those establishments carrying out only primary production, transport operations, storage of products not requiring temperature controlled storage conditions or retail operations are exempted from the approval procedure. However, some of the retailers, especially the larger ones, do need approval.

Legislation on Official Controls

This legislation is directed at the competent authorities and lays down the general principles to be respected for ensuring the official controls are objective and efficient. Furthermore, the legislation has been designed to promote a more risk-based approach to official controls.

Regulation (Ec) No 882/2004 (Official Feed and Food Controls)[4]

Regulation (EC) No 882/2004 of the European Parliament and of the Council on official controls performed to ensure the verification of compliance with feed and food law, animal health and animal welfare rules is the result of a review of the existing Community rules on the subject, which were adopted separately for the animal feed sector, the food sector and the veterinary sector. The Regulation covers the basic principles and the entire range of activities dealing with feed and food law, including animal health and animal welfare. It applies with effect from 1 January 2006, except for the provision on financing of official controls which applies with effect from 1 January 2007.

As a consequence of the new rules, the Member States have to reorganise their official controls systems so as to integrate controls at all stages of production and in all the concerned sectors, using the 'farm to fork" principles. They have to submit and annually update a general control plan for the implementation of feed and food legislation and to report annually on the implementation of that plan. National control plans and reports shall take into account guidelines drawn up by the Commission (as mentioned under guidance documents).

At present, Community controls in the Member States and in third countries are organised largely on a sectoral basis and are related to the mandates the Commission has, in different sectoral legislation. By means of this Regulation the Community approach to controls will evolve. The role of the Food and Veterinary Office as part of the EU Commission, will be essentially based on audit with the main purpose of verifying the efficiency of the control systems in the Member States and auditing the compliance or equivalence of third country legislation and control systems with EU rules. The requirement for all Member States to submit a multi-annual control plan will facilitate the carrying out of these audits. Account will also be taken of Member States' own audits and of their annual reports.

The Regulation provides for a set of general rules applicable to the official controls of all feed and food at any stage of production, processing and distribution, whether produced within the EU, exported to or imported from third countries. In addition to these rules, there are other specific control measures which are important in order to maintain a high level of protection and therefore must be kept in place. This is, for example, the case for the specific veterinary control rules on imports of animals and food of animal origin or for the specific control rules for organic products.

Regulation (EC) No 854/2004

Regulation (EC) No 854/2004 of the European Parliament and of the Council lays down specific rules for the organisation of official controls on products of animal

---

[4] OJ L 165, 30.4.2004, p. 1, corrected by OJ L 191, 28.5.2004, p. 1.

origin intended for human consumption.[5] The Regulation forms the third part of the "hygiene package" and deals, among other things, with the official controls of animals sent for slaughter, official controls with regard to fresh meat, fishery products, raw milk and dairy products and with procedures concerning imports. Modern meat inspection should be based on risk assessment and should prevent cross-contamination in the slaughter hall. In addition, meat inspection can be improved by imposing stricter hygienic measures at the farm level and by requiring the farm operator to send animals for slaughter in a clean state together with relevant management and health information called the food chain information. These principles have been introduced in the Regulation.

Other Legislation as Part of the Hygiene Package

Council Directive 2002/99/EC

Council Directive 2002/99/EC[6] lays down the animal health rules governing the production, processing, distribution and introduction of products of animal origin for human consumption. In an Annex to the Council Directive the diseases are listed of relevance to trade in products of animal origin and for which control measures have been introduced under Community legislation. Another Annex lists the treatments that are necessary to eliminate certain animal health risks linked to meat.

Directive 2004/41/EC

Directive 2004/41/EC[7] of the European Parliament and of the Council repeals the old legislation, a total of 17 Directives; each Directive dealt with a specific food item (there was a Directive for meat, fish, milk, minced meat, etc.). Consequently, the legislation has been transformed from the so called "vertical" Directives into a more horizontal approach.

Implementing Measures of the Hygiene Package

Implementing Measures

A wide range of implementing measures has been adopted on the basis of the hygiene package as foreseen in Article 12 of Regulation (EC) 852/2004,

---

[5] OJ L 139, 30.4.2004, p. 206, corrected by OJ L 226, 25.6.2004, p. 83.
[6] OJ L 18, 23.1.2003, p. 11
[7] OJ L 195, 2.6.2004, p. 12

Articles 9 and 11 of Regulation (EC) 853/2004 and Articles 16 and 18 of Regulation (EC) 854/2004.

The measures laid down in Commission Regulation (EC) No 2074/2005[8] include provisions concerning food chain information, fishery products, recognised testing methods for detecting marine biotoxins, calcium content of mechanically separated meat, lists of establishments, model health certificates for a number of products (frogs' legs, snails, gelatine, collagen, fishing products and honey), a derogation for foods with traditional characteristics and a number of amendments to Regulations (EC) No 853/2004 and (EC) No 854/2004. The amendments rectify some minor details in the Regulations.

Transitional Arrangements

Transitional arrangements with respect to certain new provisions have been taken to permit a smooth change-over from the old to the new regime. The principle of granting transitional arrangements was agreed by the European Parliament and the Council through Article 12 of Regulation (EC) No 852/2004, Article 9 of Regulation (EC) No 853/2004 and Article 16 of Regulation (EC) No 854/2004.

The measures laid down in Commission Regulation (EC) No 2076/2005[9] include provisions concerning stocks of food of animal origin, placing of food of animal origin on national markets, materials bearing pre-printed health or identification marks, marking equipment, health import conditions, food chain information, composition criteria for minced meat, use of clean water, raw milk and dairy products, eggs and egg products, training of slaughterhouse staff, certification of establishments, accreditation of laboratories carrying out official controls and some amendments to Regulations (EC) No 853/2004 and (EC) No 854/2004.

Examination of Meat for Trichinella Parasites

The adoption of Directive 2004/41/EC on 21 April 2004 by the European Parliament and the Council resulted in the repeal of Council Directive 77/96/EEC, which specified in detail the examination for *Trichinella* of carcases of swine, horses and other susceptible species. Commission Regulation (EC) No 2075/2005[10] has retained many elements from the previous legislation such as the sampling procedure, the various examination techniques in the laboratory and the derogations granted. However, at the same time the Commission Regulation has introduced a number of new elements to increase food safety for the consumer and facilitate the sampling procedure for

---

[8] OJ L 338, 22.12.2005, p. 27
[9] OJ L 338, 22.12.2005, p. 83.
[10] OJ L 338, 22.12.2005, p. 60.

those establishments where the parasite has not been encountered for a long time. The new elements are as follows:

- A larger amount of sample has to be collected and examined from those animal species that pose the greatest risk for infecting humans, mainly horses and wild boar;
- Freezing is no longer allowed to replace the examination of horsemeat (because in this host certain *Trichinella* species such as *T. spiralis*, *T. pseudospiralis* and *T. britovi* can survive freezing temperatures);
- The use of the trichinoscopic method for examining meat samples is no longer allowed, because it fails to detect *T. pseudospiralis*. A transitional arrangement for four years will give the competent authority the possibility to switch to a more reliable examination method. A number of additional requirements have to be applied whenever the trichinoscopic method is used;
- The most important regulatory change is the introduction of *Trichinella*-free holdings or a category of holdings or regions having a negligible prevalence. The competent authority can recognise a holding as free from *Trichinella* following an on-site inspection. Animals coming from a *Trichinella*-free holding or from a region with a negligible *Trichinella* risk are exempted from examination for *Trichinella*. The derogation applies only to fattening pigs. Inspection procedures can be very much simplified when the competent authority decides to recognise a category of holdings as free from *Trichinella*. Finally, the Regulation provides the possibility for a Member State to declare a region as having a negligible prevalence for *Trichinella*. Third countries will be able to apply the derogation of declaring a holding as free from *Trichinella* as well.

Regulation on Microbiological Criteria for Foodstuffs

Previously existing microbiological criteria were reviewed taking into account recent developments in food microbiology and scientific advice from the European Food Safety Authority (EFSA). Commission Regulation (EC) No 2073/2005[11] revised these criteria and introduced additional ones. The main objectives of the Commission Regulation are to ensure a high level of consumer protection with regard to food safety and to harmonize the microbiological criteria in the Member States thus facilitating international trade. In particular, the target of the Commission Regulation is to reduce the number of *Salmonella*, *Listeria* and *Enterobacter sakazakii* cases in humans. A main component of the Regulation is to set two different types of criteria for foodstuffs, which need to be complied with by the food business operator:

1. A food safety criterion defining safety of a product or a batch applicable to products placed on the market

---

[11] OJ L 338, 22.12.2005, p. 1, corrected by OJ L 278, 10.10.2006, p. 32 and by OJ L 283, 14.10.2006, p. 62.

2. A process hygiene criterion indicating the correct functioning of the manufacturing process.

Food safety criteria have been laid down for certain microorganisms which are common causes of foodborne diseases in humans, such as *Salmonella*, *Listeria monocytogenes*, *Enterobacter sakazakii*, staphylococcal enterotoxins and histamine. If food safety criteria are exceeded, the batch has to be withdrawn from the market. Food safety criteria have been set for the following combinations of food category/microorganism:

- A *Listeria monocytogenes* criterion for all ready-to-eat foods
- A *Salmonella* criterion for certain ready-to-eat foods, minced meat, meat preparations and certain meat products
- A criterion for staphylococcal enterotoxins in certain types of cheeses and milk powder
- An *Enterobacter sakazakii* criterion for dried infant formulae
- An *E. coli* criterion in live bivalve molluscs
- A histamine criterion for certain fishery products

In addition, the Commission Regulation includes process hygiene criteria, such as *Enterobacteriaceae* and *Salmonella* in carcases of slaughtered animals, *Staphylococci* in certain types of cheese, *E. coli* in pre-cut fruits and vegetables.

The sampling frequency is stipulated for a few criteria, for example, for *Salmonella* in minced meat and carcases. In other cases the food business operators have to decide the sampling frequency on a case-by-case basis taking into account the risk related to their products. Although, the Commission Regulation is directed at food business operators, competent authorities may, for various reasons, take samples to ensure that the criteria laid down are met. In the absence of Community microbiological criteria the evaluation of the food can be done in accordance with Article 14 of Regulation (EC) No 178/2002, which provides that unsafe food must not be placed on the market.

Related Food Safety Legislation

Legislation on Transmissible Spongiform Encephalopathies

The recognition of the first cases of bovine spongiform encephalopathy (BSE) in the mid eighties and the first diagnoses of the Variant Creutzfeldt-Jacob Disease (vCJD) in humans in 1996 together with the causal link between BSE and vCJD led to one of the major crises which ever affected the feed and food sectors. The key piece of legislation to protect human and animal health from the risk of BSE and other transmissible spongiform encephalopathies (TSE's) is Regulation (EC) No 999/2001[12] of the European Parliament and of the Council which lays down

---

[12] OJ L 147, 31.5.2001, p. 1.

rules for the prevention, control and eradication of certain TSE's and is commonly known as the "TSE Regulation". The TSE Regulation provides measures targeting all animal and public health risks resulting from all animal TSE, and governing the entire chain of production and placing on the market of live animals and products of animal origin. It consolidates much of the existing legislation on BSE or TSE, including rules for the monitoring of TSE in bovine, ovine and caprine animals and prohibitions concerning animal feeding. It also introduces new legislation for areas such as eradication of TSE as well as trade rules covering the domestic market, intra-community trade, import and export. Furthermore, it provides for the procedure, criteria and categories for the classification of countries according to BSE status. This very comprehensive framework is constantly evaluated through scientific review. The removal of the so-called "specified risk material" is one of the most important measures to protect the health of consumers against the risk of BSE. Specified risk materials are defined as the animal tissues being most at risk of harbouring the TSE agent. In order to prevent any recycling of possible BSE agent, these tissues are collected and completely destroyed through incineration.

Legislation on Animal By-Products

Regulation (EC) No 1774/2002[13] of the European Parliament and of the Council lays down health rules for the collection, transport, storage, handling, processing and use or disposal of all animal by-products (ABPs) not intended for human consumption. It completes the rules laid down in Regulation (EC) No 852/2004 on food waste. Its purpose is to prevent by-products from presenting a risk to animal or public health. To that end, it distinguishes three different categories of ABPs, based on risk. Category 1 material has the highest risk and is usually incinerated. Category 2 is a less risky material and can not only be incinerated, but also be composted or used for biogas production. Category 3 material can be used for animal feed under certain conditions. The last category includes parts of slaughtered animals that have been found fit for human consumption, but are not for one reason or another intended for human consumption.

Legislation on Residues

Residues are substances that can occur in foodstuffs as a side effect of using veterinary medicines or phyto-sanitary products. They are unwanted traces of medicines or plant protection products or derivatives thereof which remain in the final product[14]. Member States need to adopt and implement a plan every year to monitor live animals and products thereof, including meat, for residues of prohibited substances

---

[13] OJ L 273, 10.10.2002, p. 1.

[14] Questions and answers on residues and contaminants in foodstuffs (February 2003)

(for example hormonal substances for fattening purposes) or for substances permitted below a certain threshold, the so-called maximum residue limit (MRL). The latter group of substances includes veterinary medicinal products, pesticides and environmental contaminants. Details of the substances involved and of the residue monitoring plan can be found in Council Directive 96/22/EC[15] and Council Directive 96/23/EC[16] and their amendments. The aim of the national residue monitoring plan is to ensure that permitted levels are not exceeded and that forbidden substances are not present in food products.

Legislation on Contaminants

Contaminants are substances that can unintentionally enter the food during the various stages of its production, packaging, transport or holding or as a result of environmental contamination. Council Regulation (EC) No 315/93[17] lays down the basic principles to minimise contaminants in food, while Commission Regulation (EC) No 1881/2006[18] sets maximum levels for certain contaminants in foodstuffs. For fresh meat maximum levels have been laid down with regard to heavy metals, dioxins and polychlorinated biphenyls (PCB's).

Other Legislation

As soon as meat is being processed additional legislation will start to apply, such as rules on food additives (substances added intentionally to foodstuffs to perform certain technological functions), biocides, food contact materials, labelling requirements etc.

Animal health requirements will apply to live animals that are going to be slaughtered, while welfare requirements will apply to the management of animals on the farm, the transport of animals to the slaughterhouse and the killing of animals at the slaughterhouse.

International Aspects

Where international standards exist, they have been taken into consideration in the development or adaptation of the food safety legislation. This applies to standards developed by the Codex Alimentarius Commission, which has been created by the Food and Agriculture Organisation and the World Health Organisation to develop

---

[15] OJ L 125, 23.5.1996, p. 3.
[16] OJ L 125, 23.5.1996, p. 10.
[17] OJ L 37, 13.2.1993, p. 1.
[18] OJ L 364, 20.12.2006, p. 5.

food standards, guidelines and related texts such as codes of practice. It also applies to standards related to animal health and animal welfare developed by the World Organisation for Animal health (OIE). Similarly, ISO and CEN standards have been incorporated in the legislation as analytical reference methods as far as possible (as can be seen for example in Commission Regulation (EC) No 2073/2005).

Guidance Documents

A number of documents have been prepared to give guidance on the implementation of the food hygiene requirements and related subjects. These documents aim to assist the food business operators and the competent authorities of the Member States. The Guidance documents are not formal acts of legislation, but the Commission will defend where necessary the consensus laid down in these documents. Most of the documents (with an*) have been placed on the DG SANCO internet site[19]:

- Guidance document on Regulation (EC) No 178/2002*
- Guidance document on Regulation (EC) No 852/2004*
- Guidance document on Regulation (EC) No 853/2004*
- Guidance document on the implementation of HACCP and facilitation of the implementation of the HACCP principles in certain food businesses*
- Guidance document on community guides to good practice*
- Practical guide on food contact materials*
- Guidance document on import requirements*
- Guidance document on the preparation of multi-annual control plans as laid down in Regulation (EC) No 882/2004 (published as Commission Decision (2007/363/EC[20])
- Guidance document laying down criteria for the conduct of audits (published as Commission Decision 2006/677/EC[21])
- Guidance document on official controls, under Regulation (EC) No 882/2004, concerning microbiological sampling and testing of foodstuffs*.

Future Legislative Work

Treatment to Remove Surface Contamination

Article 3(2) of Regulation (EC) No 853/2004 provides a legal basis to permit substances other than potable water to remove surface contamination from products of animal origin. Such a legal basis did not exist in the previous legislation (Directive

---

[19] http://ec.europa.eu/food/food/index_en.htm
[20] OJ L 138, 30.5.2007, p. 24.
[21] OJ L 278, 10.10.2006, p. 15.

64/433 for red meat, Directive 71/118 for poultry meat, other Directives used to cross reference to the former Directive mentioned), but is available now that Regulation (EC) No 853/2004 is applicable.

With the adoption of the hygiene package and the introduction of the HACCP principles in the entire food chain, establishments are obliged to improve their hygiene and processing procedures. Under such circumstances the use of substances to remove surface contamination of food of animal origin can be reconsidered. It is essential that a fully integrated control programme is applied throughout the entire food chain including reduction of pathogens in water and in feed, on farms, during transport and in the processing plant. Treatment to remove surface contamination might constitute an additional element in further reducing the number of pathogens, especially with regard to *Salmonella* and *Campylobacter* provided an integrated control strategy is applied throughout the entire food chain. A draft implementing Commission Regulation is still under discussion.

Risk-Based Meat Inspection

Meat inspection has focused traditionally on the detection of the major zoonotic diseases. As stated in the introduction these diseases have either been largely eradicated from herds kept under modern management conditions or do not occur in the majority of the very young and generally healthy animals slaughtered nowadays. In order to prevent cross-contamination of subsequent carcases with microorganisms of importance in modern animal husbandry systems, like *Salmonella* and *Campylobacter*, inspection procedures without incision or palpation will have to be introduced. A detailed visual inspection without any incision or palpation of slaughter animals might be sufficient to ensure food safety. However, under these circumstances it will be necessary to take efficient preventive measures during the rearing of the animals and to provide sufficient information to the slaughterhouse on the life history of the animals. A draft Commission Regulation has been prepared, which lays down detailed requirements for risk-based meat inspection of fattening pigs and young ruminants.

Acknowledgements We wish to acknowledge the technical advice from K. van Dyck and the encouraging supervision by W. Daelman.

References

Berends, B. R. (1998). A risk assessment approach to the modernization of meat safety assurance (Doctoral dissertation, University of Utrecht, 1998).

Dorny, P., Vercammen, F., Brandt, J., Vansteenkiste, W., Berkvens, D. & Geerts, S. (2000). Sero-epidemiological study of *Taenia saginata* cysticercosis in Belgian cattle. *Veterinary Parasitology, 88*, 43–49.

Index

A

Aarestrup, F. M., 143
Aarts, H. J., 330
Aasen, I. M., 262, 263, 381, 382
Abbott, S. P., 186
Abdel-Hamid, I., 339, 342
Abee, T., 265
Abel, A. P., 344
Abildgaard, C. I., 153, 380, 381
Abraham, R. B., 265
Abriouel, H., 261, 375, 389, 390
ACE inhibitory peptides, *see* Angiotensin I-converting enzyme inhibitory peptides
Acetic acid, 303–304
  See also Beef carcasses, decontamination
Acidified sodium chlorite, 306–307
Acid production, in meat products, 135–136
Acuff, G. R., 255, 300, 304, 307, 401
Adak, G. K., 253, 403
Adams, M. R., 265, 390
Adams, N. R., 9
Adenosine monophosphate-activated protein kinase, 12
Aeschlimann, A., 234
Aggio, D., 95, 100, 104, 199
Aging, of meats, 238–239
Agrasar, A. T., 386
Agrobacterium tumefaciens, 319
Agvald-Ohman, C., 142
Ahn, C., 142, 379
Ahvenainen, R., 426, 429, 434, 436, 443
Aida, A. A., 82
Akkermans, A. D. L., 116
Alagesan, D., 339
Alakomi, H. L., 443
Alatossava, T., 107
Albano, H., 262, 379, 387
Aldissi, M., 338

Alegre, M. T., 155
Alfonso, L., 61
Alisky, J., 273
Allen, G., 406
Allen, D. M., 298
Allen, V. M., 271
Almagro, A., 175
Almakhla, H., 386
Alocilja, E., 335, 338, 339, 340, 342, 344–353, 433
Alonso, L., 234
Alpert, C. A., 153
Alston, J. M., 320
Altermann, E., 151, 220
Alter, T., 403
Altrock, A. V., 402
Alvarez, E., 233, 403
Amaral Nascimiento, A. M., 94
Ambrosiadis, I., 411
Amigo, L., 268
3-Amino-2-oxazolidinone (AOZ), 367
Ammor, M. S., 142, 222, 223
Ammor, S., 100, 103
Ampe, F., 116
AMPK, *see* Adenosine monophosphate-activated protein kinase
Amplified fragment length polymorphisms (AFLP) markers, 66, 67
Anang, D. M., 255
Ananou, S., 263, 264, 265, 387, 389, 390
Anastasi, G., 361
Andersen, L., 203, 210, 404
Anderson, A. D., 75
Anderson, E. M., 443
Anderson, G. P., 340
Anderson, L. L., 31, 33
Anderson, M. D., 302
Anderson, M. E., 295, 297, 301–303
Anderson, R. C., 268

Anderson, R. M., 68
Andrade, M. J., 169
Andrighetto, C., 95, 101, 108
Anfossi, L., 362
Angiotensin I-converting enzyme inhibitory peptides
　from meat proteins, 240, 241
Anil, M. H., 65
Animal by-products (ABPs), legislation on, 462
Annamalai, T., 268
Ann, B., 289
Antimicrobial peptides (AMPs), 267–269
Antioxidative peptides, from meat proteins, 241, 242
Anzai, K., 267
Appendini, P., 267–269, 430
Arana, A., 61, 64, 65, 68, 78
Aranda, E., 169, 171, 181
Aranishi, F., 69
Arca, P., 153
Archer, G. S., 11
Archibald, A. L., 24, 37
Ardaillon, V., 139
Arendt, E. K., 138, 152, 412
Arguello, A., 14
Arihara, K., 222–225, 231, 236–244
Aristoy, M. C., 200, 201, 203, 205, 207
Ariyapitipun, T., 265, 266
Arlindo, S., 262
Armstrong, J. F., 364
Arques, J. L., 266
Arribas-Layton, N., 347
Arrondo, J., 409
Arroyo, G., 261
Arthur, T. M., 299, 310, 311
Artigas, M. G., 385, 413
Arvanitoyannis, I. S., 217
ASC, *see* Acidified sodium chlorite
Asensio, M. A., 168, 171, 181, 186
Ashwin, H. M., 368
Aspergillus spp, 183, 186
Atanasov, P., 338, 339
Atrih, A., 386
Atterbury, R. J., 271
Aukrust, T., 155
Autio, T., 403
Avonts, L., 218
Axelsson, L., 130, 131, 135, 140, 156, 157, 223, 263, 381, 382
Ayres, J. W., 266
Ayres, K. L., 71
Azain, M. J., 234

Azcarate-Peril, M. A., 220
Azuma, T., 263

B
Baas, T. J., 34
Babacan, S., 340
Babinska, K., 232
Bachand, A., 254, 258, 259, 261, 262
Bach, S. J., 270
Bacillus spp, 108
Bacteriocin-producing bacteria and bacteriocins, 260, 261
　animals and, applications, 261, 262
　limitations, 266, 267
　meat products, applications in, 262–266
Bacteriocins
　antimicrobial activity of, 375–390
　　classification and biochemical properties, 375–379
　　hurdle theory and, 388–390
　　in vitro and in situ production of, 379–388
　BPB and, 261
　　animals and, applications, 261, 62
　　limitations, 266, 267
　　meat products, applications in, 262–266
Bacteriophage
　application of, 270
　classification and mode of action, 269
　limitations of, 272, 273
　in meat and meat products, applications, 271, 272
　preharvest control, 270, 271
Bacus, J. N., 92, 135
Badrie, N., 330
Baeten, Ch. V., 326
Baeumner, A. J., 344, 345
Bagenda, D. K., 263
Baggiani, C., 362
Bagni, A., 187
Baidini, P., 387
Bailey, D., 65
Baillie, H., 116
Baird, B., 401
Bakar, J., 255
Bakes, S. H., 390
Balaban, N., 143, 402
Baldini, P., 181, 182, 185–187, 189
Balogh, B., 270
Bang-Ce, Y., 369
Bannerman, T. L., 131
Bantleon, A., 92, 129
Bao, L., 338

BaraÂth, A., 141
Barakat, R. K., 265
Barb, C. R., 33
Barbieri, G., 185
Barbosa, J. A., 4
Barbour, W. M., 339, 340
Barbuddhe, S. B., 386
Barbuti, S., 181, 185, 387
Barcena, B. J. M., 380, 381
Barefoot, S. F., 390
Barendse, W., 24, 33–35, 37, 52
Barkholt, V., 265
Barnes, M. B., 152
Barnett, J. A., 173
Barrangou, R., 220
Barriere, C., 92, 136, 138, 200
Barrio, E., 167, 169
Barrow, P. A., 271, 273–275
Barthelmebs, L., 157
Bartholomew, D. T., 152
Baruzzi, F., 101, 108
BA, *see* Biogenic amines, in foods
Bass, J. J., 44
Bass, P., 219
Basso, A. L., 97
Battilani, P., 182, 183
Baxter, A., 368
Baxter, G. A., 368
Bayoudh, S., 362
Beach, J. C., 401
Becke-Schmid, M., 142, 155
Beck, H. C., 139, 210
Beder, I., 232
Bederova, A., 232
Beef carcasses, decontamination
    acetic acid, 303, 304
    acidified sodium chlorite, 306, 307
    acid spray washes, 303
    cetylpyridinium chloride, 299
    chemical dehairing, 299
    chlorine dioxide, 305, 306
    costs/benefit analysis for, 311–313
    electron beam irradiation (high dose-rate X-rays), 310
    fumaric acid, 304, 305
    governmental regulations and HACCP, 297, 298
    hide-on decontamination, 298
    irradiation, 309, 310
    knife trimming, 300
    lactic acid, 304
    lactoferricin B, 307
    microwave and combination technology, 311
    spray washing, 302
    steam pasteurization, 308, 309
    tri-sodium phosphate, 306
    vacuuming, 300, 301
    water sprays, 302, 303
Beef cattle slaughter process, 289–291
Bee, G., 15
Beer, J., 275
Beest, M. V., 426
Belfiore, C., 412
Belk, K. E., 291, 293, 295, 300, 305
Belkum, M. J. V., 375, 376, 377, 378
Bell, A. M., 9
Bell, J. A., 131
Bell, R. G., 295, 296, 302, 425
Bellmann, O., 47
Belloch, C., 167, 169
Benedict, R. C., 254
Benito, M. J., 171
Benkerroum, N., 256, 264, 265, 388, 412
Bennett, L. E., 244
Bensaid, M., 256, 388
Benthin, S., 381
Berben, G., 326
Berchieri, A., Jr, 273
Berdague', J. L., 139
Berdal, K. G., 327, 330, 331
Berdeaux, O., 233
Bergen, M. A. V., 270
Berger, B., 323
Berger, K. L., 443
Berggren, C., 362
Berghmans, S., 44
Bergkvist, L., 234
Bergweff, A. A., 362
Berjeaud, J. M., 153
Berney, H., 345
Berni, E., 181, 183, 188, 190–192
Bertelsen, G., 209, 425
Berthe, F., 69
Berthier, F., 94, 95, 99, 103, 108, 155
Bertrand, X., 138
Bertuzzi, T., 181
Bes, M., 107
Betthauser, J., 11
Beuchat, L. R., 168, 170–172, 297, 310
Bevan, R. M., 268
Beyatli, Y., 153
Bhaduri, S., 403
Bharathy, S., 265
Bhatia, S. K., 340, 341, 348

Bhilegaonkar, K. N., 386
Bhunia, A. K., 342, 389
Bianchi, N., 343
Bidlack, W. R., 221
Bidwell, C. A., 10
Bielke, L. R., 259
Biesalski, H. K., 232
Bigger, S. W., 426
Bimani, D., 382
Bindon, B. M., 52
Biogenic amines, in foods, 141, 142
Bioprotective cultures, 399, 406–408
    bacteriocins produced by LAB, 408
    fermented meat products, 411–413
    meat sensorial changes and, 417
    raw meat, 409
    ready-to-eat meat products and, 414, 415
    strategies for improvement of, 415, 416
Biosensors, for pathogenic bacteria detection in meat industry, 335–353
    DNA sensors, 343–345
    immunosensors, 339–343
    MEMS approach, 345
        detection principle, 345, 346
        fabrication, 347–349
        results, 350–353
        sensor design, 346, 347
        testing, 349, 350
Bishop, M. D., 37
Bisson, L. F., 106, 116
Biswas, B., 272
Biswas, S. R., 380, 381
Blaha, T., 65
Blaiotta, G., 78, 94, 95, 99, 101, 103, 107, 116, 138
Blair, I. S., 64, 78, 402
Blanco, D., 185
Blaser, M. J., 337
Blaszyk, M., 142
Blom, H., 155, 203, 239
Blott, S. C., 27, 83
Blouin, M. S., 75
Bloukas, J., 415
Board, R. G., 171, 173, 436
Bodenhammer, W. T., 433
Bodnaruk, P., 406
Boekhorst, J., 151
Boekhout, T., 168
Bogomolova, A., 338
Bohez, L., 275
Bollinger, L., 402
Bolt, D. J., 14
Boltovets, P. M., 340
Bolumar, T., 200, 201, 203, 205, 208
Bondarenko, V., 275
Bondi, M., 413
Bonet, M. E. B., 219
Bonfini, L., 329
Borgatti, M., 343
Borges, E., 107
Borsa, J., 256
Bosilevac, J. M., 298, 299
Bosworth, B. T., 274
Botta, G. A., 218
Bottari, B., 98
Bouckenooghe, T., 235
Bougle, D., 237
Bouhallab, S., 237
Bouley, C., 50
Bourke, B. J., 425
Bouttefroy, A., 416
Bover-Cid, S., 135, 141
Bovine leukocyte adhesion deficiency (BLAD), 26
Bovine meat, 6–8
Bowyer, V. L., 328
Boyd, B., 362
Boylston, T. D., 234
Boziaris, I. S., 390
BPB, see Bacteriocin-producing bacteria and bacteriocins
Brabander, H. F. D., 65
Brackett, R. E., 254, 255, 257, 304, 310, 338
Brambilla, G., 326
Brandelli, A., 382
Brashears, M. M., 256, 258, 260, 414
Bredholt, S., 264, 414
Breidt, F. Jr, 151
Bremer, V., 399
Brendehaug, J., 151
Breslin, M. F., 260, 275
Breuer, U., 175, 176
Briegel, J. R., 9
Bringel, F., 153
Broadbent, J. R., 152
Brochothrix thermosphacta, 119
Bron, P. A., 157
Brown, B. W., 8, 9
Brown, C. E., 235
Bruckner, R., 136
Bruna, J. M., 186, 187, 202, 203, 240
Brutlag, D., 95
Brynda, E., 340
Buchanan, F. C., 33, 34
Buchanan, R. L., 254, 295
Buck, B. L., 220

Index

Buckenhüskes, H. J., 130
Buckleton, J., 72
Budde, B. B., 140, 223, 225, 262, 264, 379, 411
Bukhtiyarova, M., 153
Bunch, R. J., 33
Buncic, S., 65
Bunde, R. L., 341
Buntjer, J. B., 83
Burans, J., 340
Burden, M., 78
Burgess, C. M., 253
Burpo, F. J., 326
Burriel, A., 401, 409
Burrow, H. M., 52
Burt, S. A., 256
Busboom, J. R., 10, 235
Bush, E., 403
Busolli, C., 190
Busz, H., 403
Butler, B. L., 428
Buys, E., 402
Byrne, D., 327

C

Córdoba, J. J., 168, 169, 171, 181, 186
Cacchioli, C., 181, 183, 188
Cadieu, N., 365
Caetono-Anollés, G., 95
Cagney, C., 64, 68
Cahill, S. M., 400
Cai, J., 169
Cain, T. L., 11
Cakmakci, M. L., 258
Caldera-Olivera, F., 382
Caldwell, S. L., 152
Calistri, P., 65
Callaway, T. R., 254, 259
Call, J. E., 386
Calpain 1 *(CAPN1)* gene, 34, 40
Calpastatin (CAST) gene, 10, 34, 35
Calvo, J. H., 68, 83
Calypyge allele, 44, 45
Cameron, N. D., 33
Camloh, M., 331
Campanini, M., 183, 185, 387
Campbell, D. P., 338
Campbell, H. M., 364
Campbell, K. H. S., 4
Camp, J. V., 236, 240
Campo, F., 338
Campylobacter spp, 260
Candida spp, 120

Candidate genes, 5, 32
 genetic markers characterization, for carcass and meat quality traits, 5, 6
  bovine, 6–8
  ovine, 8–11
  porcine, 11–15
 See also Meat quality traits, genetic control
Cankar, K., 329, 331
Cano, J., 169
Cantoni, C., 82, 95, 97, 100, 105, 106, 112–117, 181, 187, 199
Capece, A., 171, 172, 173
Capellas, M., 265
Caporale, V., 65
Caputo, L., 108
Carballo, J., 223, 244
Carlson, C. W., 295
Carlson, S. A., 337
Carlton, R. M., 270, 272
Carnwath, J. W., 320
Carraro, A., 96
Carrillo, C. L., 270
Carrino, J. J., 331
Carroll, J. A., 11
Carson, C. F., 256
Carter, D. B., 11
Carter, R. M., 341
Casaburi, A., 103, 138, 206, 208, 209, 266
Casaregola, S., 175
Casas, E., 34, 36, 39, 40, 45
Caseino-phosphopeptides, 237
Casser-Malek, I., 48
Castagnetti, G., 188
Castaldo, D., 64, 68, 83
Castaneda, M. P., 259
Casteele, T. V. D., 75
Castellano, P. H., 263, 264, 399, 409, 410, 412, 416, 417
Castillo, A., 255, 300, 307, 401, 409
Castillo, L. A., 265
Castro, L. P., 386
Catalase activity, 136–138
Catarame, T. M. G., 78
Catillo, G., 73
Cattivelli, D., 142
Cavani, C., 65
Cave, V. M., 81
Cavin, J. F., 157
Cayré, M. E., 404
Cenatiempo, Y., 153
Centeno, D., 136
Cerniglia, C. E., 361, 362
Cerquetti, M. C., 275

CE technology, *see* Competitive exclusion technology
Cetylpyridinium chloride, 299
Chaillou, S., 133, 135, 136, 138, 149, 151, 152, 158, 160
Chalus, T., 453
Chakrabartty, S., 342
Champomier-Vergès, M. C., 133, 135, 149, 155
Chandan, R. C., 231, 244
Chandrashekar, A., 153
Chang, Q. Q., 431
Chang, V., 401
Chang, S., 306
Chan, J. C. K., 237
Chan, W., 233
Chardigny, J. M., 233
Charlier, C., 9, 10, 43
Chartier, S., 92, 138, 200
Chartone-Souza, E., 94
Chatterjee, C., 376
Cheasty, T., 253
Chee, M., 67
Chehín, R., 409
Chemburu, S., 342
Chemical dehairing process, 299
Chen, C., 416
Chen, H., 260
Chen, H. C., 259
Chen, J., 270
Chen, J. D., 156
Chen, T. L., 330
Chen, Y., 403, 406
Chen, Y. K., 416
Chen, Z. Z., 340
Cheung, J. H., 341
Chevalet, C., 74
Chiasson, F., 256, 257
Chiba, H., 237
Chifang, P., 366
Chikindas, M. L., 261, 375
Chi, S. P., 202
Chiu, R., 24
Chlorine dioxide, 305, 306
  See also Beef carcasses, decontamination
Choi, Y. K., 156, 345
Chowdhary, B. P., 24
Chow, A. W., 131
Christine, M., 149
Chuanlai, X., 366
Chung, N., 275
Chung, T. C., 142
Church, D., 400

Church, I., 409
Chu, X. G., 366
Ciampolini, R., 77, 83
Cianci, D., 83
Cinquina, A. L., 361
Ciobanu, F., 34
Ciobanu, D. C., 12
Civera, T., 209
Clarke, A. D., 265
Clark, N., 305
CLA, *see* Conjugated linoleic acid
Clasener, H. A. L., 361
Clavero, R., 406
Clavero, M. R. S., 310
Clemens, R., 65, 80
Clemens, R. A., 221
Clement, J. M., 95
Cleveland, J., 261, 375
Clifton-Hadley, F. A., 275
Cloning, 4
Clop, A., 44
Coagulase-negative cocci (CNC), 91, 92, 95, 103–112, 118–121, 130, 133
Coakley, M., 234
Cocconcelli, P. S., 95, 97, 101, 129, 136, 142, 404, 405
Cockett, N. E., 9–11
Cocolin, L., 91–122, 170, 199
Coffey, A., 413
Cofrades, S., 223, 244
Cohen, R. N., 344
Cohn, G. E., 338
Cole, K., 261, 262
Collins, M. D., 94, 169, 220
Collins-Thompson, D., 142
Collis, M., 173
Collomb, M., 234
Colombo, F., 82
Coma, V., 430
Comi, G., 95, 100, 101, 105, 106, 109, 118, 170, 181, 199
Competitive exclusion technology, 259, 260
Compound meat products, DNA-based meat traceability and, 80–82
Cong, Y. S., 175
Conjugated linoleic acid, 14, 15, 233, 234
Connerton, I. F., 271
Connerton, P. L., 271
Connolly, L., 368
Contamination, sources of, 291–293
Conversano, M. C., 82

Cooke, H. J., 67
Cooksey, K., 266
Cooper, D. N., 67
Cooper, J., 365, 367, 368
Cooper, K. M., 364, 366, 367, 368
Coote, P. J., 256
Coppieters, W., 45
Coppola, R., 97, 103
Coppola, S., 95, 116
Coq, C. L. A.-M., 149, 153
Corcoran, K., 65
Cordoba, J. J., 186
Cordoba, M. G., 181
Corich, V., 96
Corinne, A., 319
Corkish, J. D., 275
Corn, R. M., 343
Cornuet, J. M., 77, 78
Corona, A., 103
Corredor, M., 175
Corrieu, G., 175
Corry, B., 338
Cosby, W. M., 157
Cosseddu, A. M., 103
Costs/benefit analysis, for beef carcasses decontamination, 311–313
Cotter, P. D., 258, 260, 261, 265, 375, 376, 377, 378, 408
Council Directive 2002/99/EC, 458
Cousens, S. N., 64
Couto, I., 107
Cozzani, R., 361
CPC, *see* Cetylpyridinium chloride
CPP, *see* Caseino-phosphopeptides
Crabbe, K., 381, 382
Crawford, A. M., 70
Crawley, C., 338
Crisá, A., 44
Crivori, S., 95
Crooks, S. R. H., 368
Cross, H. R., 304
Cross, K. J., 237
Crozier-Dodson, B. A., 289, 304, 311
Crudele, M. A., 142
Crum, G. A., 267
Cryolog (Gentilly, France) (eO)® adhesive TTI, 439
Cryovac Darfresh® vacuum skin packaging technology, 445
CSSM66, microsatellite marker, 35
Cuesta, E. P., 234
Cundiff, L. V., 23
Cunningham, E. P., 68, 72, 73, 78

Cunningham, F. E., 311
Cuozzo, S., 409
Cutter, C., 401
Cutter, C. N., 266, 299, 303, 306, 308, 386
Czikkova, S., 387

D

Daba, H., 381, 382
Daeschel, M. A., 386
Dahllof, I., 116
Dainty, R. H., 94, 239, 436
Dalton, H. K., 173
Daniel, C., 34
Daniel, Y. C., 289
Danielsen, M., 153, 155
Daoudi, A., 265, 412
Data in breeding, application of, 51–53
    genome selection, 53
Davenport, R. R., 173
Davey, C., 346
Davidson, C. L., 425
Davies, E., 433
Davies, E. S., 434
Davies, R. H., 260
Davila, A. M., 175
DDRT-PCR, *see* Differential Display-reverse transcribed-PCR
Deák, T., 168, 171
Deak, T., 170
Dean-Nystrom, E. A., 274
Debaryomyces hansenii, 96, 120, 200
    physiological and genetic characteristics, 173–176
Debevere, J., 387, 414, 426
Decker, E. A., 234, 235
Deegan, L. H., 258, 261, 262, 265, 375
Deisingh, A. K., 330
Dekkers, J. C. M., 23, 34
Delahaut, P., 365, 367, 368
Delaquis, P. J., 256
Deliere, E., 138
Delta-like 1 (DLK1), 10
Demain, A. L., 167
DeMarco, D., 338
Demeyer, D., 92, 199, 200, 202–204, 208–210, 404
Demnerová, Kateřina, 319
Denaturing gradient gel electrophoresis (DGGE), 93, 97
    PCR-DGGE, 104–106, 169, 170
        bacterial ecology by, 117–119
        primers, choice of, 115–117
        yeast ecology by, 119, 120

Deng, L., 338, 344
Dennis, S., 219
Dentali, S., 231
Dernburg, A., 403
Desmarchelier, P., 401, 402
Desmond, C., 219
Devlieghere, F., 387, 414, 416, 426
Dhiman, T. R., 233
Diacylglycerol-O-acyltransferase *(DGAT1)* gene, 35, 45, 46
Dickinson, D. L., 65
Dickson, J., 402, 416
Dickson, J. S., 297, 301–303, 306
Dicks, L. M., 153
Dicks, L. M. T., 388, 412
Diemen, P. M. V., 274
Dietl, G., 52
Diez-Gonzalez, F., 261
Differential Display-reverse transcribed-PCR, 47
Diles, J. J. B., 6
Dillon, V. M., 170, 171
Dilts, B. D., 255, 264, 273
Directive 2004/41/EC, 458
DNA-based meat traceability
  compound meat products and, 80–82
  DNA markers, 66, 67
  DNA as traceability tool, 67
    audit, application to, 78, 79
    breed, species/brand protection, 76–78
    DNA markers, use of, 69
    DNA sampling, 68, 69
    food safety, application to, 78
    identity, 70–74
    notation and assumptions, 69, 70
    parentage, 74, 75
    relatedness, 75, 76
  from "gate to plate", 63
    risk, 64
    stakeholders, 65
    traceability continuum, 62–64
  as part of value chain, 82–84
  species and strain differentiation, 82
DNA isolation, GMO analysis and, 325, 326
DNA markers
  multi-locus markers, 66, 67
  single locus markers, 67
DNA sampling, 68, 69
DNA sensors, 343–345
Dobrogosz, W. J., 157
Dock, L. L., 426
Dodd, C. E., 271
Dodd, E. R., 98

Dodds, K. G., 61–84
Dodson, M.V., 11
Doele, B. A., 267
Doguiet, D. D. K., 256, 388
Doherty, A., 400
Doi, K., 151
Dolci, P., 91
Dominguez, A. P. M., 171
Donkersgoed, J. V., 274
Donk, W. A. V. D., 376
Donnelly, P., 78
Donoghue, A. M., 261
Dordet-Frisoni, E., 133
Dorroch, U., 47
Dorsa, W. J., 254, 306, 308
Doskar, J., 133
Doumit, M. E., 10
Dowd, S. E., 11
Downes, F. P., 338
Doyle, M. P., 253, 254, 257, 260, 262, 295, 304, 310, 336, 338
Dragoni, I., 187
Draisci, R., 366, 368
Dransfield, E., 240
Draughton, F., 402
Dreo, T., 329
Drosinos, E. H., 95, 99, 100, 102, 104, 105, 109, 256, 375–390, 415, 436
Duan, Z. J., 366, 368
Duckett, S. K., 11
Dueger, E. L., 275, 276
Duffy, G., 78, 253
Du, L. X., 44
Dumont, V., 368
Duncan, C., 142
Dunn, B., 339
Dunne, C., 219
Dunn, N. W., 381
Dur'a, M. A., 120, 170, 173, 175, 200, 203, 206, 208, 211
Durland, P. R., 425
Durst, R. A., 341, 345
Duveneck, G. L., 344
Dwinger, R. H., 453
Dykes, G. A., 173, 271
Dziva, F., 274

E

Eastridge, J. S., 3, 14
EBV, *see* Estimated breeding value
EcoRI enzyme, 107, 108
Edlun, C., 142
Edwards, J. R., 289, 311

Eede, G. L. M. V. D., 327, 329
Eerola, S., 140, 224
Efimov, V. S., 343
EFSA, see European Food Safety Authority
Egan, A. F., 153
Eggen, A., 67
Ehlers, B., 327
Ehrat, M., 344
Ehrlich, S. D., 94, 95, 99, 103, 108, 155, 156, 157
Eijsink, V. G. H., 135, 153, 156
Eilamo, M., 434
Einspanier, R., 327, 331
Eitenmiller, R. R., 141
ElAmin, A., 64
Elder, R. O., 253, 293, 295
Electron beam irradiation (high dose-rate X-rays), decontamination of beef carcasses and, 310
Elliot, C. T., 364, 368
Elliott-Smith, W., 61
Ellis, D. F., 202
Elmi, M., 400
El-Rahman, H. A., 170
Elsen, J. M., 39
Elsser, D., 140, 223
Embryo splitting, 4
Enan, G., 263
Encinas, J. P., 175
Ender, K., 47, 52
Endo, N., 381
Ensing, K., 362
Enterobacterial repetitive intergenic consensus-PCR, 94
Enterococcus faecium, 103
Entgens, P., 142, 155
Enzyme-linked immunosorbent assays (ELISA) kits, 365–367
Ercolini, D., 78, 95, 96, 98, 101, 103, 107, 112, 116, 266
Eremin, S. A., 364
Erickson, M. C., 253, 254
ERIC-PCR, see Enterobacterial repetitive intergenic consensus-PCR
Erkkilä, S., 140, 223, 224, 225
Escherichia coli, 108
Esch, M. B., 345
Essential oils (EOs), 256–258
Esteve-Zarzoso, B., 169
Estimated breeding value, 21
Estoup, A., 77, 78
Etcheverria, A. I., 261
Ethelberg, S., 253

Etherington, D. J., 238, 240
Etienne, J., 107
European Food Safety Authority, 142, 320, 330, 455, 460
European Union (EU), meat safety and regulatory aspects in, 453
    background, 454
    future legislative work, 464–465
    general food law, 454–455
    hygiene package, 458–461
    legislation
        on hygiene, 455, 456
        on official controls, 456–458
    related food safety legislation, 461–464
Eurotium spp, 183–185
Evans, M. R., 64
Eve, L., 231
Evetoch, E. A. N. E., 261
Evett, I. W., 71
Evisceration process, 295, 296
Expression QTL (eQTL), 49
Eyer, H., 234

F

Fadda, S., 92, 138, 139, 142, 203, 205–208, 210, 388, 399, 404, 413
Fadl, A. A., 276
Faerber, P., 184
Fahrenkrug, S. C., 44
Falconer, D. S., 85
Falony, G., 217
Fang, S., 345
Fang, T. J., 303, 386
Farías, M. E., 416
Farías, R., 409
Farm animals, transgenic, 3
    carcass and meat quality traits, candidate genes/genetic markers characterization for, 5
        bovine, 6–8
        ovine, 8–11
        porcine, 11–15
Farnell, M. B., 261
Farnir, F., 46
Farnworth, E. R., 222, 243
Fawcett, S., 443
Fedtke, I., 138
Fegan, N., 401
Fehlhaber, K., 403
Ferain, T., 156
Ferguson, J., 368
Feriotto, G., 343

Fermented meats
  application of probiotics in, 222–225
  bioprotection applied to, 411–413
  flavor generation in, 202
    amino acids, transformation of, 208
    glycolysis, 204
    lipolysis, 208, 209
    oxidation, 209, 210
    proteolysis, 204–208
    volatile compounds generation, 210, 211
  microbial starters
    lactic acid bacteria (LAB), 199, 200
    Micrococcacceae, 200
    molds, 200–202
    yeasts, 200
Fermented sausages, microorganisms diversity investigation in
  culture-dependent techniques, 94–96
    strain characterization, 96–97
    strain identification, 94–96
  culture-independent techniques, 97–98
  microbial ecology of
    culture-dependent methods and, 98–104
    culture-independent methods and, 112–121
Fernández, C., 141
Fernandez-Cornejo, J., 319
Fernández-Espinar, M. T., 167–170
Fernández-Ginés, J. M., 233, 244
Fernández, M., 203, 208, 240
Fernando, R. L., 52, 53
Ferrari, M., 340
Ferrer, K., 265, 409
Ferris, M. J., 116
Ferrus, M. A., 95
Fiat, A. M., 237
Fiddler, M. B., 330
Fidotti, M., 33
Fiedler, I., 53
Figueras, M. J., 181
Fischer, M., 138, 156
Fisher, M. C., 232
FISH, *see* Fluorescence in situ hybridization
Fitzgerald, G. F., 219, 220, 267
Fitzgerald, M., 433
FitzGerald, R. J., 236
Flavor generation, in fermented meats, 202
  amino acids, transformation of, 208
  glycolysis, 204
  lipolysis, 208, 209
  oxidation, 209, 210
  proteolysis, 204–208
  volatile compounds generation, 210, 211
Fleet, G. H., 169, 170, 175
Fleming, H. P., 151
Flint, J., 51
Fliss, I., 266
Flores, M., 120, 170, 175, 199, 200–206, 211, 239
Floros, J. D., 426, 428
Fluorescence-based oxygen sensors, 432, 433
Fluorescence in situ hybridization, 93
Fodey, T. L., 365, 367
Foegeding, P. M., 386, 412
Fontana, C., 97, 101, 108, 114, 116–118, 129, 136, 404, 405
Food Sentinel System^{TM}, 442
Foods for specified health use (FOSHU), 231, 243
Forsman, P., 107
Fortes, G. G., 69
Forte, V. T., 82
Frémaux, C., 153
Francesco, C. D., 65
Fransen, G., 138
Franz, C. M. A. P., 375, 377, 378
Fratamico, P. M., 254, 255, 338
Fravalo, G., 403
Freitas, F. Z., 265
Freking, B. A., 9, 10, 11, 45
Freney, J., 107
Fresh-Check® TTI, 439
Frewer, L. J., 65
Frey, L., 153
Frias, R., 222
Friedman, R., 176
Friend, T. H., 11
Fries, R., 24
Frisvad, J. C., 187
Fritsch, E. F., 67
Fronicke, L., 24
Frost, J. A., 64
Frost, L. S., 150
Frutos, A. G., 343
Fuchu, H., 238, 241
Fujii, J., 33
Fujii, Y., 434
Fujita, H., 240, 241
Fukuhara, H., 175
Fumaric acid, 304, 305
  See also Beef carcasses, decontamination
Fungal starter cultures, in ripened meats, 186, 187

Fung, D. Y. C., 142, 171, 173, 289, 293, 294, 303–305, 308, 310, 311, 313
Fung, Z. F., 142
Fussenegger, M., 364

G
Gaeng, S., 270
Gaggiotti, O. E., 76
Gagnaire, V., 237, 243
Gaier, W., 142, 155
Gaillardin, C., 175
Galbusera, P., 75
Galdeano, C. M., 219
Gallmann, H. R., 131
Galvez, A., 261, 263, 265, 375, 387, 389, 390
Gambari, R., 343
Gandolfi, G., 69
Gannon, V. P. J., 270
Gansheroff, L. J., 274
Gantois, I., 275
Gänzle, M. G., 239, 261, 266
Gao, Z. X., 343
García, D., 82, 84
García, M. L., 171, 173, 175
García, R., 175
Garcia, M. T., 389
Garcia-Varona, M., 103
Gardini, F., 141, 142, 168
Gardner, C. D., 434
Gardner, P. I., 330
Garg, S. R., 257
Garreau, I., 242
Garrity, G. M., 131
Garro, O., 404
Gatti, M., 98
Gaudin, V., 365, 369
Gau, J. J., 339, 345
Gaya, P., 266
Gazzola, S., 142
GDF8 gene, 44
Geers, R., 65
Gehring, A. G., 338
Geisen, R., 184, 187, 254, 400
Gené, J., 181
Gene assisted selection (GAS), 42
Genetically modified organisms, 319
  analysis, basic approaches, 321, 322
    GM material, analysis of
      data quality and interpretation, 331, 332
      DNA-based methods, 324–331
      protein-based analysis, 323, 324
    sampling, 323
Gene-trap locus 2 (GTL2), 10

Geng, T., 342
Genome assisted breeding value (GAEBV), 53
Genovese, K. J., 260
Georgakis, S., 411
Georgala, D. L., 425
Georgantelis, D., 411
Georges, M., 29, 31, 36, 37, 39, 45, 66
George, T., 339, 340, 361
Georgiadou, K. G., 415
Geornaras, I., 416
Gepts, P., 319
Gerken, C. L., 6
Germain, P., 382
Gerwen, P. V., 339
Gettinby, G., 275
Gevers, D., 95, 110, 142, 153, 155
Ghalfi, H., 256, 257, 388
Ghebremedhin, B., 131
Gherardi, M. M., 275
Ghindilis, A., 338, 339
Giacomini, A., 96, 153
Gialitaki, M., 95, 99, 100, 102, 104, 105, 109
Giannetti, L., 361
Gianola, D., 53
Gibson, G. R, 219, 220, 242
Gibson, J. P., 33
Gilbert, R. O., 26
Gill, C. O., 291, 293, 295, 296
Gill, H. S., 221, 222
Gillespie, I. A., 403
Gilliland, S. E., 234, 244, 258
Gill, A. O., 256, 257, 389
Gilson, E., 95
Ginkel, L. A. V., 362
Giovannini, A., 65
Giovannoli, C., 362
Gistelink, M., 92
Giudici, P., 173
Gjerde, B., 69
Glämsta, E. L., 242
Glatz, B. A., 234
Glutahione, 235
Glycolysis, in fermented meat, 204
GMOs, *see* Genetically modified organisms
Gnadig, S., 233
Gobbetti, M., 237
Goddar, M. E., 53
Godfrey, M. A. J., 364
Godon, J. J., 99
Goebel, B. M., 92
Goldammer, T., 47
Golden, D. A., 168
Golden, T. E., 453

Gombas, D., 403, 406
Gomez, R., 339
Gonzalez, F., 261
Gonzalez, B., 153
Gonzalez, C. F., 153, 155
Gonzalez de Llano, D., 380
Goode, D., 271
Goodhue, R. E., 319
Goodridge, L., 270
Goodson, K. J., 10, 11
Gordon, J. W., 4
Gore, A., 342
Göring, H., 75
Gory, L., 99, 156
Goto, M., 434
Gotz, F., 133, 138
Govind, N. S., 176
Graham, D. C., 152
Graham, C. R., 343
Gram-positive Catalase-positive Cocci (GCC+), 129
Grande, M. J., 389
Grant, K., 406
Grant, S. A., 338
Grau, H., 138
Gravesen, A., 389
Gray, J., 82
Grazia, L., 97, 103, 173, 187
Greco, M., 103
Green, R. D., 9
Greenstein, J. L., 11
Greer, G. G., 254–256, 264, 273
Gregory, S. G., 38
Griffin, P., 402
Griffiths, M. W., 265, 270
Grimes, C. A., 343
Grisart, B., 46
Grisenti, M. S., 181
Grobet, L., 44
Grohs, C., 83
Grooms, D. L., 338, 342
Grossman, M., 52
Grosz, M. D., 36, 40
Growth hormone (GH), 8, 9
Gruber, W., 431
Gruden, K., 329, 331
Gruss, A., 157
Grutters, E., 365
Gualla, A., 181
Guangming, L., 330
Guarner, F., 219, 221, 222
Guarro, J., 181
Gueimonde, M., 222

Guerra, N. P., 386
Guerzoni, M. E., 170, 173
Guesdon, B., 237
Guglielmi, D., 187
Guilbault, G. G., 341
Guillamón, J. M., 169
Guiseppi-Elie, A., 339
Gunduz, U., 153
Guo, J., 361
Gurtler, M., 403
Gury, J., 157
Gustavsson, I., 24
Gutíerrez-Gil, B., 40

H

Hägele, G. H., 83
Haasnoot, W., 368
Haemig, P. D., 175
Hagen, B. F., 203
Haghiri, M., 65
Hagler, A. N., 169
Hahn, I., 276
Halami, P. M., 153
Hald, T., 402
Haley, C. S., 27, 39, 41, 82, 83
Haller, D., 239
Hallerman, E., 41
Hall, J. M., 343
Hall, R. J., 381
Hall, J., 268
Halothane sensitivity gene (HAL), 32
Hamada, R., 347
Hamid, A., 339, 342
Hammer, K. A., 256
Hammer, R. E., 4, 8
Hammes, W. P., 92, 129, 138, 142, 149, 152, 155, 156, 171, 173, 186, 209, 217, 239, 261, 266, 412
Hampikyan, H., 386
Hancock, D., 274
Hancock, R. E., 267
Hanekamp, T., 153
Han, J. H., 258, 426, 430
Hanlon, G. W., 269
Hansen, A., 139
Hansen, A. M., 210
Hao, Y. Y., 254, 257, 304
Hara, M., 347
Hardin, M. D., 254, 293, 300, 302
Hardt, W. D., 273
Hardy, C. L., 69, 77, 331
Hardy–Weinberg equilibrium, 69
Harms, H., 175, 176

Harris, J. J., 6
Harris, L. J., 265
Harris, N., 323
Harrison, B. E., 33
Hartley, H. A., 345
Hartman, N. F., 338
Harvey, R. B., 276
Hasler, C. M., 231
Hassan, Z., 403
Hastings, J. W., 380
Hatakka, M., 453
Haughey, S. A., 368
Hausman, J. H., 33
Havelaar, A. H., 276
Haveri, M., 402
Hawley, R. J., 11
Hayes, B. J., 53
Hayes, H., 24
Hayes, M., 268
Haynie, S. L., 267
Hazard analysis and critical control point (HACCP) plans, 289, 455, 456
    governmental regulations and, 297, 298
Heard, G. M., 224
Heasman, M., 231
Heaton, M. P., 63
Hebraud, M., 201
Hedegaard, J., 12
Heiniö, R. L., 434
Heinze, P., 329
Heithoff, D. M., 275, 276
He, F. J., 344
He, J. H., 364
Hellenas, K. E., 362, 369
Heller, C. E., 255
Heller, K. J., 331
Hemolytic uremic syndrome (HUS), 336
Henderson, C. R., 53
Hepler, A. B., 75
Hernández-Jover, T., 141
Hernandez, J., 95
Hernandez, M., 95
Herranz, B., 203, 208
Hertel, C., 99, 138, 149, 156, 217, 261
Hesseltine, C. W., 91
He, Y. H., 344
Hide-on decontamination, 298
Hierro, E. M., 202, 203, 209, 239, 240
Hiestand, D., 131
Higgins, D. G., 220
Higgins, J. P., 272
Higgins, S. E., 220
Higgs, G., 401, 402

Higgs, J. D., 232
Higgs, R., 267, 268
Hill, B. J., 69
Hill, C., 260, 261, 265, 266, 375, 408, 412
Hillebrandt, H., 347
HindIII enzyme, 108
Histidyl dipeptides, 234, 235
Hitchcock, H. L., 168
Hoagland, T. A., 268
Hoban, T. J., 64
Hobbs, J. E., 65
Hobbs, A. L., 63
Hocking, A. D., 185, 186, 191, 193
Ho, C. M., 339
Hoffman, L. C., 388, 412
Hofnung, M., 95
Hogan, S. A., 426
Ho, J. A., 342
Holck, A., 139, 203, 264, 414
Holck, A. L., 379
Holeckova, B., 402
Holgado, A. P., 388
Holgado, A. R., 413
Holley, R. A., 140, 142, 224, 225, 256, 258, 270, 389, 405
Holo, H., 375, 377, 378, 379
Holst, C. V., 331
Holst, E., 331
Holst-Jensen, A., 327, 330, 331
Holy, A. V., 173
Holzapfel, W. H., 141, 254, 258, 259, 264, 266, 375, 400, 409, 416
Homer, V., 91
Homola, J., 340
Hong, S. I., 436
Honkanen-Buzalski, T., 386, 400
Hoog, G. S. D., 181
Hooker, N. H., 335
Hoover, D. G., 260
Hornbaek, T., 262
Horn, P., 413
Hoseknechtm, K. L., 33
Hosokawa, K., 331
Hoszowski, A., 260
Hotchkiss, J. H., 267, 268, 269, 430
Houba, P. H. J., 15
Houghton, S., 275
House, J. K., 275
Houwelingen-Koukaliaroglou, M. V., 217
Hou, X. L., 364
Hovde, C. J., 270, 271
Hoyle, B., 338
Hoz, L. D. L., 202, 203, 208, 209, 239, 240

Hsu, F. C., 270
Hsu, C. W., 48
Hsu, H. W., 342
Hsu, W. H., 33
Huang, J., 381
Huang, M. R., 342
Huang, T. S., 340, 342, 364
Hubert, J. C., 153
Huet, A. C., 366, 368, 369
Huff-Lonergan, E., 11
Huffman, R. D., 254
Huff, W., 11
Hugas, M., 98, 141, 152, 263, 380, 382, 385, 386, 410, 412, 413, 415, 416
Hugenholtz, P., 92, 93
Huggins, M. B., 273
Hughes, A. L., 176
Hughes, M. C., 205
Huirne, R. B. M., 319
Huis in 't Veld, J., 221
Hu, J. M., 344
Humane Slaughter Act 1956, 289
Hume, M. E., 276
Humpheson, I., 390
Huot, E., 381
Huq, N. L., 237
Hurburgh, C. R., 331
Hurdle theory, bacteriocins and, 388–390
Hurme, E., 434, 436
Husby, K. O. V., 151
Hussein, H., 402
Huston, R. D., 33
Hutton, T., 425
Hutt, P. B., 231
Huys, G., 95, 110, 153, 155
β-Hydroxy-gamma-trimethyl amino butyric acid, 235

I

Iacumin, L., 95, 96, 100, 102, 105, 106, 110, 111
Iczkowski, K., 273
Igarashi, T., 138
Igimi, S., 403
Ihara, N., 37
Ikeda, R., 175
Immanuel, G., 265
Immunoafinity chromatography, 363–367
Immuno-modulating peptides, 237
Immunosensors, 339–343
Incze, K., 140, 217
Ingham, S., 405
Inglis, G., 403

Insulin-like growth factor (IGF1/2), 8, 45
Inter-gene spacer regions (ISR)-PCR, 107
Intimin, from *E. coli* O157:H777, 273, 274
Intramuscular fat, 39, 40
 See also Meat quality traits, genetic control
Iorizzo, M., 103
Isaac, G. E., 64
Ishikawa, S., 238, 239, 241, 242
Ishikazi, A., 381
Ismaeel, A. Y., 218
Ismail, S. A., 170
Israel, C., 82
Itoh, M., 238, 239, 241, 242
Itoh, T., 37
Ivnitski, D., 339
Iwata, T., 389
Izquierdo-Pulido, M., 135, 141
Izumi, S., 69

J

Jackson, S. P., 9, 11
Jacobsen, T., 262, 264, 411, 414
Jacoby, A., 402
Jahreis, G., 225
Jaime, I., 103, 141
Jakowski, G., 433
James, S. A., 169
James, T., 319
Jang, A., 240, 241
Jann, O. C., 37
Janssen, F. W., 83
Japanese Dietetic Information guidelines, 7
Jaroni, D., 260
Jarvi, E. J., 341
Jauhiaine, T., 243
Javier, D., 338
Jeffreys, A. J., 29, 66
Jelacic, S., 270
Jelen, P., 234
Jensen, A. H., 325, 327, 330, 331
Jensen, J. S., 425
Jensen, L., 425
Jensen, M. A., 107
Jespersen, L., 171–173, 175
Jiang, H. Y., 338, 364
Jiang, S. T., 265
Jiang, Y. L., 44
Jiang, Z.-H., 33
Jiménez-Colmenero, F., 223, 233, 244
Jimenez-Diaz, R., 152
Jin, X., 6
Joerger, M. C., 380
Joerger, R. D., 254, 272, 273

Jofre, A., 416
Johansson, M. A., 386
Johnson, M. C., 380, 381, 389
Johnston, D. J., 52
Jollès, P. S., 237, 238
Jones, B. K., 45
Jones, D. P., 235
Jones, P. J. H., 222
Jones, R. J., 404
Joosten, H. M. L. J., 381
Jordan, C. E., 343
Jordan, D., 402
Joseph, M. J., 343
Jouve, J.-L., 400
Jumriangrit, P., 265
Just, R. E., 320
Jydegaard, A.-M., 389

K

König, B., 131
König, W., 131
Kailasapathy, K., 140
Kairuz, M. N. D., 388, 413
Kaiser, A. L., 381
Kakikawa, M., 151
Kakouri, A., 168, 171, 415
Kalchayanand, N., 389
Kalinowski, R., 406
Kalischuk, L., 403
Kamal, M., 265, 412
Kambadur, R., 44
Kamps, A., 138
Kanatani, K., 153
Kanellos, T., 401, 409
Kang, S. A., 221
Kantor, A., 153
Kappes, S. M., 34, 36, 40
Kappes, S. S., 24
Karel, M., 434
Karim, L., 46
Karolewiez, A., 184
Kashi, Y., 39, 41
Kasimir, S., 403
Kastelic, J., 403
Kastner, C. L., 255, 304
Katakis, I., 345
Katayama, K., 238, 240, 241
Katikou, P., 411
Katla, T., 262, 263, 381, 410
Kato, H., 239, 243
Kato, T., 239
Kawai, Y., 263
Kay, S., 327, 329

Kearney, L., 140
Kearsey, M., 39
Kemperman, R., 376
Kemp, S., 307
Kendall, P. A., 254, 304
Kenneally, P. M., 138
Kennedy, B. W., 53
Kennedy, D. G., 366, 368
Kennedy, I., 366, 368
Kennes, Y. M., 33
Kerman, K., 345
Kerr, W. A., 64
Kerry, J. P., 425, 431–433
Kerth, C. R., 11
Khan, M. I., 268, 276, 330
Kheadr, E., 266
Khoury, C. A., 276
Kim, B., 341
Kim, D. Y., 443
Kim, H. Y., 221
Kim, J. S., 78
Kim, J. W., 443
Kim, K. S., 33
Kim, N., 341
Kim, S. B., 78
Kim, W. S., 381
Kim, W. Y., 341
Kind, A. J., 4
Kiss, I. F., 265
Kitis, F. Y. E., 390
Kittredge, C., 4
Kjeldsen, N., 65
Kjelleberg, S., 116
Klaenhammer, T. R., 151, 155, 157, 220, 261, 375, 377, 378, 386
Kleerebezem, M., 131, 149, 158
Klein, T. A., 11
Klijn, N., 116, 117
Klimant, I., 431
Klingberg, T. D., 223, 225
Klosowska, D., 13, 15
Knauf, H. J., 92, 171, 173, 186, 209
Knife trimming, of carcasses, 300
Knøchel, S., 389
Knott, S. A., 39
Koch, A. G., 262, 265, 411
Kocwin-Podsiada, M., 12, 13
Kodjo, A., 107
Koehler, P. E., 141
Kohen, F., 368
Koidis, P., 411
Kok, E. J., 330
Komarova, E., 338

Komprda, T., 141
Kondo, J. K., 152
Koning, D. J. D., 39, 41
Kono, Y., 138
Koohmaraie, M., 10, 11, 23, 34, 36, 37, 40, 238, 253, 254, 308
Koort, J., 415
Kopecny, M., 13
Koppes, L., 346
Korhonen, H., 236–238, 240, 242
Korkeala, H., 403
Korpela, R., 243
Korpimaki, T., 368
Korte, S., 11
Kortz, J., 12
Ko, S., 338
Kostrzynska, M., 254, 258, 259, 261, 262
Kotarski, S., 361, 362
Kotzekidou, P., 131, 223, 415
Koubova, V., 341
Kröckel, L., 187, 217
Krajcovicova-Kudlackova, M., 232
Kresovich, S., 338
Krier, F., 382
Krismer, B., 138
Kristiansen, B., 319
Kristoffersen, A. B., 330
Krockel, L., 265
Kruger, L. A., 389
Kruijf, N. D., 426
Krumhar, K. C., 434
Kruuk, L. E. B., 74
Krzecio, E., 12, 13
Kubelik, A. R., 66
Kuber, P. S., 11
Kubista, M., 329
Kudva, I. T., 270, 271
Kues, W. A., 320
Kuiper, H. A, 330
Kukanskis, K., 344
Kuleasan, H., 258
Kunka, B. S., 153, 155
Kurtzman, C. P., 94, 107, 167, 168, 169, 173
Kuryl, J., 13
Kwang, J., 275
Kwok, A. Y., 131

L

Lövenklev, M., 429
Lücke, F. K., 91, 92, 135, 140, 187, 223, 404, 406
López-Díaz, T. M., 175
Laan, L. J. W. V. D., 11
Labuza, T. P., 438, 443
Lachance, P. A., 232
Lacroix, C., 262, 266, 381
Lacroix, M., 256, 386
Lactic acid, 304
    See also Beef carcasses, decontamination
Lactic acid bacteria (LAB), 91, 92, 95, 98, 105, 106, 109–110, 118–121, 129, 131, 199, 200, 258
    bacteriocins produced by, 408
    genetics of
        chromosomal elements, 157–160
        mobile elements and plasmids in meat starters, 150–157
Lactobacillus curvatus Crl705, in vacuum-packaged fresh beef stored at refrigeration temperatures, 409–411
Lactobacillus spp., *see* Lactic acid bacteria (LAB)
Lactococcus garvieae, 103
Lactoferricin B, 307
Laczkova, S., 387
Laere, S. A. V., 45
Lagonigro, R., 33
Lahdenpera, S., 364
Lahti, E., 386
Lai, L., 11
Lambert, R. J. W., 256
Lammerding, A., 405
Lancastre, H. D., 107
Lanciotti, R., 170, 173
Lander, E. S., 24, 28, 50
Lan, E. H., 339
Lane, D. J., 116
Lane, W., 64
Langella, P., 155
Langseth, L., 234
LaPointe, G., 262
Larrouture, C., 139
Larsen, N., 33
Larsen, T. O., 184
Larsson, S. C., 234
Lasa, I., 61
Latouche, K., 65
Laukova, A., 387
Lauret, R., 156
Lauritsen, F. R., 139, 210
Lavoie, M. C., 262
Lawler, A. M., 13, 43
Layer, F., 131
Lazcka, O., 338, 339
L-carnitine, 235
Leahy, S. C., 220

Lebert, I., 91, 149
LeBlanc, A. D. M. D., 219
Leclercq-Perlat, M. N., 175
Lee, B. H., 152, 156
Lee, C. H., 443
Lee, G. G., 6
Lee, H., 366
Lee, H. K., 34
Lee, H. J., 366
Lee, J., 95
Lee, J. S., 345
Lee, L. P., 345
Lee, M., 240, 241
Lee, M. H., 371
Lee, S., 389
Lee, S. J., 6, 13
Lee, S.-J., 43
Lefebvre, G., 386
Le, G. W., 236
Lei, H., 344
Leiner, M. J. P., 431
Leistner, F., 199
Leistner, L., 406
LeJeune, J. T., 254, 273
Lejeune, R., 382–384
Leloup, L., 156
Lemay, M. J., 257
Lengauer, C., 24
Lenk, A., 443
Lenski, R. E., 269
Lenstra, J. A., 83
Leoncini, O., 169
Leonil, J., 243
Lepingle, A., 176
Leplae, R., 150
Lerche, M., 91
Leroy, F., 136
Leroy, S., 91, 94, 97, 111, 133, 208, 210, 382, 385, 413
Leroy-Setrin, S., 94, 136, 208, 210, 404
Letcher, S. V., 340
Leuconostoc mesenteroides, 104
Leuschner, R. G. K., 152
Leverentz, B., 273
Leveziel, H., 83
Lewis, C., 5
Lewis, A. M., 52
Leymaster, K. A., 9
Liang, C., 171, 173
Liang, P., 47
Li-Chan, E. C. Y., 237
Lieple, C., 237, 242
Li, G. H., 241

Li, M. Q., 345
Li, N., 44
Li, W. H., 169
Li, Y., 152, 345
Li, Y. B., 340, 341
Li, Y. Y., 175
Lightner, D. V., 69
Ligler, F. S., 340
Lilburn, T. G., 131
Lilly, D. M., 218
Lima-Bitterncourt, I. C., 94
Lim, D., 299, 307, 338
Lin, C., 403
Lin, C. F., 142
Lin, C. S., 48
Lindsay, D. G., 231, 234
Ling, F. H., 255
Lin, H. C., 340
Lin, I. C., 340
Linkage disequilibrium (LD), 42
Linkage mapping, QTL discovery by, 38–40
Link, N., 364, 367
Lin, L. W., 386
Linn, C. P., 343
Linnebur, M., 403
Lipolysis, in fermented meat, 208, 209
Lipp, M., 325
Lisby, M., 253
Litopoulou-Tzanetaki, E., 131, 223
Little, C., 406
Liu, L. J., 344
Liu, M. L., 152, 153
Liu, S. Q., 344
Liu, W., 275
Liu, X., 344
Livak, K. J., 66
Livisay, S. A., 234
Liying, W., 366
Llano, D. G. D., 380
Llanos, R. D., 168, 169
Locascio, L. E., 345
Loessner, M. J., 270, 272
Lohmann, M., 152
Loi, P., 8
Loizel, C., 156
Lombardi, A., 95
Lonergan, S. M., 10
Longo, F., 361
Loo, J. A. E. V., 219
Lopez-Exposito, I., 268
Lopez, R. L., 261, 390
Lopez-Villalobos, N., 64
Lorenzen, C. L., 10

Lori, D., 181
Loureiro-Dias, M. C., 175
Lovell, M. A., 273
Lu, B., 341
Luchansky, J. B., 155, 386
Lucia, L. M., 253, 255, 307
Lucke, F. K., 258, 262, 265
Luecke, F. K., 379, 386
Lues, J., 402
Luikart, G., 70, 73, 77
Lumley, T. C., 186
Lund, B., 142, 433
Lu, Z., 151

M

McAllister, T. A., 270
Macarthur, R., 323
McCarthy, L. C., 24
Macias, C. L., 386
McClane, B., 405
McClelland, R., 338
McCormick, J. K., 261
McDonnell, M. J., 253
McDowell, D. A., 78, 402
McEvoy, J., 295, 297, 402
McEwan, J. C., 70
McGown, L. B., 343
Mcguire, J., 386
Maciorowski, K. G., 270
Mackay, T. F. C., 69
McKay, L. L., 152
McKean, J. D., 61, 80
McKellar, R. C., 382
McKinstry, J., 65
MacLennan, D. H., 33
MacLeod, G., 204
McLoughlin, A., 140
McMahon, D. J., 152
McMurray, N., 431
MacNeil, M. D., 36, 40
McPherron, A. C., 13, 43, 44
MacRae, M., 402
McWhir, J., 4
Madec, F., 65
Maeda, M., 331
Maehashi, K., 345
Maggi, R. G., 176
Maguin, E., 157
Mahan, M. J., 275, 276
Maitin, V., 219
Makarova, K., 149–151, 161
Makras, L., 218, 220
Ma, L., 260

Malcata, F. X., 237
Maldini, M., 69
Malek, M., 34
Maliga, P., 319
Malignant hypothermia, 32
Malik, K. A., 221
Malik, S. V. S., 386
Malleret, C., 153, 156
Mambriani, P., 185
Manel, S., 76, 77
Maniatis, T., 67
Manicardi, G., 395, 413
Mannisto, S., 233
Man, Y. B. C., 82, 83
Manzano, M., 95, 97, 100, 104, 105,
  112–117, 199
Maqueda, M., 263, 265, 387, 389, 390
Marceau, A., 133
Marchetti, R., 170
Marchisio, E., 82
Marchitelli, C., 44
Marco, A., 120, 175, 211
Marcos, B., 266
Marcus, J. R., 205
Maria, S. D., 97, 102, 111
Mariné-Font, A., 141
Mariotti, E., 344
Maris, P., 365
Marker-assisted selection (MAS), 23, 41
Markx, G., 346
Marquie, S., 338
Marshall, K., 402
Marshall, M. R., 364
Marshall, R. T., 295, 303
Marshall, T. C., 74, 77
Martín, A., 186
Martínez Bueno, M., 263, 265, 387, 389, 390
Marteau, P., 221, 222
Martin, A., 181
Martin, B., 98, 101, 102, 107, 110, 121
Martinez-Bueno, M., 263, 265, 387, 389, 390
Martinez, Y. B. D., 265
Martinis, E. C. P. D., 265
Martinussen, J., 203, 210, 404
Martorell, P., 168, 169, 175
Martuscelli, M., 142
Marucci, P., 403
Marzoca, M., 403
Mascini, M., 343, 344
Masson, F., 139, 203, 307
Mastitis, 8
Mataragas, M., 375, 379–382, 388
Matarante, A., 108

Index 485

Maternal expressed gene 8 (MEG8), 10
Matheson, J. C., 74
Mathew, F., 338, 339
Mathias, H. C., 80, 81
Mathiesen, G., 156
Matruscelli, M., 142
Matsumoto, K., 345
Matsunaga, T., 82
Matsusaki, H., 381, 382
Matteson, P. G., 44
Matthysen, E., 75
Mattiason, B., 437
Mattiazzi, A., 96
Mattila-Sandholm, T., 140, 217, 224, 244, 434, 436
Mattsson, R., 175
Maudet, C., 76
Mauriello, G., 103, 138, 266
Maurin, F., 107
Mayer, Z., 184
Mayo, B., 153, 222, 223
Mazick, A., 253
Mazzanti, E., 83
Mazzette, R., 103
Mead, P. S., 253, 337
Mears, D. J., 339
Meat-based bioactive compounds
    conjugated linoleic acids, 233, 234
    fat and fatty acids, 232, 233
    glutahione, 235
    histidyl dipeptides, 234, 235
    L-carnitine, 235
    minerals and vitamins, 232
Meat chain, pathogens biocontrol in, 253
    antagonistic bacteria
        bacterial metabolites, 258, 259
        competitive exclusion technology, 259, 260
    antimicrobial peptides, 267–269
    bacteriocin-producing bacteria and bacteriocins, 260, 261
        animals, applications in, 261, 262
        limitations, 266, 267
        meat products, applications in, 262–266
    bacteriophage
        application of, 270, 271
        classification and mode of action, 269
        limitations of, 272, 273
        in meat and meat products, applications, 271, 272
        preharvest control, 270, 271
    organic compounds
        essential oils, 256–258

        organic acids, 254–256
    vaccines, 273–277
Meat fermentations, microbial starters in, 129
    functional features
        acid production in meat products, 135, 136
        bacteriocin and biopreservation, 139
        catalase activity, 136–138
        flavor formation, 138, 139
        meat environment, competitiveness in, 133–135
        nitrate reduction, 138
        probiotics, 139, 140
    selected bacterial starter cultures, safety of
        antibiotic resistance, 142
        biogenic amines, 141, 142
        toxigenic potential, 143
    starter cultures for fermented meats
        genomics, 132, 133
        taxonomy, 130–132
Meat and foods, and food processing environments, microbial considerations in, 293
    bung tying, 295
    carcass-to-carcass transfer, 297
    evisceration process, 295, 296
    hide removal, 295
Meat LAB starters, mobile elements and plasmids in
    bacteriophages, 151, 152
    insertion sequences and transposons, 150, 151
    plasmid content of strains, 152, 153
    plasmid-encoded functions, 153–155
    vectors and genetic tools, 155–157
Meat and meat products
    microbiology of, 401–406
    shelf stability and hurdle presence in, 406
Meat protein-derived bioactive peptides
    ACE inhibitory peptides, 240, 241
    antioxidative peptides, 241, 242
    from food proteins, 236
    functional meat products, peptides utilization for, 243, 244
    generation from, 238–240
    prebiotic peptides, 242
Meat quality traits, genetic control, 21
    application of techniques for
        candidate genes, 32
        genes and QTL mapping in livestock, 32–35
        intramuscular fat, 39
        QTL discovery by linkage mapping, 38

Meat quality traits (*Cont.*)
- QTL mapping, 35–38
- tenderness, 40–42
- trait genes starting from QTL, finding, 42–46
- data in breeding, application of, 51
  - genome selection, 53
- eQTL, 49, 50
- genetic selection, approaches and tools, 23
  - advances in knowledge, 28
  - genes controlling phenotypic variations, identifying, 30, 31
  - genetic markers, 29, 30
  - meat quality, defined, 24, 25
  - new opportunities, 27
  - traditional, 25–27
- genome sequence, 50, 51
- proteomic analysis, 49, 50

Meat safety and regulatory aspects, in EU, 453
- background, 454
- future legislative work, 464, 465
- general food law, 454, 455
- hygiene package, 458–461
- legislation
  - on hygiene, 455, 456
  - on official controls, 456–458
- related food safety legislation, 461–464

Meat tenderness, 40–42
See also Meat quality traits, genetic control

Meat traceability, DNA-based
- compound meat products and, 80, 81
- DNA markers, 66, 67
- DNA as traceability tool, 67, 68
  - audit, application to, 78, 79
  - breed, species/brand protection, 76–78
  - DNA markers, use of, 69
  - DNA sampling, 68, 69
  - food safety, application to, 78
  - identity, 70–74
  - notation and assumptions, 69, 70
  - parentage, 74, 75
  - relatedness, 75, 76
- from "gate to plate", 61
  - risk, 64
  - stakeholders, 65
  - traceability continuum, 62–64
- as part of value chain, 82–84
- species and strain differentiation, 82

Medina, M. B., 266, 338
Meenakshi, M., 276
Meester, E. D. D., 272
Meeusen, C., 338, 341

Meghen, C. M., 68, 72, 73, 78
Meghen, C. N., 68, 78
Meghrous, J., 381, 382
Meichtri, L., 402
Meinersmann, R. J., 276
Meisel, H., 236, 237, 241
Melanocyte receptor hormone (MC4R), 33
Mellander, O., 236
Mellentin, J., 231
Mellett, F. D., 388
MEMS, *see* Micro electromechanical systems, biosensor fabrication and
Mendonca, A., 416
Mendoza, M., 107
Meng, J., 336, 338
Mennecke, B., 443, 444
Menon, K. V., 257
Mera, T., 133
Mercado, I., 255, 401
Mercenier, A., 217, 218, 219, 221, 226
Mesas, J. M., 155
Messens, W., 379, 382, 385
Messi, P., 379
Meszaros, L., 265
Metaxopoulos, I., 95, 99, 100, 102, 104, 109
Metaxopoulus, J., 168, 171, 380
Metchnikoff, E., 218
Mett, A., 153
Meugnier, H., 107
Meuwissen, M. P. M., 319
Meuwissen, T. H. E., 27, 53
Mevius, D. J., 276
Meyer, A. S., 203, 208, 210
Mezzelani, A., 331
MFI, *see* Myofibriler fragmentation index
Microbial diversity, 92–94
Microbial starters, in meat fermentations, 129
- functional features
  - acid production in meat products, 135, 136
  - bacteriocin and biopreservation, 139
  - catalase activity, 136–138
  - flavor formation, 138, 139
  - meat environment, competitiveness in, 133–135
  - nitrate reduction, 138
  - probiotics, 139, 140
- selected bacterial starter cultures, safety of
  - antibiotic resistance, 142
  - biogenic amines, 141, 142
  - toxigenic potential, 143
- starter cultures for fermented meats

genomics, 132, 133
taxonomy, 130–132
Micrococcus spp, 129
Micro electromechanical systems, biosensor
  fabrication and
  detection principle, 345, 346
  fabrication, 347–349
  results, 350–353
  sensor design, 346, 347
  testing, 349, 350
Microorganisms identification in traditional
  meat products, molecular
  methods, 91
  fermented sausages
    culture-dependent techniques,
      94–97
    culture-independent techniques,
      97, 98
    microbial ecology of, 98–104, 112–121
    ISR-PCR, 107–112
    microbial diversity, approaches for study
      of, 92–94
    PCR-DGGE, 104–106
Miguez, C. B., 156
Mikami, M., 239
Mikkelsen, A., 209
Miksic, V., 344
Milan, D., 12, 67
Miller, G., 11
Miller, K. W., 155
Miller, M. F., 11
Miller, R. K., 10
Millet, L., 99
Millette, M., 386
Milliere, J. B., 416
Mills, A., 431
Mills, D. A., 106
Mills, E., 399
Mills, L., 400
Mills, M., 274
Milona, P., 268
Miltz, J., 426, 433
Minervini, F., 237, 268
Mine, Y., 236, 243
Ming, X., 266
Minnan, L., 330
Min, S., 92, 129, 344
Minunni, M., 343, 344
Miralles, M., 135
Mir, M., 345
Mirold, S., 273
Mir, P. S., 232
Mischiati, C., 343

Mitchell, A. D., 6, 14
Mittelmann, A. S., 339
Miura, H., 239
Microwave and combination technology,
  for beef carcasses
  decontamination, 311
Moioli, B., 73
Molbak, K., 253
Molds, characteristics and applications, 181,
  182, 200
  environment-contaminating species of,
    183–186
  growth and competition tests, 189–192
  growth, parameters affecting, 182, 183
  lipolytic and proteolytic activity of,
    187–189
  ripened meats, fungal starter cultures in,
    186, 187
Molle, D., 243
Møller, J. K. S., 425
Molly, K., 138, 204, 209
Mondoloni, J. A., 151
Monfort, J. M., 382, 385
Monin, G., 12, 13
3M Monitor Mark®, 439
Monounsaturated fatty acids (MUFA), 233
Monte, E., 171
Montel, M. C., 92, 99, 138, 139, 156, 200, 201,
  203, 204, 210
Montiel, V., 175
Montville, T. J., 155, 261, 375, 381
Moon, H. W., 274
Moore, L. H., 361
Moore, S. S., 52
Moore, W. E. C, 361
Moorhead, S. M., 271
Mor, A., 267
Mora, L., 235
Morea, M., 108
Morel-Deville, F., 155, 156
Morency, H., 262
Moreno, B., 175, 219
Morgan, J. B., 6, 303
Morgan, M. T., 342
Morimatsu, F., 242
Mor-Mur, M., 265
Moronne, M., 340
Morot-Bizot, S., 94, 97, 100, 102, 103,
  104, 111
Morrical, D. G., 11
Morrison, D. A., 156
Morrison, M., 115
Morris, S., 402

Morse, S. S., 398
Mortvedt-Abildgaard, C., 380, 381
Moschetti, G., 78, 95, 116
Moss, G. E., 10
Mota-Meira, M., 262
Motlagh, A. M., 153, 379
Mott, R., 51
Mount, J., 402
Mousavi, A., 443
Mo, X. T., 344
MSTN F94L gene, 6
Mueller, M. A., 270
Mueller, S., 401
Muhammad-Tahir, Z., 338, 339, 342
Mukai, T., 238
Mullis, K. B., 324
Mulvihill, B., 232
Munoz, F., 338
Murakami, Y., 345
Murani, E., 48
Murano, E.-A., 401
Muriana, P. M., 155
Murphy, B. D., 33
Murray, J. D., 8
Murray, M., 386
Murry, B., 236
Muscle regulatory factor (MRF) gene, 11
Mustapha, A., 265, 299, 307, 411, 416
Muthukumarasamy, P., 140, 224, 225, 258
Muthuswamy, J., 340
Mutti, P., 183, 185, 191, 192
Muyzer, G., 96, 112, 116, 170
Myllys, V., 402
Myofibriler fragmentation index, 41
Myostatin (MSTN), protein, 6, 43, 44

N

Núñez, F., 186
Naes, H., 203
Nagai, H., 345
Nagamine, Y., 41
Nagao, J.-I., 376
Nagao, K., 233
Nagaoka, S. Y., 237, 238
Nagao, M., 239
Nagasawa, K., 236
Naghmouchi, K., 266
Naidu, A. S., 221
Nakamura, Y., 237, 243
Nakase, T., 173, 174, 175
Nakashima, Y., 238–241
Nam, S. H., 233

Namsoo, K., 344
Nancarrow, C. D., 8
Nanninga, N., 346
Napolitano, F., 73
Narang, U., 340
Nashat, A. H., 340, 341
Naterstad, K., 140, 156, 223
Natrajan, N., 386
Naumoff, D. G., 151
Nava, G. M., 259, 261
Navarro, J. L., 205, 209
Nawata, T., 434
Neimann-Sorensen, A., 39
Nesbakken, T., 264, 414
Nes, I. F., 151, 152, 153, 261, 375, 377, 378, 379
Nesvold, H., 330
Neumeyer, B., 412
Neurater, G., 431
Neve, H., 270
Neviani, E., 98
Nevill, C. H., 11
Newell, D. G., 276
Newman, J. D., 338
Newman, L. S., 337
Nezer, C., 45
Nicolini, C., 344
Niebuhr, S., 402
Niederhausern, S. D., 413
Nie, L., 338
Nielsen, E. M., 253
Nielsen, J., 381, 382
Niemann, H., 14, 320
Nieto-Lozano, J. C., 263, 387
Nieto, P., 205
Niinivaara, F., 129
Nikolajsen, K., 382
Nilsen, A., 139, 203
Nilsson, L., 389
Nip, J. G., 199
Nishikawa, A., 175
Nitrate reduction, 138
    See also Meat fermentations, microbial starters in
Nonneman, D., 34
Nonnis Marzano, F., 69
Noonpakdee, W., 138, 265, 379
Noordman, W. H., 272
Normano, G., 402, 405
Nortje, G. L., 170
Noutomi, D., 347
Novoselova, T. A., 319
Nunez, F., 181

Nunez, M., 266
Nurmi, E., 386
Nuti, M. P., 153
Nyberg, F., 242
Nychas, G. E., 436
Nychas, G. J., 256
Nychas, G.-J. E., 256–258
Nylen, G., 64

O

Oberg, C. J., 152
Ochi, H., 237
Ockerman, H., W., 187
Oellers, K., 138, 156
O'Flynn, G., 271
Ogden, I., 402
O'Grady, M. N., 425, 426
Ogurtsov, V. I., 432
Ohata, M., 231
Okada, Y., 403
Okimoto, T., 69
Okitani, A., 239
Okumura, T., 240, 241
Okutani, A., 403
Olaizola, A., 65
Olesen, P. T., 139, 203, 208, 210
Oliver, G., 138, 207, 382, 388, 413
Olsen, M. B., 425
Olson, B., 175
Olson, D., 310
Omisakin, F., 400
Onishi, A., 11
OnVu™ TTI, 440, 441
Opioid peptides, 237
Ordonez, J. A., 239, 240
Orrú, L., 73
Osburn, W. N., 338
Oshima, G., 236
Oshimura, M., 155
Osmanagaoglu, O., 153, 379
Osta, R., 68
O'Sullivan, C. K., 262
Otani, K., 10, 13
Otero, A., 175
Ott, J., 75
Ottova-Leitmannova, A., 346
Ouattara, B., 256
Ouwehand, A. C., 222
Overall, A. D. J., 71
Ovesen, L., 233
Ovesná, J., 319
Ovine, 8–11
Owen, L., 331

Oxidation, in fermented meat, 209, 210
Oyagi, H., 389
Ozawa, T., 64, 65
Ozdemir, M., 426, 428

P

Pérez-Alvarez, J. A., 233
Pérez-Martínez, G., 135
Pace, N. R., 93
Paetkau, D., 78
Pagés, F., 412
Page, B. T., 34
Pala, T., 402
Palavesam, A., 265
Pale, soft, exudative (PSE) meat, 33
Palin, M. F., 33
Palmia, F., 190
Palmiter, R. D., 4
Pal, S., 338, 339, 342
Pancaldi, M., 68
Pangloli, P., 402
Panyim, S., 138, 265
Paoletti, C., 323
Pao, S., 270
Papa, A., 187
Papamanoli, E., 131, 140, 223
Papa, R., 69
Paparella, A., 142
Papkovskaia, N. Y., 433
Papkovsky, D. B., 425, 431–433
Papovskaia, N., 432
Paramithiotis, S., 375
Pardee, A. B., 47
Parejo, I., 209
Parente, E., 94, 103, 141, 381–383
Parker, R. B., 218
Parkes, H. C., 323
Park, I. S., 341
Park, K. H., 443
Park, S., 341
Park, W. S., 436
Park, Y. J., 235
Parma, A. E., 261, 455
Paroczay, E. W., 3, 14
Parolari, G., 181
Parrish Jr., F. C., 11
Parsons, A., 409
Parvez, S., 221
Pascual, M., 382
Pas, M. F. W., 15
Patel, J. R., 265
Paternal expressed gene, 10

Pathogens biocontrol, in meat chain, 253
  antagonistic bacteria
    bacterial metabolites, 258, 259
    competitive exclusion technology, 259, 260
  antimicrobial peptides, 267–269
  bacteriocin-producing bacteria and bacteriocins, 260
    animals, applications in, 261, 262
    limitations, 266, 267
    meat products, applications in, 262–266
  bacteriophage
    application of, 270
    classification and mode of action, 269
    limitations of, 272, 273
    in meat and meat products, applications, 271, 272
    preharvest control, 270, 271
  organic compounds
    essential oils, 256–258
    organic acids, 254–256
  vaccines, 273–277
Patton, G. C., 376
Paul, M., 376
Pavan, S., 217
Pawar, D. D., 386
Payne, R., 173
PCR, *see* Polymerase chain reaction
Pearce, P. D., 235
Peccio, A., 403
Pediococcus cerevisiae, 129, 131
Pedrazzoni, I., 387
PEG 11, *see* Paternal expressed gene 11
Pelaez-Martinez, M. C., 263, 387
Pemberton, J. M., 74
Peng, Z., 369
Penicillium spp, 183–186
Pennacchia, C., 94, 95, 97, 101–103, 107, 111, 140
Pepin, M., 139
Perdigon, G., 261
Pereira, S., 107, 141
Pérez Pulido, R., 152
Pergament, E. P., 330
Perreten, V., 142
Petaja, E., 140
Petersen, K. M., 175
Petitdemange, H., 381
Petracci, M., 65
PFGE, *see* Pulsed-field gel electrophoresis
Phaff, H. J., 169, 173
Phillips, C. A., 409

Phillips, D., 402
Phillips, R. L., 261
Piard, J. C., 152
Picard, C., 218
Pichon, L., 237
Pidcock, K., 224, 225
Piedrahita, J., 11
Piergiovanni, L., 443
Pierre, A., 243
Pierson, F. W., 272
Pietri, A., 181
Pihlanto, A., 236–238, 240, 242
Pilkington, D. H., 386, 412
Pilla, F., 33
Pillai, S. D., 270
Pinder, A., 338
Pinto, A. D., 82, 83
Pio, M., 345
Piot, J. M., 242
Piper, L. R., 9
Piry, S., 77
Pitkala, A., 402
Pitner, J. B., 343
Pitt, J. I., 185, 186, 193
Piva, G., 181
Pivarnik, P., 340
Pizzini, A., 82
Plaatjies, Z., 402
Place, R., 131
Pla, M., 98
Planchon, S., 133
Plant, A., 343, 344
Pla, R., 265
Plastow, G. S., 33
Playne, M. J., 244
Plotkin, D. J., 4
Plumpton, L., 366, 368
Podda, E., 268
Podolak, R. K., 255, 304, 305
Polejaeva, I. A., 11
Polymerase chain reaction, 92
  amplification, 29
  -DGGE, 104–106, 169, 170
    bacterial ecology by, 117–119
    primers, choice of, 115–117
    yeast ecology by, 119, 120
  for GMO analysis, 321–328
  inter-gene spacer regions (ISR), 107–112
  species-specific, 99–104
    bacterial ecology by, 120, 121
Polyunsaturated fatty acids, 10, 233
Ponomarev, G. V., 431

Ponsuksili, S., 49
Pontes, S. D., 93, 94
Porcine, 11–15
Porcine stress syndrome, 32
Pot, B., 217
Pothier, F., 33
Potter, A. A., 274
Pouchová, V., 319
Poussa, T., 243
Powell, J., 331
Power, E. G., 95
Prévost, H., 157
Prather, R. S., 11
Prebiotic peptides, from meat proteins, 242
Predki, P., 154
Prema, P., 265, 379
Pretz, J., 345
Previdi, M. P., 191
Prista, C., 175
Pritchard, J. K., 78, 82
PRKAG3 mutation, 12
Probert, H. M., 219
Probiotics, 139, 140
    applications, in fermented meat products
        as carrier for probiotic bacteria, 222, 223
        human studies, 224, 225
        screening among meat-associated bacteria, 223, 224
        technological suitability, 225
        use of known strains, 224
    history and definitions, 217–219
    microorganisms, 219, 220
    prophylactic and therapeutic effects of, 220–222
    safety considerations, 222
"Progeny test" data, 26
Pronuclear microinjection, 4
Pro-opiomelanocortin (POMC) gene, 34
Proteolysis, in fermented meat, 204–208
Proteomic analysis, 49
Pruett, W., 406
Prusa, K., 35
PSS, see Porcine stress syndrome
Publio, M. R. P., 265
Pucharoen, K., 138
PUFA, see Polyunsaturated fatty acids
Pulsed-field gel electrophoresis, 96, 97
Purchas, R. W., 235
Pursel, V. G., 14, 15
Purwati, E., 403
Putten, J. P. M. V., 276
Pyorala, S., 402

Q
QTL mapping
    families and data for, 36, 37
    genetic maps, 37, 38
    in livestock, genes and
        fat, obesity, and feed intake, 33, 34
        intramuscular fat and marbling, 35
        meat quality, 32, 33
        meat tenderization, 34, 35
Quantitative PCR (qPCR), 120–121
Quantitative trait loci (QTL), 5, 6, 23
    discovery by linkage mapping, 38–40
    finding trait genes starting from, 42
        calypyge, 44, 45
        hints from, fat in bovine milk (DGAT1), 45, 46
        insulin-like growth factor (IGF2), 45
        myostatin, 43, 44
    mapping, see QTL mapping
Quantitative trait nucleotide (QTN), 45
Quattlebaum, R. G., 390
Querol, A., 167–169
Quiberoni, A., 272
Quintavalla, S., 181, 191, 266
Quiros, M., 173
Quittelier, M., 381

R
Rabsch, W., 273
Radio frequency identification, 443, 444
Radio immunoassay, 364
Radke, S., 338, 339, 340, 342, 345–353
Radke, S. M., 433
Radosta, T., 61
Radu, S., 403
Rafalski, J. A., 66
Ragione, R. M. L., 259
Raha, A. R., 82
Rahim, R., 403
Rainelli, P., 65
Ramanaviciene, A., 345
Ramanavicius, A., 345
Ramaswamy, H. S., 426
Ramesh, A., 153
Ramos, J. P., 169, 174, 175
Ramsey, C. B., 11
Rance, K. A., 33
Randell, K., 434, 437
Rand, A. G., 340
Randolph, S. P., 270
Random amplified polymorphic DNA (RAPD)
    markers, 66
    -PCR, 93, 95, 108–111
Ransom, J., 299, 304, 307, 399, 401

Rantsiou, K., 91, 94, 95, 97, 99, 100, 102, 104–106, 109, 111, 114, 116–118
Rapid alert system for food and feed (RASFF), 454, 455
Rapoport, A., 273
Rasooly, A., 143, 402
Rastall, R. A., 219
Ratledge, C., 319
Ravnikar, M., 331
Ravula, R., 140
Raw meat, bioprotection applied to, 409–411
Raya, R., 271, 409
Ray, B., 153, 380, 381, 389
Ray, P., 153, 380
rbsABCD genes, 160
Reacher, M., 403
Ready-to-eat meat products, bioprotection and, 414, 415
Reagan, J. O., 141, 300, 302
Rebecchi, A., 95, 101, 108
Rebucci, R., 223, 224
Recio, I., 268
Reecy, J. M., 33
Reenen, C. A. V., 153
Rees, C. E., 271
Reguera-Useros, J. I., 263, 387
Regulation (EC) 852/222000, 455, 456
Regulation (EC) 854/222000, 457, 458
Regulation (EC) 882/222000, 457
Rehfeldt, C., 52
Reid, C. A., 65
Reid, G., 218, 220, 222
Reig, M., 233, 361, 362, 367
Reilly, S. S., 256
Reinders, R. D., 256
Reinheimer, J. A., 272
Reiniger, F., 431
Reitz, J., 203
Rekhif, N., 386
Remacle, C., 235
Rementzis, J., 415
Rendement Napole (RN) allele, 12
Repetitive bacterial DNA elements (Rep)-PCR, 94
Restriction fragment length polymorphism (RFLP) analysis, 29, 67, 95, 167
of rDNA, 169
Reusens, B., 235
Reuter, G., 91
Revol-Junelles, A. M., 382
Rexroad III, C. E., 34
Rexroad, Jr., C. E., 8, 14

Reynolds, E. C., 237
RFID, *see* Radio Frequency Identification
Rhue, M. R., 239
RIA, *see* Radio immunoassay
Ribosomal RNA gene sequencing, 169, 170
Ricciardi, A., 381–383
Richard, J. A., 386
Richards, H., 402
Richard, Y., 107
Ricke, S. C., 270
Riley, T. V., 256, 266
Ripened meats, fungal starter cultures in, 186, 187
Rippke, G. R., 331
Rishpon, J., 339
Ritchie, N., 338
Ritvanen, T., 443
Riva, M., 443
Rivas, L., 253
Rivera-Betancourt, M., 401
Rizzello, C. G., 237
Roberfroid, M. B., 219, 242
Roberson, A., 3
Roberts, I. N., 169
Robertson, A., 39
Roberts, P. V., 306
Robert, V., 169
Robnett, C. J., 94, 106, 168, 169
Roda, A., 364
Rodríguez, M. M., 168
Rodriguez, A., 187, 380, 402
Rodriguez, M. C., 155, 344
Rodriguez, E., 266
Rogan, D., 274
Rogers, E. K., 431
Rogers, P. J., 153
Roggiani, D., 187
Rohrer, G. A., 24, 37
Rollini, M., 382
Romano, P., 171, 172, 173, 187
Romero, D. A., 157
Ron, E. Z., 339
Rooney, M. L., 428, 429
Rosentrerer, J. J., 341
Rosmini, R., 403
Rossi, F., 94, 101, 107, 108
Ross, P., 375, 408
Ross, R., 412
Ross, R. P., 219, 256, 260, 261, 266, 267, 375
Ross, W., 403
Rotelli, D., 181
Rothschild, M. F., 33, 34
Roussel, S. A., 331

Rousset-Akrim, S., 203
Rovira, J., 103, 141
rpoB gene, 115, 118
Ruan, C. M., 342, 343
Rubner, M. F., 341
Ruddle, F. H., 4
Rudenko, L., 74
Ruiz-Barba, J. L., 152
Rule, D. C., 10
Russo, F., 78
Russo, P., 256
Rusul, G., 255, 403
Rutherfurd, S. M., 235
Ryan, M. P., 266
Ryanodyne receptor (RyR) gene, 11
Rybka, S., 140
Ryder, O. A., 8
Rydlo, T., 267

S

Sánchez, B., 169, 187
Saarela, M., 244
Sabate, J., 169
Sabia, C., 413
Sacco, C., 343
Sackmann, E., 347
Sadana, A., 339
Saeki, K., 14
Saez, R., 82
Sage, D. S., 443
Sagoo, S., 406
Sahl, H. G., 267
Saiga, A., 240, 241
Saito, K., 319
Sakai, K., 243
Sakakibara, K., 319
Sakala, R., 404
Salas, E. M., 265, 409
Salenvac®, 274, 275
Salminen, S., 219
Salmonella spp, 259, 260, 274, 275, 336, 337
Salvat, G., 403
Salzano, G., 103
Samarajeewa, U., 364
Sambrook, J., 67
Samelis, J., 168, 170–173, 182, 254, 304, 415
Sameshima, T., 224
Samsonova, J. V., 366, 368
Samson, R. A., 187, 190
Sanches, I. S., 107
Sancristobal, M., 74
Sanders, M. E., 221

Sanderson, K., 242
Sandine, W. E., 266
Sanglay, G. C., 265
Sannier, F., 242
Santarosa, P. R., 265
Santis, E. P. L. D., 103
Santivarangkna, C., 265
Santoro, E., 97
Santosa, S., 222
Santos, E. M., 103, 141
Santos, N. N., 203, 205
Santos, R. C., 203, 205
Santos, S., 222
Sanz, B., 209
Sanz, Y., 95, 101, 107, 135, 138, 139, 200–203, 205, 208
Saotta, R., 103
Sarhadi, M., 443
Saris, P. E., 156, 157
Sarra, P. G., 95, 188
Sasaki, A., 241
Sato, K., 331
Saturated fatty acids (SFA), 233
Savage, M. D., 343
Savarese, M. C., 44
Savell, J. W., 300, 304
Sayas-Barberá, E., 233
Scalari, F., 191, 192
Scangos, G. A., 4
Scannell, A., 266, 412
Schein, J. E., 24
Scheller, F. W., 341
Scherer, S., 270
Scherthan, H., 24
Schillinger, U., 254, 379, 386, 400
Schimpf, R. J., 40
Schiraldi, A., 443
Schleifer, K. H., 92
Schmidt, G., 138, 156
Schmidt, J. V., 44
Schneider, F., 47, 265
Schnieke, A. E., 4
Scholderer, J., 65
Schoppen, S., 141
Schrezenmeir, J., 259
Schrijver, I., 330
Schroeder, C. M., 68
Schultz, F. J., 254
Schulze, U., 381
Schwerin, M., 48
SCNT, *see* Somatic cell nuclear transfer
Scott, V., 403, 406
Seaton, G., 39

Sebedio, J.-L., 233
Sebranek, J., 416
Seideman, S. C., 425
Sekikawa, M., 239
Selgas, D., 209
Selgas, M. D., 171, 173
Senok, A. C., 218, 220, 222
Sentandreu, M. A., 206, 208, 235, 239
Seo, J., 345
Seo, K. H., 338
Seon, P., 344
Seppo, L., 243
Seronine, M. P., 201
Sesma, F., 409
Sevilla, A., 400
Sewalem, A., 41
Seyyedain-Ardebili, M., 204
SGMT, *see* Sperm mediated gene transfer
Shachar, D., 267
Shackelford, S. D., 10, 11, 23, 34, 36, 40
Shackell, G. H., 61, 64, 67, 68, 72, 73, 78–81
Shahidi, F., 236, 243
Shah, J., 342
Shah, N. P., 140, 231, 244
Shalaby, A., 141
Shale, K., 402
Shanahan, F., 221
Shapira, R., 153
Shareck, J., 156
Sharma, M., 265
Shaw, K. M., 273
Shay, B. J., 153
Sheldon, B. W., 386
Shen, H., 270
Shen, J. Z., 364
Sheppard, N. F., 339
Sheridan, J. J., 295, 402
Sherrington, D., 362
Shieh, Y. S., 270
Shimabukuro, H., 236
Shimada, K., 235, 249
Shimoni, E., 443
Shin, J., 306
Shin, K., 268
Shinoda, T., 175
Shi, W. M., 364, 366
Shi, Y. H., 236
Short, T., 33
Shrestha, S., 236
Shriver-Lake, L. C., 341
Shu, H. C., 437
Shutou, M., 347

Sica, M., 403
Sieber, R., 234
Sigler, L., 186
Silin, V., 343, 344
Sillence, M. N., 10
Silva, S. V., 237
Silvestri, G., 97, 114, 116, 117, 119, 120
Simard, R., 381
Simoncic, R., 232
Simoncini, N., 181, 191
Simon, L., 153
Simonova, M., 100, 103
Simon-Sarkadi, L., 141
Simpson, B. K., 426
Sinderen, D. V., 153, 220
Sineriz, F., 380
Singh, A. K., 368
Singh, V. P., 366
Single nucleotide polymorphisms, 30, 51, 67
Single strand conformation polymorphism, 93
Sinigaglia, M., 173
Siragusa, G. R., 266, 306, 308, 386
Situ, C., 365, 366, 368
Sivasubramanian, M., 265
Sivertsvik, M., 429
Si, W., 256
Siyang, S., 330
Skandamis, P. N., 256–258
Skaugen, M., 153
Skibsted, L. H., 209
Skibsted, L. S., 425
Sklar, I. B., 273
Skuridin, S. G., 343
Slade, R., 78
Slate, J., 74
Slavik, R., 340
Smagghe, G., 240
Smalla, K., 112, 170
Small, I., 319, 368, 377
Smart packaging technologies, 425
  antimicrobial packaging, 429, 430
  applications, in meat products, 444–447
  carbon dioxide scavengers and emitters, 428, 429
  indicators
    freshness, 436–438
    integrity, 434–436
    time-temperature (TTI), 438–443
  moisture control, 429
  oxygen scavengers, 426–428
  radio frequency identification, 443, 444
  sensors, 430
    biosensors, 433, 434

fluorescence-based oxygen sensors, 432, 433
gas sensors, 431, 432
Smeedsgaard, J., 184
Smerdon, W., 403
Smiddy, M. A., 425, 433
Smith, B. A., 67
Smithers, G. W., 244
Smith, G. C., 6, 254, 293, 303, 304
Smith, H. W., 273
Smith, J. A., 52
Smith, J. P., 426
Smith, P. G., 64
Smith, T. P. L., 9, 34, 44
Smolander, M., 434, 436, 438, 443
Smoragiewicz, W., 386
Smulders, F. J., 254, 256
Smyth, A., 432
Smyth, M. R., 341
Snelling, W. M., 24, 37, 50
Snowder, G. D., 10, 11
SNPs, see Single nucleotide polymorphisms
Sobsey, M. D., 270
Sofos, J. N., 170–173, 182, 254, 293, 297, 300, 301, 303, 304, 306, 309
Soldati, E., 96
Solignac, M., 77
Solinas-Toldo, S., 24, 37
Soller, M., 39, 41
Solomon, M. B., 3, 11, 14, 265
Somatic cell nuclear transfer, 319
Somer, E. B., 389
Somma, M., 325
Sonesson, A. K., 69
Sonneveld, K., 426
Sonomoto, K., 265, 381
Son, R., 82
Soothill, J. S., 273
Sorensen, D. A., 53
Soret, B., 63
Sorrentino, E., 103
Spergser, J., 131
Sperm mediated gene transfer, 319
Spescha, C., 402
Spinnler, H. E., 175
Spiriti, M. M., 343
Sportsman, J. R., 341
Spoth, B., 325
Spotti, E., 181, 183, 185, 186, 188, 190–192
Squartini, A., 153
Sriranganathan, N., 272

SSCP, see Single strand conformation polymorphism
Stahnke, L., H., 139, 200, 202–204, 210
Stanton, C., 219, 267
Staphylococcus and *Kocuria* spp., see Coagulase-negative cocci (CNC)
Stauffer, T., 434
Stavropoulos, S., 168, 171
Steam pasteurization, decontamination of beef carcasses and, 308, 309
Stebih, D., 329
Steinmetz, T., 155
Stentz, R., 156, 157
Stephan, R., 402
Stephens, M., 78
Stergiou, V. A., 265
Stern, N. J., 261, 262
Stiles, M. E., 142, 261, 375–380, 406
Stillwell, R. H., 218
Stockton, W. B., 341
Stolker, A. A. M., 362
Stone, R. T., 36, 40, 41
Stopforth, J., 301, 302
Storia, A. L., 266
Storro, I., 381, 382
Strachan, N., 400
Stratil, A., 13
Strauss, E., 325
Strauss, N., 107
Strobaugh, T. P., 338
Strunge Meyer, A., 139
Stubbings, G., 362
Suarez, J. E., 153, 380
Suarez, V. B., 272
Suehiro, J., 347
Suetsuna, K., 237
Sugita, T., 175
Suihko, M. L., 140, 224, 225
Sullivan, A., 402
Summers, A. O., 150
Sumner, J., 402
Sunesen, L. O., 200
Sung, H. R., 344
Suppakul, P., 426, 428–430
Su, X. L., 340, 341
Suzuki, M., 173
Suzzi, G., 94, 141, 142, 173
Svendsen, A., 184
Svetoch, E. A., 261, 262
Swanenburg, M., 402
Swings, J., 95, 110, 153, 155

T

Taberlet, P., 70, 73, 76
TAD Salmonella vac® E/T, 275
Takahashi, S., 6
Takala, T. M., 156, 157
Takamura, Y., 345
Takano, T., 243
Takhistov, P. T., 433
Tal'on, D., 138
Talon, R., 91, 92, 94, 97, 111, 133, 136, 139, 149, 200, 201, 203, 208, 210, 404
Tamiya, E., 345
Tanabe, S., 241, 243
Tanaka, K., 138
Tani, F., 237
Tanji, Y., 273
Tannock, G. W., 142
Tantillo, G. M., 82
Tan, W., 344
Taoukis, P. S., 438
Tapanainen, H., 233
Tarlov, M. J., 345
Tarr, P. I., 270
Tassou, C. C., 256
Tate, M. L., 68, 70, 78
Tatum, J. D., 6
Tauxe, R., 402
Taylor, J. C., 369, 389
Taylor, J. I., 389
Temperature gradient gel electrophoresis (TGGE), 93
Tenover, F. C., 97
Teratanavat, R., 335
Teresa, M., 167
Terry, C. F., 323
Teuber, M., 131, 142
Teuscher, F., 47
Thaller, G., 35, 46
Thein, S. L., 29, 66
Thermophillus aquaticus, 325
Thevenot, D., 403, 405
Thiel, A. J., 343
Thomas, A., 412
Thomas, A. B., 386
Thomas, L. V., 265
Thomas, S. C., 76
Thompson, C. S., 366, 368
Thonart, P., 256, 388
Thorberg, B., 402
Thrombotic thrombocytopenic purpura, 336
Thue, T. D., 34
Tian, X. C., 7

Tien, C. L., 386
Tien, H., 346
Tilden, J., 405
Tilghman, S. M., 44
Time-temperature indicator, 438
    enzymatic, 439
    polymer-based, 439–443
Tingey, S. V., 66
Tittiger, F., 405
Tofalo, R., 94, 142
Toldrá, F., 82, 120, 135, 170, 171, 175, 199, 200, 201, 202, 203, 204, 205, 206, 207, 208, 209, 210, 211, 233, 235, 239, 361, 362, 367
Tomé, D., 237
Tomatsu, M., 241
Tombelli, S., 343, 344
Tomlinson, J. J., 61
Tomomatsu, H., 175
Tompkin, R., 406
Torriani, S., 94
Toussaint, A., 150
Townes, C. L., 268
Townsend, A., 443, 444
Toxin Guard™, 433
Transforming growth factor (TGFβ)), 43
Transgenic farm animals, 3–15
    carcass and meat quality traits,
        candidate genes/genetic markers
        characterization for, 5, 6
    bovine, 6–8
    ovine, 8–11
    porcine, 11–15
Traynor, I. M., 368
Trettnak, W., 431
Trevisani, M., 403
Triggs, C., 72
Trimble, J., 260
Tri-sodium phosphate, 306
    See also Beef carcasses, decontamination
Troitsky, N., 273
Truszczynski, M., 260
Tsai, H. C., 303
Tsai, W. C., 340
Tsakalidou, E., 380
Tschape, H., 273
Tsigarida, E., 256–258
TTI, *see* Time-temperature indicator
TTP, *see* Thrombotic thrombocytopenic purpura
Tu, L., 411, 416
Tuomola, M., 364, 367
Turek, P., 387, 413

Turner, A. P. F., 338, 343
Työppönen, S., 140, 217
Tynkkynen, S. S., 157
Tzanetakis, N., 131, 223

U

Ugur, M., 386
Uitterlinden, A. G., 96, 116, 170
Ukeda, H., 237
Upsdell, M., 80
Upton, M., 140, 298
Ure, A. L., 233
Urso, R., 95–97, 100, 102, 105, 106, 109, 114, 116–120, 170, 199
Uruburu, F., 169
US Department of Agriculture's Food Safety and Inspection service (USDA-FSIS), 254, 291, 292
U.S. Patent and Trademark Office (USPTO), 4

V

Vázquez, J. F., 68, 71–73, 78
Vaitilingom, M., 329
Vagococcus carniphilus, 103
Vaillant, V., 403
Vainionpää, J., 443
Valdivia, E., 263, 265, 387, 389, 390
Valente, P., 169
Valentini, A., 44
Vallone, L., 187
Valsta, L. M., 233
Valyasevi, R., 138
Vandamme, E. J., 381
Vandekerckove, P., 141
Vandekinderen, I., 387, 416
Vanderlinde, P., 401, 402
Varghese, O. K., 343
Variable Number Tandem Repeat (VNTR) markers, 29, 66
Variant Creutzfeldt-Jacob Disease (vCJD), 461
Vather, R., 235
Vaughan, A., 153
Vaughan, E. E., 97
Vaughan, R. D., 341
Venkitanarayanan, K. S., 268, 276
Venter, P., 402
Ventura, M., 152
Verbeke, W., 64, 65
Vercruysse, L., 240
Vergès, C. M.-C., 149, 155
Vergnais, L., 139
Verluyten, J., 136, 223, 382, 385, 413
Vermeiren, L., 387, 415, 416, 426, 429, 430
Vermeirssen, V., 236

Vermerien, L., 414
Vermersch, D., 65
Vernozy-Rozand, C., 143, 403
Verplaetse, A., 92, 209
Verstraete, W., 236
Vescovo, G., 235
Vesseur, P., 65
Veterinary drug residues in foods, detection, 361–369
  biosensors, 368, 369
  immunoafinity chromatography, 363, 364
  immunoassay kits, 364–367
  molecular recognition, 362, 363
Vetharaniam, I., 80
Vicini, L., 266, 430
Vidal-Carou, M. C., 135, 141
Vielitz, E., 276
Viera, D. M., 270
Vignal, A., 67
Vignolo, G., 97, 101, 136, 138, 207, 263, 264, 388, 399, 404, 405, 409, 412, 413, 416, 417
Vikholm, I., 340
Viljoen, B. C., 173
Villadsen, J., 381, 382
Villani, F., 223, 226
Villanueva, J. R., 171
Villard, L., 107
Vinderola, G., 219
Virgili, R., 181
Visscher, P. M., 39, 41
VITSAB® TTI, 439
Vize, P. D., 14
Vogel, R. F., 142, 152, 153, 155, 307
Vollard, E. J., 361
Volpe, S. L., 235
Vonk, G. P., 343
Vos, P., 66
Vossen, J. M. B. M. V. D., 261
Vos, W. M. D., 116, 117
Vrese, M. D., 259
Vuyst, L. D., 136, 217, 218, 223, 265, 381, 382, 383, 384, 385, 413

W

Waal, E. C. D., 96, 116
Waal, E. D. D., 170
Waddell, T., 270
Wagenaar, J. A., 270, 276
Wagner, E., 133
Wagner, P., 347
Wagner, P. L., 269
Wagner, R. D., 259, 260

Waldor, M. K., 269
Walker, S., 11
Wall, R. J., 6
Walsh, D. J., 236
Walter, J., 116
Walz, C., 48
Wang, H. H., 368
Wang, H. L., 91
Wang, J. P., 366
Wang, L. L., 344
Wang, L. Y., 366, 368
Wang, R. H., 343
Wang, S., 367
Wang, T. T., 152
Wang, X. H., 367
Wang, X. L., 367
Wang, Y. H., 48
Wang, Z. L., 368
Waples, R. S., 76
Ward, D. M., 116
Ward, H. J. T., 64
Ward, K. A., 8, 9
Warner-Bratzler Shear Force (WBSF), 40
Warner, P. J., 34, 152
Wassenaar, T. M., 270, 276
Watkins, B. A., 233
Weber, G. H., 266
Weber, S., 266
Weber, W., 364
Webster, J. A., 107
Weeks, B. L., 139, 342
Weerkamp, A. H., 117
Wegner, J., 47
Wei, C. I., 364
Weir, B. S., 71, 75
Weis, L. J., 431
Wei, W., 338
Weller, J. I., 39, 82
Weller, M. G., 344
Weller, A. N., 152
Welt, B. A., 443
Wenijn, S., 330
Wen, Q., 405
Wesley, I., 403
Westall, S., 175
Westbrook, W., 270
West, C. A., 152
Wetzel, A. N., 254, 273
Wheeler, M. B., 320
Wheeler, T. L., 10, 11, 23, 27, 299
Whichard, J. M., 272
Wichen, P. V., 365
Widders, P. R., 276

Widmer, H. M., 344
Widstrand, C., 362
Wiegand, B. R., 11, 15
Wiegand, G., 347
Wieghart, M., 14
Wielders, C., 142
Wiener, P., 33, 52
Wildman, R. E. C., 231
Wilkins, E., 338, 342
Wilkins, M., 338, 339
Wilkinson, B. H. P., 235
Willett, J. L., 266, 386
Williams, J., 21, 27, 33
Williams, J. G. K., 66
Williams, J. L., 21, 27, 33, 52, 83
Williamson, K., 406
Williams, A. P., 193
Willopsis saturnus, 120
Will, R. G., 64
Willshaw, G., 253
Wilmut, I., 4, 8
Wilson, G. S., 341
Wilson, V., 29, 66
Wimmers, K., 48
Wingstrand, A., 402
Winkelman-Sim, D. C., 34
Woldringh, C., 346
Wolfbeis, O. S., 431
Wolk, A., 234
Wong, B., 175
Woodward, M. J., 259, 275
Woo, J. C. S., 339
Woolliams, J. A., 33, 52
Wright, M., 153
Wu, C. L., 142
Wu, C. W., 265
Wu, C. X., 44
Wu, Y. C., 202

X

Xavier, F., 338
Xiaogang, C., 366
Xiao-Juan, G., 44
Xie, L., 376
Xiraphi, N., 380
Xu, C. L., 366
Xue, Y., 233
Xu, S., 234

Y

Yamakawa, A., 151
Yamamoto, N., 243
Yamamoto, S., 403
Yamamoto, T., 263

Yamamoto, Y., 175
Yam, K. L., 433
Yanagita, T., 233
Yang, L. J., 342
Yao, S., 338
Yaron, S., 267, 268
Yarrow, D., 173, 175
Yassien, M. A., 170
Yeasts, genetic of
  biodiversity in meat products
    spoilage yeasts in meat, 170, 171
    yeast starter cultures, 171–173
  D.hansenii, physiological and genetic
    characteristics of
    genetic characteristics, 175, 176
    morphology and physiology, 173–175
  identification methods, 169
Ye, J. M., 340
Yeo, G. S., 33
Yeon, K. W., 344
Yevdokimov, Y. M., 343
Yin, L. J., 265
Yokoi, K. J., 151
Yokoyama, K., 241, 345
Yong, L., 233
Yoon, S. H., 443
Yoshikawa, M., 237, 241
Yoshikawa, Y., 434
Youderian, P., 270
Young, C. R., 276
Younts, S., 338
Yu, P., 34
Yuste, J., 265
Yu, Z., 115

Z

Zúríga, M., 135
Zagorec, M., 133, 149, 153, 155, 156, 157
Zampese, L., 95
Zanardi, E., 139, 203
Zaragoza, P., 68
Zasloff, M., 267
Zayas, J. F., 255, 304
Zel, J., 329, 331
Zeng, K. F., 343
Zhang, G., 260, 265
Zhang, S. X., 364, 366
Zhang, Z. X., 345
Zhao, H. Q., 344
Zhao, P., 260
Zhao, Q., 242
Zhao, T., 336
Zhengyu, J., 366
Zhou, A. H., 340
Zhou, S., 234
Zhou, X. D., 344
Zhou, Y. P., 344
Zhu, B., 38
Ziegler, W. E., 431
Ziprin, R. L., 276
Zoete, M. R. D., 276
Zoetendal, E. G., 116
Zoonties, P. W., 362
Zorman, T., 169
Zuckerman, H., 265
Zuo, Y., 342
Zweifel, C., 402
Zyl, W. H. V., 153

Printed in the United States
129534LV00001B/50/P